Springer Tracts in Modern Physics
Volume 232

Managing Editor: G. Höhler, Karlsruhe

Editors: A. Fujimori, Chiba
J. Kühn, Karlsruhe
Th. Müller, Karlsruhe
F. Steiner, Ulm
J. Trümper, Garching
P. Wölfle, Karlsruhe

Available online at SpringerLink.com

Starting with Volume 165, Springer Tracts in Modern Physics is part of the [SpringerLink] service. For all customers with standing orders for Springer Tracts in Modern Physics we offer the full text in electronic form via [SpringerLink] free of charge. Please contact your librarian who can receive a password for free access to the full articles by registration at:

springerlink.com

If you do not have a standing order you can nevertheless browse online through the table of contents of the volumes and the abstracts of each article and perform a full text search.

There you will also find more information about the series.

Springer Tracts in Modern Physics

Springer Tracts in Modern Physics provides comprehensive and critical reviews of topics of current interest in physics. The following fields are emphasized: elementary particle physics, solid-state physics, complex systems, and fundamental astrophysics.
Suitable reviews of other fields can also be accepted. The editors encourage prospective authors to correspond with them in advance of submitting an article. For reviews of topics belonging to the above mentioned fields, they should address the responsible editor, otherwise the managing editor.
See also springer.com

Managing Editor
Gerhard Höhler
Institut für Theoretische Teilchenphysik
Universität Karlsruhe
Postfach 69 80
76128 Karlsruhe, Germany
Phone: +49 (7 21) 6 08 33 75
Fax: +49 (7 21) 37 07 26
Email: gerhard.hoehler@physik.uni-karlsruhe.de
www-ttp.physik.uni-karlsruhe.de/

Elementary Particle Physics, Editors
Johann H. Kühn
Institut für Theoretische Teilchenphysik
Universität Karlsruhe
Postfach 69 80
76128 Karlsruhe, Germany
Phone: +49 (7 21) 6 08 33 72
Fax: +49 (7 21) 37 07 26
Email: johann.kuehn@physik.uni-karlsruhe.de
www-ttp.physik.uni-karlsruhe.de/~jk

Thomas Müller
Institut für Experimentelle Kernphysik
Fakultät für Physik
Universität Karlsruhe
Postfach 69 80
76128 Karlsruhe, Germany
Phone: +49 (7 21) 6 08 35 24
Fax: +49 (7 21) 6 07 26 21
Email: thomas.muller@physik.uni-karlsruhe.de
www-ekp.physik.uni-karlsruhe.de

Fundamental Astrophysics, Editor
Joachim Trümper
Max-Planck-Institut für Extraterrestrische Physik
Postfach 13 12
85741 Garching, Germany
Phone: +49 (89) 30 00 35 59
Fax: +49 (89) 30 00 33 15
Email: jtrumper@mpe.mpg.de
www.mpe-garching.mpg.de/index.html

Solid-State Physics, Editors
Atsushi Fujimori
Editor for The Pacific Rim
Department of Physics
University of Tokyo
7-3-1 Hongo, Bunkyo-ku
Tokyo 113-0033, Japan
Email: fujimori@wyvern.phys.s.u-tokyo.ac.jp
http://wyvern.phys.s.u-tokyo.ac.jp/welcome_en.html

Peter Wölfle
Institut für Theorie der Kondensierten Materie
Universität Karlsruhe
Postfach 69 80
76128 Karlsruhe, Germany
Phone: +49 (7 21) 6 08 35 90
Fax: +49 (7 21) 6 08 77 79
Email: woelfle@tkm.physik.uni-karlsruhe.de
www-tkm.physik.uni-karlsruhe.de

Complex Systems, Editor
Frank Steiner
Institut für Theoretische Physik
Universität Ulm
Albert-Einstein-Allee 11
89069 Ulm, Germany
Phone: +49 (7 31) 5 02 29 10
Fax: +49 (7 31) 5 02 29 24
Email: frank.steiner@uni-ulm.de
www.physik.uni-ulm.de/theo/qc/group.html

Panayotis G. Kevrekidis

The Discrete Nonlinear Schrödinger Equation

Mathematical Analysis, Numerical Computations and Physical Perspectives

With contributions by

Ricardo Carretero-González
Alan R. Champneys
Jesús Cuevas
Sergey V. Dmitriev
Dimitri J. Frantzeskakis
Ying-Ji He
Q. Enam Hoq
Avinash Khare
Kody J.H. Law
Boris A. Malomed
Thomas R.O. Melvin
Faustino Palmero
Mason A. Porter
Vassilis M. Rothos
Atanas Stefanov
Hadi Susanto

Panayotis G. Kevrekidis
University of Massachusetts, Amherst
Dept. Mathematics & Statistics
Amherst MA 01003
USA
kevrekid@math.umass.edu

P.G. Kevrekidis, *The Discrete Nonlinear Schrödinger Equation: Mathematical Analysis, Numerical Computations and Physical Perspectives*, STMP 232 (Springer, Berlin Heidelberg 2009), DOI 10.1007/ 978-3-540-89199-4

ISBN 978-3-540-89198-7　　　　　　　　　　e-ISBN 978-3-540-89199-4

DOI 10.1007/978-3-540-89199-4

Springer Tracts in Modern Physics ISSN 0081-3869　　　　e-ISSN 1615-0430

Library of Congress Control Number: 2008940668

© Springer-Verlag Berlin Heidelberg 2009

This work is subject to copyright. All rights are reserved, whether the whole or part of the material is concerned, specifically the rights of translation, reprinting, reuse of illustrations, recitation, broadcasting, reproduction on microfilm or in any other way, and storage in data banks. Duplication of this publication or parts thereof is permitted only under the provisions of the German Copyright Law of September 9, 1965, in its current version, and permission for use must always be obtained from Springer. Violations are liable to prosecution under the German Copyright Law.

The use of general descriptive names, registered names, trademarks, etc. in this publication does not imply, even in the absence of a specific statement, that such names are exempt from the relevant protective laws and regulations and therefore free for general use.

Cover design: Integra Software Services Pvt. Ltd., Puducherry

Printed on acid-free paper

9 8 7 6 5 4 3 2 1

springer.com

To:
Maria, for her love, support and patience that made this possible;
Despina and Athena, for the immense joy that they have brought in my life;
and to my parents, for always believing in me more than I do myself.

Foreword

Adventures of nonlinear science were perhaps most notably seeded at the Los Alamos National Laboratory (LANL) over half a century ago with the fundamental questions of energy equipartition in nonlinear systems, as they were posed by Fermi, Pasta, and Ulam. At the time, probably little could be imagined of the far-reaching implications that the studies of nonlinear phenomena would have, continuing to expand to this day. The Ginzburg–Landau theory of superconductivity and the order-parameter descriptions of superfluidity, the "soliton revolution" through the works of Zabusky and Kruskal on the KdV equation and the subsequent widespread applications of the nonlinear Schrödinger equation in optical fibers and Bose–Einstein condensates, the developments of bifurcation theory and chaotic dynamics and their widespread applications from climate and geophysics, to biological phenomena and chemical kinetics are only a few of the multiple arenas in which nonlinear dynamics have emerged as the appropriate description of important physical systems.

I well remember my own early days of nonlinear science appreciation, first at Cornell University in the early 1970s and then at Los Alamos where we began the Center for Nonlinear Studies (CNLS) in 1980. These were years marked by interdisciplinary discovery and by the recognition that many nonlinear equations have an inherent ability to exhibit both coherence and chaos – the beginnings of our appreciation today of spatio-temporal complexity and the functional role that this plays in multiple branches of science, technology, and engineering.

Among the many remarkable discoveries from combinations of analysis, simulation, and experimentation, the soliton (and more generically solitary wave)-bearing equations have a long and distinguished history. They connect rich histories of exactly solvable systems discovered in mathematical, statistical and many-body physics, and powerfully demonstrate the unity of nonlinear concepts across disciplines and scales – from biology to cosmology! The Nonlinear Schrödinger equation is perhaps the most ubiquitous of these equations, deriving from the familiar ingredients of wave dispersion and nonlinearity, resulting in coherent envelope profiles.

Panos Kevrekidis was a member of the family of nonlinear brothers at CNLS and Los Alamos before moving to UMass. Interestingly, CNLS was where some of the most important early steps were realized in the study of the theme of this book, namely the discrete nonlinear Schödinger equation (DNLS), appearing, e.g., through the discrete self-trapping equation. Subsequent theoretical proposals for the use of the DNLS in optical waveguide arrays, and later in Bose–Einstein condensates trapped in optical lattices, have significantly increased the impact of this model, which, in parallel, was often proposed as an envelope model in numerous other Hamiltonian nonlinear lattice settings (especially of the nonlinear Klein–Gordon type). The experimental realization of such proposals in both the optical and the atomic physics context has served to cement the relevance and importance of the study of the research theme of Hamiltonian nonlinear dynamical lattices, and especially of the DNLS as one of its prototypical realizations.

In this context the present work constitutes an important contribution to the literature and study of such systems. Despite the two decades over which this research subject has been evolving, this is the first book specifically dedicated to exploring the main physical, mathematical, and numerical aspects of the DNLS. The exposition is authored by an expert in the field with additional timely contributions from numerous highly active researchers. I expect this book to be an important contribution to the archival DNLS literature and a valuable reference for both junior and more senior researchers.

Los Alamos, NM *Alan Bishop*
July 2008

Preface

Over the past two decades, the breadth and depth of influence of nonlinear science more generally, and of dispersive lattice systems such as the discrete nonlinear Schrödinger (DNLS) equation more specifically, have grown tremendously. Starting from the speculations on Davydov's soliton in biophysics and nonlinear optical couplers and proposals for waveguide arrays in the 1970s and 1980s, the studies of the DNLS-type systems passed to a different realm in the 1990s through the experimental realization of optical waveguide arrays and the observation of key theoretical predictions including discrete solitons, diffraction, Peierls barriers, multipulse features, and diffraction management. They reemerged in yet another entirely different physical incarnation in Bose–Einstein condensates (BECs) in optical lattices in the 2000s, representing one of the most exciting aspects of nonlinear phenomena in this novel state of matter. This was even more remarkable in view of the wide range of attention that BECs garnered due to their realization being awarded the Nobel prize in Physics in 2001 and their being intimately connected to superfluidity and the Nobel Prize in Physics in 2003. In the meantime, additional related aspects arose in some of the earlier settings, including but not limited to, for instance, the realization of periodic media in photorefractive crystals.

I first came across DNLS-type equations during my time as a Ph.D. student, and started working on them during my summer visits at the Center for Nonlinear Studies, at LANL, which has always been a guiding center for work around this theme (even since its original inception!). After the end of my dissertation work, the DNLS became an even stronger focal point, among other reasons due to the increasing visibility of the physical realizations of this deceptively simple-looking mathematical model. Over the years, I have had the privilege to work with a wonderfully diverse, and physically, as well as mathematically gifted set of collaborators from whom I learnt a great deal about this topic; however, I have always been surprised by the fact that despite the level of maturity that this field has arrived at, there has not been a more comprehensive publication (i.e., a book) that focuses on this main theme at the interface of nonlinear science, lattice systems, and wave phenomena. It is in that spirit that I decided to dedicate my first sabbatical from UMass in an effort to sum up some of the main axes of the phenomenology of this equation, at least as I view it and as it has been distilled through my own research efforts on the subject over the years; undoubtedly, this brings a considerable personal flavor. However, in an effort to broaden the scope of the work, as well as to offer some of its most recent developments, I decided to partition the book into two segments. The first part, comprising the first seven chapters, consists of some of the principal and general features of the DNLS system. The second part, consisting of chapters 8–22, is a set of minireviews on more specialized, as well as often more recent topics, written by a number of friends and collaborators, who kindly offered their expert help.

At this point, I would like to thank all the contributors to this volume for their excellent coordination, and their important and informative contributions. I should also express my gratitude to all my collaborators over the years on this theme of work, for all of what they have taught me. Among them, I should especially mention Ricardo Carretero-González for his invaluable editorial and consulting assistance throughout this project. I should also thank the National Science Foundation for its support through the CAREER program and the Alexander von Humboldt Foundation for offering me the opportunity to work, as undistracted as possible, toward the completion of this book, at the University of Heidelberg, through one of its Research Fellowships. Lastly, but most importantly, I am grateful to Maria, and our two daughters, Despina and Athena, for helping me, in more ways than I can enumerate, to complete this project.

I hope that the result, despite its personal flavor and its strong parallels with my own journey through this intriguing dynamical system, will offer the reader, be they a novice, or a seasoned researcher in this field, a useful reference point where many of the fundamental ideas are explored and references are given to more specialized publications. Furthermore, I hope it will provide not only a perspective of the mathematical tools and techniques, but also a view toward the numerical computations/methods and importantly key connections to the physical realizations of this class of models. As the cycle of this work is closing, and new ones are opening up, perhaps it is relevant that I conclude this journey with one of my favorite verses of Tennyson's Ulysses:

"Tho' much is taken, much abides; and though we are not now that strength which in old days moved earth and heaven; that which we are, we are; one equal temper of heroic hearts, made weak by time and fate, but strong in will to strive, to seek, to find, and not to yield."

Heidelberg, Germany *Panayotis Kevrekidis*
June 2008

Contents

Part I General Theory

1 General Introduction and Derivation of the DNLS Equation 3
   1.1   General Introduction 3
   1.2   Prototypical Derivation of the DNLS 4
   References 8

2 The One-Dimensional Case 11
   2.1   Single-Pulse Solitary Waves 11
       2.1.1   The Continuum Approach 11
       2.1.2   The Anti-Continuum Approach 27
       2.1.3   The Variational Approach 30
   2.2   Multipulse Solitary Waves 34
       2.2.1   Multipulses Near the Anti-Continuum Limit 34
       2.2.2   A Different Approach: Perturbed Hamiltonian Systems .. 41
       2.2.3   Multipulses Close to the Continuum Limit 48
   References 52

3 The Two-Dimensional Case 55
   3.1   General Notions 55
   3.2   Single-Pulse Solitary Waves 56
   3.3   Multipulses and Discrete Vortices 60
       3.3.1   Formulation of the Bifurcation Problem Near $\epsilon = 0$ 60
       3.3.2   Persistence of Discrete Solutions 63
       3.3.3   Stability of Discrete Solutions 74
       3.3.4   Numerical Results 83
   References 98

4 The Three-Dimensional Case 99
   4.1   General Theory 100
   4.2   Discrete Solitons and Vortices 107
   References 116

5 The Defocusing Case ... 117
- 5.1 Dark Solitary Waves ... 118
  - 5.1.1 Theoretical Analysis ... 118
  - 5.1.2 Numerical Results ... 125
- 5.2 Vortex States on a Non-Zero Background ... 128
- 5.3 Gap States ... 131
  - 5.3.1 General Terminology ... 132
  - 5.3.2 Dipole Configurations ... 133
  - 5.3.3 Quadrupole Configurations ... 136
  - 5.3.4 Vortex Configuration ... 138
  - 5.3.5 General Principles Derived from Stability Considerations ... 140
- References ... 140

6 Extended Solutions and Modulational Instability ... 143
- 6.1 Continuum Modulational Instability ... 143
- 6.2 Discrete Modulational Instability ... 145
- 6.3 Some Case Examples ... 150
- References ... 152

7 Multicomponent DNLS Equations ... 153
- 7.1 Linearly Coupled ... 153
- 7.2 Nonlinearly Coupled ... 158
  - 7.2.1 One Dimension ... 159
  - 7.2.2 Higher Dimensions ... 167
- References ... 170

Part II Special Topics

8 Experimental Results Related to DNLS Equations ... 175
Mason A. Porter
- 8.1 Introduction ... 175
- 8.2 Optics ... 176
  - 8.2.1 Optical Waveguide Arrays ... 176
  - 8.2.2 Photorefractive Crystals ... 179
- 8.3 Bose–Einstein Condensation ... 182
- 8.4 Summary and Outlook ... 186
- References ... 187

9 Numerical Methods for DNLS ... 191
Kody J.H. Law and Panayotis G. Kevrekidis
- 9.1 Introduction ... 191
- 9.2 Numerical Computations Using Full Matrices ... 192

Contents

  9.3 Numerical Computations Using Sparse Matrices/Iterative Solvers . 199
  9.4 Conclusions 203
 References 203

10 The Dynamics of Unstable Waves 205
 Kody J.H. Law and Q. Enam Hoq
 10.1 Introduction 205
 10.2 Standard Scenario 206
   10.2.1 1(+1)-Dimensional Solutions 206
   10.2.2 2(+1)-Dimensional Solutions 208
   10.2.3 3(+1)-Dimensional Solutions 211
 10.3 Non-Standard Scenario 213
   10.3.1 Hexagonal Lattice 213
   10.3.2 Defocusing Nonlinearity 214
 10.4 Conclusion and Future Challenges 219
 References 219

11 A Map Approach to Stationary Solutions of the DNLS Equation 221
 Ricardo Carretero-González
 11.1 Introduction 221
 11.2 The 2D Map Approach for 1D Nonlinear Lattices 222
 11.3 Orbit Properties and Diversity in the DNLS 223
   11.3.1 Symmetries and Properties of the Cubic DNLS Steady
      States 224
   11.3.2 Homogeneous, Periodic, Modulated, and Spatially
      Chaotic Steady States 224
   11.3.3 Spatially Localized Solutions: Solitons
      and Multibreathers 226
 11.4 Bifurcations: The Road from the Anti-Continuous
   to the Continuous Limit 230
 11.5 Summary and Future Challenges 231
 References 232

12 Formation of Localized Modes in DNLS 235
 Panayotis G. Kevrekidis
 12.1 Introduction 235
 12.2 Threshold Conditions for the Integrable NLS Models 236
   12.2.1 The Continuum NLS Model 236
   12.2.2 The Ablowitz–Ladik Model 237
 12.3 Threshold Conditions for the Non-Integrable DNLS Model 240
 12.4 Conclusions and Future Challenges 245
 References 247

13 Few-Lattice-Site Systems of Discrete Self-Trapping Equations 249
Hadi Susanto
- 13.1 Introduction 249
- 13.2 Integrability 251
- 13.3 Chaos 254
- 13.4 Applications and Experimental Observations 254
- 13.5 Conclusions 256
- References 256

14 Surface Waves and Boundary Effects in DNLS Equations 259
Ying-Ji He and Boris A. Malomed
- 14.1 Introduction 259
- 14.2 Discrete Nonlinear Schrödinger Equations for Surface Waves 260
  - 14.2.1 The One-Dimensional Setting 260
  - 14.2.2 The Two-Dimensional Setting 261
  - 14.2.3 The Three-Dimensional Setting 261
- 14.3 Theoretical Investigation of Discrete Surface Waves 262
  - 14.3.1 Stable Discrete Surface Solitons in One Dimension 262
  - 14.3.2 Discrete Surface Solitons at an Interface Between Self-Defocusing and Self-Focusing Lattice Media 264
  - 14.3.3 Tamm Oscillations of Unstaggered and Staggered Solitons 265
  - 14.3.4 Discrete Surface Solitons in Two Dimensions 267
  - 14.3.5 Spatiotemporal Discrete Surface Solitons 268
  - 14.3.6 Finite Lattices and the Method of Images 268
- 14.4 Experimental Results 270
  - 14.4.1 Discrete Surface Solitons in One Dimension 270
  - 14.4.2 Staggered Modes 270
  - 14.4.3 Discrete Surface Solitons in Two Dimensions 271
- 14.5 Conclusions 273
- References 274

15 Discrete Nonlinear Schrödinger Equations with Time-Dependent Coefficients (*Management* of Lattice Solitons) 277
Jesús Cuevas and Boris A. Malomed
- 15.1 Introduction 277
- 15.2 Quiescent Solitons Under the Action of the "Management" 279
  - 15.2.1 Semi-Analytical Approximation 279
  - 15.2.2 Direct Simulations 281
- 15.3 Supporting Moving Solitons by Means of the "Management" 283
  - 15.3.1 Analytical Approximation 283
  - 15.3.2 Numerical Results 285
- 15.4 Conclusion and Future Challenges 289
- References 290

16 Exceptional Discretizations of the NLS: Exact Solutions and Conservation Laws 293
Sergey V. Dmitriev and Avinash Khare
- 16.1 Introduction 293
- 16.2 Review of Existing Works 293
  - 16.2.1 Stationary Translationally Invariant Solutions 294
  - 16.2.2 Exact Moving Solutions to DNLS 297
- 16.3 Cubic Nonlinearity 297
  - 16.3.1 Conservation Laws 298
  - 16.3.2 Two-Point Maps for Stationary Solutions 300
  - 16.3.3 Moving Pulse, Kink, and Sine Solutions 303
  - 16.3.4 Stationary TI Solutions 304
  - 16.3.5 Moving Bright Solitons 306
- 16.4 Conclusions and Future Challenges 308
- References 308

17 Solitary Wave Collisions 311
Sergey V. Dmitriev and Dimitri J. Frantzeskakis
- 17.1 Introduction and Setup 311
- 17.2 Collisions in the Weakly Discrete NLS Equation 312
- 17.3 Collisions in the Strongly Discrete NLS Equation 315
- 17.4 Strongly Discrete Nearly Integrable Case 318
- 17.5 Role of Soliton's Internal Modes 320
- 17.6 Solitary Wave Collisions in Physically Relevant Settings 322
- 17.7 Conclusions 324
- 17.8 Future Challenges 325
- References 326

18 Related Models 329
Boris A. Malomed
- 18.1 Models Beyond the Standard One 329
  - 18.1.1 Introduction 329
  - 18.1.2 Anisotropic Inter-Site Couplings 330
  - 18.1.3 Noncubic On-site Nonlinearities..................... 330
  - 18.1.4 Nonlocal Coupling 332
  - 18.1.5 The Competition Between On-site and Inter-site Nonlinearities (The Salerno Model) 333
  - 18.1.6 Semidiscrete Systems 334
- 18.2 The Anisotropic Two-Dimensional Lattice 336
  - 18.2.1 Outline of Analytical and Numerical Methods 336
  - 18.2.2 Fundamental Solitons and Vortices in the Anisotropic Model 336
- 18.3 Solitons Supported by the Cubic-Quintic On-site Nonlinearity 338

18.4 Solitons in the Salerno Model with Competing Inter-site and On-site Nonlinearities 340
    18.4.1 The 1D Model 340
    18.4.2 The 2D Model 343
18.5 One-Dimensional Solitons in the Semidiscrete System with the $\chi^{(2)}$ Nonlinearity 346
18.6 Conclusion and Perspectives 347
References 349

19 DNLS with Impurities 353
Jesús Cuevas and Faustino Palmero
19.1 Introduction 353
19.2 Stationary Solutions 354
    19.2.1 Linear Modes 355
    19.2.2 Bifurcations 356
    19.2.3 Invariant Manifold Approximation 357
19.3 Interaction of a Moving Soliton with a Single Impurity 360
19.4 Comparison with Other Related Models 365
    19.4.1 Nonlinear Impurities 365
    19.4.2 Comparison with Klein–Gordon Breathers 366
19.5 Summary and Future Challenges 366
References 367

20 Statistical Mechanics of DNLS 369
Panayotis G. Kevrekidis
20.1 Introduction 369
20.2 Theoretical Results 370
20.3 Recent Results 374
References 377

21 Traveling Solitary Waves in DNLS Equations 379
Alan R. Champneys, Vassilis M. Rothos and Thomas R.O. Melvin
21.1 Introduction 379
21.2 Mathematical Formulation 381
    21.2.1 Spatial Dynamics Formulation 382
    21.2.2 Center Manifold Reduction 385
    21.2.3 Normal Form Equations Near the Zero-Dispersion Point . 387
    21.2.4 Mel'nikov Calculations for Generalized Cubic DNLS 388
    21.2.5 Beyond-All-Orders Asymptotic Computation 389
    21.2.6 Numerical Continuation Using Pseudospectral Methods .. 390
21.3 Applications 392
    21.3.1 Saturable DNLS 392
    21.3.2 Generalized Cubic DNLS 393
    21.3.3 The Salerno Model 395

	21.4	Conclusion	397
		References	398
22	**Decay and Strichartz Estimates for DNLS**	401
	Atanas Stefanov		
	22.1	Introduction	401
	22.2	Decay and Strichartz Estimates for the Discrete Schrödinger and Klein–Gordon Equation	403
		22.2.1 A Spectral Result for Discrete Schrödinger Operators	405
		22.2.2 Application to Excitation Thresholds	406
	22.3	Decay and Strichartz Estimates for the Discrete Schrödinger Equation Perturbed by a Potential	407
		22.3.1 Spectral Theoretic Results for 1D Schrödinger Operators	410
	22.4	Challenges and Open Problems	410
		22.4.1 Does Weak Coupling Allow Eigenvalues in Three Dimensions?	410
		22.4.2 CLR-Type Bounds for Discrete Schrödinger Operators and Related Issues	411
		22.4.3 Show Analogs of Theorems 8, 9, 10 in Higher Dimensions	411
		22.4.4 Asymptotic Stability and Nucleation	411
		References	412
Index		413

Contributors

Ricardo Carretero-González Nonlinear Dynamical Systems Group, Computational Science Research Center, and Department of Mathematics and Statistics, San Diego State University, San Diego, CA, 92182-7720, USA, carreter@sciences.sdsu.edu

Alan R. Champneys Department of Engineering Mathematics, University of Bristol, UK, a.r.champneys@bristol.ac.uk

Jesús Cuevas Grupo de Física No Lineal, Departamento de Física Aplicada I, Escuela Universitaria, Politécnica, Universidad de Sevilla, C/Virgen de África, 7, 41011 Sevilla, Spain, jcuevas@us.es

Sergey V. Dmitriev Institute for Metals Superplasticity Problems RAS, 450001 Ufa, Khalturina 39, Russia, dmitriev.sergey.v@gmail.com

Dimitri J. Frantzeskakis Department of Physics, University of Athens, Panepistimiopolis, Zografos, Athens 15784, Greece, dfrantz@phys.uoa.gr

Ying-Ji He School of Electronics and Information, Guangdong Polytechnic Normal University, Guangzhou 510665; State Key Laboratory of Optoelectronic Materials and Technologies, Sun Yat-Sen University, Guangzhou 510275, China, hyj8409@sxu.edu.cn

Q. Enam Hoq Department of Mathematics, Western New England College, Springfield, MA 01119, USA, qhoq@wnec.edu

Panayotis G. Kevrekidis University of Massachusetts, Amherst, MA, 01003, USA, kevrekid@math.umass.edu

Avinash Khare Institute of Physics, Bhubaneswar, Orissa 751005, India, khare@iopb.res.in

Kody J.H. Law University of Massachusetts, Amherst, MA, 01003, USA, law@math.umass.edu

Boris A. Malomed Department of Physical Electronics, School of Electrical Engineering, Faculty of Engineering, Tel Aviv University, Tel Aviv 69978, Israel, malomed@eng.tau.ac.il

Thomas R.O. Melvin Department of Engineering Mathematics, University of Bristol, UK, thomas.melvin@bris.ac.uk

Faustino Palmero Grupo de Física No Lineal, Departamento de Física Aplicada I, Escuela Técnica, Superior de Ingeniería Informática. Universidad de Sevilla. Avda. Reina Mercedes, s/n. 41012 Sevilla, Spain, palmero@us.es

Mason A. Porter Oxford Centre for Industrial and Applied Mathematics, Mathematical Institute, University of Oxford, Oxford, England, UK, porterm@maths.ox.ac.uk

Vassilis M. Rothos Division of Mathematics, Faculty of Engineering, Aristotle University of Thessaloniki, Greece, rothos@gen.auth.gr

Atanas Stefanov The University of Kansas, Lawrence, KS, USA, stefanov@math.ku.edu

Hadi Susanto School of Mathematical Sciences, University of Nottingham, University Park, Nottingham, NG7 2RD, UK, hadi.susanto@nottingham.ac.uk

Part I
General Theory

Chapter 1
General Introduction and Derivation of the DNLS Equation

1.1 General Introduction

The discrete nonlinear Schrödinger (DNLS) equation is, arguably, one of the most fundamental nonlinear lattice dynamical models. On the one hand, this is due to its being the prototypical discretization for its famous and integrable continuum sibling, namely the nonlinear Schrödinger (NLS) equation [1, 8] which has a wide range of applications; it is the relevant dispersive envelope wave model for describing the electric field in optical fibers [13, 34], for the self-focusing and collapse of Langmuir waves in plasma physics [5, 6], or for the description of freak waves (the so-called rogue waves) in the ocean [7]. On the other hand, the DNLS is a model of particular physical interest in its own right, with a diverse host of areas where it is of physical interest; see, e.g., [38] for a relevant review. We mention a brief outline of these areas below.

Perhaps the first set of experimental investigations that triggered an intense interest in DNLS-type equations was in the area of nonlinear optics and, in particular, in fabricated AlGaAs waveguide arrays [9]. In the latter setting, a multiplicity of phenomena such as discrete diffraction, Peierls barriers (the energetic barrier that a wave needs to overcome to move over a lattice – see details below), diffraction management (the periodic alternation of the diffraction coefficient) [10, 11], and gap solitons (structures localized due to nonlinearity in the gap of the underlying linear spectrum) [12] among others [13] were experimentally observed. These phenomena, in turn, triggered a tremendous increase also on the theoretical side of the number of studies addressing such effectively discrete media; see, e.g., [14–17] for a number of relevant reviews, as well as the very recent [18].

A related area where DNLS, although it is not the prototypical model, still it yields accurate qualitative predictions both about the existence and about the stability of nonlinear localized modes is that of optically induced lattices in photorefractive media such as strontium barium niobate (SBN). Ever since the theoretical inception of such a possibility in [19], and its experimental realization in [20–23], there has been an explosive growth in the area of nonlinear waves and solitons in such periodic, predominantly two-dimensional, lattices. An ever-growing array of structures has been predicted and experimentally obtained in lattices induced with a self-focusing nonlinearity, including (but not limited to) discrete dipole [24], quadrupole [25], necklace [26], and other multipulse patterns (such as, e.g., soliton stripes [27]), discrete vortices [28, 29], and rotary solitons [30, 31]. Such structures have a definite potential to be used as carriers and conduits for data transmission and processing, in the setting of all-optical communication schemes. A recent review of this direction can be found in [32] (see also [33]).

Finally, yet another independent and completely different physical setting where such considerations and structures are relevant is that of atomic physics, where droplets of the most recently discovered state of matter, namely of Bose–Einstein condensates (BECs) may be trapped in a periodic optical lattice (OL) potential produced by counter-propagating laser beams in one, two or even all three directions [34]. The latter field has also experienced a huge growth over the past few years, including the prediction and manifestation of modulational instabilities (i.e., the instability of spatially uniform states toward spatially modulated ones) [17, 36], the observation of gap solitons [37], Landau–Zener [38], and Bloch oscillations (for matter waves subject to combined periodic and linear potentials) [39] among many other salient features; reviews of the theoretical and experimental findings in this area have also recently appeared in [40–42].

In addition to these areas more directly related to the DNLS or to NLS-type equations more generally, the DNLS serves as an envelope model to other types of lattices such as Klein–Gordon (i.e., nonlinear wave type) equations. These emerge in a number of additional applications including the oscillations of nanomechanical cantilever arrays [43], the denaturation and related phase transformations of the DNA double strand [44], or even in simple electric circuits [45].

All of the above experimental motivations from atomic, optical, nonlinear, and wave physics illustrate the relevance of acquiring a detailed understanding of a prototypical mathematical model that emerges in one form or another in the settings, namely the DNLS equation. On the other hand, more generally, this theme of interplay between nonlinearity and periodicity (that are the fundamental features of the DNLS equation, along with diffraction) is of more general and broad appeal. Nonlinearity and periodicity have been observed to introduce fundamental changes in the properties of the system. On the one hand, periodicity modifies the spectrum of the underlying linear system resulting in the potential of existence of new coherent structures, which may not exist in a homogeneous nonlinear system. On the other hand, nonlinearity renders accumulation and transmission of energy possible in "linearly" forbidden frequency domains; this, in turn, results in field localization. It is in such a prototypical setting combining these features in a Schrödinger equation context that we start our mathematical presentation, aiming at a derivation of the DNLS equation, before we subsequently focus on its mathematical properties and nonlinear wave solutions.

1.2 Prototypical Derivation of the DNLS

Our starting point for the derivation of the DNLS equation will be the continuum nonlinear Schrödinger for the wave function ψ in the presence of a periodic potential of the form

1.2 Prototypical Derivation of the DNLS

$$i\frac{\partial \psi}{\partial t} = -\frac{\partial^2 \psi}{\partial x^2} + V(x)\psi + \sigma|\psi|^2\psi \tag{1.1}$$

where $\sigma = \pm 1$ and $V(x)$ is a periodic potential $V(x+L) = V(x)$. This model is of particular relevance to (cigar-shaped) BECs in the presence of an OL potential as indicated above [40–42];[1] the sign of σ determines the nature of the interatomic interactions. The latter are attractive for negative σ, while they are repulsive for positive σ. The former case corresponds to the so-called focusing nonlinearity, while the latter to the so-called defocusing one. Note also that here we will give a one-dimensional derivation, although similar concepts can, in principle, be generalized to higher dimensions, as well.

We start by considering the linear eigenvalue problem associated with (1.1)

$$-\frac{d^2\varphi_{k,\alpha}}{dx^2} + V(x)\varphi_{k,\alpha} = E_\alpha(k)\varphi_{k,\alpha} \tag{1.2}$$

where $\varphi_{k,\alpha}$ has Bloch–Floquet functions (BFs) $\varphi_{k,\alpha} = e^{ikx}u_{k,\alpha}(x)$, with $u_{k,\alpha}(x)$ periodic with period L; α labels the energy bands $E_\alpha(k)$. It is well known [47] that $E_\alpha(k + 2\pi/L) = E_\alpha(k)$. The energy can therefore be represented as a Fourier series

$$E_\alpha(k) = \sum_n \hat{\omega}_{n,\alpha}\, e^{iknL}, \qquad \hat{\omega}_{n,\alpha} = \hat{\omega}_{-n,\alpha} = \hat{\omega}^*_{n\alpha} \tag{1.3}$$

where an asterisk stands for complex conjugation and

$$\hat{\omega}_{n,\alpha} = \frac{L}{2\pi}\int_{-\pi/L}^{\pi/L} E_\alpha(k)e^{-iknL}\,dk. \tag{1.4}$$

The BFs constitute an orthogonal basis; however, instead of that basis, we will use the Wannier function (WF) one. The WF centered around the position nL (n is an integer) and corresponding to the band α is defined as

$$w_\alpha(x - nL) = \sqrt{\frac{L}{2\pi}}\int_{-\pi/L}^{\pi/L}\varphi_{k,\alpha}(x)e^{-inkL}\,dk. \tag{1.5}$$

Conversely,

$$\varphi_{k,\alpha}(x) = \sqrt{\frac{L}{2\pi}}\sum_{n=-\infty}^{\infty}w_{n,\alpha}(x)e^{inkL}. \tag{1.6}$$

---

[1] It should be noted that it does not escape us that more elaborate reduction models have been proposed for the one-dimensional reduction of the original three-dimensional problem, e.g., in the setting of BECs, see, e.g., [46]. Here, we will nevertheless focus on the simpler cubic case, which is also the low-density limit of such models.

The WFs also form a complete orthonormal (with respect to both n and α) set of functions,

$$\int_{-\infty}^{\infty} w_{n,\alpha}^*(x) w_{n',\alpha'}(x)\, dx = \delta_{\alpha\alpha'}\delta_{nn'}$$

$$\sum_{n,\alpha} w_{n,\alpha}^*(x') w_{n,\alpha}(x) = \delta(x-x')$$

which, by properly choosing the phase of the BFs in (1.5), can be made real and exponentially decaying at infinity [47]. We therefore assume this choice: $w_{n,\alpha}^*(x) = w_{n,\alpha}(x)$. At the heart of the derivation of the DNLS equation lies the decomposition of the solution of (1.1) in the basis of WFs (given the completeness of this basis)

$$\psi(x,t) = \sum_{n\alpha} c_{n,\alpha}(t) w_{n,\alpha}(x). \tag{1.7}$$

This decomposition is then substituted in (1.1) yielding

$$i\frac{dc_{n,\alpha}}{dt} = \sum_{n_1} c_{n_1,\alpha} \hat{\omega}_{n-n_1,\alpha} + \sigma \sum_{\alpha_1,\alpha_2,\alpha_3} \sum_{n_1,n_2,n_3} c_{n_1,\alpha_1}^* c_{n_2,\alpha_2} c_{n_3,\alpha_3} W_{\alpha\alpha_1\alpha_2\alpha_3}^{nn_1n_2n_3} \tag{1.8}$$

where

$$W_{\alpha\alpha_1\alpha_2\alpha_3}^{nn_1n_2n_3} = \int_{-\infty}^{\infty} w_{n,\alpha} w_{n_1,\alpha_1} w_{n_2,\alpha_2} w_{n_3,\alpha_3}\, dx \tag{1.9}$$

are overlapping matrix elements. The expression $W_{\alpha_1\alpha_2\alpha_3\alpha_4}^{n_1n_2n_3n_4}$ is symmetric with respect to all permutations within the groups of indices $(\alpha,\alpha_1,\alpha_2,\alpha_3)$ and (n,n_1,n_2,n_3). Eq. (1.8) can be viewed as a vector form of the DNLS equation for $\mathbf{c}_n = \mathrm{col}(c_{n1}, c_{n2}, \ldots)$ with non-nearest-neighbor interactions in its general form. A key question then concerns the potential simplifications and the conditions under which it can be reduced to single-component DNLS equation.

For sufficiently rapid decay of the Fourier coefficients in (1.3) and $|\hat{\omega}_{1,\alpha}| \gg |\hat{\omega}_{n,\alpha}|$, $n > 1$ the non-nearest-neighbor coupling terms can be neglected in the linear part of Eq. (1.8), leading to a dynamical model accounting solely for nearest-neighbor interactions.

Secondly, since $w_{n,\alpha}(x)$ is localized and centered around $x = nL$, one can assume that in some cases among all the coefficients $W_{\alpha\alpha_1\alpha_2\alpha_3}^{nn_1n_2n_3}$ those with $n = n_1 = n_2 = n_3$ are dominant and other terms can be neglected, since they are exponentially weaker. Then, one arrives at the equation

$$i\frac{dc_{n,\alpha}}{dt} = \hat{\omega}_{0,\alpha} c_{n,\alpha} + \hat{\omega}_{1,\alpha} \left(c_{n-1,\alpha} + c_{n+1,\alpha} \right) + \sigma \sum_{\alpha_1,\alpha_2,\alpha_3} W_{\alpha\alpha_1\alpha_2\alpha_3}^{nnnn} c_{n,\alpha_1}^* c_{n,\alpha_2} c_{n,\alpha_3}$$

which becomes the tight-binding DNLS model

1.2 Prototypical Derivation of the DNLS

$$i\frac{dc_{n,\alpha}}{dt} = \hat{\omega}_{0,\alpha} c_{n,\alpha} + \hat{\omega}_{1,\alpha}\left(c_{n-1,\alpha} + c_{n+1,\alpha}\right) + \sigma W_{1111}^{nnnn}|c_{n,\alpha}|^2 c_{n,\alpha} \quad (1.10)$$

by restricting consideration only to the band α. One of the advantages of this derivation is that it shows directly how to generalize the single band approximation, both in the direction of including additional bands when relevant (i.e., when the interband coupling terms are comparable to the intra-band ones) and also toward that of including additional neighbors within a band when $\hat{\omega}_{n,\alpha}$ for $n > 1$ become sizeable with respect to $\hat{\omega}_{1,\alpha}$. Both of these properties are determined by the linear properties of the model, which yields $\hat{\omega}_{n,\alpha}$ and $w_{n,\alpha}(x)$ and the specific form of the nonlinearity (which determines the particular form of the overlap coefficients to be compared between bands and between sites).

A few words should be mentioned here about the history of the derivation of the DNLS equation. The equation first appeared in connection to Davydov's model in biophysics in [48, 49], as well as in a more general form, namely as the so-called discrete self-trapping (DST) equation

$$i\dot{u}_n = -\epsilon \sum_k m_{nk} u_k - |u_n|^2 u_n \quad (1.11)$$

in [50]. The DST generalization degenerates into the DNLS equation for merely nearest-neighbor couplings.

In the setting of nonlinear optics, it was first presented in [11], as stemming from a coupled mode theory approach in the case of identical, regularly spaced waveguides where the refractive index of the nonlinear material of the array increases linearly with the intensity of the optical field. A related derivation appears in the Introduction section of [8], where the starting point is the two-dimensional Maxwell's equation of the form

$$\psi_{zz} + \psi_{xx} + (f(x) + \delta|\psi|^2)\psi = 0. \quad (1.12)$$

In this setting, z is the propagation direction (as is customarily the case for the problems that we will consider in optics), $f(x)$ represents the periodically varying in x linear index of refraction, and the term proportional to the small parameter δ (in that derivation) represents the so-called Kerr nonlinear effect, whereby the refractive index is proportional to the intensity of light. The derivation then proceeds along lines parallel to those illustrated above, but also considering a slow envelope depending on $Z = \delta z$. A similar derivation can be performed even for a regular cubic nonlinearity $\propto \psi^3$, instead of $|\psi|^2\psi$, with the additional assumption of the so-called rotating wave approximation where only the terms proportional to a dominant frequency are kept, see, e.g., [12].

In the context of BECs, the equation re-emerged through the important work of [53] (see also the nearly concurrent work of [54]), where the dynamics in the presence of an OL was analyzed in the context of this equation. In the above exposition, we followed the analysis of [5], where systematic details of the derivation were provided, as well as estimates of the various coefficients (such as $\hat{\omega}$ and W above) were given for a prototypical periodic potential of the form $V(x) = A\cos(2x)$ (the interested reader is referred to that work for more details). It should also be noted that later works such as [56] also considered additional terms in the derivation and obtained more complex models that will be discussed in the last chapter of this volume. Similar results were also obtained in the optical context in the work of [57].

References

1. Sulem C., Sulem, P.L.: The Nonlinear Schrödinger Equation,Springer-Verlag, New York (1999)
2. Ablowitz, M.J., Prinari, B., Trubatch, A.D.: Discrete and Continuous Nonlinear Schrödinger Systems, Cambridge University Press, Cambridge (2004)
3. Hasegawa, A.: Solitons in Optical Communications, Clarendon Press, Oxford, NY (1995)
4. Malomed, B.A.: Progr. Opt. **43**, 71 (2002)
5. Zakharov, V.E.: Collapse and Self-focusing of Langmuir Waves, Handbook of Plasma Physics, Rosenbluth, M.N., Sagdeev R.Z. (eds), vol. 2 (Galeev, A.A., Sudan, R.N. eds.), 81–121, Elsevier, Amsterdam (1984)
6. Zakharov, V.E.: Sov. Phys. JETP **35**, 908 (1972)
7. Onorato, M., Osborne, A.R., Serio, M., Bertone, S.: Phys. Rev. Lett. **86**, 5831 (2001)
8. Kevrekidis, P.G., Rasmussen, K.Ø., Bishop, A.R.: Int. J. Mod. Phys. B **15**, 2833 (2001)
9. Eisenberg, H.S., Silberberg, Y., Morandotti, R., Boyd, A.R., Aitchison, J.S.: Phys. Rev. Lett. **81**, 3383 (1998)
10. Morandotti, R., Peschel, U., Aitchison, J.S., Eisenberg, H.S., Silberberg, Y.: Phys. Rev. Lett. **83**, 2726 (1999)
11. Eisenberg, H.S., Silberberg, Y., Morandotti, R., Aitchison, J.S.: Phys. Rev. Lett. **85**, 1863 (2000)
12. Mandelik, D., Morandotti, R., Aitchison, J.S., Silberberg, Y.: Phys. Rev. Lett. **92**, 93904 (2004)
13. Morandotti, R., Eisenberg, H.S., Silberberg, Y., Sorel, M., Aitchison, J.S.: Phys. Rev. Lett. **86**, 3296 (1999)
14. Christodoulides, D.N., Lederer, F., Silberberg, Y.: Nature **424**, 817 (2003)
15. Sukhorukov, A.A., Kivshar, Y.S., Eisenberg, H.S., Silberberg, Y.: IEEE J. Quant. Elect. **39**, 31 (2003)
16. Flach, S., Willis, C.R.: Phys. Rep. **295**, 181 (1998)
17. Campbell, D.K., Flach, S., Kivshar, Y.S.: Phys. Today **57**, 43 (2004)
18. Lederer, F., Stegeman, G.I., Christodoulides, D.N., Assanto, G., Segev, M., Silberberg, Y.: Phys. Rep. **463**, 1 (2008)
19. Efremidis, N.K., Sears, S., Christodoulides, D.N., Fleischer, J.W., Segev, M.: Phys. Rev. E **66**, 46602 (2002)
20. Fleischer, J.W., Segev, M., Efremidis, N.K., Christodoulides, D.N.: Nature **422**, 147 (2003)
21. Fleischer, J.W., Carmon, T., Segev, M., Efremidis, N.K., Christodoulides, D.N.: Phys. Rev. Lett. **90**, 23902 (2003)
22. Neshev, D., Ostrovskaya, E., Kivsharand, Yu.S., Krolikowski, W.: Opt. Lett. **28**, 710 (2003)
23. Martin, H., Eugenieva, E.D., Chen, Z., Christodoulides, D.N.: Phys. Rev. Lett. **92**, 123902 (2004)
24. Yang, J., Makasyuk, I., Bezryadina, A., Chen, Z.: Opt. Lett. **29**, 1662 (2004)
25. Yang, J., Makasyuk, I., Bezryadina, A., Chen, Z.: Stud. Appl. Math. **113**, 389(2004)
26. Yang, J., Makasyuk, I., Kevrekidis, P.G., Martin, H., Malomed, B.A., Frantzeskakis, D.J., and Zhigang C.: Phys. Rev. Lett. **94**, 113902(2005)

References

27. Neshev, D., Kivshar, Yu.S., Martin, H., Chen, Z.: Opt. Lett. **29**, 486–488 (2004)
28. Neshev, D.N., Alexander, T.J., Ostrovskaya, E.A., Kivshar, Yu.S., Martin, H., Makasyuk, I., Chen, Z.: Phys. Rev. Lett. **92**, 123903(2004)
29. Fleischer, J.W., Bartal, G., Cohen, O., Manela, O., Segev, M., Hudock, J., Christodoulides, D.N.: Phys. Rev. Lett. **92**,123904 (2004)
30. Kartashov, Y.V., Vysloukh, V.A., Torner, L.: Phys. Rev. Lett. **93**, 093904(2004)
31. Wang, X., Chen, Z., Kevrekidis, P.G.: Phys. Rev. Lett. **96**, 083904 (2006)
32. Fleischer, J.W., Bartal, G., Cohen, O., Schwartz, T., Manela, O., Freedman, B., Segev, M., Buljan, H., Efremidis, N.K.: Opt. Express **13**, 1780(2005)
33. Chen, Z., Martin, H., Bezryadina, A., Neshev, D.N., Kivshar, Yu.S., Christodoulides, D.N.: J. Opt. Soc. Am. B **22**, 1395–1405 (2005)
34. Burger, S., Cataliotti, F.S., Fort, C., Maddaloni, P., Minardi, F., Inguscio, M.: Europhys. Lett. **57**, 1 (2002)
35. Smerzi, A., Trombettoni, A., Kevrekidis, P.G., Bishop, A.R.: Phys. Rev. Lett. **89**, 170402 (2002)
36. Cataliotti, F.S., Fallani, L., Ferlaino, F., Fort, C., Maddaloni, P., Inguscio, M.: New J. Phys. **5**, 71 (2003)
37. Eiermann, B., Anker, Th., Albiez, M., Taglieber, M., Treutlein, P., Marzlin, K.-P., Oberthaler, M.K., Phys. Rev. Lett. **92**, 230401 (2004)
38. Jona-Lasinio, M., Morsch, O., Cristiani, M., Malossi, N., Müller, J.H., Courtade, E., Anderlini, M., Arimondo, E.: Phys. Rev. Lett. **91**, 230406 (2003)
39. Anderson, B.P., Kasevich, M.A.: Science **282**, 1686 (1998)
40. Brazhnyi, V.A., Konotop, V.V.: Mod. Phys. Lett. B**18**, 627 (2004)
41. Kevrekidis, P.G., Frantzeskakis, D.J.: Mod. Phys. Lett. B **18**, 173 (2004)
42. Morsch, O., Oberthaler, M.: Rev. Mod. Phys. **78**, 179 (2006)
43. Sato, M., Hubbard, B.E., Sievers, A.J.: Rev. Mod. Phys. **78**, 137 (2006)
44. Peyrard, M.: Nonlinearity **17**, R1 (2004)
45. Sato, M., Yasui, S., Kimura, M., Hikihara, T., Sievers, A.J.: Europhys. Lett. **80**, 30002 (2007)
46. Salasnich, L., Parola, A., Reatto, L.: Phys. Rev. A **65**, 043614 (2002)
47. Kohn, W.: Phys. Rev. **115**, 809 (1959)
48. Scott, A.C.: Philos. Trans. R. Soc. London Ser. A **315**, 423 (1985)
49. Scott, A.C., Macneil, L.: Phys. Lett. A **98**, 87 (1983)
50. Eilbeck, J.C., Lomdahl, P.S., Scott, A.C.: Physica D **16**, 318 (1985)
51. Christodoulides, D.N., Joseph, R.I.: Opt. Lett. **13**, 794 (1988)
52. Kivshar, Yu.S., Peyrard, M.: Phys. Rev. A **46**, 3198 (1992)
53. Trombettoni, A., Smerzi, A.: Phys. Rev. Lett. **86**, 2353 (2001)
54. Abdullaev, F.Kh., Baizakov, B.B., Darmanyan, S.A., Konotop, V.V., Salerno, M.: Phys. Rev. A **64**, 043606 (2001)
55. Alfimov, G.L., Kevrekidis, P.G., Konotop, V.V., Salerno, M.: Phys. Rev. E **66**, 046608 (2002)
56. Menotti, C., Smerzi, A., Trombettoni, A.: New J. Phys. **5**, 112 (2003)
57. Öster, M., Johansson, M., Eriksson, A.: Phys. Rev. E **67**, 056606 (2003)

Chapter 2
The One-Dimensional Case

We now focus on the analysis of the one-dimensional DNLS equation of the form

$$i\dot{u}_n = -\epsilon(u_{n+1} + u_{n-1} - 2u_n) + \beta|u_n|^2 u_n. \tag{2.1}$$

(In the above form of Eq. (2.1), one of ϵ and β can be scaled out; e.g., β can be scaled out up to a sign by $u \to u\sqrt{|\beta|}$) Note that in the next few chapters we will be considering the focusing (attractive interaction in BEC) case of $\beta < 0$; the defocusing nonlinearity of $\beta > 0$ will be treated in a separate chapter. Our analysis in both this and in the following chapters will revolve around fundamental and excited state solutions of the equation, their stability, and dynamics. Perhaps the most fundamental among these is the single-pulse solitary wave that we now turn to.

2.1 Single-Pulse Solitary Waves

2.1.1 The Continuum Approach

2.1.1.1 General Properties of the Continuum Problem

We start by considering such pulses in the continuum limit. Given the opportunity, we also present here an interlude with some fundamental features of the one-dimensional continuum NLS equation of the general form

$$iu_t = -u_{xx} - |u|^{2\sigma} u \tag{2.2}$$

(where the subscript x, t denote partial derivatives with respect to the corresponding variable). For more details, the interested reader is referred to [1].

Equation (2.2) is a Hamiltonian system, but with infinite degrees of freedom (i.e., a "field theory"). As such, we expect that it will have a Lagrangian and a Hamiltonian *density*. Indeed, the Lagrangian density for the model is

$$\mathcal{L} = \frac{i}{2}(u^\star u_t - u u_t^\star) - |u_x|^2 + \frac{1}{\sigma+1}|u|^{2\sigma+2}. \tag{2.3}$$

Then, the corresponding partial differential equation (in particular, Eq. (2.2) in this case) is derived as the Euler–Lagrange equation of the field theory [1] according to

$$0 = \frac{\delta L}{\delta u} = \frac{\partial \mathcal{L}}{\partial u} - \partial_x \left(\frac{\partial \mathcal{L}}{\partial u_x} \right) - \partial_t \left(\frac{\partial \mathcal{L}}{\partial u_t} \right). \tag{2.4}$$

where $\delta L/\delta u$ symbolizes the Fréchet derivative [2] of the Lagrangian $L = \int \mathcal{L} dx$.

It can also be shown that if the action $S = \int L dt = \int \mathcal{L} dx dt$ is invariant under a transformation $x \to x + \delta x$, $t \to t + \delta t$ and $u \to u + \delta u$, then the quantity

$$I = \int dx \left[\frac{\partial \mathcal{L}}{\partial u_t} (u_t \delta t + u_x \delta x - \delta u) + C.C. - \mathcal{L} \delta t \right] \tag{2.5}$$

is conserved. This is the celebrated Noether theorem. Its proof requires the use of calculus of variations and we omit it here (the interested reader can find a detailed derivation in Sect. 2.2 of [1]).

For the particular Hamiltonian system of interest here, we have the following invariances (and corresponding conservation laws):

- if we use the transformation $u \to v = ue^{is}$ where s is space and time independent, then the equation for v is the same as the one for u. Hence there is a phase degeneracy/invariance in the system. The generator of the corresponding invariance is found as $v \approx u + \delta u$, with $\delta u = isu$ (the leading order expansion of the above mentioned exponential phase factor). Hence, using $\delta u = isu$ and $\delta x = \delta t = 0$, we obtain that

$$P = ||u||_{L^2}^2 = \int |u|^2 dx \tag{2.6}$$

is conserved. This states that the (squared) L^2 norm is conserved by the dynamics of Eq. (2.2). This has a meaningful physical interpretation, e.g., in optics or BEC since in the former it states that the power of the beam is conserved, while in the latter it denotes the physically relevant conservation of the number of atoms in the condensate. This invariance is often referred to as the phase or gauge invariance of the NLS.
- Spatial translation $x \to x + \delta x$ also leaves Eq. (2.2) invariant. If we use $\delta t = \delta u = 0$ in Eq. (2.5), we obtain the conservation of linear momentum (just as in low-dimensional Hamiltonian systems of classical mechanics) of the form

$$M = i \int \left(u u_x^\star - u^\star u_x \right) dx. \tag{2.7}$$

Hence, translational invariance results in momentum conservation.

2.1 Single-Pulse Solitary Waves

- Finally, time translation $t \to t + \delta t$ also leaves the dynamical equation invariant, hence using $\delta x = \delta u = 0$ in Eq. (2.5) results in the conservation of the Hamiltonian (i.e., the energy) of the system

$$H = \int \left(|u_x|^2 - \frac{1}{2\sigma + 2} |u|^{\sigma+1} \right) dx. \tag{2.8}$$

The integrand of Eq. (2.8) then represents the Hamiltonian density of the system. One can then restate the problem in the Hamiltonian (as opposed to the Lagrangian) formulation by means of Hamilton's equations and/or using the structure of the Poisson brackets, e.g.,

$$i u_t = \frac{\delta H}{\delta u^\star} = \{H, u\}, \tag{2.9}$$

where the standard Poisson bracket has been used.

These are general symmetries/invariances that are present for any value of σ. We now turn to the specific, the so-called integrable case of $\sigma = 1$. The integrability of this particular case means that apart from these three above defined integrals of motion, there are infinitely many others. The unusual feature of such an integrable nonlinear partial differential equation (PDE) is that once the initial data is prescribed, we can solve the PDE for all times [3]. The nonlinear wave solutions to such PDEs are often referred to as solitons, because they are solitary coherent structures (i.e., nonlinear waves) which emerge unscathed from their interaction with other such structures.

Here we focus on the standing wave solitons of the $\sigma = 1$ case, as the main solution of the solitary pulse type. Our exposition will highlight the features of this main "building block" of the NLS equation and will show how it is modified in the presence of the non-integrable perturbation imposed by discreteness.

Such standing wave solutions can be straightforwardly obtained in an explicit form

$$u = (2\Lambda)^{1/2} \mathrm{sech}(\Lambda^{1/2}(x - ct - x_0)) e^{i\left(\frac{c}{2}x + (\Lambda - \frac{c^2}{4})t\right)}, \tag{2.10}$$

where Λ is the frequency of the wave, x_0 the initial position of its center, and c its speed. Since, there is an additional Galilean invariance, that allows us to boost a given solution to any given speed c, we will mostly focus on solutions with $c = 0$ hereafter. (Note, however, that discreteness does not preserve this invariance, hence the issue of traveling becomes an especially delicate one in the discrete case, as discussed in Part II.) These are often referred to as standing waves or occasionally as breathers (because of their periodicity in time and exponential localization in space).

Naturally, once such solutions are identified, the immediate next question concerns their stability. This can be identified at a first (but still particularly useful) step by means of linear stability analysis. Using the ansatz

$$u = e^{i\Lambda t}(u_0(x) + \varepsilon(v + iw)) \tag{2.11}$$

in Eq. (2.2), where $u_0(x) = (2\Lambda)^{1/2}\text{sech}(\Lambda^{1/2}(x - x_0))$, one obtains the linear stability equations by the $O(\varepsilon)$ expansion. It can be easily seen that the $O(1)$ equation is, by construction, identically satisfied (it is the equation satisfied by the solitary wave). The $O(\varepsilon)$ equations read as follows:

$$v_t = L_- w = \left(-\Delta + \Lambda - u_0^{2\sigma}\right) w, \tag{2.12}$$

$$w_t = -L_+ v = -\left(-\Delta + \Lambda - (2\sigma + 1)u_0^{2\sigma}\right) v. \tag{2.13}$$

In Eqs. (2.12) and (2.13), Δ denotes the second spatial derivative (per its natural higher dimensional generalization, namely the Laplacian). Note that we give the general form of the linear stability problem, even though for the time being we are interested in the particular case of $\sigma = 1$. Separating space and time variables for the solutions of the resulting equations (2.12) and (2.13) as $v(x, t) = e^{\lambda t}\tilde{v}(x)$ and $w(x, t) = e^{\lambda t}\tilde{w}(x)$, we obtain the eigenvalue problem in the form

$$\lambda^2 \tilde{v} = -L_- L_+ \tilde{v} \tag{2.14}$$

and similarly $\lambda^2 \tilde{w} = -L_+ L_- \tilde{w}$.

The invariances of the original equation are now mirrored in the zero eigenvalues of the linearization problem of Eq. (2.14). In particular, it is easy to check that for any solution of the form $u = e^{i\Lambda t} u_0(x)$, the spatial derivative du_0/dx corresponds to an eigenvector with a pair of zero eigenvalues since $L_+ du_0/dx = 0$. Similarly, the phase invariance leads to another pair of zero eigenvalues since $L_- u_0 = 0$. Hence, the linearization around the pulse-like soliton solutions of Eq. (2.10) will contain four eigenvalues at $\lambda = 0$. The algebraic multiplicity of the eigenvalues at the origin is four, but the geometric multiplicity is two. That is, each of the eigenvalues has an eigenvector and a *generalized* eigenvector associated with it. For example, the phase invariance has a generalized eigenvector $v = \partial u_0/\partial \Lambda$ [4, 5], satisfying $L_+ v = -u_0$. Similarly, there is a generalized eigenvector of translation proportional to $(x - x_0)u_0$ [4].

Furthermore, the linearization problem of Eqs. (2.12) and (2.13) will contain continuous spectrum. The latter consists of small amplitude, extended in space, plane wave eigenfunctions of the form $v + iw \propto e^{i(kx - \omega t)}$ (see [4] for their precise functional form). These satisfy the linear dispersion relation (upon substitution into the above equations) of the form

$$\omega^2 = -\lambda^2 = \pm\left(\Lambda + k^2\right). \tag{2.15}$$

Hence, in this case, the spectral plane (λ_r, λ_i) of the eigenvalues $\lambda = \lambda_r + i\lambda_i$ of the linearization around a soliton of the top panel of Fig. 2.1 will have a form such as the one given in the bottom panel of Fig. 2.1.

2.1 Single-Pulse Solitary Waves

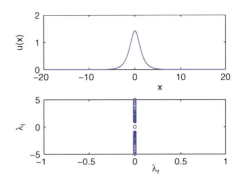

Fig. 2.1 The *top panel* shows the continuum soliton of the NLS equation for $\Lambda = 1$. The *bottom panel* shows the corresponding spectral plane of eigenvalues (λ_r, λ_i) in this integrable case. Four eigenvalues are at $\lambda = 0$ due to the invariances (see text) and the rest reside in the continuous spectrum whose band edge is at $\lambda = \pm \Lambda i$

2.1.1.2 From Continuum to Discrete

The most straightforward discretization of the NLS equation

$$i\dot{u}_n = -\epsilon \Delta_2 u_n + \beta |u_n|^{2\sigma} u_n, \qquad (2.16)$$

where $\epsilon = 1/h^2$, where h plays the role of the discrete lattice spacing and $\Delta_2 u_n = u_{n+1} + u_{n-1} - 2u_n$ is the discrete Laplacian with unit spacing. The DNLS model is also Hamiltonian with

$$H_{DNLS} = -\sum_{n=-\infty}^{\infty} \left[\epsilon |u_n - u_{n-1}|^2 + \frac{\beta}{\sigma + 1} |u_n|^{2\sigma+2} \right] \qquad (2.17)$$

and can be derived from H_{DNLS} as

$$\dot{u}_n = \{H_{DNLS}, u_n\}, \qquad (2.18)$$

using the Poisson brackets

$$\{u_m, u_n^\star\} = i\delta_{m,n}, \quad \{u_m, u_n\} = \{u_m^\star, u_n^\star\} = 0. \qquad (2.19)$$

As we indicated above, the case of $\sigma = 1$ is integrable in the continuum limit. In the discrete case, the above-mentioned DNLS is a non-integrable discretization of the continuum model. On the other hand, however, there does exist an integrable discretization, namely the so-called Ablowitz–Ladik (AL-NLS) discretization of the NLS equation [6, 7]

$$i\dot{u}_n = -\epsilon \Delta_2 u_n + \frac{\beta}{2}(u_{n+1} + u_{n-1})|u_n|^2 \qquad (2.20)$$

with similar notation as used in Eq. (2.16). In the case of the AL-NLS, the Hamiltonian is of the form

$$H_{AL\text{-}NLS} = -\sum_{n=-\infty}^{\infty} \left[\epsilon |u_n - u_{n-1}|^2 + \frac{1}{\beta} \ln\left(1 + \beta |u_n|^2\right) \right]. \tag{2.21}$$

The AL-NLS equation is derived from $H_{AL\text{-}NLS}$ using the non-standard Poisson brackets

$$\{u_m, u_n^\star\} = i\delta_{m,n}(1 + \beta |u_n|^2), \quad \{u_m, u_n\} = \{u_m^\star, u_n^\star\} = 0. \tag{2.22}$$

Additionally to the Hamiltonian, both the DNLS and the AL-NLS equation conserve a quantity that is commonly referred to as the norm or power of the solution

$$P_{DNLS} = \sum_{n=-\infty}^{\infty} |u_n|^2, \tag{2.23}$$

while for the AL-NLS equation it is

$$P_{AL\text{-}NLS} = \sum_{n=-\infty}^{\infty} \frac{1}{\beta} \log\left(1 + \beta |u_n|^2\right). \tag{2.24}$$

This conservation law, analogously to the corresponding conservation law for the continuum NLS equation, mirrors the U(1) symmetry or gauge invariance of the discrete model, i.e., its invariance with respect to an overall phase factor.

One of the fundamental differences between the DNLS and the AL-NLS model is the existence of a momentum conservation law in the latter which is *absent* in the former. In particular, the momentum

$$M = i \sum_{n=-\infty}^{\infty} \left(u_{n+1}^\star u_n - u_{n+1} u_n^\star\right) \tag{2.25}$$

is conserved in the case of the AL-NLS model; this, in turn, implies that its localized solutions can be centered *anywhere* within the discrete lattice. In fact, in the AL-NLS case, there exist exact soliton solutions which are of the form (for simplicity, setting $\epsilon = -\beta/2 = 1$)

$$u_n = \sinh(\gamma)\,\text{sech}\,(\gamma(n - \xi))\exp(i\delta(n - \xi) + \rho), \tag{2.26}$$

where $\dot{\xi} = 2\sinh(\gamma)\sin(\delta)/\gamma$ and $\dot{\rho} = -2 + 2\cos(\delta)\cosh(\gamma) + 2\delta\sin(\delta)\sinh(\gamma)/\gamma$ [8]. Then, the translational invariance is evident in the presence of an undetermined integration constant in the ordinary differential equation (ODE) for the time evolution of the position of the soliton center ξ.

In the DNLS case, the absence of translational invariance no longer permits to the solution to be arbitrarily centered anywhere along the lattice. Instead, there are only *two* stationary solutions (modulo the integer shift invariance of the lattice),

2.1 Single-Pulse Solitary Waves

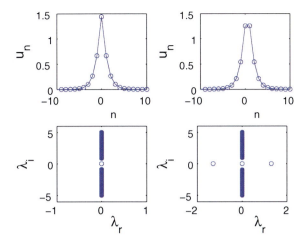

Fig. 2.2 The *top panel* shows the discrete on-site solitary wave (*left*) and inter-site solitary wave (*right*) for $\epsilon = 1$. The *bottom panels* show the corresponding linear stability eigenvalues, illustrating the linear stability of the former and linear instability of the latter (due to a real eigenvalue pair). There is only a single pair of eigenvalues at the origin, due to the phase invariance; note also the upper bound of the continuous spectrum (a feature absent in the continuum limit, where the continuous spectrum extends to $\pm i\infty$)

one centered on a lattice site (on-site) and one centered between two adjacent lattice sites (i.e., inter-site or bond-centered solution). These two solutions are shown in the top panels of Fig. 2.2. Such localized modes were initially proposed in a different dynamical lattice context in [9] and [10]; a relevant discussion of these modes in DNLS can be found in [11].

A fundamental level of understanding of this feature (although partially heuristic) can be obtained by using the continuum soliton solution of Eq. (2.10) as an *ansatz* in the formula for the *discrete energy* of Eq. (2.17). This is a "collective coordinate" type approach using as the relevant coordinate the position of the pulse center x_0. The resulting expression for the (discrete) Hamiltonian is given by

$$H_{DNLS} = \frac{16\pi^2}{h^2} \sum_{m=1}^{\infty} \frac{m \cos\left(\frac{2\pi m x_0}{h}\right)}{\sinh\left(\frac{m\pi^2}{\sqrt{\Lambda h}}\right)} \left[\epsilon - \frac{\Lambda}{3}\left(1 + \frac{m^2 \pi^2}{\Lambda h^2}\right)\right], \quad (2.27)$$

where it should be kept in mind that $\epsilon = 1/h^2$. To derive the above expression [12], terms independent of x_0 have been neglected in the energy (cf. also with the momentum invariant for the AL lattice and its relevant calculation in [13]), and the Poisson summation formula [14] has been critically used, according to which

$$\sum_{n=-\infty}^{\infty} f(\beta n) = \frac{\sqrt{2\pi}}{\beta} \sum_{m=-\infty}^{\infty} F\left(\frac{2m\pi}{\beta}\right), \quad (2.28)$$

where F is the Fourier transform of f,

$$F(k) = \frac{1}{\sqrt{2\pi}} \int_{-\infty}^{\infty} f(x)e^{ikx}dx. \quad (2.29)$$

The above calculation yields the so-called Peierls–Nabarro (PN) [15, 16] potential which is famous from the theory of crystal dislocations representing the energy landscape that a dislocation faces in a crystal lattice. Of particular importance is the so-called PN barrier, namely the energy barrier that needs to be overcome in order for the coherent structure to travel a distance of one lattice site. In the present setting, the PN barrier can be easily found from the above expression for H_{DNLS} to be

$$H_{PN} = \frac{32\pi^2}{h^2} \sum_{m=1}^{\infty} \frac{m(1-(-1)^m)}{\sinh\left(\frac{m\pi^2}{\sqrt{\Lambda}h}\right)} \left[\epsilon - \frac{\Lambda}{3}\left(1 + \frac{m^2\pi^2}{\Lambda h^2}\right)\right]. \quad (2.30)$$

(We will not attempt to evaluate this expression and compare it with numerical results for reasons that will be explained in Part II).

However, the expression of Eq. (2.30) bears an additional piece of information. In the continuum limit of $h \to 0$, the above examined energy is *independent* of the position of the pulse center due to the translational invariance of the continuum model, i.e., $H(x_0)$ is a *constant* function of x_0. Therefore, in the case where the symmetry is broken (for $h \neq 0$), the height of the periodic energy barrier represents a quantitative measure of "how much" the invariance is broken. This amount appears, in Eq. (2.30), to be *exponentially small* in the relevant parameter which is the lattice spacing h (note the dominant hyperbolic sine terms in the denominator proportional to $\sinh(m\pi^2/\sqrt{\Lambda}h)$). It is perhaps interesting to attempt to justify this exponential smallness in a qualitative way as follows: if we try to expand the operator $\Delta_2 u_n$ by means of a Taylor expansion, we obtain

$$\Delta_2 u_n = \sum_{j=1}^{\infty} \frac{2h^{2j}}{(2j)!} \frac{d^{2j}u}{dx^{2j}}. \quad (2.31)$$

However, *to all algebraic orders* in this power-law expansion of the discrete operator, the right-hand side of Eq. (2.31) contains *derivatives*. However, derivatives are translationally invariant objects. Hence, in order to be able to observe the *breaking* of the symmetry, one has to go *beyond all algebraic orders* in the power-law expansion and hence has to become *exponentially small* in h, just as the energy landscape of Eq. (2.27) suggests.

We are now in a position to discuss the linear stability problem at the discrete level of the DNLS equation. Starting with the linear stability ansatz through the discrete analog of Eq. (2.11) (where each term should be thought of as having a subscript n indexing the lattice sites), we obtain the analog of the eigenvalue equations (2.12) and (2.13) for Eq. (2.1)

2.1 Single-Pulse Solitary Waves

$$\lambda v_n = L_- w_n = (-\epsilon \Delta_2 + \Lambda - \beta u_{0,n}^2) w_n, \quad (2.32)$$

$$\lambda w_n = -L_+ v_n = -(-\epsilon \Delta_2 + \Lambda - 3\beta u_{0,n}^2) v_n. \quad (2.33)$$

In the lattice case, Eqs. (2.32) and (2.33) represent a matrix eigenvalue problem for the eigenvalues λ and the eigenvectors $(v_n, w_n)^T$ that is subsequently solved numerically.

A small remark should be added here about eigenvalue notation. In the physics literature, it is quite common to use the eigenfrequency ω, while in the more mathematically oriented texts, $\lambda = i\omega$ is used to denote eigenvalues. The two notations will be used interchangeably, with the understanding that linear instability is implied by non-zero imaginary part of ω or, equivalently, by the non-zero real part of λ.

The following features are generically observed in spectral plots analogous to the ones shown in Fig. 2.2 but for different values of the coupling strength ϵ:

- The continuous spectrum of plane wave eigenfunctions $\sim \exp(\pm i(qn - \omega t))$, exists also in the discrete problem and satisfies the following dispersion relations:

$$\omega = \Lambda + 4\epsilon \sin^2\left(\frac{q}{2}\right), \quad (2.34)$$

$$\omega = -\Lambda - 4\epsilon \sin^2\left(\frac{q}{2}\right). \quad (2.35)$$

As seen from Eqs. (2.34) and (2.35), this branch of the spectrum extends over the interval $\pm[\Lambda, \Lambda + 4\epsilon]$ (along the imaginary axis of the spectral plane).
- In addition, as indicated above, the preservation of the U(1) invariance under discretization leads to a pair of eigenvalues $\lambda^2 = 0$.
- The translational invariance breaking is, as argued above, one of the key features of the discrete problem in comparison to its continuum sibling. In the case of a site-centered (linearly stable) solution, as shown in Fig. 2.2 and justified later in this chapter, the bifurcation of the translational eigenvalues occurs along the imaginary axis, leading to linear stability. On the contrary, in the case of the bond-centered solutions, the breaking of the symmetry leads to bifurcation along the real axis, rendering such inter-site-centered solutions linearly unstable.
- Finally, as was originally illustrated in [17] and subsequently expanded in [5], as the lattice spacing increases (and the coupling strength $\epsilon = 1/h^2$ decreases), there is also a pair of eigenmodes that bifurcates from the lower edge of the continuous spectrum, becoming a point spectrum eigenvalue. This, so-called, internal mode bifurcation does not affect critically the stability of the fundamental solution (at least, in the one-dimensional problem – in higher dimensions, as we will see below in Chap. 3, Sect. 3.3.4, such bifurcations may affect the stability critically), as will be quantified in what follows.

The above information yields the full spectral information in the case of a fundamental solution (single pulse) (we will see below how this picture is modified for multipulse waveforms). In what follows, we attempt to quantify some of the features of the single-pulse linearization spectrum, namely the exponential bifurcation of the translational eigenvalue and the exponential, as well as power-law bifurcation of the internal mode from the band edge of the continuous spectrum.

We start by trying to capture the exponential bifurcation of the translational eigenmode using a more rigorous method. Specifically, we will use the discrete Evans function method developed in [18]. Here, we will outline the basic features of the method (we refer the reader to the original work of [18] for more details).

The eigenvalue problem can also be written as a first-order system

$$Y_{n+1} = A(\lambda, n) Y_n \tag{2.36}$$

with $\lambda = i\omega$. Equation (2.36) has solutions, Y^+ (Y^-) that decay exponentially as $n \to +\infty$ ($n \to -\infty$). Forming the wedge product of the two, we obtain an analytic function, the so-called Evans function, whose zeros, *by construction*, pertain to eigenvectors that span the subspace of intersection of the two spaces and hence decay for $n \to \pm\infty$. Hence, if we define

$$E(\lambda) = Y^+ \wedge Y^-, \tag{2.37}$$

the solutions of the linearization problem, λ, such that $E(\lambda) = 0$ form the *point spectrum* of eigenvalues with localized eigenfunctions in the linearization problem.

In order to evaluate the Evans function for the DNLS problem, we can use its analytic properties and Taylor expand it close to the origin of the spectral plane. To do this for the DNLS equation, we consider it as a perturbation of the AL-NLS equation. In the calculation below, for reasons of convenience, we will consider the case of $\epsilon h^2 = -\beta/2 = 1$ in Eq. (2.1); we will also denote the steady-state solution by v_n. Then, the steady-state problem can, upon setting $r_n = (v_n - v_{n-1})/h$, be rewritten as

$$v_{n+1} = \left(1 + \frac{h^2}{1 + h^2 v_n^2} \left(\Lambda - 2v_n^2\right)\right) v_n + hr_n + \varepsilon \frac{h^2 v_n^3}{1 + h^2 v_n^2} \left(\Lambda - 2v_n^2\right), \tag{2.38}$$

$$r_{n+1} = \frac{h}{1 + h^2 v_n^2} \left(\Lambda - 2v_n^2\right) v_n + r_n + \varepsilon \frac{h v_n^3}{1 + h^2 v_n^2} \left(\Lambda - 2v_n^2\right).$$

The above equation is written so that the limit of $\varepsilon = h^2 = 0$, Eq. (2.38), is the steady-state problem for the AL-NLS equation.

In the case of $\varepsilon = 0$, the exact (single soliton) solution of the AL-NLS equation is given by

$$Q_n(\xi) = \sqrt{\Lambda} \, \sinh(\alpha) \mathrm{sech}(\tilde{\alpha} n + \xi), \tag{2.39}$$

where

$$\cosh(\tilde{\alpha}) = 1 + \frac{\Lambda h^2}{2} \quad \text{and} \quad \sinh(\alpha) = \frac{\sinh(\tilde{\alpha})}{\sqrt{\Lambda} h}.$$

2.1 Single-Pulse Solitary Waves

In that case, it is also true that the steady-state problem has an infinite number of invariants $\cdots = I_n = I_{n+1} = \cdots =$ const. I_n is given by

$$I_n = h^2 v_n^2 v_{n-1}^2 + v_n^2 + v_{n-1}^2 - 2\mu v_n v_{n-1} \tag{2.40}$$

with $\mu = 1 + \Lambda h^2/2$.

Hence, following Ablowitz and Herbst [19, 20], in the perturbed (DNLS) case of $\varepsilon \neq 0$, the Melnikov function can be calculated as

$$\varepsilon M(\xi, \varepsilon) = \sum_{n=-\infty}^{\infty} \Delta_n I(\xi, \varepsilon) = \sum_{n=-\infty}^{\infty} \left[I(\mathbf{x}_{n+1}) - I(\mathbf{x}_n) \right], \tag{2.41}$$

where I is as defined above but in the $\varepsilon \neq 0$ case, it is no longer a constant. Equations (2.38) can be expressed in the form $\mathbf{x}_{n+1} = F(\mathbf{x}_n) + \varepsilon G(\mathbf{x}_n)$. Using this notation and Taylor expanding $H(F(\mathbf{x}_n) + \epsilon G(\mathbf{x}_n))$, we obtain for M the expression

$$M(\xi, \varepsilon) = \sum_{n=-\infty}^{\infty} \nabla I(\mathbf{x}_{n+1}) G(\mathbf{x}_n; \varepsilon). \tag{2.42}$$

This allows us to evaluate the Melnikov function (see [18] for details), up to corrections of $O(e^{-2\pi^2/\tilde{\alpha}})$, as

$$M(\xi, h) = C(h) e^{-\pi^2/\tilde{\alpha}} \sin\left(\frac{2\pi \xi}{\tilde{\alpha}}\right),$$

where

$$C(h) = 4\pi \frac{h}{\tilde{\alpha}} \left(\frac{2}{45} \frac{\pi^2}{\tilde{\alpha}^2} + \frac{2}{9} \frac{\pi^4}{\tilde{\alpha}^4} + \frac{8}{45} \frac{\pi^6}{\tilde{\alpha}^6} \right) \approx \frac{2147.8}{h^6}. \tag{2.43}$$

In the case of the AL-NLS equation, the "effective" (since ξ in Eq. (2.39) can take any value) translational invariance ensures that the stable and unstable manifolds of the homoclinic orbit intersect non-transversely. On the other hand, in the non-integrable case of the DNLS and generically (for $\varepsilon \neq 0$) this non-transversality will not persist. In fact, the splitting of the orbits (or effectively the angle of intersection of the manifolds) is given by the Melnikov function. Hence, in essence, the Melnikov function is a measure of the breaking of translational invariance and is expected to be transcendentally small (as in our previous calculation). This exponentially small splitting is associated with the above discussed PN barrier.

Having evaluated the Melnikov function, we can now proceed to the Taylor expansion of the Evans function of the DNLS equation, as a perturbation to the AL-NLS problem. In particular, the Evans function E is a function of the eigenvalue λ and the perturbation strength ε. Since for the AL-NLS equation $E(\lambda = 0; \epsilon = 0)$

= 0, we need to find the leading order derivatives with respect to ε and λ. As λ is an eigenvalue of algebraic multiplicity four in the AL lattice, the first derivative that will yield a non-zero contribution is $\partial_\lambda^4 E(0;0)$, which will generically be constant. The tools developed by Kapitula and co-workers [21–23] can then be used to obtain this constant. Furthermore, the preservation of the phase invariance and the bifurcation of merely the translational pair of eigenvalues in the DNLS case implies that the leading order derivative with respect to ε of non-zero contribution will be $\partial_\varepsilon \partial_\lambda^2 E(0;0)$. However, due to its relation with the splitting of the orbits of the perturbed problem, this derivative will generically be $\sim \partial_\xi M(\xi;\varepsilon)$, as can be shown on general grounds (see, e.g., [18]). The Taylor expansion of the Evans function will therefore read

$$E(\lambda;\varepsilon) = \varepsilon \lambda^2 \partial_\varepsilon \partial_\lambda^2 E(0;0) + \lambda^4 \partial_\lambda^4 E(0;0) \tag{2.44}$$

and the calculation of the details specific to the DNLS equation yields

$$E(\lambda;h^2) \sim \lambda^2(h^2 \partial_\xi M(\xi;h^2) + 2B\lambda^2), \tag{2.45}$$

where B is a constant. Substituting the expression for the Melnikov function, we conclude that for the site-centered mode (that was numerically found to be stable previously)

$$\lambda_s^\pm \sim \pm i \sqrt{\frac{\pi h^2 C(h)}{B_1 \tilde{\alpha}}}\, e^{-\pi^2/2\tilde{\alpha}}, \tag{2.46}$$

while for the inter-site-centered mode (that was previously found numerically to be unstable due to a real eigenvalue pair)

$$\lambda_u^\pm \sim \pm \sqrt{\frac{\pi h^2 C(h)}{B_1 \tilde{\alpha}}}\, e^{-\pi^2/2\tilde{\alpha}}. \tag{2.47}$$

The results of Eqs. (2.46) and (2.47) confirm the validity of the numerical findings. Furthermore, the Evans function method naturally demonstrates how the exponentially small transversality effects of (translational invariance) symmetry breaking are mirrored in the bifurcation of the translational eigenmodes away from the origin of the spectral plane. The predictions of the Evans method are compared with the results of the linear stability analysis in Fig. 2.3.

Having analyzed the bifurcation of the translational mode (note that for the inter-site-centered solution, we will discuss a different limit later in this chapter), we now turn our attention to the other interesting feature caused by discreteness as a perturbation to the continuum problem. As was first observed for kinks in a modified sine-Gordon potential [24] and later elaborated for kinks in nonlinear Klein–Gordon lattices and for standing waves in DNLS lattices in [5, 17, 18, 25–27], localized

2.1 Single-Pulse Solitary Waves

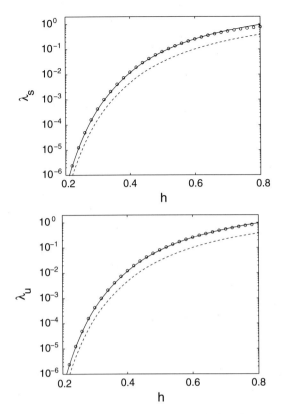

Fig. 2.3 The Evans function prediction for the imaginary eigenvalue of the site-centered mode (*top panel*) and for the real eigenvalue of the inter-site-centered mode (*bottom panel*). The theoretical results given by Eqs. (2.46) and (2.47), respectively, capture the functional dependence but not the exact prefactor. The result of Eqs. (2.46) and (2.47) without the corrected prefactor is shown by the dashed line. The result with the corrected prefactor is the solid line and it is in excellent agreement with the numerical results for the bifurcation of the corresponding translational eigenvalue given by the circles. The prefactors need to be corrected since, as explained in [18], the Evans function method captures the leading order functional dependence (on h) but higher order contributions to the prefactor are not accounted for by this method. Reprinted from [18] with permission

eigenmodes may bifurcate from the bottom edge of the continuous spectrum ($\omega_p = \pm \Lambda$ for the DNLS equation) and subsequently be present in the gap between the band edge and the origin of the spectral plane as a point spectrum eigenvalue.

This phenomenon was first tackled theoretically in the framework of intrinsic localized modes (ILMs) in [17]. In this work, discreteness was considered as a leading order power-law perturbation to the continuum problem, using the Taylor expansion of Eq. (2.31):

$$i\psi_t + \psi_{xx} + 2|\psi|^2 u + \frac{h^2}{12}\psi_{xxxx} = 0. \tag{2.48}$$

Considering the last term in Eq. (2.48) as a perturbation (of strength $\delta = h^2/12$) and using leading order perturbation theory, a corrected solution can be obtained for the perturbed problem $\psi = \exp(it)(v_0(x) + \delta v_1(x))$, with

$$v_1(x) = \frac{1}{2}\left(\frac{x\sinh(x)}{\cosh^2(x)} - \frac{7}{\cosh(x)} + \frac{8}{\cosh^3(x)}\right). \tag{2.49}$$

Then, the perturbed linearization problem of Eqs. (2.12) and (2.13) can be written as

$$U'' + \left(\frac{6}{\cosh^2 x} - 1\right)U + \omega W + (\partial_x^4 + 12v_0v_1)\delta U = 0, \tag{2.50}$$

$$W'' + \left(\frac{2}{\cosh^2 x} - 1\right)W + \omega U + (\partial_x^4 + 4v_0v_1)\delta W = 0. \tag{2.51}$$

Assuming now the eigenvalue bifurcating from the bottom of the continuous band edge to have a frequency $\omega = 1 - \delta^2\kappa^2$ (where $\delta = h^2/12$ is the measure of the perturbation), the authors of [17] project the new basis of continuous eigenfunctions $[U, W]^T$ onto the known old one [4]. Assuming that the eigenvalue bifurcating from the band edge is given by the above expression, the integral equation resulting from the projection, in the case of wave number $k = 0$ yields a solvability condition that allows to determine κ as (in the case of the DNLS equation)

$$|\kappa| = \frac{\text{sgn}(\delta)}{4} \int U(x,0)f_1(x)U(x,0) + W(x,0)f_2(x)W(x,0)\,dx, \tag{2.52}$$

where $f_{1,2}$ correspond to the perturbative operator prefactors of δU and δW in Eqs. (2.50) and (2.51), respectively. This method yields $\kappa = 4/3$ in the case of DNLS and hence

$$\omega = 1 - \frac{h^4}{81}. \tag{2.53}$$

However, as can be seen from the above methodology, only the leading order power-law correction is recovered in this way. One of the major disadvantages of this result is that it does *not* distinguish between the bifurcation of the eigenvalue pair (the "breathing" mode according to [5]), of the stable site-centered and the unstable inter-site-centered wave. Such differences, obvious in Fig. 2.4, can only be attributed to exponentially small differences between the two waves.

For this reason, in [18, 27], the Evans function methodology was developed in the vicinity of the branch point $\omega = 1$ ($\lambda = i$). However, in this case both exponential *and* power-law terms were present in the derivative of E with respect to the O(ε) perturbation in Eqs. (2.38). Evaluation of the relevant partial derivatives and Taylor expansion of $E(\lambda; \epsilon)$ near $\lambda = i; \epsilon = 0$, performed in detail in [18], yields

2.1 Single-Pulse Solitary Waves

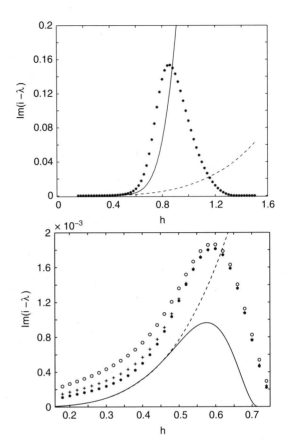

Fig. 2.4 Bifurcation of the eigenvalue λ of the breathing mode of the stable and unstable wave from the edge of the continuous spectrum ($\lambda = i$) as a function of the lattice spacing h; the dashed line shows the theory of [17], the solid line the theory of [18, 27], and the stars the results of numerical experiments on a 400-site lattice with periodic boundary conditions. In the bottom panel, circles indicate results on 200-site and plus symbols on a 300-site lattice. Reprinted from [18] with permission

$$E(\gamma, h) = 4h^{-1}\left[\gamma - \frac{h^3}{9}\left\{1 + \frac{9}{4}C(h)e^{-\pi^2/\tilde{\alpha}}\cos\left(\frac{2\pi\xi}{\tilde{\alpha}}\right)\right\}\right], \quad (2.54)$$

where

$$\gamma^2 = (1 + i\lambda)h^2\left(1 + \frac{1}{4d^2}\right). \quad (2.55)$$

Setting $E = 0$, we obtain an expression for the eigenvalue bifurcation

$$\lambda(\xi) = i\left[1 - \frac{h^4}{81}\left\{1 + \frac{9}{2}C(h)e^{-\pi^2/\tilde{\alpha}}\cos\left(\frac{2\pi\xi}{\tilde{\alpha}}\right)\right\}\right]. \quad (2.56)$$

Since ξ can take the values $n\tilde{\alpha}$ or $(n + 1/2)\tilde{\alpha}$, we obtain different eigenvalue bifurcations for the site-centered and the inter-site-centered modes. $C(h) =$

$(256\pi)/(45)(\pi/\tilde{\alpha})^7 \approx 53979.2\,h^{-7}$. It should be noted, however, that to leading (power law) order this result is the same as the one of [17]. On the other hand, as can be observed in the numerical results of Fig. 2.4, there is a significant difference between the bifurcation of the stable and the unstable wave. In fact, the numerical experiment shows that the maximal bifurcation for the former is ≈ 0.155, while for the latter it is only ≈ 0.0019. Hence, the difference between the two waves, which is discernible only at an exponentially small level of description, turns out to be important. Equation (2.56) renders this distinction clear, as it suggests that for $h > 0.45$ the exponential effects become important. This distinction has also been observed in nonlinear Klein–Gordon lattices. For instance, in the discrete sine-Gordon model, the bifurcation of the edge or breathing mode of the unstable wave is completely *suppressed* by the exponential contributions, while the bifurcation does take place for the stable wave [25].

So far, we have clarified the questions pertaining to linear stability of the site-centered and inter-site-centered modes. However, it is also important to know whether information about linear stability can be generalized to nonlinear stability of the relevant waves. This is often not possible, however, in the case of the fundamental solutions and in the vicinity of the continuum limit $h \to 0$, the work of [28] can be used to obtain nonlinear stability conclusions. Suppose that the operator L_+, defined in Eq. (2.13) has only one eigenvalue $\lambda \leq 0$, and suppose that the only eigenvalue of L_-, given in Eq. (2.12), that has the property $\lambda \leq 0$ is exactly at $\lambda = 0$. Furthermore, suppose that

$$\frac{d}{d\omega}P > 0. \tag{2.57}$$

Then it can be shown using the work of [28] that the wave is nonlinearly stable. As $h \to 0^+$ the wave is approximately given by $\sqrt{\omega}\,\mathrm{sech}(\sqrt{\omega}\,x)$. Hence,

$$\lim_{h \to 0^+} P = \int_{-\infty}^{+\infty} \left(\sqrt{\omega}\,\mathrm{sech}(\sqrt{\omega}\,x)\right)^2 dx = 2\sqrt{\omega}$$

so that

$$\lim_{h \to 0^+} \frac{d}{d\omega}P > 0.$$

Hence, the condition of Eq. (2.57), is satisfied for small h.

It is clear that $L_-(v_n) = 0$ so that $\lambda = 0$ is an eigenvalue of L_-. Since v_n is a positive solution, Sturm–Louiville theory states that $\lambda = 0$ is the minimal eigenvalue. At small h, the operator L_+ is a perturbation of the corresponding operator for the AL-NLS equation. For the AL-NLS equation the operator is such that $\partial_\xi v_n$ is an eigenfunction at $\lambda = 0$. Thus, by another application of Sturm–Liouville theory, one has that there exists one negative eigenvalue which is of O(1). Upon perturbation, both of the eigenvalues will move by an exponentially small amount. Hence, it is enough to track the eigenvalue near zero in order to determine whether the first

2.1 Single-Pulse Solitary Waves

condition above is met. However, according to the theory of [28, 29], the wave is linearly unstable if the operator L_+ has two strictly negative eigenvalues. One can conclude (by contradiction) that for the linearly unstable wave (inter-site-centered solution) the zero eigenvalue of the AL-NLS equation must move to the left in the case of the DNLS equation, while for the linearly stable wave (site-centered solution) the zero eigenvalue of the AL-NLS equation must move to the right. Thus, we have shown that if $h > 0$ is sufficiently small, for the linearly stable on-site wave, the first condition is also met and consequently the wave is also *nonlinearly stable*.

2.1.2 The Anti-Continuum Approach

We now approach the same problem, namely the existence and stability of single pulses in the DNLS equation from an entirely different perspective, that was originally proposed by MacKay and Aubry [30] and has been extensively used in the literature since. This is the initially referred to as anti-integrable, and later more appropriately termed the anti-continuum (AC) limit of $\epsilon = 0$. In this limit, the sites are *uncoupled* and it is straightforward to solve the ensuing ordinary differential equations for u_n. The key question then becomes which ones of the possible combinations of the different u_n's will persist, as soon as the coupling between the sites is turned on (i.e., $\epsilon \neq 0$). To present the relevant analysis, let us use Eq. (2.1) with $\beta = -1$ (i.e., in the focusing case) and having made a transformation of the field $u_n \to u_n \exp(-2i\epsilon t)$, as is always possible due to the gauge invariance. Then (2.1) acquires the form

$$i\dot{u}_n = -\epsilon(u_{n+1} + u_{n-1}) - |u_n|^2 u_n. \qquad (2.58)$$

As before, we look for standing waves of the form: $u_n = \exp(i\mu t)v_n$ which, in turn, satisfy the steady-state equation

$$\left(\mu - |v_n|^2\right) v_n = \epsilon \left(v_{n+1} + v_{n-1}\right). \qquad (2.59)$$

In the AC limit, it is easy to see that Eq. (2.59) is completely solvable $v_n = 0, \pm\sqrt{\mu}\exp(i\theta_n)$, where θ_n is a *free* phase parameter for *each* site. However, as indicated above, out of this huge freedom of phase selection for each site in the uncoupled limit, the important consideration is how much of it remains as soon as the cross-talking between sites is allowed. A simple way to address this question in an explicit way in the one-dimensional case of Eq. (2.58) is to multiply Eq. (2.59) by $v_n^\star$ and subtract the complex conjugate of the resulting equation, which in turn leads to

$$v_n^\star v_{n+1} - v_n v_{n+1}^\star = \text{const} \Rightarrow 2\arg(v_{n+1}) = 2\arg(v_n), \qquad (2.60)$$

where we have used the fact that the constant should be equal to 0 since we are considering solutions vanishing as $n \to \pm\infty$. The scaling freedom of the equation allows us to select $\mu = 1$ without loss of generality. Then, this yields the important conclusion that the *only* states that will persist for finite ϵ are ones containing sequences with combinations of $v_n = \pm 1$ and 0. A systematic computational classification of the simplest ones among these sequences and of their bifurcations is provided in [31]. It should be mentioned also in passing that although we considered here the focusing case of $\beta = -1$, the defocusing case of $\beta = 1$ can also be addressed based on the same considerations and using the so-called staggering transformation $w_n = (-1)^n u_n$ (which converts the defocusing nonlinearity into a focusing one).

Although we will use the above considerations later in this chapter, when considering the case of multipulses, in the present subsection, we focus on the single-pulse case. It turns out that the single pulse exists in the DNLS "all the way" between the continuum and the AC limit (a very rare feature, as can be inferred, e.g., from Fig. 14 of [31]). In fact, there are *two* waveforms at the AC limit that will both result to the continuum pulse as $\epsilon \to \infty$ ($h \to 0$). One of them is the configuration with $v_n = \delta_{n,n_0}$, i.e., the single-site excitation. The other is the configuration with $v_n = \delta_{n,n_0} + \delta_{n,n_0+1}$, i.e., a two-site, bond-centered, in-phase excitation. Since the latter is a multisite structure, it will be examined in more detail later in this chapter (where the general theory of multisite excitations will be developed). Incidentally, these two are the *only* configurations that persist throughout the continuation from the AC to the continuum limit.

We also briefly discuss the stability near the AC limit. Once again, the L_+ and L_- operators emerge when linearizing around the standing wave solutions of Eqs. (2.58) and (2.59) in the linearization problem of the form

$$\left(1 - 3v_n^2\right) a_n - \epsilon (a_{n+1} + a_{n-1}) = L_+ a_n = -\lambda b_n, \quad (2.61)$$

$$\left(1 - v_n^2\right) b_n - \epsilon (b_{n+1} + b_{n-1}) = L_- b_n = \lambda a_n. \quad (2.62)$$

It is perhaps relevant to note here that the eigenvalue problem has the general symplectic form $JLw = \lambda w$ where the J matrix has the standard symplectic structure ($J^2 = -I$), i.e.,

$$\mathcal{J} = \begin{pmatrix} 0 & I \\ -I & 0 \end{pmatrix} \quad (2.63)$$

and the operator L is defined as

$$L = \begin{pmatrix} L_+ & 0 \\ 0 & L_- \end{pmatrix}. \quad (2.64)$$

2.1 Single-Pulse Solitary Waves

In the case of the AC limit of $\epsilon = 0$, the operators L_- and L_+ simplify enormously, becoming simply multiplicative operators. It is then straightforward to solve the ensuing eigenvalue problem for each site of the AC limit. Assume a sequence for v_n with N "excited" (i.e., $\neq 0$) sites; then, it is easy to see that for $\epsilon = 0$ these sites correspond to eigenvalues $\lambda_{L_+} = -2$ for L_+ and to eigenvalues $\lambda_{L_-} = 0$ for L_-, and they result in N eigenvalue pairs with $\lambda^2 = 0$ for the full Hamiltonian problem. On the other hand, all the remaining, infinitely many (non-excited) sites in the chain with $v_n = 0$ satisfy $a_n = -\lambda b_n$ and $b_n = \lambda a_n$. This yields a pair of eigenvalues $\lambda^2 = -1$, i.e., $\lambda = \pm i$ (more generally for a frequency Λ of the solution, these will be $\lambda = \pm i\Lambda$), with infinite multiplicity.

In the case we are currently considering, namely that of a single-site excitation that will eventually give rise to the single pulse of the continuum limit, there is only a single pair of zero eigenvalues, corresponding to the excited site. There are also infinitely many pairs, corresponding to the non-excited sites, at $\lambda = \pm i$ (or $\pm i\Lambda$ more generally). As soon as ϵ becomes non-zero, the former zero eigenvalue pair will have to *remain* at 0, because of the gauge (i.e., U(1)) invariance of the equation. On the other hand, the infinitely many sites with identical eigenvalues will "expand" in accordance with Eqs. (2.34) and (2.35). As ϵ is increased, the two eigenvalues of the top panels of Figs. 2.3 and 2.4 will bifurcate from the bottom edge of the continuous spectrum. One of these eigenvalues will approach the origin exponentially as $h \to 0$, while the other, upon a maximal excursion from the continuous spectrum's lower band edge, it will return to it as $h \to 0$. This is the full spectral picture joining the results of the continuum to those of the AC limit. While we have described analytically both the $h \to 0$ limit in the previous subsection, and the $h \to \infty$ limit in this subsection, the intermediate region can only be quantified numerically, as is done in Figs. 2.3 and 2.4, as well as in Fig. 2.5 in a more conclusive way, giving information about all the point spectrum eigenvalues of the DNLS problem, as a function of the lattice spacing h.

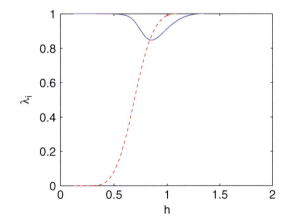

Fig. 2.5 The figure shows the (non-zero) point spectrum eigenvalues as a function of the spacing h, as they are numerically obtained (in a 400-site computation) for the linearization around a single pulse of the DNLS equation. The solid line shows the internal mode that bifurcates from the bottom edge of the continuous spectrum and returns to it, while the dashed line shows the translational mode that exponentially approaches $\lambda^2 = 0$ as $h \to 0$

2.1.3 The Variational Approach

Another approach that yields quantitative semi-analytical information about the single-pulse case *even in the intermediate regime* is the so-called variational approach (VA). In the case of the DNLS, this was originally developed in [32] and was systematically tested recently in [33]. The VA (a detailed review of which can be found in [34]) consists of selecting an a priori ansatz for the form of the solution (in the case of the DNLS a typical, quantitatively tractable choice is $u_n = A \exp(-a|n|)$). It should be understood that this is a *dramatic* oversimplification of the solution. It reduces the original *infinite dimensional* dynamical system into a two-dimensional one, in the space of "adjustable parameters" A and a (representing measures of the amplitude and the inverse width of the pulse solution, respectively). This type of *ansatz* is not substituted in the steady-state (or dynamical) equation, which it cannot, by default, satisfy, since it is typically impossible to satisfy infinitely many equations with just two free parameters, unless, fortuitously, our ansatz represents an exact solution. It is, instead, substituted in the Lagrangian (either time-dependent, or when looking for steady states, time-independent). Subsequent derivation of the Euler–Lagrange equations (i.e., extremization of the action on the restricted subspace of the adjustable parameters) is performed in the expectation of obtaining a good fit to the corresponding infinite-dimensional extremization problem. This is a reasonable expectation provided that the waveform remains close to the ansatz, but may fail considerably when that is not the case. More importantly, it is not a priori obvious *when* the method will fail, although recent efforts are starting to aim toward systematically computing the relevant error and accordingly improving the approximation, when needed [35].

As an example, we now give the steady-state version of the variational approximation for a generalized DNLS problem in line with the original presentation of [32]. Consider the steady-state problem for the discrete waveform v_n

$$\Lambda v_n = \epsilon (v_{n+1} + v_{n-1} - 2v_n) + v_n^{2\sigma+1}, \tag{2.65}$$

which is the standing wave problem for a general power nonlinearity (as discussed in the beginning of this chapter). The case of $\sigma = 1$ corresponds to our familiar cubic DNLS. Equation (2.65) can be derived from the Lagrangian

$$L = \sum_{n=-\infty}^{+\infty} \left[\epsilon(v_{n+1} + v_{n-1})v_n - (\Lambda + 2\epsilon)v_n^2 + \frac{1}{\sigma+1} v_n^{2(\sigma+1)} \right]. \tag{2.66}$$

Substituting the above-mentioned simple exponential ansatz in the Lagrangian, one can perform the summation explicitly, which yields the *effective Lagrangian*,

$$L_{\text{eff}} = 2\epsilon P \operatorname{sech} a - (\Lambda + 2\epsilon)P + \frac{P^{\sigma+1}}{\sigma+1} \frac{\coth((\sigma+1)a)}{\coth^{\sigma+1} a}. \tag{2.67}$$

2.1 Single-Pulse Solitary Waves

The (squared l^2) norm of the ansatz, which appears in Eq. (2.67), is given by

$$P \equiv \sum_{n=-\infty}^{+\infty} v_n^2 = A^2 \coth a. \qquad (2.68)$$

The Lagrangian (2.67) gives rise to the variational equations, $\partial L_{\text{eff}}/\partial P = \partial L_{\text{eff}}/\partial a = 0$, which constitute the basis of the VA toward the computation of stationary waveforms [34]. They predict relations between the norm, frequency, and width of the fundamental pulses within the framework of the VA, namely

$$P^\sigma = \frac{4\epsilon \cosh^\sigma a \sinh^2(\sigma+1)a}{\sinh^{\sigma-1} a (\sinh 2(\sigma+1)a - \sinh 2a)}, \qquad (2.69)$$

$$\Lambda = 2\epsilon(\operatorname{sech} a - 1) + P^\sigma \frac{\coth(\sigma+1)a}{\coth^{\sigma+1} a}. \qquad (2.70)$$

These ensuing transcendental equations connect the power of the pulse solution to its width (these are the two unknowns in these algebraic equations) and express these features as a function of the system parameters (such as the frequency Λ, the coupling strength ϵ, or the nonlinearity exponent σ). The equations can be solved with a standard computational mathematics package (i.e., such as the `FindRoot` routine in Mathematica) and compared to direct numerical computations, as is done in Fig. 2.6 (for $\Lambda = \sigma = 1$).

Figure 2.6 shows the continuum analog of the power of the pulse as a function of the lattice spacing h (the continuum limit of this quantity for the model considered herein can be easily seen to be 4, which is also confirmed by the relevant computations). It also shows the amplitude A of the discrete pulse as a function of the lattice spacing (the continuum limit for this quantity is $\sqrt{2}$), as well as the inverse width a as a function of h. One can easily note that for $h > 0.8$, the agreement between the variational solution of the greatly simplified 2×2 system of Eqs. (2.69) and (2.70) is truly *remarkable*. On the other hand, however, one has no a priori way to explain why the agreement starts becoming considerably worse for $h < 0.8$, other than to say that the oversimplified ansatz does not accurately describe the continuum, hyperbolic secant limit. However, a further quantification of the relevant statement cannot be a priori made. A deeper understanding of this discrepancy can, however, be partially obtained by considering the $a \to 0$ limit of the above equations (scaling out ϵ from Λ and P^σ), since we can see that $P^\sigma \approx (4 + 2\sigma + 2/\sigma)a^{2-\sigma}$, while $\Lambda \approx (1 + 2/\sigma)a^2$, as $a \to 0$. From these equations and Eq. (2.68), one can infer the dependence of the amplitude A on a as $A \approx (4 + 2\sigma + 2/\sigma)^{1/(2\sigma)} a^{1/\sigma}$. In the case of $\sigma = 1$, this yields $A \approx \sqrt{8}a \approx (2/\sqrt{3})\sqrt{2\Lambda} = (2/\sqrt{3})A_{cont}$, where A_{cont} denotes the amplitude of the continuum soliton. This indicates that in that limit the VA fails in capturing the amplitude by a factor of $2/\sqrt{3} \approx 1.15$. Nevertheless, the VA is often a useful tool in acquiring some insight on the nature of the full solution

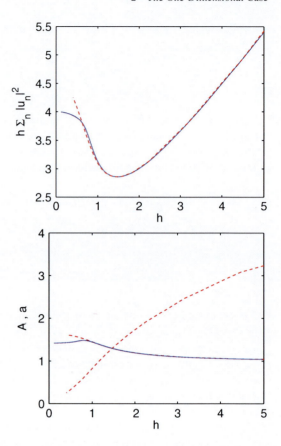

Fig. 2.6 Results of the variational approximation: the *top panel* shows the analog of the continuum squared L^2 norm as a function of the lattice spacing h. The *solid line* represents the full numerical result, while the *dashed line* shows the result of the VA. The *bottom panel* curves have the same symbolism for the amplitude A of the solution (for which the fully numerical and variational results are again compared) and for the inverse width a of the pulse (for which only the variational result is given). It is worthwhile to note how efficient the VA is in capturing the full numerical result for $h > 0.8$

of the system and an understanding of its fundamental properties. Our argument here is that it should be used with the appropriate caution; furthermore, it has been our experience that it is a method that is considerably more likely to work with fundamental solutions (given a reasonable ansatz) than with higher order excited state solutions, whose (existence and stability) properties it is often unable to track.

Before closing this discussion about single-pulse solitary waves in this generalized DNLS model, it is interesting to refer to a particular effect that this model possesses for values of σ ranging between a lower and an upper critical one, namely *bistability* [32]. In particular, within this range of values of σ, for a given ϵ, the dependence of the l^2 norm as a function of Λ is not monotonic as it would be for $\sigma = 1$, but possesses a range of powers for which two stable solutions (with $dP/d\Lambda > 0$ and an unstable solution (with $dP/d\Lambda < 0$) coexist; see Fig. 2.7. The variational prediction captures this feature of the single-pulse state fairly accurately as is shown in the figure. Equivalently, this phenomenon also arises for fixed Λ as a function of the coupling ϵ. As a measure of the accuracy of the variational prediction, we give in Fig. 2.8 the full numerical result and how it compares with the semi-analytical prediction for the lower and upper critical σ for which bistability

2.1 Single-Pulse Solitary Waves

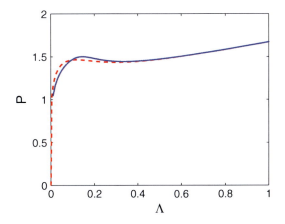

Fig. 2.7 The (squared l^2) norm of the fundamental soliton family versus Λ for $\sigma = 1.5$. *Solid lines* display numerical results, while the *dashed curves* correspond to the predictions of the variational approximation

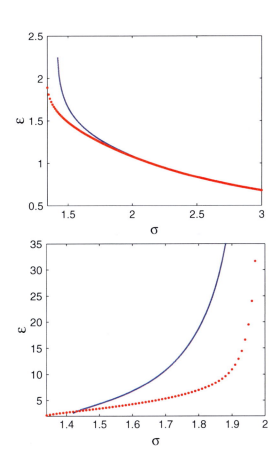

Fig. 2.8 Locations of the two bifurcations that account for the exponential destabilization and subsequent restabilization (the *top and bottom panels*, respectively) of the fundamental solitons (subject to the normalization $\Lambda \equiv 1$) in the plane of (σ, ϵ). The *line and dots* represent predictions of the variational approximation and numerical results, respectively. The restabilization corresponding to the bottom panel does not occur for $\sigma \geq 2$

occurs as concerns their dependence on the coupling ϵ (or equivalently for a fixed σ, the lower and upper couplings between which the solution is unstable). It can be seen that the variational analysis captures the general trends of the relevant behavior but the partial inaccuracy of the relevant ansatz may not allow a good quantitative comparison between the two. We should also add in passing that beyond $\sigma = 2$, for dimension $d = 1$, equivalently to the case of $\sigma = 1$ for dimension $d = 2$, the continuum version of the model is subject to catastrophic collapse-type instabilities. This issue will be discussed, along with the discrete analog of such instabilities, in more detail in the two-dimensional (cubic) setting in Chap. 3.

2.2 Multipulse Solitary Waves

2.2.1 Multipulses Near the Anti-Continuum Limit

We now turn to the consideration of multipulse solitary waves, which we are going to examine chiefly starting from the AC limit of $\epsilon = 0$. In particular, in the latter case, instead of exciting a single site (which will result, as we saw above, in a single pulse), we excite N sites in the general case. In that case, considerations similar to the ones we presented above for the case of $\epsilon \neq 0$ indicate that in the one-dimensional setting, without loss of generality, we may concern ourselves only with excitations of each site which are either $u_n = 1$ or -1 (for frequency $\mu = 1$) while the rest of the (non-excited) sites have $u_n = 0$. Then, it is straightforward to see from the structure of the L_+ and L_- operators that for $\epsilon = 0$ the N excited sites correspond to eigenvalues $\lambda_{L_+} = -2$ for L_+ and to eigenvalues $\lambda_{L_-} = 0$ for L_-, and they result in N eigenvalue pairs with $\lambda^2 = 0$ for the full Hamiltonian problem. Hence, these eigenvalues are potential sources of instability, since for $\epsilon \neq 0$, $N-1$ of those will become non-zero (there is only one symmetry, namely the U(1) invariance, persisting for $\epsilon \neq 0$, retaining one pair of eigenvalues at the origin). The key question for stability purposes is to identify the location of these $N-1$ small eigenvalue pairs.

To address the location of these eigenvalues in the presence of the perturbation induced by the inter-site coupling, one can manipulate Eqs. (2.61) and (2.62) into the form

$$\mathcal{L}_- b_n = -\lambda^2 \mathcal{L}_+^{-1} b_n \Rightarrow \lambda^2 = -\frac{(b_n, \mathcal{L}_- b_n)}{(b_n, \mathcal{L}_+^{-1} b_n)}. \tag{2.71}$$

Near the AC limit, the effect of L_+ is a multiplicative one (by -2). Hence,

$$\lim_{\epsilon \to 0}(b_n, \mathcal{L}_+^{-1} b_n) = -\frac{1}{2} \Rightarrow \lambda^2 = 2\gamma = 2(b_n, \mathcal{L}_- b_n). \tag{2.72}$$

2.2 Multipulse Solitary Waves

Therefore the problem translates into the determination of the spectrum of L_-.[1] However, using the fact that the standing wave solution v_n is an eigenfunction of L_- with $\lambda_{L_-} = 0$ and the Sturm comparison theorem for difference operators [36], one infers that if the number of sign changes in the solution at the AC limit is m (i.e., the number of times that adjacent to a $+1$ is a -1 or next to a -1 is a $+1$), then the number of negative eigenvalues $n(L_-) = m$ and therefore from Eq. (2.72), the number of imaginary eigenvalue pairs of $\mathcal{L}$ is m, while that of real eigenvalue pairs is consequently $(N-1)-m$. An immediate conclusion is that unless $m = N - 1$, or practically unless adjacent sites are out-of-phase with each other, the solution will be immediately unstable for $\epsilon \neq 0$. This eigenvalue count was presented in [37] although its stability consequences had been observed in numerous earlier works (see, e.g., [38] and references therein). It is also interesting to point out that a related count was originally presented in [39], although purely as an instability condition (rather than as a definitive count). In particular, it was recognized in that work that $n(L_-) = m$ and that that in the vicinity of the continuum limit (rather than the AC limit as here) each of the N pulses would have at least one negative eigenvalue associated with them, hence $n(L_+) \geq N$. Then, an important criterion was used that was originally developed in the work of Jones [29], namely that when $|n(L_+) - n(L_-)| > 1$, then a real eigenvalue pair will exist in the linearization. Hence, in the present case if $m < N - 1$, this criterion could be used to yield an instability, in agreement with the result presented above (but established in the vicinity of the AC limit).

An important additional realization both in the work of [37] and in that of [39] was that the m negative eigenvalues of L_-, corresponding to the m imaginary pairs of the full linearization operator JL, have negative Krein signature. The Krein signature is a fundamental topological concept in the context of nonlinear Schrödinger equations (and Hamiltonian systems more generally); the interested reader should examine [40–45] for details and examples. These eigenvalues are often also mentioned in the physical literature as negative energy modes, see, e.g., [46–48]. For our present considerations, it suffices to say that this signature is essentially the signature of the energy surface and can be found to be equivalent to the sign of $(w, L_- w)$. Hence, all of the m eigenvalue pairs bifurcating along the imaginary axis in our present calculation will be negative Krein signature (or negative energy) eigenvalues. The eigenvalues of negative Krein sign are well known to be *structurally unstable*. This means that small perturbations of the vector field can eject them off of the imaginary axis, leading to an unstable eigenvalue with a positive real part. Moreover, if eigenvalues of opposite sign collide, then they will generically form a complex conjugate pair after the collision, whereas if eigenvalues of the same sign collide, then they will pass through each other.

---

[1] It should be pointed out here that although Eq. (2.72) is particular to the cubic model, Eq. (2.71) is not and can straightforwardly be applied to any nonlinearity that depends on the field and its complex conjugate. The denominator of its right-hand side will in such cases typically provide a constant prefactor, while the spectrum of L_-, through considerations similar to those presented here, will determine the fate of small eigenvalues.

It is important to conclude from the above considerations that the only standing wave configuration of the discrete problem (starting from the AC limit) that will be structurally and nonlinearly stable is the single-site excitation. Not only did we establish nonlinear stability for that configuration above (near the continuum limit), but furthermore it is the only configuration (near the AC limit) that has neither a real eigenvalue pair (in which case it would be directly unstable) or an imaginary pair with negative Krein sign since $m = 0 = N - 1$. An additional remark worth making here concerns the case of the second solution of the AC limit reported in the previous section to eventually asymptote to the single-pulse configuration, namely the two-site, in phase excitation. In that case, $m = 0$, while $N = 2$, hence one can immediately see that for this waveform there will be $N - 1 - m = 1$ real eigenvalue pair, as soon as $\epsilon \neq 0$, indicating the instability of the relevant wave.

So far, we have used the above Eq. (2.72) in a qualitative way to obtain the relevant eigenvalue counts. In what follows, we also show how to use this equation in a quantitative manner, in order to obtain the dependence of the relevant eigenvalues on the model parameters (and, hence, wherever relevant, quantify the growth rate of the instability). To do so, we need to obtain a handle on the eigenvalues of the operator L_-. Considering the relevant eigenvalue problem $L_- \phi = \gamma \phi$, we can expand it in the vicinity of the AC limit, according to

$$L_- = L_-^{(0)} + \epsilon L_-^{(1)} + O(\epsilon^2), \tag{2.73}$$

$$\phi_n = \phi_n^{(0)} + \epsilon \phi_n^{(1)} + O(\epsilon^2), \tag{2.74}$$

$$\gamma = \epsilon \gamma_1 + O(\epsilon^2). \tag{2.75}$$

In this expansion,

$$L_-^{(0)} \phi_n = (1 - (v_n^{(0)})^2) \phi_n, \tag{2.76}$$

$$L_-^{(1)} \phi_n = -(\phi_{n+1} + \phi_{n-1}) - 2v_n^{(0)} v_n^{(1)} \phi_n, \tag{2.77}$$

where $v_n^{(0)}$ and $v_n^{(1)}$ correspond to the expansion $v_n = v_n^{(0)} + \epsilon v_n^{(1)}$ of the solution in the vicinity of the AC limit. The leading order correction $v_n^{(1)}$ will need to be computed from Eq. (2.59). It is straightforward to apply the expansion of the solution to the relevant equation (keeping in mind that $\mu = 1$ and the phases θ_n of the excited sites are $0, \pi$, corresponding to ± 1 field values). If we have the N excited sites adjacent to each other, then it is straightforward to find the leading order correction as

$$v_n^{(1)} = -\frac{1}{2} \left(\cos(\theta_{n-1} - \theta_n) + \cos(\theta_{n+1} - \theta_n) \right) e^{i\theta_n}, \quad 2 \leq n \leq N - 1, \tag{2.78}$$

$$v_1^{(1)} = -\frac{1}{2} \cos(\theta_2 - \theta_1) e^{i\theta_1}, \quad v_N^{(1)} = -\frac{1}{2} \cos(\theta_N - \theta_{N-1}) e^{i\theta_N}, \tag{2.79}$$

$$v_0^{(1)} = e^{i\theta_1}, \quad v_{N+1}^{(1)} = e^{i\theta_N}, \tag{2.80}$$

2.2 Multipulse Solitary Waves

while all other elements of $v_n^{(1)}$ are zero. Similarly, one can find the leading order corrections if the N adjacent sites are next nearest neighbor to each other (in that case, the leading order correction to the excited sites will be $v_n^{(2)}$), or for more distant initially excited sites (see [37] for such a calculation in the next-nearest-neighbor case).

If we have N excited sites at the AC limit, the relevant zero eigenvalue of the L_- operator as indicated above will have a multiplicity of N at the AC limit of $\epsilon = 0$. The corresponding linearly independent eigenvectors f_n can be conveniently selected to be $(0, \ldots, \pm 1, \ldots, 0)$ where the $\ldots$ indicate zeros and the ± 1 is located at the kth excited site for the kth eigenvector. Then the zero-order eigenvector $\phi^{(0)}$ can be expressed as a linear combination of the f_n's according to $\phi^{(0)} = \sum_{k=1}^{N} c_k f_k$, for appropriate choice of the coefficients c_k.

Our aim in this exercise is to perturbatively compute γ_1, the leading order correction to the zero eigenvalue of L_-, so that through the appropriate substitution to Eq. (2.72), we can obtain the corresponding eigenvalue of the original system. Through the above expansions, we obtain

$$\mathcal{L}_-^{(0)} \phi_n^{(1)} = \gamma_1 \phi_n^{(0)} - \mathcal{L}_-^{(1)} \phi_n^{(0)}. \tag{2.81}$$

Projecting the system of Eq. (2.81) onto the kernel of $\mathcal{L}_-^{(0)}$ eliminates the left-hand side contribution, and yields a matrix eigenvalue problem with γ_1 as its eigenvalue, namely

$$\mathcal{M}_1 c = \gamma_1 c, \tag{2.82}$$

where $c = (c_1, \ldots, c_N)$ and $\mathcal{M}_1$ is a tri-diagonal $N \times N$ matrix given by

$$(\mathcal{M}_1)_{m,n} = (f_m, L_-^{(1)} f_n). \tag{2.83}$$

Note that this matrix will only give non-trivial contributions if the excited sites are adjacent to each other (otherwise the relevant contributions will be vanishing). In the case of nearest-neighbor initially excited sites, based on the solution for $v_n^{(1)}$ above, the relevant matrix elements will be

$$\begin{aligned}
(\mathcal{M}_1)_{n,n} &= \cos(\theta_{n+1} - \theta_n) + \cos(\theta_{n-1} - \theta_n), & 1 < n < N, \\
(\mathcal{M}_1)_{n,n+1} &= (\mathcal{M}_1)_{n+1,n} = -\cos(\theta_{n+1} - \theta_n), & 1 \leq n < N, \\
(\mathcal{M}_1)_{1,1} &= \cos(\theta_2 - \theta_1), & \\
(\mathcal{M}_1)_{N,N} &= \cos(\theta_N - \theta_{N-1}).
\end{aligned} \tag{2.84}$$

Equation (2.83) more generally (or e.g. the more specific Eq. (2.84) in the simplest case of adjacent site excitations), in conjunction with Eq. (2.72) allows us to obtain definitive estimates on the eigenvalues of the linearization for multisite configurations which we can subsequently directly compare with numerical results.

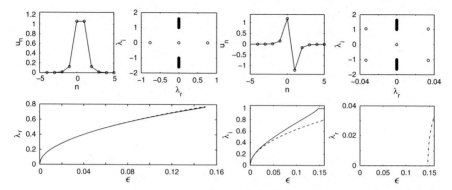

Fig. 2.9 The *top row* shows the solution profiles for the two-site in-phase mode (*first panel*) and out-of-phase mode (*third panel*) and their corresponding spectral planes (λ_r, λ_i) (*second and fourth panels*). The *bottom row* shows the relevant eigenvalue pair bifurcating from the origin. In the *left panel* it is becoming real for the in-phase mode (*solid line*: numerics, *dashed line*: theoretical prediction $\lambda = 2\sqrt{\epsilon}$). In the *right panel*, it becomes imaginary (*solid line*: numerics, *dashed line*: theoretical prediction $\lambda = 2\sqrt{\epsilon}i$). It eventually collides with the band edge of the continuous spectrum for $\epsilon > 0.146$, giving rise to a complex eigenvalue quartet (its real part is shown in the *rightmost panel*). Reprinted from [37] with permission

Such comparisons are illustrated in Fig. 2.9, for the case of two-site excitations. Fig. 2.9 presents both the case where the two excited sites are in phase (this corresponds to the unstable version of the fundamental soliton), as well as that where they are out of phase, which corresponds to the well-known example of the so-called twisted mode (see, e.g., [49–52] and references therein). In this simplest multisite case of two adjacent site excitations, the relevant matrix $\mathcal{M}_1$ becomes

$$\mathcal{M}_1 = \begin{pmatrix} \cos(\theta_1 - \theta_2) & -\cos(\theta_1 - \theta_2) \\ -\cos(\theta_1 - \theta_2) & \cos(\theta_1 - \theta_2) \end{pmatrix}, \qquad (2.85)$$

whose straightforward calculation of eigenvalues leads to $\lambda^2 = 0$ (as is expected from the U(1) invariance, one eigenvalue pair should remain at zero) and $\lambda^2 = 2\epsilon \cos(\theta_1 - \theta_2)$. Note that this result is consonant with our qualitative theory above since for same phase excitations ($\theta_1 = \theta_2$), the configuration is unstable, while the opposite is true if $\theta_1 = \theta_2 \pm \pi$. The top subplots of the figure show typical mode profiles (first and third panel) and the spectral plane $\lambda = \lambda_r + i\lambda_i$ of the corresponding linear eigenvalue problem (second and fourth panel) for $\epsilon = 0.15$. The bottom subplots indicate the corresponding real (for the in-phase mode) and imaginary (for the twisted anti-phase mode) eigenvalues from the theory (dashed line) versus the full numerical result (solid line). We find the agreement between the theory and the numerical computation to be excellent in the case of the in-phase excitation. For the twisted out-of-phase excitation, the agreement is within the 5%-error for $\epsilon < 0.0258$. For larger values of ϵ, the difference between the theory and numerics grows. The imaginary eigenvalues collide at $\epsilon \approx 0.146$ with the band

2.2 Multipulse Solitary Waves

edge of the continuous spectrum, such that the real part λ_r becomes non-zero for $\epsilon > 0.146$ (recall that these eigenvalues have negative Krein signature, hence, their collision with eigenvalues of the band edge of the continuous spectrum or ones bifurcating therefrom leads to complex quartets [40, 41] due to a Hamiltonian–Hopf bifurcation [53]).

In the case $N = 3$, the discrete three-pulse solitons consist of the three modes as follows:

$$\begin{aligned}&\text{(a) } \theta_1 = \theta_2 = \theta_3 = 0, \\ &\text{(b) } \theta_1 = \theta_2 = 0, \ \theta_3 = \pi, \\ &\text{(c) } \theta_1 = 0, \ \theta_2 = \pi, \ \theta_3 = 0.\end{aligned} \quad (2.86)$$

The eigenvalues of matrix $\mathcal{M}_1$ are given explicitly as $\gamma_1 = 0$ and

$$\gamma_{2,3} = \cos(\theta_2 - \theta_1) + \cos(\theta_3 - \theta_2)$$
$$\pm \sqrt{\cos^2(\theta_2 - \theta_1) - \cos(\theta_2 - \theta_1)\cos(\theta_3 - \theta_2) + \cos^2(\theta_3 - \theta_2)}.$$

The in-phase mode (a), which can be symbolically denoted $+ + +$ has two real unstable eigenvalues $\lambda \approx \sqrt{6\epsilon}$ and $\sqrt{2\epsilon}$ in the stability problem for small $\epsilon > 0$. The mode (b), symbolically represented as $+ + -$, has one real unstable eigenvalue pair $\lambda \approx \pm\sqrt{2\sqrt{3}\epsilon}$ and a simple pair of purely imaginary eigenvalues $\lambda \approx \pm i\sqrt{2\sqrt{3}\epsilon}$ with negative Krein signature. This pair may bifurcate to the complex plane as a result of the Hamiltonian–Hopf bifurcation. Finally mode (c) $(+ - +)$ has no unstable eigenvalues but two pairs of purely imaginary eigenvalues $\lambda \approx \pm i\sqrt{6\epsilon}$ and $\lambda \approx \pm i\sqrt{2\epsilon}$ with negative Krein signature. The two pairs may bifurcate to the complex plane as a result of the two successive Hamiltonian–Hopf bifurcations.

Figure 2.10 summarizes the results for the three modes (a–c), given in (2.86), in a presentation similar to that of Fig. 2.9. For the in-phase mode (a), two real positive eigenvalues give rise to instability for any $\epsilon \neq 0$. The error between theoretical and numerical results is within 5% for $\epsilon < 0.15$ for one real eigenvalue and for $\epsilon < 0.0865$ for the other eigenvalue. Similar results are observed for the mode (b), where the real positive eigenvalue and a pair of imaginary eigenvalues with negative Krein signature are generated for $\epsilon > 0$. The imaginary eigenvalue collides with the band edge of the continuous spectrum at $\epsilon \approx 0.169$, which results in a complex eigenvalue quartet. Finally, in the case of the out-of-phase mode (c), two pairs of imaginary eigenvalues with negative Krein signature exist for $\epsilon > 0$ and lead to the emergence of two complex quartets of eigenvalues. The first one occurs for $\epsilon \approx 0.108$, while the second one occurs for much larger values of $\epsilon \approx 0.223$.

One can do a similar calculation for the case in which the excited sites are not adjacent to each other, but are rather, e.g., one site apart. This is detailed in [37]. In that case, some of the logistic details change, most notably that we have to get to the second-order correction $\epsilon^2 \phi^{(2)}$, since the leading order correction does not

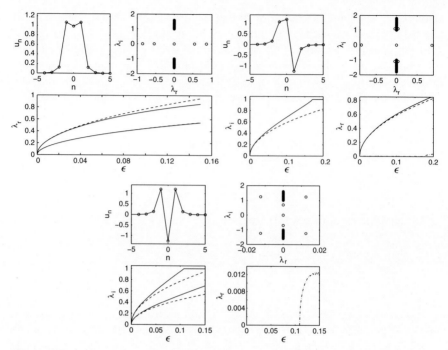

Fig. 2.10 Same as Fig. 2.9, but for the three-site branches +++ (*left panels*), ++− (*right panels*), and +−+ (*bottom panels*). Reprinted from [37] with permission

contribute to the excited sites. Furthermore, accordingly, the leading order in the expansion of the eigenvalue of L_- should be $\gamma = \epsilon^2 \gamma_2$. Then, the relevant perturbation equation becomes

$$L_-^{(0)} \phi_n^{(2)} = \gamma_2 \phi_n^{(0)} - L_-^{(1)} \phi_n^{(1)} - L_-^{(2)} \phi_n^{(0)}. \tag{2.87}$$

Subsequent projection to the kernel of $L_-^{(0)}$ yields an equation entirely analogous to Eq. (2.82) with the only difference that γ_1 and $\mathcal{M}_1$ are replaced by γ_2 and $\mathcal{M}_2$, with the latter being defined as per Eq. (2.84) but with the relevant angles being the next-nearest-neighbor excited ones. It is then straightforward to extract the analogous predictions as for the nearest-neighbor sites, but now the relevant eigenvalues, while having the same prefactors, they will be $\propto \epsilon$, rather than to $\sqrt{\epsilon}$. Note that this feature can be appropriately generalized for excitations that are k sites apart. These will "cross-talk" to each other at the kth order (and above), and the corresponding eigenvalues bifurcating from the origin will be to leading order $O(\epsilon^{k/2})$ (or higher).

The corresponding numerical results are shown in Fig. 2.11 for two sites and Fig. 2.12 for three sites, which are entirely analogous to Figs. 2.9 and 2.10, respectively. In the in-phase, two-site case, the agreement with the theory is excellent for $\epsilon < 0.2$. For the twisted mode, we also have very good agreement for $\epsilon < 0.415$;

2.2 Multipulse Solitary Waves

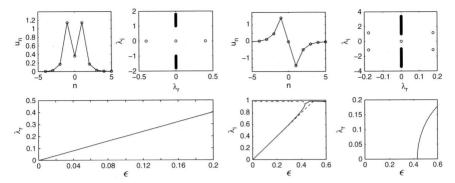

Fig. 2.11 A direct analog to Fig. 2.9, but now for the case of next-nearest-neighbor two-site excitations. The resulting eigenvalues are (approximately) linear in ϵ in this case. Reprinted from [37] with permission

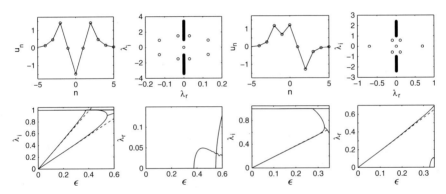

Fig. 2.12 A direct analog to Fig. 2.10, but now for the case of next-nearest-neighbor three-site excitations. The resulting eigenvalues are (approximately) linear in ϵ in this case. Reprinted from [37] with permission

the Hamiltonian–Hopf bifurcation occurs at $\epsilon \approx 0.431$ (in this case, the collision occurs with an eigenvalue that has bifurcated from the band edge of the continuous spectrum). Similarly, for the three-site excitations, we observe excellent agreement in the examined range between the numerical results and the corresponding theoretical predictions. Here, the quartets emerge at $\epsilon \approx 0.328$ for the $++-$ mode, while for $+-+$, there are two such bifurcations arising at $\epsilon \approx 0.375$ and 0.548, respectively.

2.2.2 A Different Approach: Perturbed Hamiltonian Systems

We now take a detour to provide a different (and more general) approach to the stability problem of the one-dimensional DNLS system, while discussing some recent general results for this class of systems developed in [54, 55, 57]. These recent results were obtained on the basis of Lyapunov–Schmidt reductions, as well as through the earlier, important functional analytic work of [28, 42].

Our starting point will be a perturbed system of the general form

$$\frac{du}{dt} = JE'(u), \qquad (2.88)$$

where J is the usual invertible skew-symmetric operator with bounded inverse and $E(u) = E_0(u) + \epsilon E_1(u)$, with $0 < \epsilon \ll 1$. Here $E(u)$ represents the total energy of the system. The underlying assumption is that the perturbation breaks some of the symmetries of the unperturbed problem. The aim of these results is to relate the spectrum (denoted henceforth by σ) $\sigma(E''(\Phi))$ to $\sigma(JE''(\Phi))$, where Φ represents a solution to the steady-state problem $E'(u) = 0$. The operator $E''(\Phi)$ is self-adjoint; hence, $\sigma(E''(\Phi)) \subset \mathbb{R}$. Since $JE''(\Phi)$ is the composition of a skew-symmetric operator with a self-adjoint operator, if $\lambda \in \sigma(JE''(\Phi))$, then $-\lambda, \pm\lambda^* \in \sigma(JE''(\Phi))$. Thus, eigenvalues for $JE''(\Phi)$ come in quartets. Below one sees the manner in which the negative directions for $E''(\Phi)$ influence the unstable spectrum of $JE''(\Phi)$. It is of particular relevance to note that negative directions for $E''(\Phi)$ do *not* necessarily lead to an exponential instability of the wave. A detailed discussion of the proof of these results can be found in [55]. The more epigrammatic discussion of these results below follows the work of [57].

2.2.2.1 The Unperturbed Problem

Let H be a Hilbert space with inner product $\langle \cdot, \cdot \rangle$; also denote by G a finite-dimensional Abelian connected Lie group with Lie algebra $\mathfrak{g}$, setting $\dim(\mathfrak{g}) = n$. We use $e^\omega := \exp(\omega)$ to denote the exponential map from $\mathfrak{g}$ to G. Let T be a unitary representation of G on H, so that $T'(e)$ maps $\mathfrak{g}$ into the space of closed skew-symmetric operators. Denote $T_\omega := T'(e)\omega$ as the generator of the semigroup $T(e^{\omega t})$, and note that T_ω is linear in $\omega \in \mathfrak{g}$. The group orbit Gu is defined by $Gu := \{T(g)u : g \in G\}$. It is assumed that E is invariant under a group orbit, i.e., $E(T(g)u) = E(u)$ for all $g \in G$ and $u \in H$. Define the functional $Q_\omega(u) := \frac{1}{2}\langle J^{-1}T_\omega u, u \rangle$, and note that $Q''_\omega = J^{-1}T_\omega$ is a symmetric linear operator. Furthermore, Q_ω is invariant under a group orbit.

The Hamiltonian system of interest is given by

$$\frac{dv}{dt} = JE'(v).$$

We are interested in relative equilibria of this system, i.e., stationary solutions which satisfy $u(t) \in Gu(0)$ for all t. A relative equilibrium satisfies $u(t) = T(e^{\omega t})u(0)$ for some $\omega \in \mathfrak{g}$. Changing variables via

$$v(t) = T(\exp(\omega t))u(t),$$

2.2 Multipulse Solitary Waves

leads to the system

$$\frac{du}{dt} = JE'_0(u;\omega), \tag{2.89}$$

where

$$E'_0(u;\omega) := E'(u) - J^{-1}T_\omega u.$$

We are therefore seeking critical points of the functional $E_0(u;\omega) := E(u) - Q_\omega(u)$ for some $\omega \in \mathfrak{g}$.

The steady-state equation is

$$E'_0(u;\omega) = 0$$

and we assume that it has a smooth family $\Phi(\omega)$ of solutions, where ω varies in $\mathfrak{g}$. Furthermore, we assume that the isotropy subgroups $\{g \in G : T(g)\Phi(\omega) = \Phi(\omega)\}$ are discrete. This assumption implies that the group orbits $G\Phi(\omega)$ have dimension n for each fixed $\omega \in \mathfrak{g}$. Since G is Abelian, for each fixed $\omega \in \mathfrak{g}$ the entire group orbit $T(g)\Phi(\omega)$ consists of relative equilibria with time evolution $T(e^{\omega t})$.

We denote the linearization operator around the wave by JE''_0. Fix a basis $\{\omega_1, \ldots, \omega_n\}$ that satisfies the property that the set $\{T_{\omega_1}\Phi, \ldots, T_{\omega_n}\Phi\}$ is orthogonal. One has that $E''_0 T_{\omega_j}\Phi = 0$ for $j = 1, \ldots, n$; see Section 2 of [55]. Since G is Abelian, under the non-degeneracy condition that D_0 is non-singular, where D_0 is defined in (2.91), it is known that the operator JE''_0 will have a non-trivial kernel

$$JE''_0(\Phi)T_{\omega_i}\Phi = 0, \quad JE''_0(\Phi)\partial_{\omega_i}\Phi = T_{\omega_i}\Phi \tag{2.90}$$

for $i = 1, \ldots, n$, with $\partial_\omega := \partial/\partial\omega$. Furthermore, this set is a basis for the kernel. Note that solutions to the above linear system yield not only a basis for the tangent space to the group orbit, but also a basis for the tangent space of the manifold of relative equilibria. We will assume that the linear operator E''_0 is Fredholm of index zero. If one sets

$$Z = \text{Span}\{T_{\omega_1}\Phi, \ldots, T_{\omega_n}\Phi\},$$

then $H = N \oplus Z \oplus P$, where N is the finite-dimensional subspace

$$N = \{u \in H : \langle u, E''_0 u \rangle < 0\}$$

and $P \subset H$ is a closed subspace such that

$$\langle u, E''_0 u \rangle > \delta \langle u, u \rangle, \quad u \in P$$

for some constant $\delta > 0$.

We set

$$H_1 := \{u \in H : \langle u, E_0'' \partial_{\omega_i} \Phi \rangle = 0, \, i = 1, \ldots, n\}.$$

It is shown in [58] that when solving the linear eigenvalue problem $JE_0''(\Phi)u = \lambda u$, it is sufficient to consider only those $u \in H_1$. This also follows from standard solvability theory, as $J^{-1}T_{\omega_i}\Phi = E_0''\partial_{\omega_i}\Phi$ are solutions to the adjoint eigenvalue problem at $\lambda = 0$ for $i = 1, \ldots, n$. We now define the symmetric matrix $D_0 \in \mathbb{R}^{n \times n}$ by

$$(D_0)_{ij} = \langle \partial_{\omega_j} \Phi, E_0'' \partial_{\omega_i} \Phi \rangle. \tag{2.91}$$

For a given self-adjoint operator A, we denote the number of negative eigenvalues by n(A), while p(A) will be the number of positive eigenvalues, and z(A) the number of zero eigenvalues.

The following was proved in [8]. Suppose that z(D_0) = 0. The operator E_0'' restricted to the space H_1 has the negative index

$$\mathrm{n}\left(E_0''|_{H_1}\right) = \mathrm{n}\left(E_0''\right) - \mathrm{n}(D_0).$$

If n($E_0''|_{H_1}$) = 0, then the wave is a local minimizer for the energy $E_0(u)$, and is therefore stable. The interpretation of this statement can be made as follows. Suppose that the operator E_0'' satisfies n(E_0'') = $k \geq 1$. One then has that the wave is not a local minimizer for E_0. However, there are conserved quantities associated with the evolution equation, and it is possible that these quantities may "prohibit" accessing some or all of the unstable eigendirections. The dim(g) conserved quantities are given by

$$Q_i(u) := \frac{1}{2}\langle J^{-1}T_{\omega_i}u, u \rangle, \quad i = 1, \ldots, n.$$

The quantity n(D_0) precisely determines the number of directions which are rendered inaccessible by the conserved quantities. Hence, n(E_0'') − n(D_0) determines the number of unstable directions for the energy after the constraints have been taken into account.

2.2.2.2 The Perturbed Problem

We now turn to the perturbed problem, where the energy is of the form $E_0(u) + \epsilon E_1(u)$, with $0 < \epsilon \ll 1$. It is assumed that the perturbation breaks $1 \leq k_s \leq n$ of the original symmetries, so that the perturbed system will have $n - k_s$ symmetries. Furthermore, it is assumed that the problem is well-understood for $\epsilon = 0$, as per the above discussion. The existence question is settled by the work of [54], through the following condition, based on Lyapunov–Schmidt reductions: a necessary condition for persistence of the wave is

2.2 Multipulse Solitary Waves

$$\langle E_1'(\Phi(\omega)), T_{\omega_j}\Phi\rangle = 0, \quad j = 1, \ldots, n \tag{2.92}$$

for some $\omega \in \mathfrak{g}$. The condition is sufficient if $z(M) = n - k_s$, where the symmetric matrix M satisfies

$$M_{ij} := \langle T_{\omega_i}\Phi, E_1''(\Phi(\omega)) T_{\omega_j}\Phi\rangle.$$

Since the perturbation breaks k_s symmetries, and the system is Hamiltonian, $2k_s$ eigenvalues will leave the origin. The following lemma, which tracks these small eigenvalues, was proven in [55] via a Lyapunov–Schmidt reduction: the $O(\sqrt{\epsilon})$ eigenvalues and associated eigenfunctions for the perturbed problem are given by

$$\lambda = \sqrt{\epsilon}\lambda_1 + O(\epsilon), \quad u = \sum_{i=1}^{n} v_i \left(T_{\omega_i}\Phi + \sqrt{\epsilon}\lambda_1 \partial_{\omega_i}\Phi\right) + O(\epsilon),$$

where λ_1 is the eigenvalue and $\mathbf{v}$ is the associated eigenvector for the generalized eigenvalue problem

$$\left(D_0 \lambda_1^2 + M\right) \mathbf{v} = \mathbf{0}.$$

It should be noted that the above eigenvalue problem will have $2(n - k_s)$ zero eigenvalues, due to the fact that this many symmetries are assumed to be preserved under the perturbation.

If an eigenvalue has non-zero real part, the Krein signature is zero [46, 59]. The Krein signature of a purely imaginary $O(\sqrt{\epsilon})$ eigenvalue given above is

$$K = \text{sign}\left(\mathbf{v}^T M \mathbf{v}\right) = \text{sign}(\mathbf{v}^T D_0 \mathbf{v}), \tag{2.93}$$

where $\mathbf{v}$ is the associated eigenvector [55]. It may also be possible for eigenvalues to emerge out of the continuous spectrum, creating internal modes, as discussed above. Since these eigenvalues will be of $O(1)$, they will not be captured by the perturbation expansion given in the above lemma. However, this is not problematic (at least in models of the DNLS type) since any $O(1)$ eigenvalues will be purely imaginary with positive Krein sign, and hence for small ϵ do not contribute to an instability.

In the statement of the theorem below, the symmetric matrix D_ϵ is defined by

$$(D_\epsilon)_{ij} := \mathbf{w}_i^T D_0 \mathbf{w}_j, \tag{2.94}$$

where the set $\{\mathbf{w}_1, \ldots, \mathbf{w}_{n-k_s}\}$ is a basis for $\ker(M)$. The following is proved in [55] regarding $\sigma(J(E_0'' + \epsilon E_1''))$ for $0 < \epsilon \ll 1$.

Theorem 1. *Suppose that the unperturbed wave is stable, i.e., $n(E_0'') = n(D_0)$. Let k_r represent the number of real negative eigenvalues, $2k_c$ the number of complex eigenvalues with negative real part, and $2k_i$ the number of purely imaginary eigenvalues with negative Krein signature for the perturbed problem (counting multiplicity). Assume that $z(D_\epsilon) = 0$. Then*

$$k_r + 2k_c + 2k_i = \text{n}(E_0'') + \text{n}(M) - \text{n}(D_\epsilon). \tag{2.95}$$

Furthermore, all of these eigenvalues are of $\text{O}(\sqrt{\epsilon})$, *and*

$$k_s \geq k_r \geq |\text{n}(M) - (\text{n}(D_0) - \text{n}(D_\epsilon))|.$$

Any eigenvalues arising from an edge bifurcation will be purely imaginary with positive Krein signature.

The following remarks can be made about the count of eigenvalues:

1. The upper bound on k_r arises from the facts that there are only $2k_s$ eigenvalues of $\text{O}(\sqrt{\epsilon})$ and the system is Hamiltonian.
2. Since $\text{n}(D_\epsilon) \leq \text{n}(D_0) = \text{n}(E_0'')$, the perturbed wave cannot be a minimizer unless $\text{n}(M) = 0$ and that $\text{n}(D_\epsilon) = \text{n}(D_0)$.

One possible interpretation of 1 is as follows. As previously mentioned, for the unperturbed problem each unstable direction associated with E_0'' is neutralized by an invariance, which in turn are each generated by a symmetry. Now, D_ϵ is the representation of D_0 when restricted to the symmetry group which persists upon the perturbation. The quantity

$$\text{n}(E_0'') - \text{n}(D_\epsilon) = \text{n}(D_0) - \text{n}(D_\epsilon)$$

then precisely details the number of unstable directions associated with E_0'' which are no longer neutralized by the invariances. The quantity $\text{n}(M)$ is the number of additional unstable directions generated by the symmetry-breaking perturbation E_1. The theorem essentially illustrates that the number of potentially unstable eigenvalue pairs in the system is obtained by keeping track of these eigendirections.

2.2.2.3 Case Example: DNLS from the Anti-Continuum Limit

One can consider the DNLS equation near the AC limit as such a perturbed problem. In fact, one can do this even for a general interaction matrix between sites (that is not restricted to nearest-neighbor interactions) as follows:

$$i\dot{u}_n + u_n - |u_n|^2 u_n = -\epsilon \sum_{m=1}^{N} k_{nm} u_m. \tag{2.96}$$

Then, one can label the unperturbed energy at the AC limit as

$$E_0(u) = \sum_n |u_n|^2 - \frac{1}{2}|u_n|^4, \tag{2.97}$$

while the relevant perturbation will be of the form

2.2 Multipulse Solitary Waves

$$E_1(u) = -\epsilon \sum_{n,m=1}^{N} k_{nm} \left(u_n^* u_m + u_n u_m^*\right). \tag{2.98}$$

At the AC limit, the solutions for the excited sites will be $u_n = e^{i\theta_n}$, where the θ_n are free arbitrary phase parameters.

To determine the persistence of the waves, one has to evaluate the perturbed energy at the unperturbed limit solution, in which case, we can straightforwardly evaluate it to be

$$E_1 = -\sum_{n,m=1}^{N} 2k_{nm} \cos(\theta_n - \theta_m). \tag{2.99}$$

Based on the discussion of the previous subsection, the persistence conditions then read $\partial_{\theta_n} E_1 = 0$, which leads to an equivalent condition as Eq. (2.60) derived previously, namely

$$\sum_{m \neq n} k_{nm} \sin(\theta_n - \theta_m) = 0. \tag{2.100}$$

One can subsequently based on the above theory evaluate the relevant matrices D_0 and M_{ij} that enter the stability calculation, in order to obtain information for the relevant eigenvalues that will leave the origin of the spectral plane, upon deviation from the AC limit of $\epsilon = 0$. One can thus find that

$$D_0 = -I_N, \tag{2.101}$$
$$M_{ij} = \partial^2_{\theta_i \theta_j} E_1. \tag{2.102}$$

I_N is the unit matrix of size N (the number of excited sites). Then the correction λ_1 to the eigenvalues will be obtained from the reduced eigenvalue problem $(D_0 \lambda_1^2 + M)v = 0$ which leads in our case to

$$\left(-I_N \lambda_1^2 + 2\mathcal{M}_1\right) v = 0, \tag{2.103}$$

since $M = 2\mathcal{M}_1$, where $\mathcal{M}_1$ is defined by the Eq. (2.84) above (in the nearest-neighbor case; in fact the result obtained here is also true for a more general interaction matrix). This confirms the validity of the direct calculation given above, but also places these results in a broader context of perturbed Hamiltonian dynamics.

Before closing this subsection, we should also note that the eigenvalue count given above confirms the closure relation of [55] (see also [60]), since $n(E_0'') = N + m$ (where N is the number of excited sites and m the number of sign changes between them), $k_r = N - m - 1$, $k_c = 0$, and $k_i = m$, with the imaginary eigenvalues bearing negative Krein signature. A direct calculation shows that $n(M) - n(D_\epsilon) = -1$ and therefore the relevant relation is satisfied.

2.2.3 Multipulses Close to the Continuum Limit

We close the discussion of the one-dimensional problem by briefly considering the case of multipulses in the vicinity of the continuum limit, following closely the discussion of [39] (for this reason, we also use $\epsilon = 1/(2h^2)$, with $h \to 0$ and $\beta = -1$ in Eq. (2.1)). In the quasi-continuum approximation, it can be obtained that the interaction between two solitons is given by the potential energy

$$U_{int}(\xi_i - \xi_2, \Delta\phi) = -8\eta^3 \exp(-\eta|\xi_1 - \xi_2|)\cos(\Delta\phi). \tag{2.104}$$

In this expression $\Delta\xi = \xi_1 - \xi_2$ is the relative separation between the wave centers and $\Delta\phi$ is the relative phase which also plays an important role in their interaction. In particular, note that the interaction is *attractive* for in-phase solitons, while exactly the opposite (i.e., repulsion) is true for out-of-phase solitary waves. Although this result can be derived using the perturbation technique of Karpman and Solov'ev [61] or the variational approximation [34], here we present a different and fairly direct approach of obtaining it, due to Manton [63] (who pioneered it in Klein–Gordon-type equations). Here, we are following the relevant discussion of [64].

In the case of the NLS equation, we have defined previously the mass (whose role is played by the squared L^2 norm, given by Eq. (2.6), while the momentum is defined in Eq. (2.7) (although, in the present setting a factor of 1/2 would be multiplying the right-hand side). Assuming then that we have a soliton centered at $\xi = 0$ and one at $\xi = \Delta\xi$, we can find the derivative of the momentum (the "force") evaluated between $a \ll 0$ and $0 \ll b \ll \Delta\xi$. (This brings in the assumption of sufficiently large separation between the waves for this approach to work.) We thus obtain

$$\frac{dM}{dt} = \frac{1}{4}\left[uu^\star_{xx} + u_{xx}u^\star - 2|u_x|^2\right]_a^b. \tag{2.105}$$

Note that if integrating between $-\infty$ and ∞, Eq. (2.105) would yield a vanishing right-hand side, due to the total conservation of the momentum. However, in the present setting, it yields a non-vanishing contribution to the solitary wave in this interval (a non-vanishing force) due to the solitary wave outside of the interval. Hence, we can use this approach to infer the force exerted from one soliton to the other (and also their respective equations of motion). We use the standard two-soliton decomposition, $u = u^{(1)} + u^{(2)}$ where $u^{(1)} = \eta\operatorname{sech}(\eta x)\exp(i\eta^2 t/2)$, $u^{(2)} = \eta\operatorname{sech}(\eta(x - \Delta\xi))\exp(i\eta^2 t/2)\exp(i\phi)$ are the standing waves, and the relative phase ϕ between them has been incorporated in u_2. Then, one obtains

$$\frac{dM}{dt} = 8\eta^4 \exp(-\eta\Delta\xi), \tag{2.106}$$

which results in the dynamical equation for the separation (using the mass of the soliton; see also the details of the discussion of [64]) of the form

2.2 Multipulse Solitary Waves

$$\ddot{\Delta\xi} = -8\eta^3 \exp(-\eta\Delta\xi)\cos\phi. \tag{2.107}$$

The anti-derivative of the right-hand side of Eq. (2.106) yields the potential of interaction between the solitary waves, coinciding with the result of Eq. (2.104).

In the continuum, this interaction does not allow for the formation of (stationary) bound states between the solitary waves. However, in the realm of the lattice (near the continuum limit), the idea of [39] was that each of the solitary waves will face the energetic contributions of two distinct factors: on the one hand, there will be the interaction with the second solitary wave. On the other hand, each of the waves will be subject to the PN potential due to the existence of the lattice. The latter energetic contribution has been described previously. An asymptotic lowest order approximation of the relevant formula was used in [39] in the form

$$H_{PN}(\xi) \approx -\frac{8\pi^4}{3h^3} \exp\left(-\frac{\pi^2}{\eta h}\right) \cos\left(\frac{2\pi}{h}\xi\right). \tag{2.108}$$

Then, the full energy landscape can be described as

$$H \approx H_{PN}(\xi_1) + H_{PN}(\xi_2) + U_{int}(\xi_1 - \xi_2, \Delta\phi) \tag{2.109}$$

and it is expected that the locations of the centers of the relevant solitary waves can be obtained from extremization of the energy of Eq. (2.109).

While the expression of Eq. (2.109) gives a nice intuitive way to understand the balance of interactions for multipulses in the lattice setting (see also the relevant sketch of Fig. 2.13), for practical purposes, it is perhaps less useful. This is because if h is small, the H_{PN} terms are exponentially weak and hence are practically negligible in comparison to the interaction energy, while for h large so that these terms are sizeable, the pulse deviates from its continuous form and the calculation of H_{PN} is less accurate based on the quasi-continuum expression. Hence, given the nature of the approximations in the calculation, we do not attempt to test it quantitatively herein, although we acknowledge its qualitative usefulness in elucidating the relevant energy landscape (see also the relevant discussions in [65]).

It is interesting to compare/contrast this picture with the integrable analog of the DNLS, namely the AL-NLS model, where the above-mentioned Manton calculation can be carried through [66]. In the latter case, as discussed in the beginning of the chapter, the conserved momentum is given by Eq. (2.25). As in the continuum case, we now consider two solitons, one centered at 0 and one centered at $s \gg 0$, i.e., *widely separated*. We compute dM/dt by performing the summation over n not for the infinite lattice (when the result would be zero due to the relevant conservation law), but rather from $n = L$ to $n = N$, with $L \ll 0$, and $0 \ll N \ll \Delta\xi$. The idea behind this calculation is that, in fact, the force in this interval is not going to be zero, but rather would be *finite* due to the soliton–soliton interaction. For a finite interval encompassing only one soliton, the amount of momentum gain is finite, due to the fact that the one soliton experiences the pull (or push) of the other soliton at

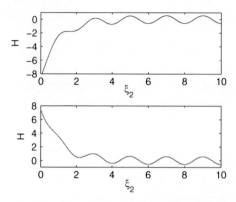

Fig. 2.13 The energy landscape of Eq. (2.109), as it is given for the interaction of two identical solitary waves with $\eta = 1$, for $h = 2$. The first wave is located at $\xi_1 = 0$ and the contributions to the energy landscape affecting the second solitary wave are shown as a function of its center location ξ_2. The *top panel* shows the case of attractive interaction for $\Delta\phi = 0$, while the *bottom panel* shows the case of repulsive interaction for $\Delta\phi = \pi$

the boundary of the interval where we perform the calculation. Specifically, we can evaluate

$$\frac{dM}{dt} = -2 \sum_{n=L}^{N} \left(|u_{n+1}|^2 - |u_n|^2 \right)$$
$$+ \sum_{n=L}^{N} \left(u_n u_{n+2}^\star + u_n^\star u_{n+2} \right) \left(1 + |u_{n+1}|^2 \right)$$
$$- \sum_{n=L}^{N} \left(u_{n-1} u_{n+1}^\star + u_{n-1}^\star u_{n+1} \right) \left(1 + |u_n|^2 \right). \tag{2.110}$$

However, observing the telescopic nature of the sums in the right-hand side of Eq. (2.110), we infer that

$$\frac{dM}{dt} = -2 \left(|u_{N+1}|^2 - |u_L|^2 \right)$$
$$+ \left(u_N u_{N+2}^\star + u_N^\star u_{N+2} \right) \left(1 + |u_{N+1}|^2 \right)$$
$$- \left(u_{L-1} u_{L+1}^\star + u_{L-1}^\star u_{L+1} \right) \left(1 + |u_L|^2 \right). \tag{2.111}$$

As usual in Manton's method, and based on intuitive physical arguments, the main contribution in this asymptotic calculation stems from the boundary between the two solitons. Hence, we drop the terms with subscript L and only consider the contributions with subscript N in what follows.

2.2 Multipulse Solitary Waves

We then select a two-soliton ansatz

$$u_n = u_n^{(1)} + u_n^{(2)} \qquad (2.112)$$

with $u_n^{(1)} = \sinh(\gamma)\text{sech}(\gamma n)\exp(i\sigma)$ and $u_n^{(2)} = \sinh(\gamma)\text{sech}(\gamma(n - \Delta\xi))\exp(i\sigma)$ (i.e., two in-phase solitons). Since $0 \ll N \ll \Delta\xi$, we can use the asymptotic form of the soliton tail at $n = N$, according to

$$u_n^{(1)} = 2\sinh(\gamma)\exp(-\gamma N)\exp(i\sigma), \qquad (2.113)$$

$$u_n^{(2)} = 2\sinh(\gamma)\exp(\gamma(N - \Delta\xi))\exp(i\sigma). \qquad (2.114)$$

Through direct substitution of Eq. (2.112) and the expressions in Eqs. (2.113) and (2.114) into Eq. (2.111), we obtain that

$$\frac{dM}{dt} \approx 32\sinh^4(\gamma)\exp(-\gamma\Delta\xi). \qquad (2.115)$$

Using Newton's equation of motion for the solitons we obtain

$$P_s\ddot{\Delta\xi} = -2\frac{dM}{dt}, \qquad (2.116)$$

where M_s is the mass (power) of the soliton; the factor "2" comes from the fact that there is an equal and opposite pull (or push) on the second soliton, and hence their relative distance decreases by twice the contribution of dM/dt to each of them; and finally the "–" sign originates from the fact that a positive boundary contribution to dM/dt decreases the soliton distance, while the opposite is true for a negative dM/dt. In this case,

$$P_s = \sum_{n=-\infty}^{\infty} \ln\left(1 + |\psi_n|^2\right) = 2\gamma. \qquad (2.117)$$

Thus, the equation for the $s(t)$ becomes

$$\ddot{\Delta\xi} = -\frac{32}{\gamma}\sinh^4(\gamma)\exp(-\gamma\Delta\xi), \qquad (2.118)$$

while the relevant effective soliton interaction potential (for a unit mass particle) is

$$V(\Delta\xi) = -\frac{32}{\gamma^2}\sinh^4(\gamma)\exp(-\gamma\Delta\xi). \qquad (2.119)$$

If the solitons additionally possess a phase difference ϕ, the above calculations gives a factor of $\cos(\phi)$ in Eqs. (2.118) and (2.119). Relevant results illustrating the attraction of in-phase and repulsion of out-of-phase solitons are shown in Fig. 2.14.

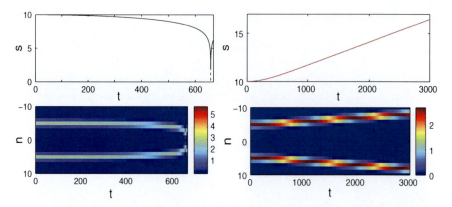

Fig. 2.14 Example of an in-phase (*left*) and an out-of-phase (*right*) collision of two discrete bright solitons of the AL-NLS model. The *top panels* show the distance $s(t) \equiv \Delta \xi$ numerically and from the ODE of Eq. (2.118), while the *bottom* shows the space-time contour plot of the AL-NLS evolution. The quality of the agreement of the ODE result with the full numerical computation is such that the two lines of the *top panels* can not be distinguished. Reprinted from [66] with permission

It should be pointed out that, as expected, the AL-NLS solitons do *not* face a PN barrier when traveling through the lattice. For this reason, the only contribution to their potential energy stems from the exponential tail–tail interactions, contrary to what we saw is the case in the DNLS model. Finally, accounting for a factor of 1/2 in the equation (and also another such factor in the definition of the momentum), as well as taking the limit of $\sinh(\gamma) \to \gamma$, we can derive the continuum analog (2.107) of Eq. (2.118).

References

1. Sulem, C., Sulem, P.L.: The Nonlinear Schrödinger Equation. Springer-Verlag, New York, (1999)
2. Morrison, P.J.: Rev. Mod. Phys. **70**, 467–521 (1998)
3. Ablowitz, M.J., Segur, H.: Solitons and the Inverse Scattering Transform. SIAM, Philadelphia (1981)
4. Kaup, D.J.: Phys. Rev. A **42**, 5689 (1990)
5. Johansson, M., Aubry, S.: Phys. Rev. E **61**, 5864 (2000)
6. Ablowitz, M.J., Ladik, J.F.: J. Math. Phys. **16**, 598 (1975)
7. Ablowitz, M.J., Ladik, J.F.: J. Math. Phys. **17**, 1011 (1976)
8. Ablowitz, M.J., Prinari, B., Trubatch, A.D.: Discrete and Continuous Nonlinear Schrödinger Systems. Cambridge University Press, Cambridge (2004)
9. Sievers, A.J., Takeno, S.: Phys. Rev. Lett. **61**, 970 (1988)
10. Page, J.B.: Phys. Rev. B **41**, 7835 (1990)
11. Kivshar, Yu.S., Campbell, D.K.: Phys. Rev. E **48**, 3077 (1993)
12. Kevrekidis, P.G., Kivshar, Yu.S., Kovalev, A.S.: Phys. Rev. E **67**, 046604 (2003)
13. Cai, D. Bishop, A.R., Grønbech-Jensen, N.: Phys. Rev. E **53**, 4131 (1996)
14. Morse, P.M., Feshbach, H.: Methods of Theoretical Physics. McGraw-Hill, New York (1953)

References

15. Peierls, R.F.: Proc. R. Soc. London **52**, 34 (1940)
16. Nabarro, F.R.N.: Proc. R. Soc. London **59**, 256 (1947)
17. Kivshar, Yu.S., Pelinovsky, D.E., Cretegny, T., Peyrard, M.: Phys. Rev. Lett. **80**, 5032 (1998)
18. Kapitula, T., Kevrekidis, P.G.: Nonlinearity **14**, 533 (2001)
19. Ablowitz, M.J., Herbst, B.M.: In: Important Developments in Soliton Theory, Fokas, A., Zakharov, V. (eds.) Springer-Verlag, Berlin (1993)
20. Herbst, B.M., Ablowitz, M.J.: J. Comp. Phys. **105**, 122 (1993)
21. Kapitula, T.: SIAM J. Math. Anal. **30**, 273 (1999)
22. Kapitula, T., Rubin, J.: Nonlinearity, **13**, 77 (2000)
23. Kapitula, T., Sandstede, B.: Physica **124D**, 58 (1998)
24. Braun, O.M., Kivshar, Yu.S., Peyrard, M.: Phys. Rev. E **56**, 6050 (1997)
25. Balmforth, N.J., Craster, R.V., Kevrekidis, P.G.: Physica **135D**, 212 (2000)
26. Kevrekidis, P.G., Jones, C.K.R.T.: Phys. Rev. E **61**, 3116 (2000)
27. Kapitula, T., Kevrekidis, P.G., Jones, C.K.R.T.: Phys. Rev. E, **63** 036602 (2001)
28. Grillakis, M., Shatah, J., Strauss W.: J. Funct. Anal. **74**, 160 (1987)
29. Jones, C.K.R.T.: Ergod. Theor. Dynam. Syst. **8**, 119 (1988)
30. MacKay, R.S., Aubry, S.: Nonlinearity **7**, 1623 (1994)
31. Alfimov, G.L., Brazhnyi, V.A., Konotop, V.V.: Physica D **194**, 127 (2004)
32. Malomed, B.A., Weinstein, M.I.: Phys. Lett. A **220** 91 (1996)
33. Cuevas, J., Kevrekidis, P.G., Frantzeskakis, D.J., Malomed, B.A.: Physica D **238**, 67 (2009)
34. Malomed, B.A.: Progr. Opt. **43**, 71 (2002)
35. Kaup, D.J., Vogel, T.K.: Phys. Lett. A **362** 289 (2007)
36. Levy, H., Lessman, F.: Finite DIfference Equations, Dover, New York (1992)
37. Pelinovsky, D.E., Kevrekidis, P.G., Frantzeskakis, D.J.: Physica D **212**, 1 (2005)
38. Kevrekidis, P.G., Rasmussen, K.Ø., Bishop, A.R.: Int. J. Mod. Phys. B **15**, 2833 (2001)
39. Kapitula, T., Kevrekidis, P.G., Malomed, B.A.: Phys. Rev. E **63**, 036604 (2001)
40. MacKay, R.S., Stability of equilibria of Hamiltonian systems. In: MacKay, R.S., Meiss, J. (eds.) Hamiltonian Dynamical Systems, Adam Hilger, London (1987)
41. MacKay, R.S.: Phys. Lett. A **155**, 266 (1991)
42. Grillakis, M.: Commun. Pure Appl. Math. **43**, 299 (1990)
43. Grillakis, M.: Commun. Pure Appl. Math. **41**, 745 (1988)
44. Li, Y., Promislow, K.: Physica D **124**, 137 (1998)
45. Li, Y., Promislow, K.: SIAM J. Math. Anal. **31**, 1351(2000)
46. Skryabin, D.V.: Phys. Rev. E **64**, 055601(R) (2001)
47. Skryabin, D.V.: Phys. Rev. A **63**, 013602 (2001)
48. Kawaguchi, Y., Ohmi, T.: Phys. Rev. A **70**,043610 (2004)
49. Laedke, E.W., Kluth, O., Spatschek, K.H.: Phys. Rev. E **54**, 4299 (1996)
50. Johansson, M., Aubry, S.: Nonlinearity **10**, 1151(1997)
51. Darmanyan, S., Kobyakov, A., Lederer, F.: Sov. Phys. JETP **86**, 682 (1998)
52. Kevrekidis, P.G., Bishop, A.R., Rasmussen, K.Ø.: Phys. Rev. E **63**, 036603 (2001)
53. van der Meer, J.-C.: Nonlinearity **3**, 1041 (1990)
54. Kapitula, T.: Physica D **156**, 186 (2001)
55. Kapitula, T., Kevrekidis, P.G., Sandstede, B.: Physica D **195**, 263 (2004)
56. Kapitula, T., Kevrekidis, P.G., Sandstede, B.: Physica D **201**, 199 (2005)
57. Kapitula, T., Kevrekidis, P.G.: J. Phys. A **37**, 7509 (2004)
58. Grillakis, M., Shatah, J., Strauss, W.: J. Func. Anal. **94**, 308 (1990)
59. Aubry, S.: Physica **103D**, 201 (1997)
60. Pelinovsky, D.E.: Proc. Roy. Soc. Lond. A **461**, 783 (2005)
61. Karpman, V.I., Solov'ev, V.V.: Physica D **3**, 142 (1981)
62. Karpman, V.I., Solov'ev, V.V.: Physica D **3**, 487 (1981)
63. Manton, N.S.: Nucl. Phys. B **150**, 397 (1979)
64. Kevrekidis, P.G., Khare, A., Saxena, A.: Phys. Rev. E **70**, 057603 (2004)
65. Kevrekidis, P.G., Phys. Rev. E **64**, 026611 (2001)
66. Kevrekidis, P.G., Khare, A., Saxena, A., Bena, I.: Math. Comp. Simul. **74**, 405 (2007)

Chapter 3
The Two-Dimensional Case

3.1 General Notions

We now turn to the examination of the two-dimensional DNLS equation, and of the type of excitations that can emerge in that context.

The dynamical equation can be written in the form

$$i\dot{u}_{n,m} + \epsilon \left(u_{n+1,m} + u_{n-1,m} + u_{n,m+1} + u_{n,m-1} - 4u_{n,m}\right) + |u_{n,m}|^2 u_{n,m} = 0, \quad (3.1)$$

where $u_{n,m}$ represents the two-dimensional complex field. The corresponding Hamiltonian function of this Hamiltonian system can be expressed as

$$H = \sum_{(n,m) \in \mathbb{Z}^2} \epsilon |u_{n+1,m} - u_{n,m}|^2 + \epsilon |u_{n,m+1} - u_{n,m}|^2 - \frac{1}{2}|u_{n,m}|^4. \quad (3.2)$$

In addition to the time translational invariance inducing the conservation of the Hamiltonian, this infinite dimensional dynamical system also has the U(1) invariance, analogously to its one-dimensional sibling, hence it also preserves the squared l^2 norm or power $P = \sum_{m,n} |u_{n,m}|^2$. These are the two fundamental conservation laws that are known for the discrete case. In the continuum analog of the model, there exist additional conservation laws; a natural one among these corresponds to the vector form of the momentum

$$M = i \int \left(u \nabla u^\star - u^\star \nabla u\right) dx dy. \quad (3.3)$$

A far less obvious invariance of the two-dimensional setting is the so-called pseudo-conformal invariance which is *particular* to the two-dimensional case (the so-called critical case for the cubic nonlinearity). If one defines $l(t) = (t_\star - t)/t_0$, then the transformation $\mathbf{x}'(t) = \mathbf{x}(t)/l(t)$, $t' = \int_0^t ds/l^2(s)$ and $u'(\mathbf{x}', t') = lu(\mathbf{x}, t) \exp(ia|\mathbf{x}|^2/(4l^2))$, leaves the equation unchanged. In the above expressions, $a = -ldl/dt$ and $\mathbf{x}$ is used to denote the spatial vector. The corresponding conserved quantity is

$$C = \int \left(|\mathbf{x}u + 2it\nabla u|^2 - 2t^2|u|^4\right) d\mathbf{x}. \tag{3.4}$$

It is important to note that neither of these last two conservation laws is preserved in the discrete case. Furthermore, the continuum case is well known to lead to collapse in its dynamical evolution; see a detailed analysis of the relevant phenomena in [1].

In fact, the two-dimensional case is special as it is the "critical dimension" beyond which collapse is occurring. The pseudo-conformal invariance allows the rescaling of the amplitude and the width of the solution in a self-similar way, without costing energy and in this way gives rise to the possibility of the solution to collapse along this group orbit in finite time. In fact, if the power of the solution exceeds that of the fundamental, single-humped radial solution of the equation (often referred to as the Townes soliton [2]):

$$\Delta R + R - R^3 = 0, \tag{3.5}$$

then the initial condition leads to self-focusing and collapse, while if it is lower than that, it instead leads to dispersion. This collapse-type effect is of course no longer possible in the discrete case, since the l^2 conservation prevents any particular site from acquiring infinite amplitude (at best, the whole power of the initial condition may be concentrated on a single site, in a phenomenon referred to as *quasi-collapse*). Since the treatise of [1] addresses this issue in considerable detail (see also references therein), we will not discuss it further here. Instead we will focus on the stationary states of the discrete problem and their stability analysis.

As in the previous chapter, we will start by briefly discussing the single-pulse case, and we will then turn to more complex multisite solutions such as multi-pulse solitary waves and discrete vortices. The latter are a new feature of the two-dimensional discrete system with no direct analog in the one-dimensional infinite lattice case.

3.2 Single-Pulse Solitary Waves

The fundamental standing wave ($u_n = e^{i\Lambda t}v_n$) solution of the two-dimensional discrete equation should be a single-humped solitary wave that asymptotes to the Townes soliton [2] as the continuum limit is approached. Its stationary profile should satisfy

$$\Lambda v_n = \epsilon \Delta_2 v_n + v_n^3. \tag{3.6}$$

In view of the scaling property of the equation (under $\tilde{\Lambda} = \Lambda/\epsilon$ and the $\tilde{u}_n = u_n/\sqrt{\epsilon}$), ϵ (or Λ) can be scaled out, and we will consider the relevant problem of Eq. (3.6) monoparametrically.

3.2 Single-Pulse Solitary Waves

In fact, we can consider this monoparametric problem at any dimension, if Δ_2 represents the corresponding d-dimensional discrete Laplacian (whereby site n has $2d$ neighbors). To obtain a semi-analytical understanding of the properties of the relevant ground state, we will analyze the problem of Eq. (2.65) starting with a variational approximation, which can be carried out at any dimension d. This approach for the single-pulse solutions of interest can be obtained using

$$v_n = A e^{-a|n|_{l_1}}. \tag{3.7}$$

Then the Lagrangian from which Eq. (3.6) can be derived, namely

$$\mathcal{L}_{\text{eff}} = \sum_n \left(\sum_j 2\epsilon u_{n+e_j} u_n - (\Lambda + 2d\epsilon) u_n^2 + \frac{1}{2} u_n^4 \right), \tag{3.8}$$

(where e_j is the unit vector along the jth direction) can be explicitly evaluated as

$$\mathcal{L}_{\text{eff}} = 2d\epsilon P \operatorname{sech} a - (\Lambda + 2d\epsilon) P + \frac{P^2}{2^{d+1}} \frac{\cosh^d(2a) \sinh^d(a)}{\cosh^{3d}(a)}, \tag{3.9}$$

where the power $P = \sum_n u_n^2$ can be evaluated as

$$P = A^2 \coth^d(a). \tag{3.10}$$

From Eq. (3.9) and the extremization conditions of this effective Lagrangian:

$$\frac{\partial \mathcal{L}_{\text{eff}}}{\partial P} = \frac{\partial \mathcal{L}_{\text{eff}}}{\partial a} = 0, \tag{3.11}$$

one can obtain the $P = P(\Lambda)$ (in any dimension, and for different values of ϵ).

One can compare the results of Eq. (3.11) with direct numerical computations identifying the ground state solutions of the DNLS equation (again, in any dimension). As relevant examples, we present in Fig. 3.1 the cases of $d = 2$ that we focus on in this chapter, but also for comparison those of $d = 1, 3$. The power of the solutions is given as a function of Λ (for $\epsilon = 1$) in Fig. 3.1. One of the key observations of the figure is the difference in the stationary state properties between the case of $d = 1$ (so-called subcritical case) and the cases of $d = 2$ (critical) and $d = 3$ (supercritical). In fact, as was originally demonstrated in the work of [11], through scaling arguments, and was later proved more rigorously in [12] (see also the recent discussion of [13]), for dimensions larger than the critical dimension (which is given by $d\sigma = 2$ for nonlinearities $|u|^{2\sigma} u$), there is a *power threshold* for the excitation of localized solitary waves. That is, contrary to what is the case for one dimension such excitations where there is such a solution *for any value of the power*, for higher dimensional problems, such a solution exists *only* for powers $P \geq P_{cr}$, where P_{cr} denotes the relevant threshold.

Fig. 3.1 The plot shows the one-, two-, and three-dimensional results for the power P of the stationary solutions for different Λ in Eq. (3.6). The *solid lines* denote the full numerical results, while the *dashed* ones the results of the variational approximation described in the text

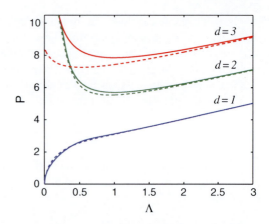

Another important comment to make here concerns the accuracy of the variational approximation in characterizing the stationary solutions. We can see that in the one- and two-dimensional settings, the VA is fairly accurate in capturing the trends of the full numerical solution, however, in the three-dimensional context it clearly misses in its quantitative description (although it does share some qualitative trends with the actual solution). As an example, we note that in that case, the minimum power occurs for $\Lambda \approx 1$ in the numerical results, while it happens for $\Lambda \approx 0.5234$ in the variational approximation.

The existence of the power thresholds discussed above also bears important information for the stability of the localized solutions of the discrete model. In particular, as discussed in the previous chapter and as stems from the original work of [6–8] and later from the work of [9, 10], the change of monotonicity of the $P = P(\Lambda)$ curve has an important consequence in the stability of the structure, in particular, the single-humped solution is stable when $dP/d\Lambda > 0$, while it is unstable due to a real eigenvalue pair when $dP/d\Lambda < 0$. Hence, as the continuum limit is approached through decreasing Λ (or equivalently through increasing ϵ), the originally stable discrete single-humped soliton has to become unstable, and this happens precisely at the point where $dP/d\Lambda = 0$. This instability was observed in [14] and the relevant criterion was originally proposed in the work of Vakhitov and Kolokolov and therefore is often referred to as the VK criterion [12].

In Fig. 3.2, we show two examples of the relevant solution, as obtained numerically, one deeply in the discrete regime for $\Lambda = 1.5$ (linearly stable) and one very close to the continuum regime for $\Lambda = 0.05$ (linearly unstable, since $\Lambda < \epsilon = 1$). The figure also shows the corresponding spectral planes (λ_r, λ_i) of the relevant linearization eigenvalues $\lambda = \lambda_r + i\lambda_i$, verifying the stability of the former and instability of the latter (due to a real eigenvalue pair). Note how the latter solution appears to approach its radially symmetric continuum limit (the Townes soliton of [2]). It should also be pointed out in that regard that as $\Lambda \to 0$ (see also the relevant trend in Fig. 3.1), the squared l^2 norm of the solution can be seen to approach $P \approx 11.7$ which is the well-known critical mass of the Townes solution [13].

3.2 Single-Pulse Solitary Waves

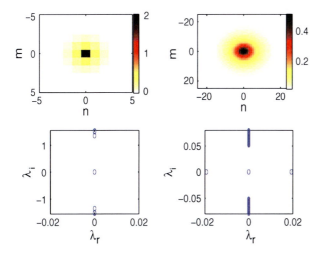

Fig. 3.2 The *top left panel* shows a contour plot of the fundamental solution of the two-dimensional DNLS for $\Lambda = 1.5$, while the *top right* shows the same solution for $\Lambda = 0.05$ (when it approaches its continuum profile). The former solution is stable as is shown in the *bottom left panel* illustrating the spectral plane (λ_r, λ_i) of its linearization eigenvalues $\lambda = \lambda_r + i\lambda_i$, while the latter solution is unstable (having $\Lambda < 1$), as is shown in the *bottom right panel*

In closing, it would be interesting to highlight one important open problem arising in the computations of this fundamental solution for $d = 2$ and 3. In particular, our numerical results indicate that for such solutions, the instability due to the sign change of $dP/d\Lambda$ occurs *precisely* at $\Lambda = 1$ (or, when ϵ is present in the equation, when $\Lambda = \epsilon$). Hence, a relevant mathematical question arises as to whether indeed for dimensions $d \geq 2$, one can more precisely quantify the location of the instability and whether in particular it indeed occurs for $\Lambda = \epsilon$ more generally or not. This is a conjecture that it would be worthwhile to settle.

One direction toward proving this conjecture is to try to derive properties based on Eq. (3.6), connecting the power of the solution P, the frequency Λ, and the coupling strength ϵ. Such a property can be obtained, e.g., by multiplying (3.6) by u_n and summing over n, resulting in the identity

$$(2d\epsilon + \Lambda)P = 2\epsilon L + N, \tag{3.12}$$

where $L = \sum_n \sum_j u_{n+e_j} u_n$ and $N = \sum_n u_n^4$. Similarly, multiplying Eq. (3.6) by $\partial u_n / \partial \Lambda$ and summing over n, along with using Eq. (3.12), one can also derive the identity

$$P = \frac{1}{2} \frac{dN}{d\Lambda}. \tag{3.13}$$

Unfortunately, in addition to Eqs. (3.12) and (3.13), one needs one more equation to eliminate both N and L and derive the dependence of P on ϵ and Λ that would permit an explicit calculation of the relevant critical point $dP/d\Lambda = 0$.

3.3 Multipulses and Discrete Vortices

We now try to address the existence of more complex multisite structures, starting from the AC limit of $\epsilon = 0$ in Eq. (3.6). Note that similarly to our one-dimensional exposition of the previous chapter, for notational simplicity, we set $\Lambda = 1$ in what follows. Our discussion will closely follow [14].

3.3.1 Formulation of the Bifurcation Problem Near $\epsilon = 0$

The relevant stationary equation of interest then reads

$$\left(1 - |v_{n,m}|^2\right) \phi_{n,m} = \epsilon \left(v_{n+1,m} + v_{n-1,m} + v_{n,m+1} + v_{n,m-1}\right). \tag{3.14}$$

In the $\epsilon = 0$, similarly to the one-dimensional case, the solutions of this equation can be fully characterized

$$v_{n,m}^{(0)} = \begin{cases} e^{i\theta_{n,m}}, & (n,m) \in S, \\ 0, & (n,m) \in \mathbb{Z}^2 \backslash S, \end{cases} \tag{3.15}$$

where S is a finite set of nodes on the lattice and $\theta_{n,m}$ are parameters for these *excited sites*. Since θ_0 is arbitrary, we can set $\theta_{n_0,m_0} = 0$ for a particular node $(n_0, m_0) \in S$. Using this convention, we can define two special types of localized modes, called discrete solitons and vortices.

The localized solution of the difference equations (3.14) with $\epsilon > 0$ is called a *discrete soliton* when it has all real-valued amplitudes $v_{n,m}$, $\forall (n,m) \in S$ and at the limit (3.15), $\theta_{n,m} = \{0, \pi\}$ for $(n,m) \in S$. On the other hand, if S is a simple closed discrete contour on the plane and the localized solution has complex valued $v_{n,m}$ that satisfies the limit (3.15) with $\theta_{n,m} \in [0, 2\pi]$, $(n,m) \in S$, then we call such a solution a *discrete vortex*.

Discrete vortices can be partitioned into symmetric and asymmetric ones as follows. If S is a simple closed discrete contour on the plane, such that each node $(n,m) \in S$ has exactly two adjacent nodes in vertical or horizontal directions along S. Let $\Delta\theta_j$ be the phase difference between two successive nodes in the contour S, defined according to the enumeration $j = 1, 2, \ldots, \dim(S)$, such that $|\Delta\theta_j| \leq \pi$. If the phase differences $\Delta\theta_j$ are constant along S, the discrete vortex is called symmetric. Otherwise, it is called asymmetric. The total number of 2π phase shifts across the closed contour S is called the vortex charge. More specifically, we consider discrete contours

3.3 Multipulses and Discrete Vortices

$$S_M = \{(1,1), (2,1), \ldots, (M+1,1), (M+1,2), \ldots, (M+1, M+1),$$
$$(M, M+1), \ldots, (1, M+1), (1, M), \ldots, (1,2)\}, \quad (3.16)$$

containing $4M$ sites. Given the above definition, the contour S_M for a fixed M could support symmetric and asymmetric vortices with some charge L. Arguably, the simplest vortex is the symmetric charge-one vortex cell ($M = L = 1$: $\theta_{1,1} = 0$, $\theta_{2,1} = \pi/2$, $\theta_{2,2} = \pi$, $\theta_{1,2} = 3\pi/2$) [16, 29]. Although the main formalism below is developed for any $M \geq 1$, we obtain a complete set of results on persistence and stability of discrete vortices only in the cases $M = 1, 2, 3$, which are of most physical interest.

It then follows directly from the general method [17] that the discrete solitons of the two-dimensional NLS lattice (3.14) can be continued for $0 < \epsilon < \epsilon_0$ for some $\epsilon_0 > 0$. It is more complicated to find a configuration of $\theta_{n,m}$ for $(n,m) \in S$ that allows us to continue the discrete vortices for $\epsilon > 0$. The continuation of the discrete solitons and vortices is based on the Implicit Function Theorem and the Lyapunov–Schmidt Reduction Theorem [18, 19].

We denote by $\mathcal{O}(0)$ a small neighborhood of $\epsilon = 0$, such that $\mathcal{O}(0) = (-\epsilon_0, \epsilon_0)$ for some $\epsilon_0 > 0$. Let $N = \dim(S)$ and $\mathcal{T}$ be the torus on $[0, 2\pi]^N$, such that $\theta_{n,m}$ for $(n,m) \in S$ form a vector $\boldsymbol{\theta} \in \mathcal{T}$. Let $\Omega = L^2(\mathbb{Z}^2, \mathbb{C})$ be the Hilbert space of square-summable complex-valued sequences $\{\phi_{n,m}\}_{(n,m) \in \mathbb{Z}^2}$, equipped with the standard inner product and l^2 norm.

It can then be proved that there exists a unique (discrete soliton) solution of the difference equations (3.14) in the domain $\epsilon \in \mathcal{O}(0)$ with a real profile satisfying $\lim_{\epsilon \to 0} v_{n,m} = v_{n,m}^{(0)}$, and $v_{n,m}^{(0)}$ is given by (3.15) with $\theta_{n,m} = \{0, \pi\}$, $(n,m) \in S$. The solution is analytic in ϵ in this neighborhood. This can be proved by considering the equations for the stationary solution of Eq. (3.14) as the zeros of a vector valued function $f_{n,m}$. Then this mapping has a bounded and continuous Fréchet derivative

$$\mathcal{L}_{n,m} = \left(1 - 3v_{n,m}^2\right) - \epsilon\left(s_{+1,0} + s_{-1,0} + s_{0,+1} + s_{0,-1}\right), \quad (3.17)$$

where $s_{n',m'}$ is the shift operator, such that $s_{n',m'} u_{n,m} = u_{n+n',m+m'}$. The kernel of $\mathcal{L}_{n,m}$ is empty for $\epsilon = 0$. Therefore, for $\epsilon = 0$, $\mathcal{L}_{n,m}^{(0)}$ has a bounded inverse, which implies by the implicit function theorem (see Appendix 1 in [19] and Chap. 2.2 in [18]) that there is a continuous (in fact, analytic) in ϵ solution $v_{n,m}(\epsilon)$ in our case.

Now for a general profile $v_{n,m}^{(0)}$ at the AC limit, the continuation of such a solution for $\epsilon \in \mathcal{O}(0)$ requires that some conditions, constituting a vector valued function which we will denote by $\mathbf{g}(\boldsymbol{\theta}, \epsilon)$, be satisfied. Moreover, the function $\mathbf{g}(\boldsymbol{\theta}, \epsilon)$ is analytic in $\epsilon \in \mathcal{O}(0)$ and $\mathbf{g}(\boldsymbol{\theta}, 0) = \mathbf{0}$ for any $\boldsymbol{\theta}$. This can be shown for a general solution $v_{n,m}$, by considering the vector equations (3.14) and their complex conjugate (denote the vector valued function by $\mathbf{f}(\mathbf{v}, \bar{\mathbf{v}}, \epsilon)$). Then, taking the Fréchet derivative of $\mathbf{f}(\mathbf{v}, \bar{\mathbf{v}}, \epsilon)$ with respect to $\mathbf{v}$ and $\bar{\mathbf{v}}$, we compute the linearization operator $\mathcal{H}$ for the difference Eq. (3.14):

$$\mathcal{H}_{n,m} = \begin{pmatrix} 1 - 2|v_{n,m}|^2 & -v_{n,m}^2 \\ -\bar{v}_{n,m}^2 & 1 - 2|v_{n,m}|^2 \end{pmatrix}$$

$$-\epsilon \left(s_{+1,0} + s_{-1,0} + s_{0,+1} + s_{0,-1} \right) \begin{pmatrix} 1 & 0 \\ 0 & 1 \end{pmatrix}. \qquad (3.18)$$

Let $\mathcal{H}^{(0)} = \mathcal{H}(\boldsymbol{\phi}^{(0)}, 0)$. Note that dim $\ker(\mathcal{H}^{(0)}) = N$. Moreover, eigenvectors of $\ker(\mathcal{H}^{(0)})$ re-normalize the parameters $\theta_{n,m}$ for $(n,m) \in S$ in the limiting solution (3.15). By the Lyapunov Reduction Theorem [19, Chap. 7.1], there exists a decomposition $\Omega = \ker(\mathcal{H}^{(0)}) \oplus \omega$, such that $\mathbf{g}(\boldsymbol{\theta}, \epsilon)$ is defined in terms of the projections to $\ker(\mathcal{H}^{(0)})$. Let $\{\mathbf{e}_{n,m}\}_{(n,m) \in S}$ be a set of N linearly independent eigenvectors in the kernel of $\mathcal{H}^{(0)}$. It follows from the representation,

$$\mathcal{H}_{n,m}^{(0)} = -\begin{pmatrix} 1 & e^{2i\theta_{n,m}} \\ e^{-2i\theta_{n,m}} & 1 \end{pmatrix}, \quad (n,m) \in S \qquad (3.19)$$

that each eigenvector $\mathbf{e}_{n,m}$ in the set $\{\mathbf{e}_{n,m}\}_{(n,m) \in S}$ has the only non-zero element $(e^{i\theta_{n,m}}, -e^{-i\theta_{n,m}})^T$ at the (n,m)th position of $(\mathbf{u}, \mathbf{w}) \in \Omega \times \Omega$. By projections of the nonlinear equations to $\ker(\mathcal{H}^{(0)})$, we derive an implicit representation for the functions $\mathbf{g}(\boldsymbol{\theta}, \epsilon)$:

$$(n,m) \in S: \ 2ig_{n,m}(\boldsymbol{\theta}, \epsilon) = (1 - |v_{n,m}|^2)\left(e^{-i\theta_{n,m}} v_{n,m} - e^{i\theta_{n,m}} \bar{v}_{n,m} \right)$$
$$- \epsilon e^{-i\theta_{n,m}} \left(v_{n+1,m} + v_{n-1,m} + v_{n,m+1} + v_{n,m-1} \right)$$
$$+ \epsilon e^{i\theta_{n,m}} \left(\bar{v}_{n+1,m} + \bar{v}_{n-1,m} + \bar{v}_{n,m+1} + \bar{v}_{n,m-1} \right), \quad (3.20)$$

where the factor $(2i)$ is introduced for convenience. By setting $v_{n,m} = e^{i\theta_{n,m}} \phi_{n,m}$ for $(n,m) \in S$ and renaming $\phi_{n,m} \to v_{n,m}$ we end up obtaining the solvability conditions

$$(n,m) \in S: \ -2ig_{n,m}(\boldsymbol{\theta}, \epsilon) = \epsilon e^{-i\theta_{n,m}} \left(v_{n+1,m} + v_{n-1,m} + v_{n,m+1} + v_{n,m-1} \right)$$
$$- \epsilon e^{i\theta_{n,m}} \left(\bar{v}_{n+1,m} + \bar{v}_{n-1,m} + \bar{v}_{n,m+1} + \bar{v}_{n,m-1} \right). \ (3.21)$$

Note that these are the same conditions to leading order to the solvability conditions directly inferred by Eq. (3.14), by multiplying the equation by $\bar{v}_{n,m}$ and subtracting the complex conjugate, namely

$$(n,m) \in S: \ -2ig_{n,m}(\boldsymbol{\theta}, \epsilon) = \epsilon \bar{v}_{n,m} \left(v_{n+1,m} + v_{n-1,m} + v_{n,m+1} + v_{n,m-1} \right)$$
$$- \epsilon v_{n,m} \left(\bar{v}_{n+1,m} + \bar{v}_{n-1,m} + \bar{v}_{n,m+1} + \bar{v}_{n,m-1} \right). \ (3.22)$$

Note, also, that given the analyticity of these solvability conditions, they can be Taylor expanded in ϵ, and so can the solution $v_{n,m}$. Furthermore, in this reformulation

3.3 Multipulses and Discrete Vortices   63

of the problem for the angles $\theta_{m,n}$ (as a function of ϵ), the gauge or U(1) invariance of the original problem can be translated into a shift of the angle $\theta_{n,m} \to \theta_{n,m} + \theta_0$, which yields a one parameter family of roots of $\mathbf{g}(\boldsymbol{\theta}, \epsilon)$. This implies that the Jacobian matrix $(\mathcal{M}_1)_{jk} = \partial g_j^{(1)}/\partial \theta_k$ of the first-order expansion $g^{(1)}$ will have a non-empty kernel with an eigenvector $\mathbf{p}_0 = (1, 1, \ldots, 1)$ due to the gauge transformation. However, if we let X_0 be the constrained subspace of $\mathbb{C}^N$:

$$X_0 = \{\mathbf{u} \in \mathbb{C}^N : (\mathbf{p}_0, \mathbf{u}) = 0\}. \tag{3.23}$$

and the matrix $\mathcal{M}_1$ is non-singular in the subspace X_0, then there exists a unique (modulo the shift) analytic continuation of the root of the bifurcation equations for $\epsilon \in \mathcal{O}(0)$ by the Implicit Function Theorem, applied to the nonlinear equation $\mathbf{g}(\boldsymbol{\theta}, \epsilon) = \mathbf{0}$ [19, Appendix 1].

An important generalization of the above continuation is the following (regarding conditions under which a solution family cannot be continued for $\epsilon \in \mathcal{O}(0)$: let $\boldsymbol{\theta}_*$ be a $(1+d)$-parameter solution of $\mathbf{g}^{(1)}(\boldsymbol{\theta}) = \mathbf{0}$ and $\mathcal{M}_1$ have a zero eigenvalue of multiplicity $(1+d)$, where $1 \leq d \leq N-1$. Let $\mathbf{g}^{(2)}(\boldsymbol{\theta}_*) = \cdots = \mathbf{g}^{(K-1)}(\boldsymbol{\theta}_*) = \mathbf{0}$ but $\mathbf{g}^{(K)}(\boldsymbol{\theta}_*) \neq \mathbf{0}$. The limiting solution (3.15) can be continued in the domain $\epsilon \in \mathcal{O}(0)$ only if $\mathbf{g}^{(K)}(\boldsymbol{\theta}_*)$ is orthogonal to $\ker(\mathcal{M}_1)$. If $\mathbf{g}^{(K)}(\boldsymbol{\theta}_*) \notin X_d$, where

$$X_d = \{\mathbf{u} \in X_0 : (\mathbf{p}_l, \mathbf{u}) = 0,\ l = 1, \ldots, d\}, \tag{3.24}$$

then the solution can not be continued in $\epsilon \in \mathcal{O}(0)$, according to Chap. 1.3 of [19].

3.3.2 Persistence of Discrete Solutions

We now consider discrete soliton and vortex solutions over the contours S_M and order the angles of which the contour consists as $\theta_1, \theta_2, \ldots, \theta_N$. Given the nature of the considered (closed, square) contours, periodic boundary conditions are applied ($\theta_0 = \theta_N$, $\theta_1 = \theta_{N+1}$). As per the definition above, a discrete vortex has the charge L if the phase difference $\Delta \theta_j$ between two successive nodes changes by $2\pi L$ along the discrete contour S_M, where $\Delta \theta_j$ is defined within the fundamental branch $|\Delta \theta_j| \leq \pi$. By gauge transformation, we can always set $\theta_1 = 0$ for convenience. We will also choose $\theta_2 = \theta$ with $0 \leq \theta \leq \pi$ for convenience, which corresponds to discrete vortices with $L \geq 0$ (the existence and stability of their negative charge counterparts is the same).

To identify the leading order persistence conditions, we substitute the limiting AC solution $v_{n,m}^{(0)}$ solution in the bifurcation equations to obtain $g^{(1)}$ in the form

$$\mathbf{g}_j^{(1)}(\boldsymbol{\theta}) = \sin\left(\theta_j - \theta_{j+1}\right) + \sin\left(\theta_j - \theta_{j-1}\right), \quad 1 \leq j \leq N. \tag{3.25}$$

The bifurcation equations $\mathbf{g}^{(1)}(\boldsymbol{\theta}) = \mathbf{0}$ are rewritten as a system of N nonlinear equations for N parameters $\theta_1, \theta_2, \ldots, \theta_N$ as follows:

$$\sin(\theta_2 - \theta_1) = \sin(\theta_3 - \theta_2) = \cdots = \sin(\theta_N - \theta_{N-1}) = \sin(\theta_1 - \theta_N). \quad (3.26)$$

These types of conditions also arose in the work of [20, 21]. We now attempt to classify all solutions of the bifurcation equations.

If we let $a_j = \cos(\theta_{j+1} - \theta_j)$ for $1 \leq j \leq N$, such that $\theta_1 = 0$, $\theta_2 = \theta$, and $\theta_{N+1} = 2\pi L$, where $N = 4M$, $0 \leq \theta \leq \pi$ and L is the vortex charge. All solutions of the bifurcation equations (3.26) reduce to the following four families:

(i) discrete solitons with $\theta = \{0, \pi\}$ and

$$\theta_j = \{0, \pi\}, \quad 3 \leq j \leq N, \quad (3.27)$$

such that the set $\{a_j\}_{j=1}^N$ includes l coefficients $a_j = 1$ and $N - l$ coefficients $a_j = -1$, where $0 \leq l \leq N$.

(ii) symmetric vortices of charge L with $\theta = \pi L/2M$, where $1 \leq L \leq 2M - 1$, and

$$\theta_j = \frac{\pi L(j-1)}{2M}, \quad 3 \leq j \leq N, \quad (3.28)$$

such that all N coefficients are the same: $a_j = a = \cos(\pi L/2M)$.

(iii) one-parameter families of asymmetric vortices of charge $L = M$ with $0 < \theta < \pi$ and

$$\theta_{j+1} - \theta_j = \left\{ \begin{array}{c} \theta \\ \pi - \theta \end{array} \right\} \mod(2\pi), \quad 2 \leq j \leq N, \quad (3.29)$$

such that the set $\{a_j\}_{j=1}^N$ includes $2M$ coefficients $a_j = \cos\theta$ and $2M$ coefficients $a_j = -\cos\theta$.

(iv) zero-parameter asymmetric vortices of charge $L \neq M$ and

$$\theta = \theta_* = \frac{\pi}{2}\left(\frac{n + 2L - 4M}{n - 2M}\right), \quad 1 \leq n \leq N - 1, \; n \neq 2M, \quad (3.30)$$

such that the set $\{a_j\}_{j=1}^N$ includes n coefficients $a_j = \cos\theta_*$ and $N - n$ coefficients $a_j = -\cos\theta_*$ and the family (iv) does not reduce to any of the families (i)–(iii).

This can be seen because essentially there are only two roots of the sine function that permit simultaneously satisfying the bifurcation equations. These are the choices of Eq. (3.29), leading, respectively, to either $a_j = \cos(\theta)$ or to $a_j = -\cos(\theta)$. If we generically assume that there are totally n choices of the former type and $N - n$ ones of the latter type within the contour, then

$$\theta_{N+1} = n\theta + (N - n)(\pi - \theta) = (2n - N)\theta + (N - n)\pi = 2\pi L,$$

3.3 Multipulses and Discrete Vortices

where L is the integer charge of the discrete vortex. There are only two solutions of the above equation. When θ is arbitrary parameter, we have $n = N/2 = 2M$ and $L = M$, which gives the one-parameter family (iii). When $\theta = \theta_*$ is fixed, we have

$$\theta_* = \frac{\pi}{2}\left(\frac{n + 2L - 4M}{n - 2M}\right).$$

When $n = N - 2L$, we have the family (i) with $N - 2L$ phases $\theta_j = 0$ and $2L$ phases $\theta_j = \pi$. Since the charge is not assigned to discrete solitons, the parameter L could be half-integer: $L = (N - l)/2$, where $0 \le l \le N$. When $n = 4M$, we have the family (ii) for any $1 \le L \le 2M - 1$. Other choices of n, which are irreducible to the families (i)–(iii), produce the family (iv). Furthermore, it is worthwhile to note that there are special cases where family (iii) reduces to families (ii) and (i); these will be dubbed supersymmetric cases. In particular, when $\theta = 0$ and π, the family (iii) reduces to the family (i) with $l = 2M$. When $\theta = \pi/2$, the family (iii) reduces to the family (ii) with $L = M$. We shall call the corresponding solutions of family (i) the supersymmetric soliton and of family (ii) the supersymmetric vortex.

One can make a simple combinatorial enumeration of the solutions of families (i)–(iii). In the case of (i), there are $N_1 = 2^{N-1}$, since aside from the first site, all others can be either 0 or π. There are also $N_2 = 2M - 1$ solutions of family (ii), and N_3 solutions of family (iii), where

$$N_3 = 2^{N-1} - \sum_{k=0}^{2M-1} \frac{N!}{k!(N-k)!}. \qquad (3.31)$$

As special case examples that will be of relevance to our discussion below, we mention the contours with $M = 1$ (four sites) and $M = 2$ (eight sites). In the first case, there are eight solutions of type (i), one of type (ii), three solutions of type (iii), and no solutions of the family (iv). The three one-parameter asymmetric solutions are

(a) $\theta_1 = 0,\ \theta_2 = \theta,\ \theta_3 = \pi,\ \theta_4 = \pi + \theta,$ \hfill (3.32)

(b) $\theta_1 = 0,\ \theta_2 = \theta,\ \theta_3 = 2\theta,\ \theta_4 = \pi + \theta,$ \hfill (3.33)

(c) $\theta_1 = 0,\ \theta_2 = \theta,\ \theta_3 = \pi,\ \theta_4 = 2\pi - \theta.$ \hfill (3.34)

Similarly in the case with $M = 2$, there are 128 solutions of family (i), 3 solutions of family (ii), 35 solutions of family (iii), and 14 of family (iv). The three symmetric vortices have charge $L = 1$ ($\theta = \pi/4$), $L = 2$ ($\theta = \pi/2$), and $L = 3$ ($\theta = 3\pi/4$). The one-parameter asymmetric vortices include 35 combinations of 4 upper choices and 4 lower choices in (3.29). Finally, the zero parameter asymmetric vortices include seven combinations of vortices with $L = 1$ for seven phase differences $\pi/6$

and one phase difference $5\pi/6$ and seven combinations of vortices with $L = 3$ for one phase difference $\pi/6$ and seven phase differences $5\pi/6$.

3.3.2.1 First-Order Reductions

The Jacobian $\mathcal{M}_1$ of the first-order bifurcation equations $\mathbf{g}^{(1)}(\boldsymbol{\theta})$ can be obtained from Eq. (3.25) as

$$(\mathcal{M}_1)_{i,j} = \begin{cases} \cos(\theta_{j+1} - \theta_j) + \cos(\theta_{j-1} - \theta_j), & i = j, \\ -\cos(\theta_j - \theta_i), & i = j \pm 1, \\ 0, & |i - j| \geq 2, \end{cases} \quad (3.35)$$

subject to the periodic boundary conditions. It is interesting that this is the same type of structure as we encountered in the one-dimensional configurations of the previous chapter, and it is the one also encountered in the perturbation theory of continuous multipulse solitons in coupled NLS equations [20].

If we let n_0, z_0, and p_0 be the numbers of negative, zero, and positive terms of $a_j = \cos(\theta_{j+1} - \theta_j)$, $1 \leq j \leq N$ (therefore $n_0 + z_0 + p_0 = N$), it is important based on this information to infer how many eigenvalues of $\mathcal{M}_1$ are negative, zero, or positive. Assuming $z_0 = 0$ (note that this is not true in supersymmetric cases), it turns out (see the appendix of [20] for a relevant proof by induction arguments) that the eigenvalues are intimately connected with the quantity

$$A_1 = \sum_{i=1}^{N} \prod_{j \neq i} a_j = \left(\prod_{i=1}^{N} a_i\right) \left(\sum_{i=1}^{N} \frac{1}{a_i}\right). \quad (3.36)$$

In particular, as expected there exists a zero eigenvalue in $\mathcal{M}_1$ due to the gauge invariance. Therefore, denoting the characteristic polynomial of $\mathcal{M}_1$ as $D(\lambda)$, it is clear that $D(0) = 0$. It is then important to evaluate $D'(0) = -\lambda_1 \lambda_2 \ldots \lambda_{N-1}$. A key result in that regard is that for this matrix $D'(0) = -NA_1$; therefore, an immediate consequence is that if $A_1 \neq 0$, then the multiplicity of the zero eigenvalue is $z(\mathcal{M}_1) = 1$. If $A_1 = 0$, then $z(\mathcal{M}_1) \geq 2$. Note also that from the above results (comparing $D'(0)$) $(-1)^{n(\mathcal{M}_1)} = \text{sign}(A_1)$, where $n(\mathcal{M}_1)$ denotes the number of negative eigenvalues of the matrix; this implies that $n(\mathcal{M}_1)$ is even if $A_1 > 0$ and is odd if $A_1 < 0$. Finally, if $p(\mathcal{M}_1)$ will be used to denote the number of positive eigenvalues, then as indicated in [20], there are two possible scenario (if $A_1 \neq 0$) for the related eigenvalues: either $n(\mathcal{M}_1) = n_0 - 1$, $p(\mathcal{M}_1) = p_0$ or $n(\mathcal{M}_1) = n_0$, $p(\mathcal{M}_1) = p_0 - 1$.

In some important special cases (for what follows), the eigenvalues of the matrix can be computed analytically. More specifically, if all coefficients $a_j =$

3.3 Multipulses and Discrete Vortices

$\cos(\theta_{j+1} - \theta_j)$, $1 \leq j \leq N$ are equal $a_j = a$, then, the eigenvalues of the matrix can be computed as

$$\lambda_n = 4a \sin^2 \frac{\pi n}{N}, \quad 1 \leq n \leq N. \tag{3.37}$$

This is straightforward to see since the eigenfunction equation becomes

$$a\left(2x_j - x_{j+1} - x_{j-1}\right) = \lambda x_j, \quad x_0 = x_N, \; x_1 = x_{N+1}, \tag{3.38}$$

which is solvable by the using the discrete Fourier modes $x_j = \exp(i(2\pi jn/N))$ for $1 \leq j, n \leq N$, yielding the above eigenvalue expression.

On the other hand, if the elements of the matrix alternate in sign $a_j = (-1)^j a$, $1 \leq j \leq 4M$, then the eigenvalue problem can be written as a system of two coupled linear difference equations of the form

$$a\left(y_j - y_{j-1}\right) = \lambda x_j, \quad a\left(x_j - x_{j+1}\right) = \lambda y_j, \quad 1 \leq j \leq 2M, \tag{3.39}$$

subject to the periodic boundary conditions: $x_1 = x_{2M+1}$ and $y_0 = y_{2M}$. Once again, the discrete Fourier transform can be used according to $x_j = x_0 \exp(i(2\pi jn/2M))$ and $y_j = y_0 \exp(i(2\pi jn/2M))$ for $1 \leq j, n \leq 2M$ and this yields the eigenvalues

$$\lambda_n = -\lambda_{n+2M} = 2a \sin \frac{\pi n}{2M}, \quad 1 \leq n \leq 2M, \tag{3.40}$$

such that $n(\mathcal{M}_1) = 2M - 1$, $z(\mathcal{M}_1) = 2$, and $p(\mathcal{M}_1) = 2M - 1$. These numbers do not change if the set $\{a_j\}_{j=1}^N$ is obtained from the sign-alternating set $\{(-1)^j a\}_{j=1}^N$ by permutations (see [14] for a proof of the last statement).

The above results indicate that discrete solitons can be typically continued uniquely for finite ϵ, since $z(\mathcal{M}_1) = 1$. This is with the notable exception of supersymmetric solitons where the number of positive and negative a_j's is equal. On the other hand, for family (ii), all coefficients a_j are the same: $a_j = a = \cos(\pi L/2M)$, $1 \leq j \leq N$. The above calculation for equal a_j's yields the presence of a zero eigenvalue λ_N; the remaining $(N - 1)$ eigenvalues are all positive for $a > 0$ (when $1 \leq L \leq M - 1$), negative for $a < 0$ (when $M + 1 \leq L \leq 2M - 1$), and zero for $a = 0$ (in the supersymmetric case of $L = M$). Therefore, states other than the supersymmetric ones are also guaranteed to have a unique continuation also in the case of discrete *symmetric* vortices. The first-order reductions are less informative in the case of asymmetric discrete vortices of family (iii), whereby there are $2M$ coefficients $a_j = \cos\theta$ and $2M$ coefficients $a_j = -\cos\theta$, which are non-zero for $\theta \neq \pi/2$. The count of eigenvalues of the matrix $\mathcal{M}_1$ yields $n(\mathcal{M}_1) = 2M - 1$, $z(\mathcal{M}_1) = 2$, and $p(\mathcal{M}_1) = 2M - 1$. Therefore in this case, the higher multiplicity of the zero eigenvalue (related, at heart, with the additional freedom in the selection of the angular parameter θ) leads to an inconclusive result for such solutions. Lastly, for family (iv), the fact that $A_1 \neq 0$, again preserves the zero eigenvalue multiplicity as $z(\mathcal{M}_1) = 1$, permitting a unique continuation of such zero parameter asymmetric vortices.

3.3.2.2 Second-Order Reductions

To determine the fate of supersymmetric solitons and vortices, as well as that of mono-parametric, asymmetric vortices, we now expand the bifurcation function $\mathbf{g}(\boldsymbol{\theta}, \epsilon)$ to second order. Starting with the first-order correction in the solution, we have

$$\left(1 - 2|v_{n,m}^{(0)}|^2\right) v_{n,m}^{(1)} - v_{n,m}^{(0)2} \bar{v}_{n,m}^{(1)} = v_{n+1,m}^{(0)} + v_{n-1,m}^{(0)} + v_{n,m+1}^{(0)} + v_{n,m-1}^{(0)}. \quad (3.41)$$

To solve this more complicated equation, we will distinguish the cases of the different contours; we will consider, in particular, the contours with $M = 1$, $M = 2$, and $M \geq 3$.

In the case of $M = 1$, the inhomogeneous equation (3.41) has a solution of the form

$$v_{n,m}^{(1)} = -\frac{1}{2} \left[\cos\left(\theta_{j-1} - \theta_j\right) + \cos\left(\theta_{j+1} - \theta_j\right)\right] e^{i\theta_j}, \quad (3.42)$$

for sites in the contour S_M, while for their non-contour neighbors,

$$v_{n,m}^{(1)} = e^{i\theta_j} \quad (3.43)$$

and every other site vanishes at this order. Substituting this first order correction within the bifurcation equations to deduce $\mathbf{g}^{(2)}(\boldsymbol{\theta})$, we find the form

$$\mathbf{g}_j^{(2)}(\boldsymbol{\theta}) = \frac{1}{2} \sin\left(\theta_{j+1} - \theta_j\right) \left[\cos(\theta_j - \theta_{j+1}) + \cos\left(\theta_{j+2} - \theta_{j+1}\right)\right] \quad (3.44)$$

$$+ \frac{1}{2} \sin\left(\theta_{j-1} - \theta_j\right) \left[\cos\left(\theta_j - \theta_{j-1}\right) + \cos(\theta_{j-2} - \theta_{j-1})\right], \quad 1 \leq j \leq N.$$

One can then straightforwardly compute the vector $\mathbf{g}^{(2)}(\boldsymbol{\theta})$ for the asymmetric solutions (3.32), (3.33), and (3.34):

(a) $\mathbf{g}_2 = \begin{pmatrix} 0 \\ 0 \\ 0 \\ 0 \end{pmatrix}$, (b) $\mathbf{g}_2 = \begin{pmatrix} 2 \\ 0 \\ -2 \\ 0 \end{pmatrix} \sin\theta \cos\theta$, (c) $\mathbf{g}_2 = \begin{pmatrix} 0 \\ -2 \\ 0 \\ 2 \end{pmatrix} \sin\theta \cos\theta.$

The key observation, however, concerns the kernel of $\mathcal{M}_1$, which as illustrated above has a second element (in addition to the gauge invariance eigenvector $(1, 1, 1, 1)^T$). This element is evaluated as

3.3 Multipulses and Discrete Vortices

(a) $\mathbf{p}_1 = \begin{pmatrix} 0 \\ 1 \\ 0 \\ 1 \end{pmatrix}$, (b) $\mathbf{p}_1 = \begin{pmatrix} 0 \\ 1 \\ 2 \\ 1 \end{pmatrix}$, (c) $\mathbf{p}_1 = \begin{pmatrix} 0 \\ 1 \\ 0 \\ -1 \end{pmatrix}$.

The Fredholm alternative $(\mathbf{p}_1, \mathbf{g}_2) = 0$ is satisfied for the solution (a) but fails for the solutions (b) and (c), unless $\theta = \{0, \pi/2, \pi\}$. Therefore, the important conclusion from this exercise is that solutions (b) and (c) *cannot* be continued for $\epsilon \neq 0$.

In the case of $M = 2$, the correction is the same as in the previous case, as given by Eqs. (3.42) and (3.43), except for the central node (2, 2), where contributions from its four neighboring sites yield

$$v_{2,2}^{(1)} = e^{i\theta_2} + e^{i\theta_4} + e^{i\theta_6} + e^{i\theta_8}. \tag{3.45}$$

This, in turn, modifies the entries of the bifurcation function according to

$$\mathbf{g}_j^{(2)}(\boldsymbol{\theta}) \to \mathbf{g}_j^{(2)}(\boldsymbol{\theta}) + \sin(\theta_j - \theta_{j-2}) + \sin(\theta_j - \theta_{j+2}) + \sin(\theta_j - \theta_{j+4}), \quad j = 2, 4, 6, 8. \tag{3.46}$$

In that case, there are 35 one-parameter asymmetric vortex solitons, each of which has a corresponding second eigenvector $\mathbf{p}_1$ in the kernel of $\mathcal{M}_1$. For all but one of these solutions (assuming that $\theta \neq \{0, \pi/2, \pi\}$), the condition for continuation of the solution, namely $(\mathbf{p}_1, \mathbf{g}_2) = 0$ fails, hence the solutions cannot exist. The only solution that can be continued in this case is the one with alternating signs of $a_j = \cos(\theta_{j+1} - \theta_j)$.

Finally, in the case of $M \geq 3$, the first-order corrections to the solution still obey (3.42) and (3.43), except for the four corner nodes $(2, 2), (M, 2), (M, M)$, and $(2, M)$ each of which have two neighbors which lead to

$$v_{n,m}^{(1)} = e^{i\theta_{j-1}} + e^{i\theta_{j+1}}, \quad j = 1, M+1, 2M+1, 3M+1. \tag{3.47}$$

The correction term $\mathbf{g}^{(2)}(\boldsymbol{\theta})$ is given by (3.45), except for the adjacent entries to the four corner nodes on the contour S_M: $(1, 1), (1, M+1), (M+1, M+1)$, and $(M+1, 1)$, which are modified by

$$\mathbf{g}_j^{(2)}(\boldsymbol{\theta}) \to \mathbf{g}_j^{(2)}(\boldsymbol{\theta}) + \sin(\theta_j - \theta_{j-2}), \quad j = 2, M+2, 2M+2, 3M+2,$$

$$\mathbf{g}_j^{(2)}(\boldsymbol{\theta}) \to \mathbf{g}_j^{(2)}(\boldsymbol{\theta}) + \sin(\theta_j - \theta_{j+2}), \quad j = M, 2M, 3M, 4M. \tag{3.48}$$

For any $M \geq 3$, there is a solution of family (iii), where $\mathbf{g}_2 = \mathbf{0}$, which is characterized by the alternating signs of coefficients $a_j = \cos(\theta_{j+1} - \theta_j)$ for $1 \leq j \leq N$. In the case $M = 3$, all other solutions of family (iii) have $(\mathbf{p}_1, \mathbf{g}_2) \neq 0$ and hence terminate at the second-order reductions.

Generalizing the results of this section, we have that all asymmetric vortices of family (iii), except for the sign-alternating set $a_j = \cos(\theta_{j+1} - \theta_j) = (-1)^{j+1}\cos\theta$, $1 \leq j \leq N$, cannot be continued to $\epsilon \neq 0$ for $M = 1, 2, 3$. The only solution of this type which can be continued has the explicit form

$$\theta_{4j-3} = 2\pi(j-1), \quad \theta_{4j-2} = \theta_{4j-3}+\theta, \quad \theta_{4j-1} = \theta_{4j-3}+\pi, \quad \theta_{4j} = \theta_{4j-3}+\pi+\theta, \tag{3.49}$$

where $1 \leq j \leq M$ and $0 \leq \theta \leq \pi$. This solution includes two particular cases of supersymmetric solitons of family (i) for $\theta = 0$ and π and supersymmetric vortices of family (ii) for $\theta = \pi/2$. Continuation of the solution (3.49) must be considered beyond the second-order reductions.

In the case of supersymmetric solitons, and considering the matrix $\mathcal{M}_1 + \epsilon \mathcal{M}_2$, it can be found that the second zero eigenvalue of $\mathcal{M}_1$ bifurcates off zero. As a result, the supersymmetric solutions of family (i) can be uniquely continued to discrete solitons.

On the other hand, we need to consider the Jacobian matrix $\mathcal{M}_2$ and its eigenvalues for discrete supersymmetric vortices of different charges (we consider the cases $M = 1$, $M = 2$, and $M \geq 3$). In the case of $M = 1$, the elements of $\mathcal{M}_2$ are given by

$$(\mathcal{M}_2)_{i,j} = \begin{cases} +1, & i = j, \\ -\frac{1}{2}, & i = j \pm 2, \\ 0, & |i-j| \neq 0, 2 \end{cases} \tag{3.50}$$

or explicitly

$$\mathcal{M}_2 = \begin{pmatrix} 1 & 0 & -1 & 0 \\ 0 & 1 & 0 & -1 \\ -1 & 0 & 1 & 0 \\ 0 & -1 & 0 & 1 \end{pmatrix}. \tag{3.51}$$

The matrix $\mathcal{M}_2$ has four eigenvalues: $\lambda_1 = \lambda_2 = 2$ and $\lambda_3 = \lambda_4 = 0$. The two eigenvectors for the zero eigenvalue are $\mathbf{p}_3 = (1, 0, 1, 0)^T$ and $\mathbf{p}_4 = (0, 1, 0, 1)^T$. The eigenvector $\mathbf{p}_4$ corresponds to the derivative of the asymmetric vortex (3.32) with respect to parameter θ, while the eigenvector $\mathbf{p}_0 = \mathbf{p}_3 + \mathbf{p}_4$ corresponds to the shift due to gauge invariance.

3.3 Multipulses and Discrete Vortices

In the case of $M = 2$, the second-order Jacobian matrix is given by the form

$$\tilde{\mathcal{M}}_2 = \begin{pmatrix} 1 & 0 & -\frac{1}{2} & 0 & 0 & 0 & -\frac{1}{2} & 0 \\ 0 & 0 & 0 & \frac{1}{2} & 0 & -1 & 0 & \frac{1}{2} \\ -\frac{1}{2} & 0 & 1 & 0 & -\frac{1}{2} & 0 & 0 & 0 \\ 0 & \frac{1}{2} & 0 & 0 & 0 & \frac{1}{2} & 0 & -1 \\ 0 & 0 & -\frac{1}{2} & 0 & 1 & 0 & -\frac{1}{2} & 0 \\ 0 & -1 & 0 & \frac{1}{2} & 0 & 0 & 0 & \frac{1}{2} \\ -\frac{1}{2} & 0 & 0 & 0 & -\frac{1}{2} & 0 & 1 & 0 \\ 0 & \frac{1}{2} & 0 & -1 & 0 & \frac{1}{2} & 0 & 0 \end{pmatrix}. \qquad (3.52)$$

The corresponding eigenvalue problem can be decoupled into two linear difference equations with constant coefficients, as follows:

$$2x_j - x_{j+1} - x_{j-1} = 2\lambda x_j, \qquad j = 1, 2, 3, 4$$
$$-2y_{j+2} + y_{j+1} + y_{j-1} = 2\lambda y_j, \qquad j = 1, 2, 3, 4,$$

subject to the periodic boundary conditions for x_j and y_j; this can again be solved by discrete Fourier transform, yielding the eigenvalues $\lambda_1 = 1$, $\lambda_2 = 2$, $\lambda_3 = 1$, and $\lambda_4 = 0$; $\lambda_5 = 1$, $\lambda_6 = -2$, $\lambda_7 = 1$, and $\lambda_8 = 0$ (the first four are obtained from the first problem, while the latter four from the second problem). In this case also, there are two eigenvectors with zero eigenvalue, namely $\mathbf{p}_4 = (1, 0, 1, 0, 1, 0, 1, 0)^T$ and $\mathbf{p}_8 = (0, 1, 0, 1, 0, 1, 0, 1)^T$, where the eigenvector $\mathbf{p}_8$ corresponds to the derivative of the asymmetric vortex (3.49) with respect to parameter θ and the eigenvector $\mathbf{p}_0 = \mathbf{p}_4 + \mathbf{p}_8$ corresponds to the shift due to gauge invariance.

Finally, in the case of $M \geq 3$, the Jacobian matrix still resembles that of Eq. (3.50), but now the additional entries stem from the four corner nodes of the contours, namely $(1, 1)$, $(1, M + 1)$, $(M + 1, M + 1)$, and $(M + 1, 1)$. The second-order Jacobian in this case reads

$$\tilde{\mathcal{M}}_2 = \mathcal{M}_2 + \Delta\mathcal{M}_2, \qquad (3.53)$$

where $\Delta\mathcal{M}_2$ is a rank-four non-positive matrix with the elements

$$(\Delta\mathcal{M}_2)_{i,j} = \begin{cases} -1, & i = j = 2, M, M+2, 2M, 2M+2, 3M, 3M+2, 4M, \\ +1, & i = j - 2 = M, 2M, 3M, 4M, \\ +1, & i = j + 2 = 2, M+2, 2M+2, 3M+2 \end{cases} \qquad (3.54)$$

and all other elements are zeros. The explicit form for the modified matrix $\tilde{\mathcal{M}}_2$ in the case $M = 3$ is

$$\tilde{\mathcal{M}}_2 = \begin{pmatrix} 1 & 0 & -\frac{1}{2} & 0 & 0 & 0 & 0 & 0 & 0 & 0 & -\frac{1}{2} & 0 \\ 0 & 0 & 0 & -\frac{1}{2} & 0 & 0 & 0 & 0 & 0 & 0 & 0 & \frac{1}{2} \\ -\frac{1}{2} & 0 & 0 & 0 & \frac{1}{2} & 0 & 0 & 0 & 0 & 0 & 0 & 0 \\ 0 & -\frac{1}{2} & 0 & 1 & 0 & -\frac{1}{2} & 0 & 0 & 0 & 0 & 0 & 0 \\ 0 & 0 & \frac{1}{2} & 0 & 0 & 0 & -\frac{1}{2} & 0 & 0 & 0 & 0 & 0 \\ 0 & 0 & 0 & -\frac{1}{2} & 0 & 0 & 0 & \frac{1}{2} & 0 & 0 & 0 & 0 \\ 0 & 0 & 0 & 0 & -\frac{1}{2} & 0 & 1 & 0 & -\frac{1}{2} & 0 & 0 & 0 \\ 0 & 0 & 0 & 0 & 0 & \frac{1}{2} & 0 & 0 & 0 & -\frac{1}{2} & 0 & 0 \\ 0 & 0 & 0 & 0 & 0 & 0 & -\frac{1}{2} & 0 & 0 & 0 & \frac{1}{2} & 0 \\ 0 & 0 & 0 & 0 & 0 & 0 & 0 & -\frac{1}{2} & 0 & 1 & 0 & -\frac{1}{2} \\ -\frac{1}{2} & 0 & 0 & 0 & 0 & 0 & 0 & 0 & \frac{1}{2} & 0 & 0 & 0 \\ 0 & \frac{1}{2} & 0 & 0 & 0 & 0 & 0 & 0 & 0 & -\frac{1}{2} & 0 & 0 \end{pmatrix}.$$

(3.55)

The computation of the eigenvalues of $\tilde{\mathcal{M}}_2$ again decouples into eigenvalue problems for two 6×6 matrices and the resulting spectra can be computed as

$$\lambda_1 = \lambda_7 = -0.780776, \quad \lambda_2 = \lambda_8 = -0.5, \quad \lambda_3 = \lambda_9 = 0,$$
$$\lambda_4 = \lambda_{10} = 0.5, \quad \lambda_5 = \lambda_{11} = 1.28078, \quad \lambda_6 = \lambda_{12} = 1.5.$$

Just as in the previous two cases, also in this case the matrix has exactly two zero eigenvalues, one of which is related to the derivative of the asymmetric vortex and one of which is related to the shift of the gauge invariance.

3.3.2.3 Higher Order Reductions

Since the family of solutions (3.49) survives up to second-order reductions, one needs to consider higher order reductions in order to examine the potential persistence or non-existence of such solutions. Intuitively speaking, the presence of the arbitrary parameter θ in this family of asymmetric vortices appears not to be supported by the symmetry of the corresponding discrete contour, or that of the original dynamical equation. One therefore has to attempt to algorithmically expand

3.3 Multipulses and Discrete Vortices

the considerations of the above subsections to higher orders (as direct calculations become extremely cumbersome), to address the issue. We do so as follows.

Let M be the index of the discrete contour S_M and K be the truncation order of the Lyapunov–Schmidt reduction. We construct a squared domain $(n, m) \in D(M, K)$ which includes $N_0 \times N_0$ lattice nodes, where $N_0 = 2K + M + 1$. Corrections of the power series for a given configuration of $\boldsymbol{\theta}$ in (3.49) solve the set of inhomogeneous equations

$$\mathcal{H}^{(0)} \begin{pmatrix} \boldsymbol{\phi}^{(k)} \\ \bar{\boldsymbol{\phi}}^{(k)} \end{pmatrix} = \begin{pmatrix} \mathbf{f}^{(k)} \\ \bar{\mathbf{f}}^{(k)} \end{pmatrix}, \qquad 1 \leq k \leq K,$$

where $\mathcal{H}^{(0)}$ is given by (3.19) and $\mathbf{f}^{(k)}$ represents the right-hand side terms, which are defined recursively from the nonlinear Eq. (3.14). When $\boldsymbol{\phi}^{(k)} \in \omega \subset \Omega$, we have a unique solution of the inhomogeneous equations for any $1 \leq k \leq K$:

$$\phi_{n,m}^{(k)} = -\frac{1}{2} f_{n,m}^{(k)}, \quad (n, m) \in S_M, \qquad \phi_{n,m}^{(k)} = f_{n,m}^{(k)}, \quad (n, m) \in \mathbb{Z}^2 \setminus S_M,$$

provided that

$$g_{n,m}^{(k)} = -\mathrm{Im}(f_{n,m}^{(k)} e^{-i\theta_{n,m}}) = 0, \qquad (n, m) \in S_M, \quad 1 \leq k \leq K.$$

Now, if all $\mathbf{g}^{(k)} = 0$ for $1 \leq k \leq K - 1$, but $(\mathbf{p}_1, \mathbf{g}^{(K)}) \neq 0$, where $\mathbf{p}_1$ is the derivative vector of (3.49) with respect to parameter θ, then the family (3.49) terminates at the Kth order of the Lyapunov–Schmidt reduction.

Following this algorithm, one can find that for $M = 1$, when $\mathbf{p}_1 = (0, 1, 0, 1)^T$, the vector $\mathbf{g}^{(k)}$ is zero for $k = 1, 2, 3, 4, 5$ and non-zero for $k = K = 6$. Moreover, $(\mathbf{p}_1, \mathbf{g}^{(6)}) \neq 0$ for any $\theta \neq \{0, \pi/2, \pi\}$. Similarly, in the case $M = 2$, we have also found that $K = 6$ and $(\mathbf{p}_1, \mathbf{g}^{(6)}) \neq 0$ for any $\theta \neq \{0, \pi/2, \pi\}$. Therefore, indeed, such solutions cannot be continued for the cases of $M = 1$ and 2. Finally, for $M = 3$, one obtains similar conclusions through numerical computations; it is therefore natural to conjecture that such a solution (asymmetric, one parameter family) cannot be continued to finite ϵ, for any value of M.

Summarizing our conclusions for the persistence of the different classified families of discrete solitons and discrete symmetric and asymmetric vortices, we have the following:

- discrete solitons of family (i) in (3.27)
- symmetric vortices of family (ii) in (3.28)
- asymmetric vortices of family (iii) in (3.29) cannot be continued to the domain $\epsilon \in \mathcal{O}(0)$ for $M = 1, 2, 3$.
- zero-parameter asymmetric vortices of family (iv) in (3.30)

It is now natural to turn to the examination of the stability of the relevant persisting solutions.

3.3.3 Stability of Discrete Solutions

To examine the spectral stability of discrete solitons and vortices, we use the linearization

$$u_{n,m}(t) = e^{i(1-4\epsilon)t}\left(\phi_{n,m} + a_{n,m}e^{\lambda t} + \bar{b}_{n,m}e^{\bar{\lambda}t}\right), \quad (n,m) \in \mathbb{Z}^2, \quad (3.56)$$

where $\lambda \in \mathbb{C}$ and $(a_{n,m}, b_{n,m}) \in \mathbb{C}^2$ are the eigenvalues and eigenfunctions, respectively, satisfying

$$i\lambda a_{n,m} = \left(1 - 2|\phi_{n,m}|^2\right) a_{n,m} - \phi_{n,m}^2 b_{n,m}$$
$$- \epsilon \left(a_{n+1,m} + a_{n-1,m} + a_{n,m+1} + a_{n,m-1}\right),$$

$$-i\lambda b_{n,m} = -\bar{\phi}_{n,m}^2 a_{n,m} + \left(1 - 2|\phi_{n,m}|^2\right) b_{n,m}$$
$$- \epsilon \left(b_{n+1,m} + b_{n-1,m} + b_{n,m+1} + b_{n,m-1}\right).$$

This stability problem can be rephrased as

$$i\lambda \psi = \sigma \mathcal{H} \psi, \quad (3.57)$$

where $\psi = (a_{n,m}, b_{n,m})^T$ (the ^{T} denotes transpose), $\mathcal{H}$ is defined by the linearization operator (2.13), and σ consists of 2×2 blocks of Pauli matrices σ_3 (σ_3 is the diagonal matrix with elements $(1, -1)$ along the diagonal). In the eigenvalue problem of Eq. (3.57), the presence of λ with non-zero real part illustrates the presence of instability.

The Taylor expansion of the matrix $\mathcal{H}$ will play a central role in our stability considerations below and is as follows:

$$\mathcal{H} = \mathcal{H}^{(0)} + \sum_{k=1}^{\infty} \epsilon^k \mathcal{H}^{(k)}, \quad (3.58)$$

where $\mathcal{H}^{(0)}$ is defined in (3.19), while the first- and second-order corrections are given by

$$\mathcal{H}_{n,m}^{(1)} = -2 \begin{pmatrix} \bar{\phi}_{n,m}^{(0)} \phi_{n,m}^{(1)} + \phi_{n,m}^{(0)} \bar{\phi}_{n,m}^{(1)} & \phi_{n,m}^{(0)} \phi_{n,m}^{(1)} \\ \bar{\phi}_{n,m}^{(0)} \bar{\phi}_{n,m}^{(1)} & \bar{\phi}_{n,m}^{(0)} \phi_{n,m}^{(1)} + \phi_{n,m}^{(0)} \bar{\phi}_{n,m}^{(1)} \end{pmatrix}$$

$$- \left(s_{+1,0} + s_{-1,0} + s_{0,+1} + s_{0,-1}\right) \begin{pmatrix} 1 & 0 \\ 0 & 1 \end{pmatrix}$$

and

$$\mathcal{H}_{n,m}^{(2)} = -2 \begin{pmatrix} \bar{\phi}_{n,m}^{(0)} \phi_{n,m}^{(2)} + \phi_{n,m}^{(0)} \bar{\phi}_{n,m}^{(2)} & \phi_{n,m}^{(0)} \phi_{n,m}^{(2)} \\ \bar{\phi}_{n,m}^{(0)} \bar{\phi}_{n,m}^{(2)} & \bar{\phi}_{n,m}^{(0)} \phi_{n,m}^{(2)} + \phi_{n,m}^{(0)} \bar{\phi}_{n,m}^{(2)} \end{pmatrix} - \begin{pmatrix} 2|\phi_{n,m}^{(1)}|^2 & \phi_{n,m}^{(1)2} \\ \bar{\phi}_{n,m}^{(1)2} & 2|\phi_{n,m}^{(1)}|^2 \end{pmatrix}.$$

3.3 Multipulses and Discrete Vortices

It is important to consider again the starting point of $\epsilon = 0$ and the eigenvalue count at that AC limit. There, $\mathcal{H}_{n,m} \equiv \mathcal{H}_{n,m}^{(0)}$ has exactly N negative eigenvalues $\gamma = -2$, N zero eigenvalues (these two sets will constitute the point spectrum of the solution with N excited sites), as well as infinitely many positive eigenvalues $\gamma = 1$ (these will become the continuous spectrum of the solution). In connection to the full eigenvalue problem of the stability operator $\sigma \mathcal{H}_{n,m}^{(0)}$, both the negative and zero eigenvalues correspond to $\lambda = 0$, while the $\gamma = 1$ positive eigenvalues correspond to $\lambda = \pm i$. As before in the one-dimensional problem (cf. Eqs. (2.34) and (2.35)), the latter part will develop a continuous spectral band $\lambda = \pm i[1 + 4\epsilon(\sin^2(q_n/2) + \sin^2(q_m/2))]$, extending in the interval $\pm i[1, 1 + 8\epsilon]$, which will be bounded away from the origin and will not produce instabilities for small ϵ. On the other hand, it is important to examine how the zero eigenvalues will move in the presence of the coupling-induced perturbation.

Focusing now on the zero eigenvalues of the operator $\mathcal{H}$ in the eigenvalue problem $\mathcal{H}\varphi = \gamma\varphi$, we can use the expansion

$$\varphi = \varphi^{(0)} + \epsilon\varphi^{(1)} + \epsilon^2\varphi^{(2)} + O(\epsilon^3), \qquad \gamma = \epsilon\gamma_1 + \epsilon^2\gamma_2 + O(\epsilon^3), \qquad (3.59)$$

where $\varphi^{(0)} = \sum_{j=1}^{N} c_j \mathbf{e}_j$ and $\mathbf{e}_j(\theta)$, $j = 1, \ldots, N$ are eigenvectors of the kernel of $\mathcal{H}^{(0)}$. These eigenvectors contain a single non-zero block $i(e^{i\theta_j}, -e^{-i\theta_j})^T$ at the jth position, which corresponds to the node (n, m) on the contour S_M and are orthogonal according to

$$(\mathbf{e}_i(\theta), \mathbf{e}_j(\theta)) = 2\delta_{i,j}, \qquad 1 \leq i, j \leq N. \qquad (3.60)$$

The corresponding generalized eigenvectors are $\hat{\mathbf{e}}_j(\theta)$, $j = 1, \ldots, N$, such that each eigenvector $\hat{\mathbf{e}}_j(\theta)$ contains the only non-zero block $(e^{i\theta_j}, e^{-i\theta_j})^T$ at the jth position. Direct computations show that

$$\sigma\mathcal{H}^{(0)}\hat{\mathbf{e}}_j(\theta) = 2i\mathbf{e}_j(\theta), \qquad 1 \leq j \leq N. \qquad (3.61)$$

Then, the first-order correction in the eigenvalue equation for the matrix $\mathcal{H}$, $\varphi^{(1)}$, satisfies the inhomogeneous equation

$$\mathcal{H}^{(0)}\varphi^{(1)} + \mathcal{H}^{(1)}\varphi^{(0)} = \gamma_1\varphi^{(0)}. \qquad (3.62)$$

Projection to the kernel of $\mathcal{H}^{(0)}$ gives the eigenvalue problem for γ_1:

$$\frac{1}{2}\sum_{i=1}^{N} \left(\mathbf{e}_j, \mathcal{H}^{(1)}\mathbf{e}_i\right) c_i = \gamma_1 c_j. \qquad (3.63)$$

If one represents the operator $\mathcal{H}$ as $\mathcal{H} = \mathcal{H}_p + \epsilon\mathcal{H}_s$, then $\mathcal{H}^{(1)} = \mathcal{H}_p^{(1)} + \mathcal{H}_s$. On the other hand, the bifurcation conditions to leading order can be represented as

$$g_j^{(1)}(\boldsymbol{\theta}) = \frac{1}{2}\left(\mathbf{e}_j(\boldsymbol{\theta}), \mathcal{H}_s \boldsymbol{\phi}^{(0)}(\boldsymbol{\theta})\right).$$

A direct calculation then [14], of the left-hand side of Eq. (3.63) yields it to be equal to $\cos(\theta_j - \theta_{j+1}) + \cos(\theta_j - \theta_{j-1})$, which is also equal to the Jacobian element $(\mathcal{M}_1)_{ij} = \partial g_i^{(1)}/\partial \theta_j$. Hence for the N small eigenvalues of the eigenvalue problem $\mathcal{H}\varphi = \gamma\varphi$, we have

$$\lim_{\epsilon \to 0} \frac{\gamma_j}{\epsilon} = \mu_j^{(1)}, \qquad 1 \le j \le N, \tag{3.64}$$

where $\mu_j^{(1)}$ are the eigenvalues of $(\mathcal{M}_1)$.

It is then relevant to connect the eigenvalues of the Jacobian $\mathcal{M}_1$ (and of the matrix $\mathcal{H}$) to those of the full stability problem $\sigma \mathcal{H} \boldsymbol{\psi} = i\lambda \boldsymbol{\psi}$. The corresponding statement will be of the form

$$\lim_{\epsilon \to 0} \frac{\lambda_j^2}{\epsilon} = 2\mu_j^{(1)}, \qquad 1 \le j \le N. \tag{3.65}$$

This can be established by using the regular perturbation series

$$\boldsymbol{\psi} = \boldsymbol{\psi}^{(0)} + \sqrt{\epsilon}\boldsymbol{\psi}^{(1)} + \epsilon \boldsymbol{\psi}^{(2)} + \epsilon\sqrt{\epsilon}\boldsymbol{\psi}^{(3)} + O(\epsilon^2), \tag{3.66}$$

$$\lambda = \sqrt{\epsilon}\lambda_1 + \epsilon \lambda_2 + \epsilon\sqrt{\epsilon}\lambda_3 + O(\epsilon^2), \tag{3.67}$$

where, due to the relations (3.60) and (3.61), we have

$$\boldsymbol{\psi}^{(0)} = \sum_{j=1}^{N} c_j \mathbf{e}_j, \qquad \boldsymbol{\psi}^{(1)} = \frac{\lambda_1}{2} \sum_{j=1}^{N} c_j \hat{\mathbf{e}}_j, \tag{3.68}$$

according to the kernel and generalized kernel of $\sigma \mathcal{H}^{(0)}$. The second-order correction term $\boldsymbol{\psi}^{(2)}$ satisfies the inhomogeneous equation

$$\mathcal{H}^{(0)} \boldsymbol{\psi}^{(2)} + \mathcal{H}^{(1)} \boldsymbol{\psi}^{(0)} = i\lambda_1 \sigma \boldsymbol{\psi}^{(1)} + i\lambda_2 \sigma \boldsymbol{\psi}^{(0)}. \tag{3.69}$$

Projection to the kernel of $\mathcal{H}^{(0)}$ gives the eigenvalue problem for λ_1:

$$\mathcal{M}_1 \mathbf{c} = \frac{\lambda_1^2}{2} \mathbf{c}, \tag{3.70}$$

where $\mathbf{c} = (c_1, c_2, \ldots, c_N)^T$ and the matrix $\mathcal{M}_1$ is the same as in the eigenvalue problem (3.63). Relation (3.65) follows from (3.70).

Based on these results, we can quantify the number of eigenvalues of different types in the first-order reductions for the different families of solutions (except for the supersymmetric vortices that we will need to study at the second-order reduc-

3.3 Multipulses and Discrete Vortices

tions since $\mathcal{M}_1 = 0$. In particular, considering the quantity A_1 defined in Eq. (3.36), we have that for the family (i), it is $A_1 = (-1)^{N-l}(2l - N)$, where l is the number of $+1$'s in the configuration. In that case $n(\mathcal{M}_1) = N - l - 1$, $z(\mathcal{M}_1) = 1$ and $p(\mathcal{M}_1) = l$ for $0 \leq l \leq 2M - 1$, while if $2M + 1 \leq l \leq 4M$, there is one more negative $(N - l)$ and one less positive $(l - 1)$ eigenvalue. In the supersymmetric case of $l = 2M$, $n(\mathcal{M}_1) = 2M - 1$, $z(\mathcal{M}_1) = 2$, and $p(\mathcal{M}_1) = 2M - 1$. This implies that for the number of real eigenvalue pairs N_r, imaginary ones with negative Krein signature N_i^-, and zero eigenvalues N_0, in this case we have $N_i^- = N - l - 1$, $N_0 = 1$, and $N_r = l$ for $0 \leq l \leq 2M - 1$; $N_i^- = N - l - 1$, $N_0 = 2$, and $N_r = l - 1$ for $l = 2M$; and $N_i^- = N - l$, $N_0 = 1$, and $N_r = l - 1$ for $2M + 1 \leq l \leq N$.

In the case of family (ii), the corresponding counts for $\mathcal{M}_1$ are $n(\mathcal{M}_1) = 0$, $z(\mathcal{M}_1) = 1$, and $p(\mathcal{M}_1) = N - 1$ for $1 \leq L \leq M - 1$ and $n(\mathcal{M}_1) = N - 1$, $z(\mathcal{M}_1) = 1$, and $p(\mathcal{M}_1) = 0$ for $M + 1 \leq L \leq 2M - 1$, for discrete vortices of charge L within the contour S_M. This, in turn, implies that the full eigenvalue problem will have $N_i^- = 0$, $N_0 = 1$, and $N_r = N - 1$ for $1 \leq L \leq M - 1$; $N_i^- = 0$, $N_0 = N$, and $N_r = 0$ for $L = M$; and $N_i^- = N - 1$, $N_0 = 1$, and $N_r = 0$ for $M + 1 \leq L \leq 2M - 1$.

Having eliminated the potential for the monoparametric asymmetric vortices of family (iii), we lastly examine the zero parameter asymmetric vortices of family (iv) in the realm of first-order reductions. We find there for $\cos\theta_* \neq 0$, $L \neq M$ and $1 \leq n \leq N - 1$, $n \neq 2M$, that the parameter $A_1 = (-1)^{N-n}(\cos\theta_*)^{N-1}(2n - N)$, such that $z(\mathcal{M}_1) = 1$ in all cases. For $\cos\theta_* > 0$, $n(\mathcal{M}_1) = N - n - 1$ and $p(\mathcal{M}_1) = n$ for $1 \leq n \leq 2M - 1$ and $n(\mathcal{M}_1) = N - n$ and $p(\mathcal{M}_1) = n - 1$ for $2M + 1 \leq n \leq N - 1$. In the opposite case of $\cos\theta_* < 0$, we have $n(\mathcal{M}_1) = n$ and $p(\mathcal{M}_1) = N - n - 1$ for $1 \leq n \leq 2M - 1$ and $n(\mathcal{M}_1) = n - 1$ and $p(\mathcal{M}_1) = N - n$ for $2M + 1 \leq n \leq N - 1$. These results lead to the full eigenvalue problem counts: for $\cos\theta_* > 0$, we have $N_i^- = N - n - 1$, $N_0 = 1$, and $N_r = n$ for $1 \leq n \leq 2M - 1$ and $N_i^- = N - n$, $N_0 = 1$, and $N_r = n - 1$ for $2M + 1 \leq n \leq N - 1$; for $\cos\theta_* < 0$, we have $N_i^- = n$, $N_0 = 1$, and $N_r = N - n - 1$ for $1 \leq n \leq 2M - 1$ and $N_i^- = n - 1$, $N_0 = 1$, and $N_r = N - n$ for $2M + 1 \leq n \leq N - 1$.

Despite the considerable wealth of information provided by the first-order reductions, there are still features that need to be clarified at the second-order reductions. Among them are the second zero eigenvalue of supersymmetric solitons (that should bifurcate away from the origin at a higher order), the potential splitting of real eigenvalues of the first-order reductions in the complex plane for solutions of family (ii), or the analysis of the stability of supersymmetric vortices with $L = M$ within family (ii).

3.3.3.1 Eigenvalue Splitting at Second-Order Reductions

In the case of family (i), for $l = 2M$ (supersymmetric solitons) and $a_j \neq (-1)^j a$, as indicated above the Jacobian has two zero eigenvalues with eigenvectors $\mathbf{p}_0$ and $\mathbf{p}_1$. On the other hand, the matrix $\mathcal{M}_1 + \epsilon \mathcal{M}_2$ has only one zero eigenvalue with eigenvector $\mathbf{p}_0$ (due to the gauge invariance). Therefore, the second eigenvalues should bifurcate away from zero at the second-order reduction. We can therefore expand the perturbation theory at the next order using $\mathbf{c} = \mathbf{p}_1$ to derive

$$\gamma_2 = \frac{(\mathbf{p}_1, \mathcal{M}_2 \mathbf{p}_1)}{(\mathbf{p}_1, \mathbf{p}_1)}.$$

One can therefore find at the second order that

$$\lambda_2^2 = 2\frac{(\mathbf{p}_1, \mathcal{M}_2 \mathbf{p}_1)}{(\mathbf{p}_1, \mathbf{p}_1)} = 2\gamma_2.$$

It can therefore be concluded that the splitting of the additional zero eigenvalue in the second-order reduction resembles that of the zero eigenvalues in the first-order reductions. If $\gamma_2 > 0$, then the eigenvalues $\epsilon \lambda_2$ will be real, while if $\gamma_2 < 0$, they will be imaginary with negative Krein sign.

The next topic of interest that needs to be addressed at the level of second-order reductions is the potential splitting of non-zero eigenvalues (of the first order). In particular, if we use the explicit solutions for $\boldsymbol{\phi}^{(1)}$, required in $\mathcal{H}^{(1)}$, it is possible to compute the explicit solution of the inhomogeneous equation (3.69) for $\boldsymbol{\psi}^{(2)}$ as

$$\boldsymbol{\psi}^{(2)} = \frac{\lambda_2}{2} \sum_{j=1}^{N} c_j \hat{\mathbf{e}}_j + \frac{1}{2} \sum_{j=1}^{N} (\sin(\theta_{j+1} - \theta_j) c_{j+1} + \sin(\theta_{j-1} - \theta_j) c_{j-1}) \hat{\mathbf{e}}_j$$
$$+ \sum_{j=1}^{N} c_j (\mathcal{S}_+ + \mathcal{S}_-) \mathbf{e}_j, \tag{3.71}$$

where the operators $\mathcal{S}_\pm$ shift elements of $\mathbf{e}_j$ from the node $(n,m) \in S_M$ to the adjacent nodes outside of S_M. Then, for the third-order correction $\boldsymbol{\psi}^{(3)}$, one has the subsequent order inhomogeneous equation of the form

$$\mathcal{H}^{(0)} \boldsymbol{\psi}^{(3)} + \mathcal{H}^{(1)} \boldsymbol{\psi}^{(1)} = i\lambda_1 \sigma \boldsymbol{\psi}^{(2)} + i\lambda_2 \sigma \boldsymbol{\psi}^{(1)} + i\lambda_3 \sigma \boldsymbol{\psi}^{(0)}. \tag{3.72}$$

To obtain an expression for the second-order correction λ_2, we project the inhomogeneous problem (3.72) also to the kernel of $\mathcal{H}^{(0)}$ (as before to obtain Eq. (3.70)), which, in turn, leads to

$$\mathcal{M}_1 \mathbf{c} = \frac{\lambda_1^2}{2} \mathbf{c} + \sqrt{\epsilon} (\lambda_1 \lambda_2 \mathbf{c} + \lambda_1 \mathcal{L}_1 \mathbf{c}), \tag{3.73}$$

where the matrix $\mathcal{L}_1$ is defined by

$$(\mathcal{L}_1)_{i,j} = \begin{cases} \sin(\theta_j - \theta_i), & i = j \pm 1, \\ 0, & |i-j| \neq 1, \end{cases} \tag{3.74}$$

subject to the periodic boundary conditions. If we now label the eigenvalue of the first-order Jacobian matrix $\mathcal{M}_1$ as $\mu_j^{(1)}$ (with eigenvector $\mathbf{c}_j$), then the two leading

3.3 Multipulses and Discrete Vortices

order expressions for the bifurcation of the eigenvalue from zero in Eq. (3.66) are given by

$$\lambda_1 = \pm\sqrt{2\mu_j^{(1)}}, \qquad \lambda_2 = -\frac{(\mathbf{c}_j, \mathcal{L}_1\mathbf{c}_j)}{(\mathbf{c}_j, \mathbf{c}_j)}. \tag{3.75}$$

Now, given the skew-symmetric nature of the operator $\mathcal{L}_1$, we infer that the second-order correction term λ_2 is purely imaginary or zero. In the case of discrete solitons of family (i), since $\sin(\theta_{j+1} - \theta_j) = 0$, the elements of $\mathcal{L}_1$ (and hence λ_2) will be vanishing. On the other hand, in the case of symmetric vortices of family (ii) with $L \neq M$, the matrix $\mathcal{M}_1$ has double eigenvalues, according to the roots of $\sin^2 \pi n/N$ in the explicit solution (3.37). In that case, all the coefficients $a_j = \cos(\theta_{j+1} - \theta_j)$ and $b_j = \sin(\theta_{j+1} - \theta_j)$, $1 \leq j \leq N$ will be the same: $a_j = a$ and $b_j = b$. Then, one can again compute both the first, as well as the second-order correction for the eigenvalues explicitly as

$$\lambda_1 = \pm\sqrt{8a}\sin\frac{\pi n}{N}, \qquad \lambda_2 = -2ib\sin\frac{2\pi n}{N}, \qquad 1 \leq n \leq N. \tag{3.76}$$

This implies that all double roots of λ_1 with $n \neq N/2$ and N split along the imaginary axis in λ_2. When $a > 0$, the splitting occurs in the transverse directions to the real values of λ_1. When $a < 0$, the splitting occurs in the longitudinal directions to the imaginary values of λ_1. The simple roots at $n = N/2$ and N are not affected, since $\lambda_2 = 0$ for $n = N/2$ and N. This implies, e.g., that for discrete vortices of family (ii) with $1 \leq L \leq M - 1$, their positive double roots will, in fact, split and become into complex quartets, as we will see below for example in the case of $M = 2$ and $L = 1$.

We now turn to the case of supersymmetric vortices (where the first-order reductions are completely degenerate and yield no information) in order to use the second-order reductions to address the splitting of their zero eigenvalues. We extend the results of the regular perturbation series (3.59) and (3.66) to the case $\mathcal{M}_1 = 0$, which occurs for supersymmetric vortices of family (ii) with charge $L = M$. More specifically, when $\mathcal{M}_1 = 0$ (and hence $\gamma_1 = 0$), then the first-order correction is given by

$$\varphi^{(1)} = \frac{1}{2}\sum_{j=1}^{N}(c_{j+1} - c_{j-1})\hat{\mathbf{e}}_j + \sum_{j=1}^{N} c_j\,(\mathcal{S}_+ + \mathcal{S}_-)\,\mathbf{e}_j, \tag{3.77}$$

where the meaning of the operators $\mathcal{S}_\pm$ is the same as in Eq. (3.71). Going to the next order in perturbation theory, we obtain the inhomogeneous equation

$$\mathcal{H}^{(0)}\varphi^{(2)} + \mathcal{H}^{(1)}\varphi^{(1)} + \mathcal{H}^{(2)}\varphi^{(0)} = \gamma_2\varphi^{(0)}. \tag{3.78}$$

Hence, projecting to the kernel of $\mathcal{H}^{(0)}$ gives the eigenvalue problem for γ_2:

$$\frac{1}{2}\left(\mathbf{e}_j, \mathcal{H}^{(1)}\boldsymbol{\varphi}^{(1)}\right) + \frac{1}{2}\sum_{i=1}^{N}\left(\mathbf{e}_j, \mathcal{H}^{(2)}\mathbf{e}_i\right)c_i = \gamma_2 c_j. \tag{3.79}$$

Through direct computation, one can verify that the matrix on the left-hand side of (3.79) is identical to the Jacobian matrix $\mathcal{M}_2$ (defined as $(\mathcal{M}_2)_{ij} = \partial \mathbf{g}_i^{(2)}/\partial \theta_j$). Using our results from the Jacobian of the second-order reductions in the previous section (for $M = 1, 2,$ and 3), we have that $n(\mathcal{M}_2) = 0$, $z(\mathcal{M}_2) = 2$, and $p(\mathcal{M}_2) = 2$ for $M = 1$; $n(\mathcal{M}_2) = 1$, $z(\mathcal{M}_2) = 2$, and $p(\mathcal{M}_2) = 5$ for $M = 2$; and $n(\mathcal{M}_2) = 4$, $z(\mathcal{M}_2) = 2$, and $p(\mathcal{M}_2) = 6$ for $M = 3$. We now need to connect the eigenvalues of the Jacobian matrix $\mathcal{M}_2$ with those of the full eigenvalue problem λ. This is done again by using the perturbation series but now in the form

$$\boldsymbol{\psi} = \boldsymbol{\psi}^{(0)} + \epsilon \boldsymbol{\psi}^{(1)} + \epsilon^2 \boldsymbol{\psi}^{(2)} + O(\epsilon^3), \qquad \lambda = \epsilon \lambda_1 + \epsilon^2 \lambda_2 + O(\epsilon^3), \tag{3.80}$$

where

$$\boldsymbol{\psi}^{(0)} = \sum_{j=1}^{N} c_j \mathbf{e}_j, \qquad \boldsymbol{\psi}^{(1)} = \boldsymbol{\varphi}^{(1)} + \frac{\lambda_1}{2}\sum_{j=1}^{N} c_j \hat{\mathbf{e}}_j, \tag{3.81}$$

and $\boldsymbol{\varphi}^{(1)}$ is given by (3.77). The second-order correction term $\boldsymbol{\psi}^{(2)}$ is found from the inhomogeneous equation

$$\mathcal{H}^{(0)}\boldsymbol{\psi}^{(2)} + \mathcal{H}^{(1)}\boldsymbol{\psi}^{(1)} + \mathcal{H}^{(2)}\boldsymbol{\psi}^{(0)} = i\lambda_1\sigma\boldsymbol{\psi}^{(1)} + i\lambda_2\sigma\boldsymbol{\psi}^{(0)}. \tag{3.82}$$

In this case, the projection of the inhomogeneous equation to the kernel of $\mathcal{H}^{(0)}$ gives the eigenvalue problem for λ_1:

$$\mathcal{M}_2 \mathbf{c} = \lambda_1 \mathcal{L}_2 \mathbf{c} + \frac{\lambda_1^2}{2}\mathbf{c}, \tag{3.83}$$

where $\mathbf{c} = (c_1, c_2, \ldots, c_N)^T$, the matrix $\mathcal{M}_2$ is the same as in the eigenvalue problem (3.79), and the matrix $\mathcal{L}_2$ follows from the matrix $\mathcal{L}_1$ in the form (3.74) with $\sin(\theta_{j+1} - \theta_j) = 1$, or explicitly:

$$(\mathcal{L}_2)_{i,j} = \begin{cases} +1, & i = j-1, \\ -1, & i = j+1, \\ 0, & |i-j| \neq 1, \end{cases} \tag{3.84}$$

subject to the periodic boundary conditions. Given that $\mathcal{M}_2$ is symmetric and $\mathcal{L}_2$ is skew-symmetric, the eigenvalues of the problem (3.83) arise in pairs $(\lambda_1, -\lambda_1)$. A direct comparison of the matrices $\mathcal{M}_2$ in (3.50) and $\mathcal{L}_2$ in (3.84), leads to the conclusion $\mathcal{M}_2 = -(1/2)\mathcal{L}_2^2$. However, the Jacobian matrices $\tilde{\mathcal{M}}_2$ are modified in the case $M = 2$ and $M \geq 3$ by the rank-one and rank-four non-positive matrices $\Delta \mathcal{M}_2$. As a result, the eigenvalue problem (3.83) can be factorized as follows:

3.3 Multipulses and Discrete Vortices

$$\frac{1}{2}(\mathcal{L}_2 + \lambda_1)^2 \mathbf{c} = \Delta \mathcal{M}_2 \mathbf{c}. \tag{3.85}$$

One can now consider the reduced eigenvalue problem of Eq. (3.83) in the cases of $M = 1, 2,$ and 3. In the case of $M = 1$, the problem takes the form of the following constant-coefficient difference equation:

$$-c_{j+2} + 2c_j - c_{j-2} = \lambda_1^2 c_j + 2\lambda_1 (c_{j+1} - c_{j-1}), \quad 1 \leq j \leq 4M, \tag{3.86}$$

subject to the periodic boundary conditions, which, as usual, can be tackled by the discrete Fourier transform yielding

$$\left(\lambda_1 + 2i \sin \frac{\pi n}{2M}\right)^2 = 0, \quad 1 \leq n \leq 4M. \tag{3.87}$$

This implies that the eigenvalue problem will have two eigenvalue pairs with $\lambda_1 = \pm 2i$, and two more pairs of eigenvalues with $\lambda_1 = 0$ (only one of which will persist to higher order reductions).

On the other hand, in the case of $M = 2$, the problem of Eq. (3.83) can be reduced to two constant-coefficient difference equations

$$-x_{j+1} + 2x_j - x_{j-1} = \lambda_1^2 x_j + 2\lambda_1 (y_j - y_{j-1}), \quad j = 1, 2, 3, 4,$$

$$y_{j+1} - 2y_{j+2} + y_{j-1} = \lambda_1^2 y_j + 2\lambda_1 (x_{j+1} - x_j), \quad j = 1, 2, 3, 4,$$

where $x_j = c_{2j-1}$ and $y_j = c_{2j}$ are subject to the periodic boundary conditions. In this case, the characteristic equation has the explicit form

$$\lambda_1^4 - 2\lambda_1^2 \left(1 - (-1)^n - 8 \sin^2 \frac{\pi n}{4}\right) + 8 \sin^2 \frac{\pi n}{4} \left(1 - (-1)^n - 2 \sin^2 \frac{\pi n}{4}\right) = 0$$

for $n = 1, 2, 3, 4$, leading to four pairs of eigenvalues with $\lambda_1 = \pm\sqrt{2}i$, a single pair with $\lambda_1 = \pm\sqrt{\sqrt{80} - 8}$ (which renders the configuration with $L = M = 2$ *immediately* unstable for $\epsilon \neq 0$), a single pair with $\lambda_1 = \pm i\sqrt{\sqrt{80} + 8}$ and finally two pairs with $\lambda_1 = 0$, only one of which will survive for higher order reductions.

Finally, in the case of $M = 3$, it is less straightforward to compute the eigenvalues explicitly via Fourier decomposition. For this reason, we instead obtain them from a numerical linear algebra package as

$$\lambda_{1,2} = \pm 3.68497i, \quad \lambda_{3,4} = \lambda_{5,6} = \pm 3.20804i, \quad \lambda_{7,8} = \pm 2.25068i,$$

$$\lambda_{9,10} = \lambda_{11,12} = \pm i, \lambda_{13,14} = \lambda_{15,16} = \pm 0.53991,$$

$$\lambda_{17,18,19,20} = \pm 0.634263 \pm 0.282851i, \lambda_{21,22,23,24} = 0.$$

Therefore, in the case of the $L = M = 3$, we expect a double real eigenvalue pair and a complex eigenvalue quartet to immediately destabilize the relevant configuration (although additional potentially unstable eigendirections may exist, since the algebraic multiplicity of the zero eigenvalue is larger than two).

It is interesting to slightly expand here on the reasons for the destabilization of the $L = M = 2$ and $L = M = 3$ supersymmetric vortex solutions. More specifically, the destabilization of the supersymmetric vortex with $M = 2$ occurs due to the center node $(2, 2)$, which couples its nearest-neighbors of the contour S_2 in the second-order reductions. This modifies the Jacobian matrix $\tilde{\mathcal{M}}_2$ (due to the non-zero nature of the rank-one non-positive matrix $\Delta\mathcal{M}_2$) in a way such as to produce a simple negative eigenvalue, while none such exists for the non-negative matrix $\mathcal{M}_2$. A similar feature arises in the destabilization of the $L = M = 3$ supersymmetric vortex, whereby the coupling of eight nodes of the contour S_3 with four interior corner points $(2, 2)$, $(2, M)$, (M, M), and $(M, 2)$ induces the destabilization. In the latter case, the rank-four non-positive matrix $\Delta\mathcal{M}_2$ leads to four negative eigenvalues in the Jacobian matrix $\tilde{\mathcal{M}}_2$ and to four unstable eigenvalues in the reduced eigenvalue problem (3.85). It is interesting that this mathematical quantification leads to an intuitive understanding of the origin of the instability and, therefore, to insights as to how to avoid it. In particular, the latter can be achieved, for instance, if a hole is drilled at the central node (2,2) of the $M = 2$ contour, or four such holes at the sites $(2, 2)$, $(2, M)$, (M, M), and $(M, 2)$ of the $M = 3$ contour, then the matrix $\Delta\mathcal{M}_2 = 0$; then, all the relevant eigenvalues of the Jacobian would be positive leading to imaginary eigenvalues for the full eigenvalue problem, similarly to the case of $M = 1$. We will test this type of insight numerically in what follows.

3.3.3.2 Eigenvalue Splitting at Higher Order Reductions

It is important to note that in all the above supersymmetric cases, there are additional (to the ones stemming from the gauge invariance) zero eigenvalues at the level of the second-order corrections, which need to be resolved at the level of higher order reductions. In particular, we consider the splitting of the double zero eigenvalue of $\mathcal{M}_2$ which corresponds to the eigenvectors $\mathbf{p}_0$ and $\mathbf{p}_1$, where $\mathbf{p}_0 = (1, 1, \ldots, 1, 1)^T$ and $\mathbf{p}_1 = (0, 1, \ldots, 0, 1)^T$. To this effect, we set $\mathbf{c} = (c_1, \ldots, c_N) = \mathbf{p}_1 + \alpha \mathbf{p}_0$, where α is a parameter and generally assume that the splitting occurs at the Kth order of reductions. Then, the perturbation series (3.59) needs to be extended to that order, leading to the following inhomogeneous equation:

$$\mathcal{H}^{(0)}\boldsymbol{\varphi}^{(k)} = -\sum_{m=1}^{k} \mathcal{H}^{(m)}\boldsymbol{\varphi}^{(k-m)}, \qquad 1 \leq k \leq K-1$$

and

$$\mathcal{H}^{(0)}\boldsymbol{\varphi}^{(K)} = -\sum_{m=1}^{K} \mathcal{H}^{(m)}\boldsymbol{\varphi}^{(K-m)} + \gamma_K \boldsymbol{\varphi}^{(0)},$$

where the zeroth order $\boldsymbol{\varphi}^{(0)} = \sum_{j=1}^{N} c_j \mathbf{e}_j$, and $\gamma = \gamma_K \epsilon^K + O(\epsilon^{K+1})$ is the leading order approximation for the smallest non-zero eigenvalue of $\mathcal{H}$. From the projection

3.3 Multipulses and Discrete Vortices

formulas on to the kernel of $\mathcal{H}^{(0)}$, then the equations for the correction γ_K and α can be obtained.

This algorithmic procedure can be used for the supersymmetric vortices with $M = 1$, leading to $K = 6$, $\alpha = -1/2$, and $\gamma_6 = -16$ (and, hence, to the conclusion that the zero eigenvalue becomes a small negative eigenvalue for small $\epsilon \neq 0$). A similar conclusion is obtained for $M = 2$, where $K = 6$, $\alpha = -1/2$, and $\gamma_6 = -8$. This allows us to develop the regular perturbation series for the eigenvalue problem $\sigma \mathcal{H} \psi = i\lambda \psi$ starting with the zeroth order $\psi^{(0)} = \sum_{j=1}^{N} c_j \mathbf{e}_j$ and $\mathbf{c} = \mathbf{p}_1 + \alpha \mathbf{p}_0$, where $\alpha = -1/2$. This leads in the case of $M = 1$ to the conclusion that $\lambda_1 = \lambda_2 = 0$ but $\lambda_3 \neq 0$, such that $\lambda_3^2 = -32 = 2\gamma_6$; this results in a small imaginary eigenvalue of the linear stability matrix with negative Krein signature. Similarly, for the case of $M = 2$, we find that $\lambda_3^2 = -16 = 2\gamma_6$.

Based on the above results, we can summarize our stability conclusions as follows. We expect in the vicinity of $\epsilon \in \mathcal{O}(0)$ to have stable solutions of the following forms:

- discrete solitons,
- discrete symmetric vortices of family (ii) over contours S_M with charge $M + 1 \leq L \leq 2M - 1$,
- discrete supersymmetric vortices with $L = M = 1$.

We now turn to a numerical examination of the above findings.

3.3.4 Numerical Results

3.3.4.1 Discrete Solitons

Based on the above considerations, we can firstly construct any discrete soliton configuration that we would like (comprising essentially of $+1$'s and -1's on the lattice) at the AC limit. We can subsequently continue the relevant configuration to finite values of the coupling ϵ, by solving Eq. (3.6) and finally obtain the corresponding linearization eigenvalues, by solving the linear stability problem $\sigma \mathcal{H} \psi = i\lambda \psi$ numerically.

For the case of discrete solitons, we will only consider some illustrative cases to highlight the comparison of theoretical and numerical results, although it should be stressed that the same approach can essentially be used for any configuration of interest. Any two-site configuration in the two-dimensional problem can be effectively thought of as a quasi-one-dimensional one along the line of sight connecting the two sites. Keep in mind, however, that this is genuinely true only when the sites are connected by a lattice direction; when they are not, the relevant eigenvalues are expected to be non-zero to leading order at $\epsilon^{d_{min}/2}$ where d_{min} is the minimal distance between the sites along the lattice directions. Hence, to consider genuinely non-quasi-one-dimensional properties, we need to examine configurations with three or more sites. As a prototypical three-site example, we will consider the following configuration:

$$\begin{pmatrix} 1 & 0 & -1 \\ 0 & 0 & 0 \\ 0 & 0 & 1 \end{pmatrix}. \tag{3.88}$$

We will also consider two prototypical four site configurations, namely

$$\begin{pmatrix} 1 & 0 & -1 \\ 0 & 0 & 0 \\ -1 & 0 & 1 \end{pmatrix} \text{ and } \begin{pmatrix} 1 & -1 & -1 \\ 0 & 1 & 0 \\ 0 & 0 & 0 \end{pmatrix}, \tag{3.89}$$

as well as two five-site configurations

$$\begin{pmatrix} 1 & 0 & 1 \\ 0 & -1 & 0 \\ 1 & 0 & 1 \end{pmatrix} \text{ and } \begin{pmatrix} 0 & -1 & 1 \\ 0 & 1 & 0 \\ 1 & -1 & 0 \end{pmatrix}. \tag{3.90}$$

The above matrices yield the spatial form of the (real) field at the anti-continuum limit, in the vicinity of the spatially excited sites. For completeness/comparison, we also consider a configuration with many more sites such as the nine-site configuration of the form

$$\begin{pmatrix} 1 & -1 & 1 \\ -1 & 1 & -1 \\ 1 & -1 & 1 \end{pmatrix}. \tag{3.91}$$

For the three-site configuration, for bifurcation equation purposes, the structure is similar to the three-site one-dimensional structures of Fig. 2.12 of the previous chapter. Hence, the relevant Jacobian and eigenvalues in this case also will be $\lambda = \pm\sqrt{2\epsilon}$ and $\pm\sqrt{6\epsilon}$ (in addition to the zero eigenvalue of the phase invariance); see the relevant discussion around Eq. (2.86).

For the four-site configurations of (3.89), as regards the first configuration, the bifurcation equations for all four sites are $g_j = \sin(\theta_j - \theta_{j+1}) + \sin(\theta_j - \theta_{j-1})$, where θ_j is the phase (0 or π depending on whether the AC limit is $+1$ or -1 for each site) and $\theta_{j\pm 1}$ is the phase of their closest two neighbors in the configuration. From this, once again the first-order reduction Jacobian can be computed, leading to the eigenvalues $\lambda = \pm 2\epsilon i$ (a double eigenvalue pair) and $\lambda = \pm 2\sqrt{2}\epsilon i$ (a single pair) in addition to the zero eigenvalue of the phase invariance. For the second four-site configuration of (3.89), the bifurcation equations are slightly more complicated in that the site that has three neighbors has $g_2 = \sin(\theta_2 - \theta_1) + \sin(\theta_2 - \theta_3) + \sin(\theta_2 - \theta_4)$, where θ_2 is the phase of that site and $\theta_{1,3,4}$ those of its neighbors, while the rest of the sites have $g_j = \sin(\theta_j - \theta_2)$ for $j = 1, 3, 4$. From the corresponding Jacobian one can extract the relevant eigenvalues to be $\lambda = \pm\sqrt{2\epsilon}i$ (a double eigenvalue) and $\lambda = \pm\sqrt{8\epsilon}i$ (a single eigenvalue), as well as the zero eigenvalue pair.

For the first five-site configurations of (3.90), we have the bifurcation equation $g_0 = 2\sin(\theta_0 - \theta_1) + 2\sin(\theta_0 - \theta_2) + 2\sin(\theta_0 - \theta_3) + 2\sin(\theta_0 - \theta_4)$, where we

3.3 Multipulses and Discrete Vortices

have labeled as θ_0 the phase of the central site, while θ_j, with $j = 1, 2, 3, 4$, denote the phases of the four corner sites. We also have for the corner sites $g_j = \sin(\theta_j - \theta_{j+1}) + \sin(\theta_j - \theta_{j-1}) + 2\sin(\theta_j - \theta_0)$ for $j = 1, 2, 3, 4$. In this case, in addition to the zero eigenvalue, there is a real eigenvalue pair $\lambda = \pm 2\epsilon$, an eigenvalue of $\mathcal{O}(\epsilon^2)$ and an imaginary pair $\lambda = \pm\sqrt{20\epsilon} i$. Hence, this configuration is predicted to *always* be linearly unstable. Finally, for the second five-site configuration, one can similarly write down the bifurcation equations and accordingly obtain the eigenvalues: $\lambda = \pm 0.874i\sqrt{\epsilon}$, $\pm 1.6625i\sqrt{\epsilon}$, $\pm 2.288i\sqrt{\epsilon}$, and $\pm 2.69i\sqrt{\epsilon}$, as well as the zero eigenvalue.

Finally, for the configuration with nine sites of (3.91), labeling θ_0 the phase of the central site of the contour, $\theta_{1,3,5,7}$ the phases of the four corners, and $\theta_{2,4,6,8}$ those of the four sites adjacent to the central one, we have the bifurcation equations $g_j = \sin(\theta_j - \theta_{j+1}) + \sin(\theta_j - \theta_{j-1})$ for $j = 1, 3, 5, 7$, while $g_j = \sin(\theta_j - \theta_{j+1}) + \sin(\theta_j - \theta_{j-1}) + \sin(\theta_j - \theta_0)$ for $j = 2, 4, 6, 8$, and $g_0 = \sum_{k=1}^{4} \sin(\theta_0 - \theta_{2k})$. From the corresponding Jacobian the first-order reduction for the eigenvalues yields $\lambda = \pm\sqrt{2\epsilon} i$ (double), $\lambda = \pm 2\sqrt{\epsilon} i$ (single), $\lambda = \pm\sqrt{6\epsilon} i$ (double), $\lambda = \pm\sqrt{8\epsilon} i$ (double), and $\lambda = \pm\sqrt{12\epsilon} i$ (single), with the parenthesis denoting in each case the multiplicity of the relevant eigenvalue.

The results for the three-site configuration of (3.88) are shown in Fig. 3.3. We illustrate the configuration for two different values of the coupling (one stable and one unstable, while the comparison of the theoretically predicted and numerically obtained eigenvalues is shown in the bottom panel. It can be seen that the agreement is very good between the two. The configuration is, in fact, found to be unstable for $\epsilon > 0.295$, when the larger of the two imaginary eigenvalues of negative Krein sign collides with an eigenvalue bifurcating from the lower band edge of the continuous spectrum, leading to an oscillatory instability through the generation of a quartet of eigenvalues.

Figure 3.4 shows the two cases of (3.89) concerning four-site solutions. The top left panel shows the first configuration which is linearly stable for small values of

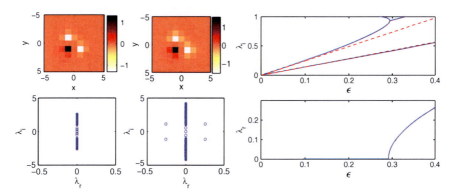

Fig. 3.3 The *left two rows* show the three-site solution of (3.88) for $\epsilon = 0.2$ and 0.4 and the corresponding linear stability eigenvalues (the former case is stable, while the latter is unstable). The *right row* shows the evolution of the imaginary (*top*) and real (*bottom*) eigenvalues as a function of ϵ. The *solid lines* show the full numerical results, while the dashed lines indicate the corresponding theoretical predictions

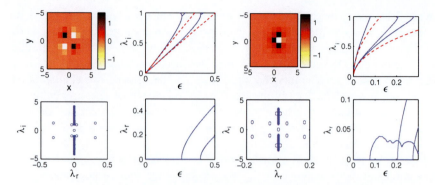

Fig. 3.4 The *four left panels* concern the first solution of (3.89), while the *four right panels* concern the second one. The *top left panel* in each case shows a typical solution profile (for $\epsilon = 0.4$ in the *left rows*, and for $\epsilon = 0.3$ in the *right rows*), the *bottom left* the corresponding eigenvalues in the spectral plane, while the *top* and *bottom right* in each case show the dependence on ϵ of the imaginary and real parts of the relevant eigenvalues (again the *solid lines* correspond to the numerical results, while *dashed* ones to theoretical approximations discussed above)

ϵ, with a double and a single imaginary eigenvalue pairs with negative Krein sign. These eigenvalues eventually destabilize the solution upon collision with eigenvalues bifurcating from the band edge of the continuous spectrum. The relevant collisions occur for $\epsilon = 0.262$ and 0.396. The right panels show the second configuration whose three imaginary negative Krein sign eigenvalues collide with the band edge at $\epsilon = 0.076, 0.206$, and 0.276 giving rise to three eigenvalue quartets.

The five-site configurations of (3.90) are, in turn, shown in Fig. 3.5. Note that the first one of these configurations is immediately unstable, as soon as $\epsilon \neq 0$, due to a very accurately captured real eigenvalue pair $\lambda = \pm 2\epsilon$. Additionally, there is complex quartet emerging from the collision of the $\mathcal{O}(\epsilon)$ imaginary eigenvalue of

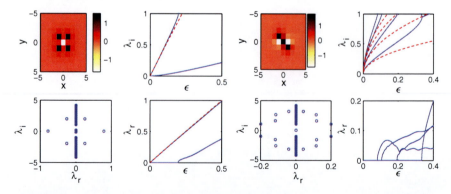

Fig. 3.5 Same as Fig. 3.4, but now for the five-site configurations of (3.90). The four left panels correspond to the first configuration, while the right four panels to the second configuration. Note the immediate instability of the former. The numerical solutions and their linear stability are for $\epsilon = 0.4$

3.3 Multipulses and Discrete Vortices

the configuration with the continuous spectrum at $\epsilon = 0.198$. On the other hand, the double pair of $\mathcal{O}(\epsilon^2)$ eigenvalues moves more slowly and does not collide with the band edge for the parameter values considered. On the other hand, the second configuration is stable for small ϵ, but becomes increasingly unstable as ϵ is increased due to a sequence of four collisions with the band edge (or eigenvalues bifurcating from the band edge) of the continuous spectrum occurring at $\epsilon = 0.081, 0.106, 0.195$, and $\epsilon = 0.334$.

Finally, the nine-site waveform of (3.91) is demonstrated in Fig. 3.6, along with the dependence of its eigenvalues on ϵ. Once again as predicted by the theory, the solution is found to be linearly stable with eight imaginary eigenvalue pairs (three of which are double) for small ϵ. However, for $\epsilon > 0.054$, a complex web of oscillatory instabilities is initiated (which is also affected by finite size effects in the figure, discussed in more detail in the chapter on dark solitons), rendering the solution unstable thereafter.

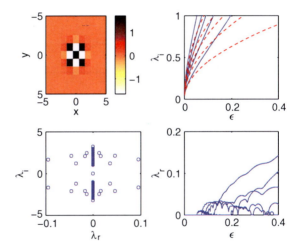

Fig. 3.6 Same as Fig. 3.4, but now for the nine-site configurations of (3.91). The *left panels* are for $\epsilon = 0.25$. Note that the solution is stable for small ϵ, but the negative Krein sign of the imaginary eigenvalues leads to a complex web of oscillatory instabilities for $\epsilon > 0.054$

3.3.4.2 Discrete Vortices

We now turn to the examination of discrete vortex solutions of families (ii) (both symmetric and supersymmetric) and (iv) that were previously considered theoretically in this chapter. The relevant features will be presented in a unified way for these solutions in Figs. 3.7, 3.8, 3.9, 3.10, 3.11, and 3.12. The top left panel will, in each case, show the profile of the vortex solution for a specific value of ϵ by means of contour plots of the real (top left), imaginary (top right), modulus (bottom left), and phase (bottom right) two-dimensional profiles. The right panels will in each case show the spectral plane of the linearization eigenvalues for the corresponding value of ϵ. The bottom panel shows the dependence of small eigenvalues as a function of ϵ, obtained via continuation methods from the AC limit of $\epsilon = 0$. In these graphs,

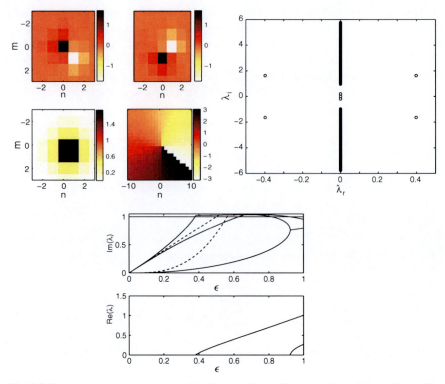

Fig. 3.7 The (supersymmetric) vortex cell with $L = M = 1$. The *top left panel* shows the profile of the solution for $\epsilon = 0.6$. The *subplots* show the real (*top left*), imaginary (*top right*), modulus (*bottom left*) and phase (*bottom right*) fields. The *top right panel* shows the spectral plane (λ_r, λ_i) of the linear eigenvalue problem (3.57). The *bottom panel* shows the small eigenvalues versus ϵ (the *top subplot* shows the imaginary part, while the *bottom* shows the real part). The *solid lines* show the numerical results, while the *dashed lines* show the results of the Lyapunov–Schmidt reductions. Reprinted from [14] with permission

as before, the solid lines will denote theoretical results, while the dashed ones, the result of the first-, second-, and higher order reductions presented above.

Figure 3.7 concerns the case of the supersymmetric vortex of charge $L = 1$ on the contour S_M with $M = 1$. In the second- and sixth-order reductions, the stability spectrum of the vortex solution has a pair of imaginary eigenvalues $\lambda \approx \pm i\sqrt{32}\epsilon^3$ and two pairs of imaginary eigenvalues $\lambda \approx \pm 2\epsilon i$. The latter pairs split along the imaginary axis beyond the second-order reductions. The larger pair of negative Krein signature becomes subject to a Hamiltonian–Hopf bifurcation for larger values of $\epsilon \approx 0.38$ upon collision with the continuous spectrum. The smaller pair of positive Krein signature disappears in the continuous spectrum for $\epsilon > 0.66$. The smallest pair of imaginary eigenvalues has negative Krein signature and a Hamiltonian–Hopf bifurcation occurs in the case for $\epsilon \approx 0.92$ due to collision with another pair of positive Krein signature bifurcating from the continuous spectrum.

3.3 Multipulses and Discrete Vortices

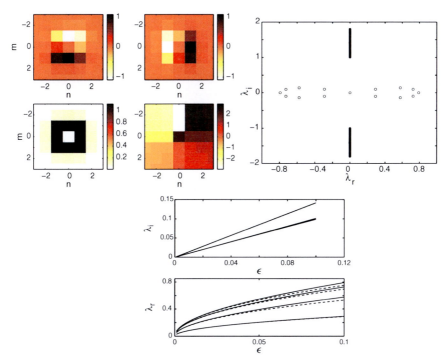

Fig. 3.8 The symmetric vortex with $L = 1$ and $M = 2$ for $\epsilon = 0.1$. Reprinted from [14] with permission

Figure 3.8 presents results for the symmetric vortex of charge $L = 1$ on the contour S_M with $M = 2$. There are three double and one simple real unstable eigenvalues in the first-order reductions, but all double eigenvalues split into the complex plane in the second-order reductions. The asymptotic result of Eq. (3.76) for eigenvalues $\lambda \approx \sqrt{\epsilon}\lambda_1 + \epsilon\lambda_2$ with $N = 8$, $a = \cos(\pi/4)$ and $b = \sin(\pi/4)$ is shown on Fig. 3.8 in very good agreement with numerical results.

Figure 3.9 shows results for the supersymmetric vortex with $L = M = 2$. The non-zero eigenvalues of the second- and sixth-order reductions consist of a pair of simple real eigenvalues $\lambda \approx \pm\epsilon\sqrt{\sqrt{80}-8}$, a pair of simple imaginary eigenvalues $\lambda \approx \pm i\epsilon\sqrt{\sqrt{80}+8}$, a pair of simple imaginary eigenvalues $\lambda \approx \pm 4i\epsilon^3$, and a pair of imaginary eigenvalues of algebraic multiplicity four at $\lambda \approx \pm i\epsilon\sqrt{2}$. The bottom right panel of Fig. 3.9 shows the splitting of multiple imaginary eigenvalues beyond the second-order reductions along the imaginary axis and also four subsequent oscillatory instabilities for larger values of ϵ ($\epsilon \approx 0.23, 0.5, 0.5$, and 1.45). The other two pairs of purely imaginary eigenvalues collide with the band edge of the continuous spectrum at $\epsilon \approx 1.315$ and 1.395 and disappear into the continuous spectrum. Note, once again, the level of accuracy of our theoretical predictions in comparison with the direct numerical results, especially for the cases of small ϵ illustrated at the bottom left panel of Fig. 3.9. We will return to this structure below,

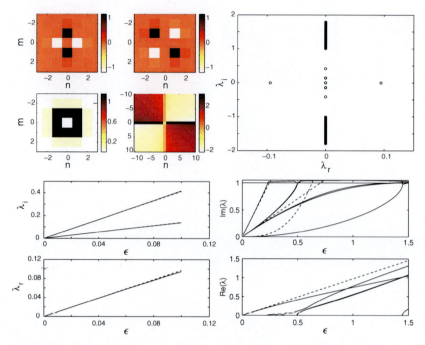

Fig. 3.9 The supersymmetric vortex with $L = M = 2$ for $\epsilon = 0.1$. The *bottom right panel* is an extension of the *bottom left panel* to larger values of ϵ. Reprinted from [14] with permission

using it as a case example of the "mathematical intuition" that our reductions offer on the origin of the observed dynamical instability and how this can be avoided in such a case.

Figure 3.10 shows results for the symmetric vortex with $L = 3$ and $M = 2$. The first-order reductions predict three pairs of double imaginary eigenvalues, a pair of simple imaginary eigenvalues and a double zero eigenvalue. The double eigenvalues split in the second-order reductions along the imaginary axis, given by (3.76) with $N = 8$, $a = \cos(3\pi/4)$ and $b = \sin(3\pi/4)$. The seven pairs of imaginary eigenvalues lead to a cascade of seven complex quartets of eigenvalues emerging for larger values of ϵ due to their collisions with the continuous spectrum. The first bifurcation when the symmetric vortex becomes unstable occurs for $\epsilon \approx 0.096$. It is interesting to note in connection to this solution the sharp contrast between this result (i.e., the fact that a solution with $L = 3$ may be *stable*, while the lower charge $L = 2$ solution is *always unstable* over the same contour) and the continuum NLS intuition; see, e.g., [22] for relevant analytical considerations and [23, 24] for numerical results. The latter indicates that over this discrete contour higher charge vortices are more prone to instability than the lower charge ones. On the contrary, the stability of the discrete $L = 3$ structure was first observed in [25].

Zero parameter asymmetric vortices of family (iv) on the contour S_M with $M = 2$ are shown in Fig. 3.11 for $L = 1$ and in Fig. 3.12 for $L = 3$. In the case of Fig. 3.11, all the phase differences between adjacent sites in the contour are $\pi/6$, except for

3.3 Multipulses and Discrete Vortices

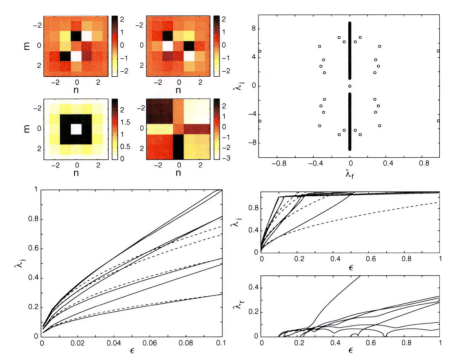

Fig. 3.10 The symmetric vortex with $L = 3$ and $M = 2$ for $\epsilon = 1$. Reprinted from [14] with permission

the last one which is $5\pi/6$, completing a phase trip of 2π for a vortex of topological charge $L = 1$. Eigenvalues of the matrix $\mathcal{M}_1$ in the first-order reductions can be computed numerically as follows: $\mu_1^{(1)} = -1.154$, $\mu_2^{(1)} = 0$, $\mu_3^{(1)} = 0.507$, $\mu_4^{(1)} = 0.784$, $\mu_5^{(1)} = 1.732$, $\mu_6^{(1)} = 2.252$, $\mu_7^{(1)} = 2.957$, and $\mu_8^{(1)} = 3.314$. As a result, the corresponding eigenvalues $\lambda \approx \pm\sqrt{2\mu^{(1)}\epsilon}$ yield one pair of imaginary eigenvalues and six pairs of real eigenvalues, in agreement with our numerical results. The bottom panel of Fig. 3.11 shows that two pairs of real eigenvalues collide for $\epsilon \approx 0.047$ and 0.057 and lead to two quartets of eigenvalues.

In the case of Fig. 3.12, all the phase differences in the contour are $5\pi/6$, except for the last one which is $\pi/6$, resulting in a vortex of topological charge $L = 3$. Eigenvalues of the matrix $\mathcal{M}_1$ are found numerically as follows: $\mu_1^{(1)} = -3.314$, $\mu_2^{(1)} = -2.957$, $\mu_3^{(1)} = -2.252$, $\mu_4^{(1)} = -1.732$, $\mu_5^{(1)} = -0.784$, $\mu_6^{(1)} = -0.507$, $\mu_7^{(1)} = 0$, and $\mu_8^{(1)} = 1.154$. Consequently, this solution has six pairs of imaginary eigenvalues and one pair of real eigenvalues. The first Hamiltonian–Hopf bifurcation in this case occurs for $\epsilon \approx 0.086$.

It is interesting to note in passing that this approach is equally well-suited to address not just "isotropic" square lattices, where the x and y directions are equivalent, but also even *anisotropic* such lattices, where for instance the coefficient of the discrete Laplacian has a different prefactor (e.g., ϵ and $\epsilon\alpha$, respectively) in the two

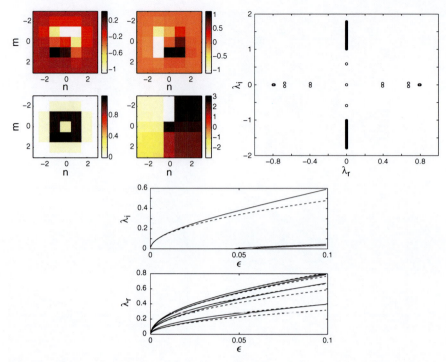

Fig. 3.11 The asymmetric vortex with $L = 1$ and $M = 2$ for $\epsilon = 0.1$. Reprinted from [14] with permission

different directions. Although, we won't analyze this possibility in more detail here, the relevant details can be found by the interested reader in the work of [26].

3.3.4.3 Stabilization of Unstable Waves

One of the most remarkable features of the above developed technology for the detection of the coherent structures and their linear stability in dynamical lattices of the DNLS type is its "mathematical intuition" about the nature of the encountered instabilities and how to potentially eliminate them.

As a case example of this type, we consider one of the most prototypical unstable vortex configurations considered above, namely the vortex with $L = M = 2$. The examination of the real eigenmode leading to the direct instability of the $S = 2$ vortex (that has support over the central site that we denote by a double zero subscript in what follows), as well as the apparent mediation of the instability by means of the central site (see below), lead us to consider the possibility of having an "impurity" at the central site, e.g., a strong localized potential such as a laser beam in BECs or an inhomogeneity in a photorefractive crystal, enforcing $\phi_{0,0} = 0$. More specifically, what we observe mathematically is that the additional terms in the second order reductions that appear to mediate the instability are the terms in the corresponding

3.3 Multipulses and Discrete Vortices

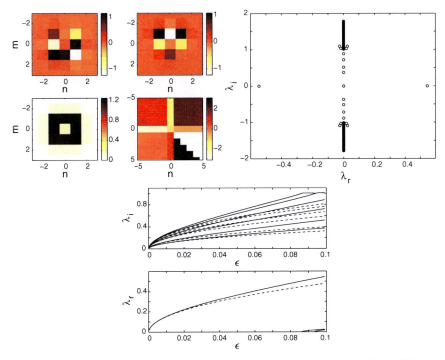

Fig. 3.12 The asymmetric vortex with $L = 3$ and $M = 2$ for $\epsilon = 0.1$. Reprinted from [14] with permission

reduction matrix $\mathcal{M}_2$ coupling the sites adjacent to $(0, 0)$. If we therefore eliminate this central site, disallowing the "communication" between its neighboring sites, then the bifurcation function g_j^2 of the form

$$g_j^2 = \frac{1}{2} \sin(\theta_{j+1} - \theta_j) \left[\cos(\theta_j - \theta_{j+1}) + \cos(\theta_{j+2} - \theta_{j+1})\right]$$

$$+ \frac{1}{2} \sin(\theta_{j-1} - \theta_j) \left[\cos(\theta_j - \theta_{j-1}) + \cos(\theta_{j-2} - \theta_{j-1})\right]$$

$$+ \left[\sin(\theta_j - \theta_{j+2}) + \sin(\theta_j - \theta_{j+4}) + \sin(\theta_j - \theta_{j-2})\right](\delta_{j,2} + \delta_{j,4} + \delta_{j,6} + \delta_{j,8})$$

(with $1 \leq j \leq 8$ and where δ denotes the Kronecker symbol) lacks the last term, since these are interactions "mediated" by the now inert site. The second-order Jacobian is then much simpler and acquires the form $(\mathcal{M}_2)_{j,k} = 1$ for $j = k$, $-1/2$ for $j = k \pm 2$, and 0 for $|j - k| \neq 0, 2$. One can then repeat the calculation of the corresponding eigenvalues, via the discrete Fourier transform, to obtain the characteristic equation

$$\left(\lambda_1 + 2i \sin\left(\frac{j\pi}{4}\right)\right)^2 = 0, \qquad j = 1, \ldots, 8. \tag{3.92}$$

This results into three eigenvalues of algebraic multiplicity four, namely $\lambda = 0$ and $\lambda = \pm \epsilon i \sqrt{2}/2$. There are also two double eigenvalues $\lambda = \pm 2\epsilon i$. The crucial observation, however, is that in this case, there are no real eigenvalues immediately present as $\epsilon \neq 0$ and hence the discrete vortex with $S = 2$ will be *linearly stable*, due to the stabilizing effect of the impurity (or, to be more precise, due to the absence of the instability mediated by the central site).

The stabilization of the relevant structure is clearly shown in Fig. 3.13. The figure illustrates the principal relevant eigenvalues in the case of the inert central site in the left panels, clearly demonstrating not only the validity of the theoretical reduction results presented above, but most importantly the absence of any eigenvalues with non-zero real part for small ϵ. Note that the instability only sets in due to a complex quartet in this case for $\epsilon > 0.36$, which indicates that under the present conditions the stability range of the vortex of $L = 2$ is comparable to that of $L = 1$ (which is stable for $\epsilon < 0.38$). As a result, the right panel shows that for $\epsilon = 0.2$, the same structure with the same perturbation that would have clear instability dynamics for the uniform chain, would no longer be subject to such an instability in the chain with the inert central defect site. These results were first presented in [27].

Interestingly, this suggestion has motivated further studies on this topic such as the work of [28], which suggested on a purely numerical basis the consideration of a cross-like vortex of $L = 2$, such as the one illustrated in the left panels of Fig. 3.14. Note that in this case, once again "reduced communication" is achieved between the four sites previously cross-talking through the central node of the contour; however, instead of this being realized through the central site being inert, here, it is achieved geometrically through increasing the distance between these sites (in the

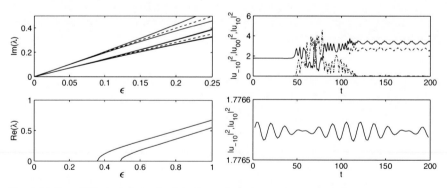

Fig. 3.13 From [27]: the *left panels* of the figure show the imaginary (*top panels*) and real (*bottom panels*) parts of the eigenvalues initially at the origin of the spectral plane for the vortex with $L = M = 2$, but with the central site inert. We can see that, contrary to the case where the central site is present, the structure is linearly stable for small ϵ; note again the agreement between the theoretical prediction of the reductions (*dashed line*) and the full numerics (*solid line*). This results in a dynamical evolution shown in the *right panels* for $\epsilon = 0.2$, where some of the main sites of the configuration are shown in the presence (*top panels*) or absence (*bottom panels*) of the central site. Note how the same configuration which is unstable in the *top* (for the same initial perturbation) becomes stabilized in the *bottom*

3.3 Multipulses and Discrete Vortices

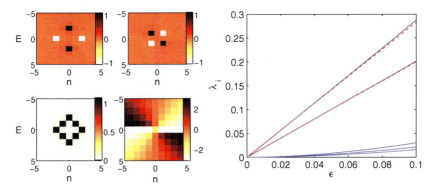

Fig. 3.14 The *left panels* show a typical example of the vortex cross of $L = 2$ which is stable for small ϵ according to the *right panel* illustrating the relevant eigenvalues close to $\lambda = 0$ (which are all imaginary). Note once again the agreement between the result of the analytical reductions (*dashed lines*) and the numerical findings (*solid lines*)

resulting rhombic pattern). In fact, we have studied this structure at the level of the analytical reductions, finding from the relevant Jacobian of the bifurcation equations that it should have a single eigenvalue pair $\lambda = \pm\sqrt{8\epsilon}i$, a double eigenvalue pair $\lambda = \pm 2\epsilon i$, as well as four pairs of eigenvalues of a higher order (not discussed in detail here) which are also imaginary (see the right panel of Fig. 3.14). The relevant theoretical predictions are compared to full numerical results for small ϵ in Fig. 3.14, illustrating, once again, the usefulness of the method in providing quantitative information about the stability (as well as the stabilization) of the various configurations.

3.3.4.4 Solitons and Vortices in Non-Square Lattices

The considerations presented above in the case of square lattices can be straightforwardly generalized to different types of lattices (such as hexagonal or honeycomb ones), where the number of nearest neighbors is different (six and three, respectively) and hence we expect quantitative, and perhaps even qualitative changes in the relevant phenomenology. As a concrete example of this type, we will consider for definiteness the DNLS in a hexagonal geometry

$$i\frac{du_{m,n}}{dz} = -\epsilon\left(\sum_{<m',n'>} u_{m',n'} - 6u_{m,n}\right) - |u_{m,n}|^2 u_{m,n}, \quad (3.93)$$

where the summation is meant over the six nearest neighbors (denoted by $\langle m', n'\rangle$) of the site (m, n). This type of setting was originally considered in [29], while subsequent works such as [30] extended it also to Klein–Gordon lattices and the examination of breather states therein.

In this context, selecting a simple hexagonal contour with a central inert site, it is straightforward to construct a configuration with topological charge S over the contour, provided that we select the AC limit solutions in the form $u_j = \exp(i\theta_j)\exp(it)$

(normalizing, without loss of generality, the propagation constant to unity), where $\theta_j = 2\pi j S/6$ and $j = 1, \ldots, 6$ for the six sites constituting the relevant contour. It is straightforward to see that this configuration yields non-trivial phase profiles for $S = 1$ and 2, while for $S = 3$ it yields a "discrete hexapole" (i.e., not a genuine $S = 3$ vortex structure, but instead a real configuration emulating that waveform in the discrete setting), which can nonetheless also be considered within the analytical framework provided below; also $S = 0$ corresponds to the case of an in-phase structure, that we should expect to be unstable. It should also be noted that this framework works not only for contours of "size" $N = 6$ as will be considered here, but also for ones with $N = 3$ (in that case, e.g., the $\theta_j = 2\pi j S/N$ with $N = 3$), as in [30]. A brief discussion of the results in that context is given below. The results below follow closely the presentation of [31].

We can straightforwardly adapt the calculations presented above to formulate the persistence conditions for the configuration in the presence of finite coupling as:

$$g_j \equiv \sin(\theta_j - \theta_{j+1}) + \sin(\theta_j - \theta_{j-1}) = 0 \tag{3.94}$$

for all $j = 1, \ldots, 6$. Then one can also adapt the stability conditions obtained previously based on the *Jacobian* $\mathcal{M}_{jk} = \partial g_j/\partial \theta_k$ and its eigenvalues γ_j in connection to the eigenvalues of the full linearization $\lambda_j = \sqrt{2\gamma_j \epsilon}$ (to leading order). In the present setting, the Jacobian matrix is given by the expression of (3.35), where the factor $a \equiv \cos(\theta_{j+1} - \theta_j) = \cos(\pi S/3)$ appears in all the elements of the matrix (multiplied by 2 for the diagonal elements and by -1 for the off-diagonal ones). As a result, the eigenvalue problem for the γ's is equivalent to

$$a(2x_n - x_{n+1} - x_{n-1}) = \gamma x_n, \tag{3.95}$$

which can be solved by discrete Fourier transform (i.e., using for the eigenvector $x_n \sim \exp(i\pi j n/3)$), yielding $\gamma_j = 4a \sin^2(\pi j/6)$ and hence, finally,

$$\lambda_j = \pm \sqrt{8\epsilon \cos\left(\frac{\pi S}{3}\right) \sin^2\left(\frac{\pi j}{6}\right)}. \tag{3.96}$$

More specifically, in the case of $S = 1$ this predicts that the fundamental vortex solution will be *unstable* due to two double real eigenvalue pairs with $\lambda = \pm\sqrt{\epsilon}$ and $\lambda = \pm\sqrt{3\epsilon}$ and a single real eigenvalue pair of $\lambda = \pm 2\sqrt{\epsilon}$ (one of the six eigenvalues of the Jacobian is zero due to the phase invariance of the equation), while on the other hand, the $S = 2$ configuration will be *stable* because its eigenvalues will be those of $S = 1$ multiplied by the complex unity (and hence will be all imaginary). It is interesting to note in passing that for $S = 0$ and 3 the above theoretical prediction encompasses the instability and stability, respectively, of a hexagonal discrete soliton with in-phase and out-of-phase nearest-neighbor excitations.

Note that these results can be straightforwardly extended to the three-site contour of the hexagonal lattice, in which case $N = 3$, and therefore the corresponding

3.3 Multipulses and Discrete Vortices

expression will become

$$\lambda_j = \pm \sqrt{8\epsilon \cos\left(\frac{2\pi S}{3}\right) \sin^2\left(\frac{\pi j}{3}\right)}. \tag{3.97}$$

It is also interesting to point out that the results would not change to this leading order for a six-site contour of a honeycomb lattice, since the (absent in that case) inert central site of the hexagonal contour is not accounted for in the above leading order calculation. Finally, it should be pointed out that the stability conclusions obtained above should be expected to be reversed for $\epsilon < 0$ (the defocusing case that we will examine in more detail in Chap. 5). In particular, the in-phase solution will be stable, while the out-of-phase hexapole will be unstable, and similarly the charge $S = 1$ vortex will be the stable one, while the $S = 2$ vortex will be unstable.

We test these predictions in a prototypical case, namely for the $S = 1$ and 2 vortices in the six-site hexagonal lattice contour in Fig. 3.15. The four left panels of the figure represent a typical example of an $S = 1$ vortex (for $\epsilon = 0.025$, close to the AC limit). The second row illustrates the eigenvalues of the associated linearization of Eq. (3.93) around the vortex solution, revealing the presence of five unstable eigenmodes (with non-zero real parts), in agreement with the theoretical prediction. Interestingly, the double eigenvalues of the above theoretical prediction split into complex quartets (a similar feature was observed in the case of $S = 1$ vortices on eight-site square contours earlier in this chapter). Note the quality of the comparison of the theoretical prediction of the modes' growth rates with respect to the corresponding numerical results for small ϵ. The right four panels of Fig. 3.15 represent the case of $S = 2$ which, again in accordance with our theoretical prediction, is indeed found to be linearly stable for small coupling (no eigenvalues with non-zero real part). While in that case the double eigenvalues split, they still follow fairly accurately the trends of the relevant theoretical predictions. These results illustrate

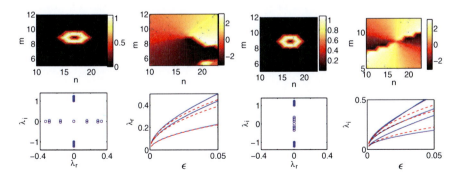

Fig. 3.15 Amplitude and phase of the $S = 1$ vortex (*left panels*) and $S = 2$ vortex (*right panels*), for $\epsilon = 0.025$. The corresponding spectral planes (λ_r, λ_i) of the linearization eigenvalues $\lambda = \lambda_r + i\lambda_i$ are shown in the corresponding *bottom left panels*. The respective *bottom right panels* show the eigenvalues bifurcating from the spectral plane origin as numerically obtained (*solid*) and theoretically predicted from Eq. (3.96) (*dashed*)

that the techniques presented in this chapter are by no means restricted to square lattices, but rather can be directly adapted to address more general lattices, as well as potentially DNLS equations on graphs with different types of connectivities.

References

1. Sulem, C., Sulem, P.L.: The Nonlinear Schrödinger Equation. Springer-Verlag, New York (1999)
2. Chiao, R.Y., Garmire, E., Townes, C.H.: Phys. Rev. Lett. **13**, 479 (1964)
3. Flach, S., Kladko, K., MacKay, R.S.: Phys. Rev. Lett. **78**, 1207 (1997)
4. Weinstein, M.I.: Nonlinearity **12**, 673 (1999)
5. Kastner, M.: Phys. Rev. Lett. **93**, 150601 (2004)
6. Grillakis, M.: Commun. Pure Appl. Math. **43**, 299 (1990)
7. Grillakis, M.: Commun. Pure Appl. Math. **41**, 745 (1988)
8. Grillakis, M., Shatah, J., Strauss, W.: J. Func. Anal. **94**, 308 (1990)
9. Kapitula, T., Kevrekidis, P.G., Sandstede, B.: Physica D **195**, 263 (2004)
10. Kapitula, T., Kevrekidis, P.G., Sandstede, B.: Physica D **201**, 199 (2005)
11. Kevrekidis, P.G., Rasmussen, K.Ø., Bishop, A.R.: Phys. Rev. E **61**, 2006 (2000)
12. Vakhitov, M.G., Kolokolov, A.A.: Radiophys. Quantum Electron. **16**, 783 (1973)
13. Fibich, G., Gaeta, A.L.: Opt. Lett. **25**, 335 (2000)
14. Pelinovsky, D.E., Kevrekidis, P.G., Frantzeskakis, D.J.: Physica D **212**, 20 (2005)
15. Malomed, B.A., Kevrekidis, P.G.: Phys. Rev. E **64**, 026601 (2001)
16. Yang, J., Musslimani, Z.: Opt. Lett. **28** 2094–2096, (2003)
17. MacKay, R.S., Aubry, S.: Nonlinearity **7**, 1623 (1994)
18. Chow, S.N., Hale, J.K.: Methods of Bifurcation Theory. Springer-Verlag, Heidelberg (1982)
19. Golubitsky, M., Schaeffer, D.G.: Singularities and Groups in Bifurcation Theory. vol. 1, Springer-Verlag, New York (1985)
20. Kapitula, T., Kevrekidis, P.G., J. Phys. A **37**, 7509 (2004)
21. Alexander, T.J., Sukhorukov, A.A., Kivshar, Yu.S.: Phys. Rev. Lett. **93**, 063901 (2004)
22. Pego, R.L., Warchall, H.: J. Nonlin. Sci. **12**, 347 (2002)
23. Carr, L.D., Clark, C.W.: Phys. Rev. A **74**, 043613 (2006)
24. Herring, G., Carr, L.D., Carretero-González, R., Kevrekidis, P.G., Frantzeskakis, D.J.: Phys. Rev. A **77**, 023625 (2008)
25. Kevrekidis, P.G., Malomed, B.A., Zhigang, C., Frantzeskakis, D.J.: Phys. Rev. E **70**, 056612 (2004)
26. Kevrekidis, P.G., Frantzeskakis, D.J., Carretero-González, R., Malomed, B.A., Bishop, A.R.: Phys. Rev. E **72**, 046613 (2005)
27. Kevrekidis, P.G., Frantzeskakis, D.J.: Phys. Rev. E **72**, 016606 (2005)
28. Öster, M., Johansson, M.: Phys. Rev. E **73**, 066608 (2006)
29. Kevrekidis, P.G., Malomed, B.A., Gaididei, Yu.B.: Phys. Rev. E **66**, 016609 (2002)
30. Koukouloyannis, V., MacKay, R.S.: J. Phys. A: Math. Gen. **38**, 1021–1030 (2005)
31. Law, K.J.H., Kevrekidis, P.G., Koukouloyannis, V., Kourakis, I., Frantzeskakis, D.J., Bishop, A.R.: Phys. Rev. E **78**, 066610 (2008)

Chapter 4
The Three-Dimensional Case

We now turn to the case of the fully three-dimensional dynamical lattices of the DNLS type, described by the model of the form

$$i\dot{u}_n + \epsilon \Delta_2 u_n + |u_n|^2 u_n = 0, \qquad n \in \mathbb{Z}^3, \ t \in \mathbb{R}_+, \ u_n \in \mathbb{C}, \qquad (4.1)$$

where, now, $\Delta_2 u_n$ is the discrete three-dimensional Laplacian

$$\Delta_2 u_n = u_{n+e_1} + u_{n-e_1} + u_{n+e_2} + u_{n-e_2} + u_{n+e_3} + u_{n-e_3} - 6u_n \qquad n \in \mathbb{Z}^3,$$

with $\{e_1, e_2, e_3\}$ being standard unit vectors in $\mathbb{Z}^3$. As usual, this is a Hamiltonian infinite dimensional (in the infinite lattice) dynamical system preserving the Hamiltonian function

$$H = \epsilon \left(\|u_{n+e_1} - u_n\|^2_{l^2(\mathbb{Z}^3)} + \|u_{n+e_2} - u_n\|^2_{l^2(\mathbb{Z}^3)} + \|u_{n+e_3} - u_n\|^2_{l^2(\mathbb{Z}^3)} \right) \qquad (4.2)$$
$$- \frac{1}{2} \|u_n\|^4_{l^4(\mathbb{Z}^3)},$$

due to the time translation invariance. On the other hand, the l^2-norm is also, as usual, conserved by this dynamical model $P(t) = \|u_n(t)\|^2_{l^2(\mathbb{Z}^3)} = P(0)$.

From the point of view of physical applications, the three-dimensional DNLS remains relevant to understanding the superfluid dynamics of BECs in three-dimensional optical lattices, which have been realized in a variety of experiments [1–3] (although it cannot capture genuinely quantum phenomena within these lattices which require a Bose–Hubbard-type model [4]). On the other hand, there exists an additional possibility for the physical realization of such a system, offered by a three-dimensional crystal built of microresonators [5].

Naturally, in this three-dimensional context, it is straightforward to "embed" lower dimensional structures [7, 21]. A series of such examples is shown in Fig. 4.1. In particular the top left panel shows a fundamental soliton solution; for this solution, as we saw in Fig. 3.1, we expect it to be stable for $\epsilon < \Lambda$, while for $\epsilon > \Lambda$, as the continuum limit is approached, the solution becomes destabilized. In addition to that solution, there exist multipulse solutions such as the dipole and quadrupole

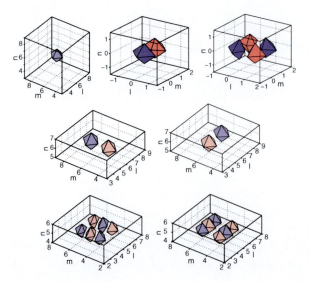

Fig. 4.1 The *top row* shows a typical case-example of the fundamental (single-site) solution (*left panel*), a dipole solution consisting of two excited sites (*middle panel*), and a quadrupole solution of four excited sites (*right panel*). The second row shows the real part (*left panel*) and the imaginary part (*right panel*) of a discrete vortex with $S = 1$. The third row shows the real (*left*) and imaginary (*right*) panels of a vortex with $S = 3$. In all the cases shown here from [21] and [7], the darker surfaces indicate a large positive value of the contour, while the lighter ones a large negative value thereof, illustrating the excited sites of the configuration

(top middle and top right panels in Fig. 4.1). These solutions are principally two-dimensional along the relevant embedded plane and as a result the corresponding Lyapunov–Schmidt equations, and linearization eigenvalues remain to leading order the same as in the corresponding discussions of Chaps. 2 and 3, respectively. The same, in fact, holds true for the vortices of topological charge $S = 1$ (second row) and even of topological charge $S = 3$ (third row) which can both be stable for sufficiently weak coupling, as in the corresponding two-dimensional case of Chap. 3.

However, in addition to these lower dimensional structures, it is possible to construct genuinely three-dimensional structures with no-direct analog in lower dimensional settings. It is to the general theory of these structures that we now turn, subsequently followed by corresponding direct calculations and numerical comparisons for some of the most fundamental among these structures, namely ones lying on a cube, on an eight-site double cross, or on a six-site, diamond-like structure. Our exposition will follow closely the detailed analysis of [8].

4.1 General Theory

As before, the DNLS can be written as a dynamical system of the form $d\mathbf{u}/dt = J\nabla H[\mathbf{u}]$, where the operators (J, ∇) are block-diagonal, being given at each node by

4.1 General Theory

$$J_n = \begin{bmatrix} 0 & 1 \\ -1 & 0 \end{bmatrix}, \quad \nabla_n = \begin{bmatrix} \partial_{\text{Re}(u_n)} \\ \partial_{\text{Im}(u_n)} \end{bmatrix}.$$

Our aim in the present chapter is to seek stationary (i.e., standing wave) solutions to this problem of the form

$$u_n(t) = \phi_n e^{i(1-6\epsilon)t}. \tag{4.3}$$

Then, the ϕ_n satisfies the corresponding steady-state equation of the form

$$(1 - |\phi_n|^2)\phi_n = \epsilon \Sigma \phi_n, \quad \Sigma = \Delta_2 + 6, \quad n \in \mathbb{Z}^3. \tag{4.4}$$

Once the stationary state ϕ_n has been obtained, it will be of interest to us to identify its linear stability. This will be done by means of the linearization

$$u_n(t) = e^{i(1-6\epsilon)t} \left(\phi_n + a_n e^{\lambda t} + \bar{b}_n e^{\bar{\lambda} t} \right), \tag{4.5}$$

where λ is the eigenvalue parameter and the sequence $\{(a_n, b_n)\}$ solves the linear eigenvalue problem for the difference operators

$$\left(1 - 2|\phi_n|^2\right) a_n - \phi_n^2 b_n - \epsilon \Sigma a_n = i\lambda a_n,$$
$$-\bar{\phi}_n^2 a_n + \left(1 - 2|\phi_n|^2\right) b_n - \epsilon \Sigma b_n = -i\lambda b_n. \tag{4.6}$$

As in the previous chapters, the existence of eigenvalues λ, such that $\text{Re}(\lambda) > 0$ indicates the exponential instability of the corresponding waveform.

As in Chaps. 2 and 3, our starting point for the exploration of the existence and stability problems will be the AC limit of $\epsilon = 0$. Considering a bounded set of nodes S, which are excited on the lattice and denoting its complement by $S^\perp$, given our selection of frequency $\Lambda = 1$ (again, without loss of generality), we have the following solution at the AC limit

$$\phi_n^{(0)} = \begin{cases} e^{i\theta_n}, & n \in S, \\ 0, & n \in S^\perp. \end{cases} \tag{4.7}$$

We consider the nonlinear vector field (whose zeros are the solutions of the stationary problem) in the form

$$\mathbf{F}_n(\boldsymbol{\phi}, \epsilon) = \begin{bmatrix} \left(1 - |\phi_n|^2\right) \phi_n - \epsilon \Sigma \phi_n \\ \left(1 - |\phi_n|^2\right) \bar{\phi}_n - \epsilon \Sigma \bar{\phi}_n \end{bmatrix}. \tag{4.8}$$

The Jacobian $D_{\boldsymbol{\phi}} \mathbf{F}(\boldsymbol{\phi}, \epsilon)$ of the nonlinear vector field $\mathbf{F}(\boldsymbol{\phi}, \epsilon)$ at the solution $\boldsymbol{\phi}$ for any ϵ coincides with the *linearized energy operator* $\mathcal{H}$, which defines the quadratic form for the Lyapunov function $\Lambda[\mathbf{u}] = H[\mathbf{u}] + (1 - 6\epsilon)Q[\mathbf{u}]$ in the form

$$\Lambda[\mathbf{u}] = \Lambda[\boldsymbol{\phi}] + \frac{1}{2}(\boldsymbol{\psi}, \mathcal{H}\boldsymbol{\psi}) + O(\|\boldsymbol{\psi}\|^3),$$

where $\text{Re}\, u_n = \phi_n + a_n$, $\text{Im}\, u_n = b_n$, and the two-block of $\boldsymbol{\psi}$ is defined at the node n as

$$\boldsymbol{\psi}_n = \begin{bmatrix} a_n \\ b_n \end{bmatrix}.$$

The matrix operator $\mathcal{H}$ on $\boldsymbol{\psi}$ is not block-diagonal due to the presence of the shift operator Σ. We can still use a formal notation $\mathcal{H}_n$ for the "two-block" of $\mathcal{H}$ at the node n in the form

$$\mathcal{H}_n = \begin{pmatrix} 1 - 2|\phi_n|^2 & -\phi_n^2 \\ -\bar{\phi}_n^2 & 1 - 2|\phi_n|^2 \end{pmatrix} \tag{4.9}$$

$$- \epsilon \left(s_{+e_1} + s_{-e_1} + s_{+e_2} + s_{-e_2} + s_{+e_3} + s_{-e_3} \right) \begin{pmatrix} 1 & 0 \\ 0 & 1 \end{pmatrix},$$

where $s_{e_j} u_n = u_{n+e_j}$ for $j = 1, 2, 3$. In this notation, the linear eigenvalue problem can be rewritten in the form

$$\sigma \mathcal{H} \boldsymbol{\psi} = i \lambda \boldsymbol{\psi}, \tag{4.10}$$

where the corresponding two-block of σ is a diagonal matrix of $(1, -1)$ at each node.

If we now denote the relevant solution at the AC limit by $\boldsymbol{\phi}^{(0)} = \boldsymbol{\phi}^{(0)}(\boldsymbol{\theta})$, then we have that $\mathbf{F}(\boldsymbol{\phi}^{(0)}, 0) = \mathbf{0}$ is satisfied and $\mathcal{H}^{(0)} = D_{\boldsymbol{\phi}} \mathbf{F}(\boldsymbol{\phi}^{(0)}, 0)$ is block-diagonal with the two-block at the node $n \in \mathbb{Z}^3$ given by

$$(\mathcal{H}^{(0)})_n = \begin{bmatrix} 1 & 0 \\ 0 & 1 \end{bmatrix}, \ n \in S^\perp, \quad (\mathcal{H}^{(0)})_n = \begin{bmatrix} -1 & -e^{2i\theta_n} \\ -e^{-2i\theta_n} & -1 \end{bmatrix}, \ n \in S.$$

It is clear in this case, by means of explicit calculation, that $\mathcal{H}^{(0)} \mathbf{e}_n = \mathbf{0}$ and $\mathcal{H}^{(0)} \hat{\mathbf{e}}_n = -2\hat{\mathbf{e}}_n$, where the two-blocks of eigenvectors $\mathbf{e}_n$ and $\hat{\mathbf{e}}_n$ at the node k are given by

$$(\mathbf{e}_n)_k = i \begin{bmatrix} e^{i\theta_n} \\ -e^{-i\theta_n} \end{bmatrix} \delta_{k,n}, \quad (\hat{\mathbf{e}}_n)_k = \begin{bmatrix} e^{i\theta_n} \\ e^{-i\theta_n} \end{bmatrix} \delta_{k,n}$$

with $\delta_{k,n}$ being a standard Kronecker symbol. Therefore, $\text{Ker}(\mathcal{H}^{(0)}) = \text{Span}(\{\mathbf{e}_n\}_{n \in S})$. If we now denote by $\mathcal{P} : X \mapsto \text{Ker}(\mathcal{H}^{(0)})$ be the orthogonal projection operator to the N-dimensional kernel of $\mathcal{H}^{(0)}$, then $\mathcal{P}$ is expressed by

4.1 General Theory

$$(\mathcal{P}\mathbf{f})_n = \frac{1}{2i}\left(e^{-i\theta_n}(\mathbf{f})_n - e^{i\theta_n}(\bar{\mathbf{f}})_n\right). \qquad n \in S. \qquad (4.11)$$

In this case, the decomposition of ϕ can be given as

$$\boldsymbol{\phi} = \boldsymbol{\phi}^{(0)}(\boldsymbol{\theta}) + \sum_{n \in S} \alpha_n \mathbf{e}_n + \boldsymbol{\varphi}, \qquad (4.12)$$

where $\boldsymbol{\varphi} \in \text{Range}(\mathcal{H}^{(0)})$ and $\alpha_n \in \mathbb{R}$ for each $n \in S$. We note that

$$\forall \boldsymbol{\theta}_0 \in \mathcal{T}: \quad \boldsymbol{\phi}^{(0)}(\boldsymbol{\theta}_0) + \sum_{n \in S} \alpha_n \mathbf{e}_n = \boldsymbol{\phi}^{(0)}(\boldsymbol{\theta}_0 + \boldsymbol{\alpha}) + \mathrm{O}\left(\|\boldsymbol{\alpha}\|_{\mathbb{R}^n}^2\right). \qquad (4.13)$$

Since the values of $\boldsymbol{\theta}$ in $\boldsymbol{\phi}^{(0)}(\boldsymbol{\theta})$ have not been defined yet, we can set $\alpha_n = 0$, $n \in S$ without loss of generality. The splitting equations in the Lyapunov–Schmidt reduction algorithm are

$$\mathcal{P}\mathbf{F}(\boldsymbol{\phi}^{(0)}(\boldsymbol{\theta}) + \boldsymbol{\varphi}, \epsilon) = 0, \qquad (\mathcal{I} - \mathcal{P})\mathbf{F}(\boldsymbol{\phi}^{(0)}(\boldsymbol{\theta}) + \boldsymbol{\varphi}, \epsilon) = 0.$$

We note that $(\mathcal{I} - \mathcal{P})\mathcal{H}(\mathcal{I} - \mathcal{P}) : \text{Range}(\mathcal{H}^{(0)}) \mapsto \text{Range}(\mathcal{H}^{(0)})$ is analytic in $\epsilon \in \mathcal{O}(0)$ and invertible at $\epsilon = 0$, while $\mathbf{F}(\boldsymbol{\phi}, \epsilon)$ is analytic in $\epsilon \in \mathcal{O}(0)$. By the implicit function theorem for analytic vector fields, there exists a unique solution $\boldsymbol{\varphi}$ analytic in $\epsilon \in \mathcal{O}(0)$ and dependent on $\boldsymbol{\theta}$, such that $\boldsymbol{\varphi} \equiv \boldsymbol{\varphi}(\boldsymbol{\theta}, \epsilon)$ and $\|\boldsymbol{\varphi}\| = \mathrm{O}(\epsilon)$ as $\epsilon \to 0$. As a result, there exists the nonlinear vector field $\mathbf{g}$, such that the Lyapunov–Schmidt bifurcation equations are

$$\mathbf{g}(\boldsymbol{\theta}, \epsilon) = \mathcal{P}\mathbf{F}(\boldsymbol{\phi}^{(0)}(\boldsymbol{\theta}) + \boldsymbol{\varphi}(\boldsymbol{\theta}, \epsilon), \epsilon) = 0. \qquad (4.14)$$

By the construction, the function $\mathbf{g}(\boldsymbol{\theta}, \epsilon)$ is analytic in $\epsilon \in \mathcal{O}(0)$ and $\mathbf{g}(\boldsymbol{\theta}, 0) = \mathbf{0}$ for any $\boldsymbol{\theta}$. Therefore, $\mathbf{g}(\boldsymbol{\theta}, \epsilon)$ can be represented for $\epsilon \in \mathrm{O}(0)$ by the Taylor series

$$\mathbf{g}(\boldsymbol{\theta}, \epsilon) = \sum_{k=1}^{\infty} \epsilon^k \mathbf{g}^{(k)}(\boldsymbol{\theta}). \qquad (4.15)$$

In this case, the presence of the U(1) invariance leads to

$$\forall \alpha_0 \in \mathbb{R}: \qquad \mathbf{g}(\boldsymbol{\theta} + \alpha_0 \mathbf{p}_0, \epsilon) = \mathbf{g}(\boldsymbol{\theta}, \epsilon), \qquad (4.16)$$

where $\mathbf{p}_0 = (1, 1, ..., 1)^T \in \mathbb{R}^N$. The Lyapunov–Schmidt decomposition can then be summarized as follows.

The configuration $\boldsymbol{\phi}^{(0)}(\boldsymbol{\theta})$ can be continued to the domain $\epsilon \in \mathcal{O}(0)$ if and only if there exists a root $\boldsymbol{\theta}_*$ of the vector field $\mathbf{g}(\boldsymbol{\theta}, \epsilon)$. Moreover, if the root $\boldsymbol{\theta}_*$ is analytic in $\epsilon \in \mathcal{O}(0)$ and $\boldsymbol{\theta}_* = \boldsymbol{\theta}_0 + \mathcal{O}(\epsilon)$, the solution $\boldsymbol{\phi}$ of the difference equation is analytic in $\epsilon \in \mathcal{O}(0)$, such that

$$\phi = \phi^{(0)}(\theta_*) + \varphi(\theta_*, \epsilon) = \phi^{(0)}(\theta_0) + \sum_{k=1}^{\infty} \epsilon^k \phi^{(k)}(\theta_0). \tag{4.17}$$

This suggests the following algorithm for testing the persistence of a configuration at different orders in ϵ. We assume that the order κ is the leading order for which $\mathbf{g}^{(\kappa)}(\boldsymbol{\theta}) \neq \mathbf{0}$ and the corresponding Jacobian is denoted by $\mathcal{M}^{(k)} = D_{\boldsymbol{\theta}} \mathbf{g}^{(k)}(\boldsymbol{\theta}_0)$ for $k \geq \kappa$. Then,

1. If $\mathrm{Ker}(\mathcal{M}^{(\kappa)}) = \mathrm{Span}(\mathbf{p}_0) \subset \mathbb{R}^N$, then the configuration (4.7) is uniquely continued in $\epsilon \in \mathcal{O}(0)$ modulo the gauge transformation (4.16).
2. Let $\mathrm{Ker}(\mathcal{M}^{(\kappa)}) = \mathrm{Span}(\mathbf{p}_0, \mathbf{p}_1, ..., \mathbf{p}_{d_\kappa}) \subset \mathbb{R}^N$ with $1 \leq d_\kappa \leq N - 1$ and $P^{(\kappa)} : \mathbb{R}^N \mapsto \mathrm{Ker}(\mathcal{M}^{(\kappa)})$ be the projection matrix. Then,

   a) If $\mathbf{g}^{(\kappa+1)}(\boldsymbol{\theta}_0) \notin \mathrm{Range}(\mathcal{M}^{(\kappa)})$, the configuration (4.7) does not persist for any $\epsilon \neq 0$.
   b) If $\mathbf{g}^{(\kappa+1)}(\boldsymbol{\theta}_0) \in \mathrm{Range}(\mathcal{M}^{(\kappa)})$, the configuration (4.7) is continued to the next order. Replace

$$\mathcal{M}^{(\kappa)} \mapsto P^{(\kappa)} \mathcal{M}^{(\kappa+1)} P^{(\kappa)},$$
$$P^{(\kappa)} \mapsto P^{(\kappa+1)} : \mathbb{R}^N \mapsto \mathrm{Ker}(P^{(\kappa)} \mathcal{M}^{(\kappa+1)} P^{(\kappa)}),$$
$$\boldsymbol{\theta}_0 \mapsto \boldsymbol{\theta}_0 - \epsilon \left(\mathcal{M}^{(\kappa)}\right)^{-1} \mathbf{g}^{(\kappa+1)}(\boldsymbol{\theta}_0),$$
$$\mathbf{g}^{(\kappa+1)} \mapsto \mathbf{g}^{(\kappa+2)}$$

and then repeat the above two steps.

This algorithm allows us to determine (if it stops in a finite number of iterations), whether a configuration continues in ϵ or whether it only exists at the very special AC limit. Hereafter we will assume that the algorithm converges to an existing solution after a finite number of iterations, and we will focus on the stability of the resulting structure.

First, we consider the eigenvalue problem of the operator $\mathcal{H}$. In particular, we use the Taylor series expansion

$$\mathcal{H} = \mathcal{H}^{(0)} + \sum_{k=1}^{\infty} \epsilon^k \mathcal{H}^{(k)} \tag{4.18}$$

and focus on the truncated eigenvalue problem for the spectrum of $\mathcal{H}$:

$$\left[\mathcal{H}^{(0)} + \epsilon \mathcal{H}^{(1)} + \cdots + \epsilon^{k-1} \mathcal{H}^{(k-1)} + \epsilon^k \mathcal{H}^{(k)} + O(\epsilon^{k+1})\right] \boldsymbol{\psi} = \mu \boldsymbol{\psi},$$

where μ is eigenvalue and $\boldsymbol{\psi}$ is the corresponding eigenvector.

We now assume that the above persistence algorithm produces multidimensional kernels of matrices $\mathcal{M}^{(\kappa)}, \mathcal{M}^{(\kappa+1)}, ..., \mathcal{M}^{(k-1)}$, such that $\dim \mathrm{Ker}\, \mathcal{M}^{(k)} < \dim \mathrm{Ker}\, \mathcal{M}^{(k-1)}$ for some $\kappa \leq k \leq K$, where κ and K are the starting and termination orders of the algorithm. The kernel of $M^{(K)}$ is one-dimensional. We also let $\boldsymbol{\alpha}$

4.1 General Theory

be an element of $\text{Ker}(\mathcal{M}^{(\kappa)}) \cap \text{Ker}(\mathcal{M}^{(\kappa+1)}) \cap \ldots \cap \text{Ker}(\mathcal{M}^{(k-1)}) \subset \mathbb{R}^N$, such that $\alpha \notin \text{Ker}(\mathcal{M}^{(k)})$ for some $\kappa \leq k \leq K$. Then, α has $(d_{k-1} + 1)$ arbitrary parameters, where $d_{k-1} \leq N - 1$. By using the projection operator $\mathcal{P}$ in (4.11) and the relation (4.13), we obtain that

$$\alpha = \mathcal{P}\left(\sum_{n \in S} \alpha_n \mathbf{e}_n\right), \quad \left(\sum_{n \in S} \alpha_n \mathbf{e}_n\right) = D_{\boldsymbol{\theta}} \boldsymbol{\phi}^{(0)}(\boldsymbol{\theta}_0) \alpha,$$

where $D_{\boldsymbol{\theta}} \boldsymbol{\phi}^{(0)}(\boldsymbol{\theta}_0)$ is the Jacobian matrix of the infinite-dimensional vector $\boldsymbol{\phi}^{(0)}(\boldsymbol{\theta})$ with respect to the N-dimensional vector $\boldsymbol{\theta}$. It is clear that

$$\mathcal{H}^{(0)} \boldsymbol{\psi}^{(0)} = 0, \quad \text{where} \quad \boldsymbol{\psi}^{(0)} = \sum_{n \in S} \alpha_n \mathbf{e}_n = D_{\boldsymbol{\theta}} \boldsymbol{\phi}^{(0)}(\boldsymbol{\theta}_0) \alpha.$$

Furthermore, the partial $(k-1)$th sum of the power series (4.17) yields the zero of the nonlinear vector field (4.8) up to the order $O(\epsilon^k)$ and has $(d_{k-1} + 1)$ arbitrary parameters if $\boldsymbol{\theta}_0$ is shifted in the direction of the vector α. The linear inhomogeneous system

$$\mathcal{H}^{(0)} \boldsymbol{\psi}^{(m)} + \mathcal{H}^{(1)} \boldsymbol{\psi}^{(m-1)} + \cdots + \mathcal{H}^{(m)} \boldsymbol{\psi}^{(0)} = \mathbf{0}$$

then has a particular solution in the form $\boldsymbol{\psi}^{(m)} = D_{\boldsymbol{\theta}} \boldsymbol{\phi}^{(m)}(\boldsymbol{\theta}_0) \alpha$ for $m = 1, 2, \ldots, k - 1$. By extending the regular perturbation series for isolated zero eigenvalues of $\mathcal{H}^{(0)}$,

$$\boldsymbol{\psi} = \boldsymbol{\psi}^{(0)} + \epsilon \boldsymbol{\psi}^{(1)} + \cdots + \epsilon^k \boldsymbol{\psi}^{(k)} + O\left(\epsilon^{k+1}\right), \quad \mu = \mu_k \epsilon^k + O\left(\epsilon^{k+1}\right), \quad (4.19)$$

we obtain the linear inhomogeneous equation

$$\mathcal{H}^{(0)} \boldsymbol{\psi}^{(k)} + \mathcal{H}^{(1)} \boldsymbol{\psi}^{(k-1)} + \cdots + \mathcal{H}^{(k)} \boldsymbol{\psi}^{(0)} = \mu_k \boldsymbol{\psi}^{(0)}.$$

The projection operator $\mathcal{P}$ may now be applied recalling the definition (4.14); this yields for the left-hand side of the linear equation

$$\mathcal{P}\left[\mathcal{H}^{(1)} D_{\boldsymbol{\theta}} \boldsymbol{\phi}^{(k-1)}(\boldsymbol{\theta}_0) + \cdots + \mathcal{H}^{(k)} D_{\boldsymbol{\theta}} \boldsymbol{\phi}^{(0)}(\boldsymbol{\theta}_0)\right] \alpha = D_{\boldsymbol{\theta}} \mathbf{g}^{(k)}(\boldsymbol{\theta}_0) \alpha = \mathcal{M}^{(k)} \alpha,$$

where we have used that $\mathcal{P} \boldsymbol{\psi}^{(0)} = \alpha$. Therefore, μ_k is an eigenvalue of the Jacobian matrix $\mathcal{M}^{(k)}$ and α is the corresponding eigenvector. The equivalence between non-zero small eigenvalues of $\mathcal{H}$ and non-zero eigenvalues of $\mathcal{M}$ can be summarized as follows.

Let us assume that the above persistence algorithm converges at the Kth order and the solution $\boldsymbol{\phi}$ persists for $\epsilon \neq 0$. Then,

$$\lambda_{\neq 0}(\mathcal{H}) = \lambda_{\neq 0}\left(P^{(m-1)} \mathcal{M}^{(m)} P^{(m-1)}\right) \epsilon^m + O\left(\epsilon^{m+1}\right), \quad m = \kappa, \kappa + 1, \ldots, K,$$

where $\lambda_{\neq 0}(\mathcal{A})$ is a non-zero eigenvalue of operator $\mathcal{A}$.

The next step is then to consider eigenvalues of the spectral problem (3.57) truncated at the kth-order approximation

$$\left[\mathcal{H}^{(0)} + \epsilon\mathcal{H}^{(1)} + \cdots + \epsilon^{k-1}\mathcal{H}^{(k-1)} + \epsilon^{k}\mathcal{H}^{(k)} + O\left(\epsilon^{k+1}\right)\right]\psi = i\lambda\sigma\psi.$$

Using the above obtained relations $\hat{\mathbf{e}}_n = -i\sigma\mathbf{e}_n$ and $\mathcal{H}^{(0)}\hat{\mathbf{e}}_n = -2\hat{\mathbf{e}}_n$ for all $n \in S$, it can be seen that the linear inhomogeneous equation

$$\mathcal{H}^{(0)}\varphi^{(0)} = 2i\sigma D_{\theta}\phi^{(0)}(\theta_0)\alpha$$

has a solution

$$\varphi^{(0)} = \sum_{n\in S}\alpha_n\hat{\mathbf{e}}_n = \Phi^{(0)}(\theta_0)\alpha,$$

where $\Phi^{(0)}(\theta_0)$ is the matrix extension of $\phi^{(0)}(\theta_0)$, which consists of vector columns $\hat{\mathbf{e}}_n$, $n \in S$. Similarly, there exists a special solution of the inhomogeneous problem

$$\mathcal{H}^{(0)}\varphi^{(m)} + \mathcal{H}^{(1)}\varphi^{(m-1)} + \cdots + \mathcal{H}^{(m)}\varphi^{(0)} = 2i\sigma D_{\theta}\phi^{(m)}(\theta_0)\alpha, \qquad m = 1, 2, \ldots, k',$$

in the form $\varphi^{(m)} = \Phi^{(m)}(\theta_0)\alpha$, where $k' = (k-1)/2$ if k is odd and $k' = k/2 - 1$ if k is even. By extending the regular perturbation series for isolated zero eigenvalue of $\sigma\mathcal{H}^{(0)}$,

$$\psi = \psi^{(0)} + \epsilon\psi^{(1)} + \cdots + \epsilon^{k-1}\psi^{(k-1)} \qquad (4.20)$$
$$+\frac{1}{2}\lambda\left(\varphi^{(0)} + \epsilon\varphi^{(1)} + \cdots + \epsilon^{k'}\varphi^{(k')}\right) + \epsilon^{k}\psi^{(k)} + O\left(\epsilon^{k+1}\right),$$

where $\psi^{(m)} = D_{\theta}\phi^{(m)}(\theta_0)\alpha$ for $m = 0, 1, \ldots, k-1$, $\varphi^{(m)} = \Phi^{(m)}(\theta_0)\alpha$ for $m = 0, 1, \ldots, k'$, and $\lambda = \epsilon^{k/2}\lambda_{k/2} + O(\epsilon^{k/2+1})$, we obtain a linear inhomogeneous problem for $\psi^{(k)}$ at the order $O(\epsilon^k)$. When k is odd, the linear problem takes the form

$$\mathcal{H}^{(0)}\psi^{(k)} + \mathcal{H}^{(1)}\psi^{(k-1)} + \cdots + \mathcal{H}^{(k)}\psi^{(0)} = \frac{i}{2}\lambda_{k/2}^2\sigma\varphi^{(0)}. \qquad (4.21)$$

On the other hand, when k is even, the linear problem reads

$$\mathcal{H}^{(0)}\psi^{(k)} + \mathcal{H}^{(1)}\psi^{(k-1)} + \cdots + \mathcal{H}^{(k)}\psi^{(0)}$$
$$+\frac{1}{2}\lambda_{k/2}\left(\mathcal{H}^{(1)}\varphi^{(k')} + \cdots + \mathcal{H}^{(k'+1)}\varphi^{(0)}\right) = \frac{i}{2}\lambda_{k/2}^2\sigma\varphi^{(0)}. \qquad (4.22)$$

We can then connect the small non-zero eigenvalues of the full spectral problem of the operator $\sigma\mathcal{H}$ and the reduced one, as follows.

Assume that we can identify a solution (at the Kth order) ϕ persisting for $\epsilon \neq 0$. Let operator $\mathcal{H}$ have a small eigenvalue μ of multiplicity d, such that

4.2 Discrete Solitons and Vortices

$\mu = \epsilon^k \mu_k + O(\epsilon^{k+1})$. Then, the eigenvalue problem (3.57) admits $(2d)$ small eigenvalues λ, such that $\lambda = \epsilon^{k/2}\lambda_{k/2} + O(\epsilon^{k/2+1})$, where non-zero values $\lambda_{k/2}$ are found from the quadratic eigenvalue problems

$$\text{odd } k: \quad \mathcal{M}^{(k)}\boldsymbol{\alpha} = \frac{1}{2}\lambda_{k/2}^2\boldsymbol{\alpha}, \tag{4.23}$$

$$\text{even } k: \quad \mathcal{M}^{(k)}\boldsymbol{\alpha} + \frac{1}{2}\lambda_{k/2}\mathcal{L}^{(k)}\boldsymbol{\alpha} = \frac{1}{2}\lambda_{k/2}^2\boldsymbol{\alpha}, \tag{4.24}$$

where $\mathcal{L}^{(k)} = \mathcal{P}\left[\mathcal{H}^{(1)}\Phi^{(k')}(\boldsymbol{\theta}_0) + \cdots + \mathcal{H}^{(k'+1)}\Phi^{(0)}(\boldsymbol{\theta}_0)\right]$.

Based on the above, in order to identify the eigenvalues of the linearization around a solution ϕ, we first compute $\mathcal{M}^{(k)} = D_{\boldsymbol{\theta}}\mathbf{g}^{(k)}(\boldsymbol{\theta}_0)$ for $\kappa \le k \le K$. For each order k, where eigenvalues of $\mathcal{M}^{(k)}$ are non-zero, we compute matrices $\mathcal{L}^{(k)}$. We then identify the roots $\lambda_{k/2}$ of the determinant equation for the quadratic eigenvalue problems (4.23) and (4.24).

We now turn to specific examples where we can showcase this approach.

4.2 Discrete Solitons and Vortices

The three main types of waveforms that will be considered in this section are the cube, the diamond and the double cross, an example of each of which can be found in Fig. 4.2. We will focus on the fundamental (and more complex) vortex waveforms on these contours, although naturally solitonic configurations with phases of 0 and π can also be excited in these (see a relevant brief discussion at the end of the chapter).

Perhaps the most fundamental example that can be considered as a genuinely three-dimensional building block is a cube consisting of the sites of two adjacent planes: $S = S_0 \oplus S_1$, where

$$S_l = \{(0, 0, l), (1, 0, l), (1, 1, l), (0, 1, l)\}, \quad l = 0, 1, \tag{4.25}$$

such that $N = \dim(S) = 8$; see top left panel of Fig. 4.2 We will enumerate the relevant angles of excitation at the AC limit as $\theta_{l,j}$, with $l = 0, 1$ and $j = 1, 2, 3, 4$ for the eight excited sites of the cube configuration and we will focus on vortex solutions (although, as indicated above, also octupole solutions are possible, among others, in this contour). The bifurcation equations are then straightforward to obtain as

$$g_{l,j}^{(1)} = \sin(\theta_{l,j+1} - \theta_{l,j}) + \sin(\theta_{l,j-1} - \theta_{l,j}) + \sin(\theta_{l+1,j} - \theta_{l,j}), \quad l = 0, 1, \; j = 1, 2, 3, 4, \tag{4.26}$$

where $\theta_{2,j} = \theta_{0,j}$. Roots of $\mathbf{g}^{(1)}(\boldsymbol{\theta})$ occur for vortex configurations with

$$\theta_{0,j} = \frac{\pi(j-1)}{2}, \quad \theta_{1,j} = \theta_0 + s_0\frac{\pi(j-1)}{2}, \quad j = 1, 2, 3, 4, \tag{4.27}$$

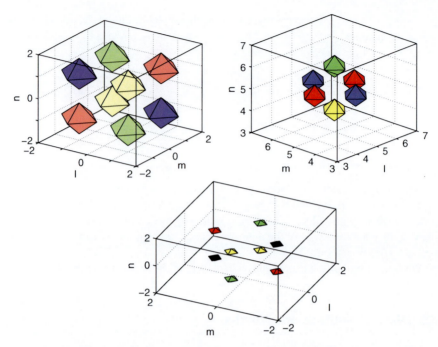

Fig. 4.2 Prototypical examples of the stable three-dimensional configurations. The *top left* shows a stable eight-site vortex cube, the *top right* a stable six-site diamond, while the *bottom* shows a stable eight-site double vortex cross. See the text for the relevant distributions. Reprinted from [8] with permission

where $\theta_0 = \{0, \pi/2, \pi, 3\pi/2\}$ and $s_0 = \{+1, -1\}$. Two out of the eight possible vortex configurations of this "vortex cube" are redundant for $s_0 = -1$ and $\theta_0 = \pi$ or $3\pi/2$, as they can be obtained from the cases with $\theta_0 = 0$ and $\pi/2$, respectively, by means of rotation of the whole structure through multiplication by i.

Detailed application of the above algorithms for persistence and stability [8] illustrates that the only configurations that persist are the ones with $\theta_0 = \{0, \pi\}$ and $s_0 = \{1, -1\}$ (for which the algorithm converges at order $K = 6$). On the other hand, solutions with $\theta_0 = \{\pi/2, 3\pi/2\}$ and $s_0 = \{1, -1\}$ terminate at the $k = 1$ order, and therefore cannot be continued to finite values of $\epsilon \in \mathcal{O}(0)$. The results for the persistence, the eigenvalues of $\mathcal{H}$, as well as the eigenvalues of the full linear stability problem (i.e., the eigenvalues of $i\sigma\mathcal{H}$) for this structure are given in Table 4.1.

In summarizing the results of Table 4.1, we infer that the *only* vortex cube configuration that may be stable is the one with $\theta_0 = \pi$ and $s_0 = 1$, while the other two irreducible configurations, namely the ones with $\theta_0 = 0$ and $s_0 = \pm 1$ have multiple real eigenvalues in each case. We now examine how these analytical predictions compare with full numerical computations.

Figure 4.3 presents the simple cube configuration with $S_1 = \{0, \pi/2, \pi, 3\pi/2\}$. This configuration should be unstable due to an eigenvalue $\lambda \approx 2\epsilon^{1/2}$, of multiplicity

4.2 Discrete Solitons and Vortices

Table 4.1 Vortex Cube Configurations: the first column shows the configuration, the next its persistence (Per), the following ones the eigenvalues of $\mathcal{H}$ and then those of the full stability problem (i.e., of $i\sigma\mathcal{H}$), and the last column determines whether the structure may be stable (St)

S_1	Per	$\mathcal{O}(\epsilon)\,\mathcal{H}$	$\mathcal{O}(\epsilon^2)\,\mathcal{H}$	$\mathcal{O}(\epsilon^6)\,\mathcal{H}$	$\mathcal{O}(\epsilon^{1/2})\,i\sigma\mathcal{H}$	$\mathcal{O}(\epsilon)\,i\sigma\mathcal{H}$	$\mathcal{O}(\epsilon^3)\,i\sigma\mathcal{H}$	St
$\{0,\frac{\pi}{2},\pi,\frac{3\pi}{2}\}$	Y	$\{2\times 4\}$	$\{2\times 2\}$	$\{-16\}$	$\{\pm 2\times 4\}$	$\{\pm 2i\times 2\}$	$\{\pm 4i\sqrt{2}\}$	N
$\{0,\frac{3\pi}{2},\pi,\frac{\pi}{2}\}$	Y	$\{-2\times 2,\ 2\times 2\}$	$\{2\times 2\}$	$\{-16\}$	$\{\pm 2\times 2,\ \pm 2i\times 2\}$	$\{\pm 2\times 2\}$	$\{\pm 4i\sqrt{2}\}$	N
$\{\frac{\pi}{2},\pi,\frac{3\pi}{2},0\}$	N							
$\{\frac{\pi}{2},0,\frac{3\pi}{2},\pi\}$	N							
$\{\pi,\frac{3\pi}{2},0,\frac{\pi}{2}\}$	Y	$\{-2\times 4\}$	$\{2\times 2\}$	$\{-16\}$	$\{\pm 2i\times 4\}$	$\{\pm 2i\times 2\}$	$\{\pm 4i\sqrt{2}\}$	Y
$\{\frac{3\pi}{2},0,\frac{\pi}{2},\pi\}$	N							

four. In the numerical computations, the pair of multiple real eigenvalue splits into two identical pairs of real eigenvalues and a quartet of complex eigenvalues. Nevertheless, all four eigenvalues in the right half-plane have the same real part denoted by the very thick solid line in the left panel of Fig. 4.3. The imaginary part of the quartet of complex eigenvalues is denoted by dash–dotted line in the right panel of the figure. While the real part of all four eigenvalues is of order $O(\epsilon^{1/2})$, the imaginary part of complex eigenvalues arises at order $O(\epsilon)$ with a numerical approximation $\lambda \approx 2\epsilon^{1/2} \pm 2i\epsilon$. Additionally, there exists a pair of double imaginary eigenvalues at $O(\epsilon)$ and a pair of simple imaginary eigenvalues at $O(\epsilon^3)$. The latter pairs are shown on the right panel of Fig. 4.3 by thin lines, since the pair of double imaginary

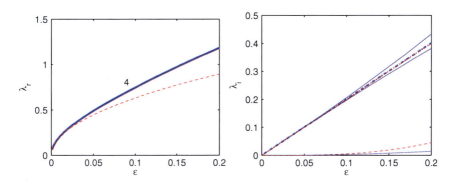

Fig. 4.3 The real (*left*) and imaginary (*right*) parts of small eigenvalues of the linearized problem of Eq. (4.6) associated with the vortex cube configuration with $S_1 = \{0, \pi/2, \pi, 3\pi/2\}$ versus ϵ. Numerically computed eigenvalues are denoted by *solid lines*, while their counterparts from symbolic computations are plotted by *dashed lines*. Multiple real or imaginary eigenvalues are denoted by *thick solid lines* and their corresponding multiplicity is shown beside the relevant line, while complex eigenvalues are denoted by *thick dash–dotted lines*. Reprinted from [8] with permission

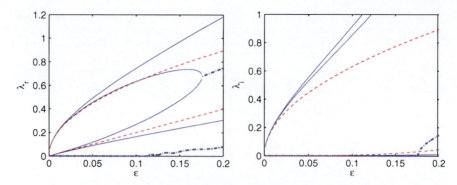

Fig. 4.4 Same as in Fig. 4.3, but for the simple cube vortex configuration with $S_1 = \{0, 3\pi/2, \pi, \pi/2\}$. Reprinted from [8] with permission

eigenvalues splits into two pairs of simple imaginary eigenvalues. It is clear from these comparisons that the leading order predictions of Table 4.1 represent adequately the pattern of unstable and neutrally stable eigenvalues of the relevant configuration.

Figure 4.4 describes the vortex cube configuration with $S_1 = \{0, 3\pi/2, \pi, \pi/2\}$. This is also immediately unstable, due to a double pair of real eigenvalues at the order $O(\epsilon^{1/2})$ and another double pair of real eigenvalues of $O(\epsilon)$. Both pairs split for small values of ϵ but remain simple pairs of real eigenvalues for sufficiently small values of ϵ. Then, a pair of the former and one of the latter collide for $\epsilon \approx 0.175$, leading to a quartet of complex eigenvalues. Another double pair of imaginary eigenvalues exists at $O(\epsilon^{1/2})$ and it splits into simple pairs of imaginary eigenvalues. When these eigenvalues meet the continuous spectrum located at $\pm i[1, 1+6\epsilon]$, the pairs of imaginary eigenvalues generate additional quartets of complex eigenvalues for $\epsilon > 0.113$ and $\epsilon > 0.125$. Finally, one more pair of imaginary eigenvalues exists at the order $O(\epsilon^3)$ and it remains small for $0 < \epsilon < 0.2$.

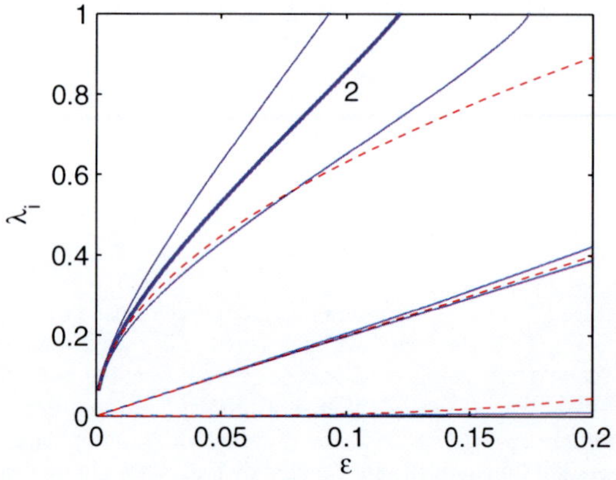

Fig. 4.5 Same as in Fig. 4.3, but for the stable simple cube vortex configuration with $S_1 = \{\pi, 3\pi/2, 0, \pi/2\}$. Reprinted from [8] with permission

4.2 Discrete Solitons and Vortices

Lastly, as regards possible vortex cube configurations, Fig. 4.5 describes the spectrally stable (at least for small ϵ) case with $S_1 = \{\pi, 3\pi/2, 0, \pi/2\}$. The quadruple pair of imaginary eigenvalues at $O(\epsilon^{1/2})$ splits for small ϵ into a double pair and two simple pairs of imaginary eigenvalues. All these pairs generate quartets of complex eigenvalues upon collision with the continuous spectrum for $\epsilon > 0.1$, $\epsilon > 0.125$, and $\epsilon > 0.174$. Therefore, the vortex configuration becomes unstable for sufficiently large ϵ. A double pair of imaginary eigenvalues at $O(\epsilon)$ splits for small ϵ into simple pairs of imaginary eigenvalues. The additional pair of imaginary eigenvalues at the order $O(\epsilon^3)$ remains small for $0 < \epsilon < 0.2$.

We now turn to the vortex diamond configurations (see top right panel of Fig. 4.2). One can straightforwardly again apply the results of our reductions to obtain the relevant bifurcation conditions and eigenvalues, which are summarized in Table 4.2, for the six irreducible configurations of this type that exist when there is a quadrupole structure in the central plane, surrounded by two symmetric central off-peak nodes. Describing the configuration more precisely in mathematical terms, we can write $S = S_{-1} \oplus S_0 \oplus S_1$, where

$$S_0 = \{(-1, 0, 0), (0, -1, 0), (1, 0, 0), (0, 1, 0)\}, \quad S_{\pm 1} = \{(0, 0, \pm 1)\}, \quad (4.28)$$

such that $N = \dim(S) = 6$. The vortex diamonds can be expressed by

$$\theta_{0,j} = \pi(j-1), \quad j = 1, 2, 3, 4, \quad \theta_{\pm 1, 0} = \theta_0^{\pm}, \quad (4.29)$$

where $\theta_0^{\pm} = \{0, \pi/2, \pi, 3\pi/2\}$. Six configurations with $\theta_0^- > \theta_0^+$ are redundant as they can be obtained from the corresponding configurations with $\theta_0^- < \theta_0^+$ by reflection: $\theta_0^{\pm} \mapsto \theta_0^{\mp}$. Three other configurations with $\theta_0^- = \theta_0^+ = \{\pi, 3\pi/2\}$ and $\theta_0^- = \pi$, $\theta_0^+ = 3\pi/2$ can be obtained from the configurations $\theta_0^- = \theta_0^+ = \{0, \pi/2\}$ and $\theta_0^- = 0$, $\theta_0^+ = \pi/2$, respectively by multiplication of u_n by -1 and rotation of the whole structure by $180°$. One more configuration with $\theta_0^- = 0$, $\theta_0^+ = 3\pi/2$ can be obtained from the configuration with $\theta_0^- = 0$, $\theta_0^+ = \pi/2$ by complex conjugation.

Table 4.2 Same as Table 4.1 but for the vortex diamond configurations

S_{-1}	S_1	Per	$O(\epsilon^2)\mathcal{H}$	$O(\epsilon^4)\mathcal{H}$	$O(\epsilon)i\sigma\mathcal{H}$	$O(\epsilon^2)i\sigma\mathcal{H}$	St
0	0	Y	$\{-12, -6, 2\times 2, 4\}$		$\{\pm 2\times 2, \pm 2\sqrt{2}, \pm 2i\sqrt{3}, \pm 2i\sqrt{6}\}$		N
0	$\frac{\pi}{2}$	N					
0	π	Y	$\{-2\times 2, -5\pm\sqrt{41}\}$	12	$\{\pm 2i\times 2, \pm\sqrt{-10+2\sqrt{41}}, \pm i\sqrt{10+2\sqrt{41}}\}$	$\pm 2\sqrt{6}$	N
$\frac{\pi}{2}$	$\frac{\pi}{2}$	N					
$\frac{\pi}{2}$	π	N					
$\frac{\pi}{2}$	$\frac{3\pi}{2}$	Y	$\{-8, -2\times 3\}$	-12	$\{\pm 2i\times 3, \pm 4i\}$	$\pm 2i\sqrt{6}$	Y

Therefore, we only consider the six irreducible vortex configurations. As can be seen from the table [8], only three configurations persist for $\epsilon \neq 0$, namely the ones with $(\theta_0^-, \theta_0^+) = (0, 0), (0, \pi)$, and $(\pi/2, 3\pi/2)$. Furthermore, among these only one configuration with $\theta_0^- = \pi/2$ and $\theta_0^+ = 3\pi/2$ is spectrally stable. Note that in this case the bifurcation conditions are

$$g_{0,j}^{(1)} = 2\sin(\theta_{0,j} - \theta_{0,j+1}) + 2\sin(\theta_{0,j} - \theta_{0,j-1}) + \sin(\theta_{0,j} - \theta_{0,j+2}) \quad (4.30)$$
$$+ 2\sin(\theta_{0,j} - \theta_{1,0}) + 2\sin(\theta_{0,j} - \theta_{-1,0})$$

for each of the four sites of the plane, while they are

$$g_{\pm 1, j} = \sin(\theta_{\pm 1, j} - \theta_{\mp 1, j}) + 2 \sum_j \sin(\theta_{\pm 1, j} - \theta_{0, j}). \quad (4.31)$$

We now compare the analytical results with numerical computations for these configurations.

The case of $S_{-1} = 0$ and $S_1 = 0$ is shown in Fig. 4.6. The configuration is unstable due to a simple and a double pair of real eigenvalues, both of $O(\epsilon)$, captured very accurately by our theoretical approximation. In addition, a complex quartet emerges because of the collision of a pair of imaginary eigenvalues with the continuous spectrum for $\epsilon > 0.175$. The second diamond configuration with $S_{-1} = 0$ and $S_1 = \pi$ is shown in Fig. 4.7; the latter is found to be unstable due to two simple pairs of real eigenvalues, one at the order $O(\epsilon)$ and one at the order $O(\epsilon^2)$. The simple pair of imaginary eigenvalues becomes a quartet of complex eigenvalues upon collision with the continuous spectrum for $\epsilon > 0.179$. The double pair of imaginary eigenvalues remains double for $0 < \epsilon < 0.2$. Lastly, the third diamond vortex configuration with $S_{-1} = \pi/2$ and $S_1 = 3\pi/2$, shown in Fig. 4.8, is spectrally stable for $0 < \epsilon < 0.2$. We can again see that the theoretical prediction accurately reflects the simple pair of imaginary eigenvalues at the order $O(\epsilon)$, the triple pair of

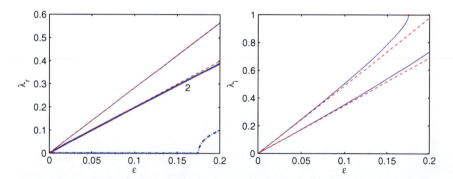

Fig. 4.6 The real and imaginary parts of the pertinent eigenvalues of the diamond configuration with $S_{-1} = 0$ and $S_1 = 0$ as a function of ϵ. Reprinted from [8] with permission

4.2 Discrete Solitons and Vortices

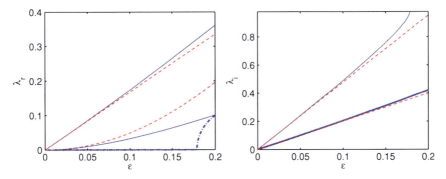

Fig. 4.7 Same as in Fig. 4.6, but for the diamond configuration with $S_{-1} = 0$ and $S_1 = \pi$. Reprinted from [8] with permission

imaginary eigenvalues (splitting into a double pair and a simple one) at the order $O(\epsilon)$, and the simple pair of imaginary eigenvalues at the order $O(\epsilon^2)$.

Finally, we turn to the case of the double vortex cross (see the bottom panel of Fig. 4.2). This is a configuration that consists of two symmetric planes of aligned vortex crosses separated by an empty plane. More precisely, we set $S = S_{-1} \oplus S_1$, where

$$S_l = \{(-1, 0, l), (0, -1, l), (1, 0, l), (0, 1, l)\}, \qquad l = -1, 1, \tag{4.32}$$

such that $N = \dim(S) = 8$. By using the same convention as in (i), the double cross vortex configurations are expressed by

$$\theta_{-1,j} = \frac{\pi(j-1)}{2}, \qquad \theta_{1,j} = \theta_0 + s_0 \frac{\pi(j-1)}{2}, \qquad j = 1, 2, 3, 4, \tag{4.33}$$

where $\theta_0 = \{0, \pi/2, \pi, 3\pi/2\}$ and $s_0 = \{+1, -1\}$. In this case as well, there are only six irreducible representations and among them, three persist, while only one (namely, the one with $\theta_0 = \pi$ and $s_0 = 1$) is spectrally stable. The analysis can

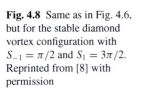

Fig. 4.8 Same as in Fig. 4.6, but for the stable diamond vortex configuration with $S_{-1} = \pi/2$ and $S_1 = 3\pi/2$. Reprinted from [8] with permission

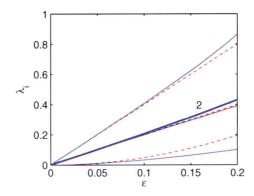

Table 4.3 Same as Table 4.1 but for the double cross vortex configurations

S_1	Per	$\mathcal{O}(\epsilon^2)\mathcal{H}$	$\mathcal{O}(\epsilon^4)\mathcal{H}$	$\mathcal{O}(\epsilon)i\sigma\mathcal{H}$	$\mathcal{O}(\epsilon^2)i\sigma\mathcal{H}$	St
$\{0, \frac{\pi}{2}, \pi, \frac{3\pi}{2}\}$	Y	$\{-2\times 2, 2\times 2\}$	$\{-8, 28\times 2\}$	$\{\pm 2\times 2, \pm 2i\times 2\}$	$\{\pm 4i, \pm 2\sqrt{14}\times 2\}$	N
$\{0, \frac{3\pi}{2}, \pi, \frac{\pi}{2}\}$	Y	$\{-4, -2\times 3, 2\}$	$\{-8, 28\}$	$\{\pm 2, \pm 2i\times 3, \pm 2i\sqrt{2}\}$	$\{\pm 4i, \pm 2\sqrt{14}\}$	N
$\{\frac{\pi}{2}, \pi, \frac{3\pi}{2}, 0\}$	N					
$\{\frac{\pi}{2}, 0, \frac{3\pi}{2}, \pi\}$	N					
$\{\pi, \frac{3\pi}{2}, 0, \frac{\pi}{2}\}$	Y	$\{-4\times 2, -2\times 4\}$	$\{-8\}$	$\{\pm 2i\times 4, \pm 2i\sqrt{2}\times 2\}$	$\{\pm 4i\}$	Y
$\{\frac{3\pi}{2}, 0, \frac{\pi}{2}, \pi\}$	N					

proceed in a similar way as in the cases outlined above and its results are summarized in Table 4.3.

The comparison of these findings with full numerical computations yields the following. The double vortex cross with $S_1 = \{0, \pi/2, \pi, 3\pi/2\}$ is shown in Fig. 4.9. The results are found to be consistent with the eigenvalue approximations. Note that both double pairs of real and imaginary eigenvalues at $O(\epsilon)$ split for small ϵ into simple pairs of real and imaginary eigenvalues, while the double pair of real eigenvalues at the order $O(\epsilon^2)$ remains double for small ϵ. The case of $S_1 = \{0, 3\pi/2, \pi, \pi/2\}$ is shown in Fig. 4.10. All the pairs are numerically found to be simple in this case, including the triple pair of imaginary eigenvalues at the order $O(\epsilon)$ which splits for small ϵ into three simple imaginary pairs. Finally, the case of $S_1 = \{\pi, 3\pi/2, 0, \pi/2\}$ is the only spectrally stable one for $\epsilon < 0.2$ and is shown in Fig. 4.11. The double and quadruple pairs of imaginary eigenvalues at $O(\epsilon)$ split for small ϵ into individual simple pairs of imaginary eigenvalues.

It should also be noted that in addition to the above-mentioned cases of the discrete vortices, one can straightforwardly consider by means of the same methods the

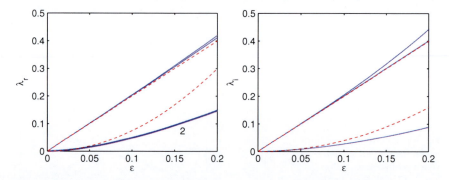

Fig. 4.9 Same as Fig. 4.3, but for the double cross vortex configuration with $S_1 = \{0, \pi/2, \pi, 3\pi/2\}$. Reprinted from [8] with permission

4.2 Discrete Solitons and Vortices

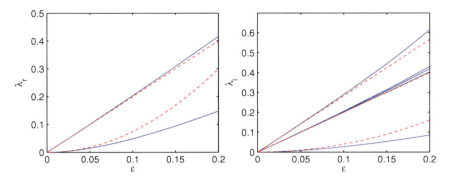

Fig. 4.10 Same as Fig. 4.3, but for the double cross vortex configuration with $S_1 = \{0, 3\pi/2, \pi, \pi/2\}$. Reprinted from [8] with permission

stability of discrete solitons on similar (or other) three-dimensional contours. As a prototypical example, we mention the discrete solitons that can be formed on a cube, among which we consider the potentially stable octupole structure of Fig. 4.12. In this case also the bifurcation equations will be given, as above, by

$$g_{l,j}^{(1)} = \sin(\theta_{l,j+1}-\theta_{l,j})+\sin(\theta_{l,j-1}-\theta_{l,j})+\sin(\theta_{l+1,j}-\theta_{l,j}), \quad l = 0, 1, \quad j = 1, 2, 3, 4,$$

however, now the relevant angles will alternate between 0 and π. The octupole with adjacent sites being out of phase with each other is, in fact, stable. Its leading order eigenvalues can be found to be $\pm 2\sqrt{\epsilon}i$ (a triple eigenvalue pair), $\pm\sqrt{8\epsilon}i$ (another triple eigenvalue pair), and finally $\lambda = \pm\sqrt{12\epsilon}i$, which is a single eigenvalue pair. As before these findings tested against full numerical computations in Fig. 4.12 are found to be in good agreement between theory and numerics. The configuration is stable for small ϵ, but becomes unstable for $\epsilon > 0.051$ (the largest eigenvalue collides with the band edge of the continuous spectrum, producing an eigenvalue quartet). Additional quartets emerge, e.g., for $\epsilon > 0.071$, when the first triplet of eigenvalue pairs collides with the band edge and the configuration becomes

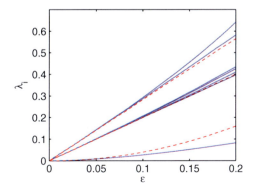

Fig. 4.11 Same as Fig. 4.3, but for the stable double cross vortex configuration with $S_1 = \{\pi, 3\pi/2, 0, \pi/2\}$. Reprinted from [8] with permission

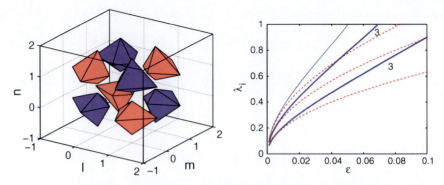

Fig. 4.12 The figure shows the discrete solitary wave (octupole) with an eight-site out-of-phase structure between nearest neighbors (*left panel*). This configuration is a stable one with seven non-zero imaginary eigenvalue pairs (two triple and a single pair) quantified in the text (*right panel*)

increasingly unstable thereafter. Other genuinely 3d configurations can be tackled in a similar fashion.

Having, thus, completed our examination of fundamental structures in the focusing, single component case, we now turn to the defocusing regime and highlight the analogies and the differences between the two.

References

1. Gericke, T., Gerbier, F., Widera, A., Fölling, S., Mandel, O., Bloch, I.: J. Mod. Opt **54**, 735 (2007) (for a recent discussion)
2. Bloch, I.: Nat. Phys. **1**, 23 (2005) (for a relevant review article)
3. Schori, C., Stöferle, T., Moritz, H., Köhl, M., Esslinger, T.: Phys. Rev. Lett. **93**, 240402 (2004)
4. Bruder, C., Fazio, R., Schön, G.: Ann. Phys (Leipzig) **14**, 566 (2005)
5. Heebner, J.E., Boyd, R.W.: J. Mod. Opt. **49**, 2629–2636 (2002)
6. Kevrekidis, P.G., Malomed, B.A., Frantzeskakis, D.J., Carretero-González, R.: Phys. Rev. Lett. **93**, 080403 (2004)
7. Carretero-González, R., Kevrekidis, P.G., Malomed, B.A., Frantzeskakis, D.J.: Phys. Rev. Lett. **94**, 203901 (2005)
8. Lukas, M., Pelinovsky, D.E., Kevrekidis, P.G.: Phys. D **237**, 339 (2008)

Chapter 5
The Defocusing Case

In this chapter, we will turn to the defocusing version of the DNLS equation, setting $\beta = 1$ in Eq. (2.1) of Chap. 2. Part of the reason for considering the latter setting stems from recent experimental results both in the area of optical waveguide arrays and also in the related area of optically induced lattices in photorefractive crystals.

More specifically, the work of [1] considered the anomalous diffraction regime of AlGaAs waveguide arrays, which feature the Kerr-type cubic nonlinearity. In that context, the formation of fundamental dark soliton discrete excitations was observed. Experiments revealing the same type of structures have more recently taken place in defocusing lithium niobate waveguide arrays, which exhibit a different type of nonlinearity, namely a saturable, defocusing one due to the photovoltaic effect [2]. The dark soliton is, arguably, the fundamental solution of the NLS equation of (2.2) with the defocusing nonlinearity. In the stationary case (it can also be boosted similarly to the bright soliton to a finite speed, but here we will be concerned with stationary solutions) it assumes the form of a heteroclinic connection with

$$u(x,t) = \exp(-i\Lambda t)\sqrt{\Lambda}\tanh\left[\sqrt{\frac{\Lambda}{2}}(x-x_0)\right] \tag{5.1}$$

for the cubic nonlinearity case. Naturally, we will seek to identify such a solution in the discrete case in what follows. Note, by the way, that the propagation constant here has the opposite sign in comparison to the focusing case.

However, on the other hand, the discrete case allows the potential for solutions that cannot exist in the continuum limit. These solutions once again stem from the AC limit in the form of localized excitations on a few sites. Perhaps the easiest way to see this is by substituting $u_n = \exp(-i\Lambda t)v_n$ and observing that the resulting stationary state equation

$$(\Lambda - |v_n|^2)v_n = -\epsilon(v_{n+1} + v_{n-1}) \tag{5.2}$$

is identical to that of the focusing equation, upon the so-called staggering transformation $v_n = (-1)^n w_n$. This very simple transformation will allow us on numerous occasions to convert our knowledge of the focusing problem into an understanding of the corresponding defocusing setting. However, note that this transformation does *not* survive the continuum limit. Nevertheless, it indicates that single-site excitations with $v_n = \sqrt{\Lambda}\delta_{nn_0}$, as well as multisite (multipulse) solutions $v_n = \sum_k \pm\sqrt{\Lambda}\delta_{nk}$ can naturally arise in the vicinity of this limit. These bright solitary excitations of the defocusing case are collectively termed *gap* solitons, because the alternating phase structure at their tails, as $\epsilon \neq 0$, indicates that they emanate from states with wave number $k = \pi$, rather than $k = 0$, as regular bright solitons of the focusing case do. Furthermore, the staggering transformation can be applied to the setting of the linear stability analysis, in the defocusing analog of Eqs. (2.61) and (2.62), converting them, as well, to the focusing case. This has a very interesting implication for the gap states. In particular, it is easy to note that if we excite the simplest multipulse state, namely a two-site excitation, if the sites of the excitation are an *odd* number of sites apart, then the staggering transformation will convert such an in-phase state of the defocusing case into an out-of-phase (and hence linearly stable near the AC limit) focusing state, and an out-of-phase gap state into an in-phase focusing state (hence linearly unstable). On the other hand, if they are an *even* number of sites apart, then the staggering transformation preserves the phase structure in the focusing case. This implies that, say, nearest-neighbor in-phase solitons in the defocusing case will be stable, while next-nearest-neighbors will *not*, and that out-of-phase nearest-neighbor solitary waves will be unstable, while next-nearest-neighbor ones will be stable. These predictions will be tested in the two-dimensional setting, in connection with the very recent experimental work of [3].

We start our presentation of this chapter with the fundamental dark solitary wave structures of the one-dimensional case, and then we turn to gap states and vortices in the two-dimensional setting.

5.1 Dark Solitary Waves

The first works revealing the particularities and interesting instabilities of dark solitary waves in the DNLS equation were those of [4] and [5]. Here, we will follow in our presentation the recent work of [6].

5.1.1 Theoretical Analysis

There are two fundamental discrete dark soliton states, namely the on-site and the inter-site dark soliton (similar to the corresponding single-pulse bright structures). The former waveform will be at the AC limit as follows:

$$v_{n\leq -1} = \sqrt{\Lambda}, \tag{5.3}$$
$$v_{n=0} = 0, \tag{5.4}$$
$$v_{n\geq 1} = -\sqrt{\Lambda_1}, \tag{5.5}$$

5.1 Dark Solitary Waves

while the latter type of dark soliton reads

$$v_{n \leq 0} = \sqrt{\Lambda_1}, \tag{5.6}$$

$$v_{n \geq 1} = -\sqrt{\Lambda_1}. \tag{5.7}$$

One can then examine the linear stability of these prototypical configurations, as a starting point for the finite ϵ case. As before, this can be done through the linear stability ansatz

$$u_n = \exp(-i\Lambda t) \left[v_n + \delta \left(\exp(\lambda t) p_n + \exp(\lambda^* t) q_n \right) \right]. \tag{5.8}$$

As before, the linear stability of the configuration will be determined by the nature of the eigenvalues λ of the ensuing matrix eigenvalue problem for λ and its corresponding eigenvector $(p_n, q_n^*)^T$. A configuration will be (neutrally) stable for this Hamiltonian system if $\forall \lambda$, the real part λ_r of the eigenvalue ($\lambda = \lambda_r + i\lambda_i$) is such that $\lambda_r = 0$. The resulting matrix eigenvalue problem reads

$$i\lambda \begin{pmatrix} p_n \\ q_n^* \end{pmatrix} = \begin{pmatrix} 2|v_n|^2 - \Lambda - \epsilon \Delta_2 & v_n^2 \\ -(v_n^2)^* & \Lambda - 2|v_n|^2 + \epsilon \Delta_2 \end{pmatrix} \begin{pmatrix} p_n \\ q_n^* \end{pmatrix}. \tag{5.9}$$

In the case of the AC limit where the sites become uncoupled, the relevant stability matrix for all non-zero sites is identical and assumes the form

$$\Lambda_1 \begin{pmatrix} 1 & 1 \\ -1 & -1 \end{pmatrix}. \tag{5.10}$$

The matrix of Eq. (5.10) yields a pair of zero eigenvalues for each of these non-zero sites. Hence, in an inter-site configuration at the AC limit, the linearization would only result in (infinitely many) zero eigenvalues.

The only difference of an on-site configuration lies in the existence of the central $v_0 = 0$ site. This site produces a 2×2 stability matrix of the form

$$\Lambda_1 \begin{pmatrix} -1 & 0 \\ 0 & 1 \end{pmatrix} \tag{5.11}$$

and, therefore, an eigenvalue pair $\lambda = \pm i\Lambda$.

We now turn to the finite coupling case with $\epsilon \neq 0$. We start by considering the solution profile. The solutions will be deformed from their AC limit profile of Eqs. (5.3), (5.4), (5.5), (5.6), and (5.7). To address this deformation, the solution can be expanded into a power series

$$v_n = v_n^{(0)} + \epsilon v_n^{(1)} + O(\epsilon^2). \tag{5.12}$$

The leading order correction can be straightforwardly computed by using the expansion into the stationary state equation as

$$v_n^{(1)} = \frac{\Delta_2 v_n^{(0)}}{2\Lambda} \qquad (5.13)$$

for all excited sites. For the zeroth site of the on-site configuration, the symmetry of the profile yields a zero correction (to all relevant orders). It is easy to see that the correction of Eq. (5.13) only contributes to leading order to the sites with $n \in \{1, -1\}$ for the on-site and to those with $n \in \{0, 1\}$ for the inter-site configuration. These corrections amount to

$$v_1^{(1)} = \frac{1}{2\sqrt{\Lambda}}, \qquad (5.14)$$

$$v_{-1}^{(1)} = -v_1^{(1)} \qquad (5.15)$$

for the on-site and to

$$v_1^{(1)} = \frac{1}{\sqrt{\Lambda}}, \qquad (5.16)$$

$$v_0^{(1)} = -v_1^{(1)} \qquad (5.17)$$

for the inter-site case.

We subsequently focus on the linear stability problem. The eigenvalues of the latter will be of two types, namely the continuous spectrum that will emerge from the background (and correspond to the finite coupling generalization of the zero eigenvalues of the AC limit) and the point spectrum resulting from the vicinity of the center of the dark soliton configuration.

The continuous spectrum corresponds to plane wave eigenfunctions of the form $\{p_n, q_n\} \sim \exp(ikn)$. These, in turn, result into a matrix eigenvalue problem

$$\begin{pmatrix} s_1 & \Lambda \\ -\Lambda & -s_1 \end{pmatrix}, \qquad (5.18)$$

where $s_1 = \Lambda + 4\epsilon \sin^2(k/2)$. This leads to a continuous eigenvalue spectrum described by the dispersion relation

$$\omega \equiv i\lambda = \pm\sqrt{s_1^2 - \Lambda^2}, \qquad (5.19)$$

which is associated with the eigenvalue band $i\lambda \in [-\sqrt{16\epsilon^2 + 8\epsilon\Lambda}, \sqrt{16\epsilon^2 + 8\epsilon\Lambda}]$.

On the other hand, the point spectrum eigenvalues will result from the central part of the excitation. For the inter-site configuration, we consider a leading order approximation only for the two sites ($n = 0$ and 1) participating in the dark soliton (as they are the only ones modified to leading order in perturbation theory). Using

5.1 Dark Solitary Waves

the perturbative expansion of Eqs. (5.6) and (5.7) for the relevant part of the eigenvalue problem, we obtain the 4 × 4 matrix

$$\begin{pmatrix} \Lambda - 2\epsilon & -\epsilon & \Lambda - 2\epsilon & 0 \\ -\epsilon & \Lambda - 2\epsilon & 0 & \Lambda - 2\epsilon \\ -\Lambda + 2\epsilon & 0 & -\Lambda + 2\epsilon & \epsilon \\ 0 & -\Lambda + 2\epsilon & \epsilon & -\Lambda + 2\epsilon \end{pmatrix}. \tag{5.20}$$

This stability matrix leads to a pair of real eigenvalues

$$\lambda = \pm\sqrt{2\epsilon\Lambda - 5\epsilon^2}. \tag{5.21}$$

As a result, the configuration will be immediately unstable, for $\epsilon \neq 0$. This prediction will be directly compared with the numerical results of the following section.

One can use a similar argument for the on-site configuration by considering the three central sites of the solitary structure and constructing a 6 × 6 matrix whose eigenvalues can, in principle, be computed. However, the resulting expressions are too cumbersome; hence, we study two alternative arguments that describe very accurately the behavior of the relevant point spectrum eigenvalue originally at $\lambda = \pm i\Lambda_1$, associated with the on-site dark solitary wave.

The first approach is a rigorous one and is based on the so-called Gerschgorin's theorem (see, e.g., [7]). Let us consider matrices $A = [a_{lj}]$ of order N and define the radii $r_l = \sum_{j=1, j \neq l}^{N} |a_{lj}|$ and denote the circles in the complex spectral plane $Z_l = \{z \in C : |z - a_{ll}| < r_l\}$. Then, Gerschgorin's theorem states that the eigenvalues of the matrix belong to these circles and, in fact, its refined version states that if m of these circles form a connected set, S, disjoint from the remaining $N-m$ circles, then exactly m eigenvalues are contained in S. The above setting of the on-site solution is an excellent testbed for the application of Gerschgorin's theorem because the sole eigenvalue discussed above is at $\pm i\Lambda$ in the AC limit (of zero radius for the Gerschgorin circles), while all others are located at the origin. Hence, for small $\epsilon \ll \Lambda$, the (single) relevant point spectrum eigenvalue remains in the corresponding Gerschgorin circle which can be easily computed. In fact, considering that only the diagonal and super- and subdiagonal elements are present for a site with $v_0 = 0$, the Gerschgorin estimate for the relevant eigenvalue emerges immediately as

$$|i\lambda \pm (\Lambda - 2\epsilon)| \leq 2\epsilon, \tag{5.22}$$

which, in turn, necessitates that the relevant eigenvalue lies between $\Lambda - 4\epsilon \leq i\lambda \leq \Lambda$. This is a rigorous result based on the above theorem, which provides a linear (in ϵ) bound on the growth of the relevant eigenvalue.

On the other hand, there is also an even more successful approach put forth in [5], which, however, is to a certain degree an approximation. According to the latter, one considers the eigenvalue equations for the eigenvectors p_n and q_n^*. Considering the anti-symmetry of the central site one may use the approximation of $p_{n+1} + p_{n-1} \approx 0$.

Then, the eigenvalue equations (5.9) decouple and provide a relevant estimate for this eigenvalue as

$$i\lambda = \pm(\Lambda_1 - 2\epsilon). \tag{5.23}$$

This prediction will also be compared with our numerical results in the following section. It is also worth noting here that this eigenvalue, moving toward the spectral plane origin, will eventually collide with the (growing) band edge of the continuous spectrum when the predictions of Eq. (5.23) and of the band edge of the continuous spectrum coincide, which occurs for

$$\epsilon_{cr} = \frac{2\sqrt{3}-3}{6}\Lambda_1 \equiv 0.07735\Lambda_1. \tag{5.24}$$

This collision, per the opposite Krein signature [8, 9] of the relevant eigenvalues, will lead to a Hamiltonian–Hopf bifurcation and a quartet of eigenvalues, as was originally observed in [5] (see also the numerical results below).

We now examine an alternative method to tackle the linear stability analysis of these structures on the basis of a formal perturbative expansion, as proposed in [6]. In particular, we introduce the following linearization ansatz:

$$u_n = v_n + \delta C_n.$$

Substituting this into the equation of motion yields to $\mathcal{O}(\delta)$

$$i\dot{C}_n = -\epsilon \Delta_2 C_n + \beta_1 \left(2|v_n|^2 C_n + v_n^2 C_n^\star\right) - \Lambda_1 C_n. \tag{5.25}$$

Decomposing $C_n(t) = \eta_n + i\xi_n$ and assuming that v_n is real, Eq. (5.25) gives

$$\begin{pmatrix}\dot{\eta}_n \\ \dot{\xi}_n\end{pmatrix} = \begin{pmatrix} 0 & \mathcal{L}_-(\epsilon) \\ -\mathcal{L}_+(\epsilon) & 0 \end{pmatrix}\begin{pmatrix}\eta_n \\ \xi_n\end{pmatrix} = \mathcal{H}\begin{pmatrix}\eta_n \\ \xi_n\end{pmatrix}, \tag{5.26}$$

where the operators $\mathcal{L}_-(\epsilon)$ and $\mathcal{L}_+(\epsilon)$ are defined as $\mathcal{L}_-(\epsilon) \equiv -\epsilon\Delta_2 + v_n^2 - \Lambda$ and $\mathcal{L}_+(\epsilon) \equiv -\epsilon\Delta_2 + 3v_n^2 - \Lambda$. The stability of v_n is then determined by the eigenvalues λ of $\mathcal{H}$.

Since (5.26) is linear, we can eliminate one of the eigenvectors, for instance ξ_n, from which we obtain the following eigenvalue problem:

$$\mathcal{L}_-(\epsilon)\mathcal{L}_+(\epsilon)\eta_n = -\lambda^2 \eta_n = \Xi\eta_n. \tag{5.27}$$

As before, we expand the eigenvector η_n and the eigenvalue Ξ as

$$\eta_n = \eta_n^{(0)} + \epsilon\eta_n^{(1)} + \mathcal{O}(\epsilon^2), \quad \Xi = \Xi^{(0)} + \epsilon\Xi^{(1)} + \mathcal{O}(\epsilon^2).$$

5.1 Dark Solitary Waves

Substituting into Eq. (5.27) and identifying coefficients for consecutive powers of ϵ yields

$$[\mathcal{L}_-(0)\mathcal{L}_+(0) - \Xi^{(0)}]\eta_n^{(0)} = 0, \qquad (5.28)$$

$$[\mathcal{L}_-(0)\mathcal{L}_+(0) - \Xi^{(0)}]\eta_n^{(1)} = f \qquad (5.29)$$

with

$$f = \left[(\Delta_2 - 2v_n^{(0)}v_n^{(1)})\mathcal{L}_+(0) + \mathcal{L}_-(0)(\Delta_2 - 6v_n^{(0)}v_n^{(1)}) + \Xi^{(1)}\right]\eta_n^{(0)}. \qquad (5.30)$$

First, let us consider the order $\mathcal{O}(1)$ equation (5.28). One can do a simple analysis as above to show that there is one eigenvalue, i.e., $\Xi^{(0)} = 0$ for the inter-site configuration and two eigenvalues $\Xi^{(0)} = 0$ and $\Xi^{(0)} = \Lambda^2$ for the on-site one. The zero eigenvalue has infinite multiplicity and is related to the continuous spectrum, as discussed previously.

For the inter-site configuration, there is an eigenvalue bifurcating from the continuous spectrum as soon as the coupling is turned on. Therefore, this zero eigenvalue is the crucial eigenvalue for its stability. The normalized eigenvector of this eigenvalue is $\eta_n^{(0)} = 1/\sqrt{2}$, for $n = 0, 1$ and $\eta_n^{(0)} = 0$ otherwise. For the on-site configuration, the crucial eigenvalue for the stability is $\Xi^{(0)} = \Lambda_1^2$ with the normalized eigenvector $\eta_n^{(0)} = 1$, for $n = 0$ and $\eta_n^{(0)} = 0$ otherwise.

The dependence of the relevant eigenvalue on ϵ can be calculated from Eq. (5.29). Due to the fact that the corresponding eigenvector is zero almost everywhere, we only need to consider the site with non-zero component eigenvector, i.e., $n = 0, 1$ for the inter-site and $n = 0$ for the on-site.

It is simple to show that the solvability condition of Eq. (5.29) using, e.g., the Fredholm alternative requires $f = 0$ from which one immediately obtains that $\Xi^{(1)} = -2\Lambda$ for the inter-site and $\Xi^{(1)} = -4\Lambda$ for the inter-site.

Hence, the critical eigenvalue is

$$\lambda = \pm\sqrt{2\Lambda\epsilon} + \mathcal{O}(\epsilon^2) \qquad (5.31)$$

for the inter-site configuration and

$$\lambda = \pm i\sqrt{\Lambda^2 - 4\Lambda_1\epsilon} + \mathcal{O}(\epsilon^2) \qquad (5.32)$$

for the on-site one.

These two results agree with the leading order findings of our earlier analysis. In particular, Eq. (5.31) can be immediately seen to agree with the leading order prediction of Eq. (5.21). On the other hand, as regards Eq. (5.32), a Taylor expansion to leading order yields $\lambda = \pm i(\Lambda - 2\epsilon)$, again in agreement with the findings of Eq. (5.23).

It is important to highlight here that at the qualitative level the staggering transformation allows us to infer the stability of either on-site dark soliton or inter-site dark soliton states, or even any kind of multiple hole (or multiple "domain walls" between the uniform stationary states ± 1) such as the ones considered in [10]. In particular for all the sites in the tail of the structure, either as $v_n \to 1$ or as $v_n \to -1$ or $\pm\infty$, the staggering transformation results into out of phase excitations in the focusing regime, which per the analysis of Chap. 2 are associated with small imaginary eigenvalues (in this defocusing case with positive Krein signature). Then, the key question is what happens at locations where there is a jump from ± 1 to ∓ 1 (either through a number of zeros or directly). What the staggering transformation illustrates (and which can be proved based on the Sturm theory arguments that we gave for the focusing case in Chap. 2 is the following:

1. *In-phase* excitation separated by an *odd* number of sites gives rise to purely imaginary eigenvalues.
2. *In-phase* excitation separated by an *even* number of sites gives rise to a real eigenvalue (and, hence, instability).
3. *Out-of-phase* excitation separated by an *odd* number of sites gives rise to a real eigenvalue (and, hence, instability).
4. *Out-of-phase* excitation separated by an *even* number of sites gives rise to purely imaginary eigenvalues.

On the basis of the above principle, we can quantify the qualitative features of the stability of not only any staggered state (as mentioned above), but also even of any dark soliton state (it can be easily checked, for instance, that the above corollary of our considerations in Chap. 2 and the staggering transformation, can immediately explain the linear stability of the on-site dark soliton structure, while it can also justify the instability of the corresponding inter-site structure).

A final comment should be made in passing here, regarding the very recent work of [11]. Based on the recent interest in the excitation of more complex, potentially multihole states [10], the work of [11] provided a systematic framework to address the stability of such configurations. An example of this type is shown in the next section. More specifically, the eigenvalue problem for the operator L_- can be solved (see the discussion in Chap. 2) in the form

$$V_n \psi_n - \epsilon (\psi_{n+1} + \psi_{n-1} - 2\psi_n) = \gamma \psi_n, \qquad V_n = V_n^{(0)} + \sum_{k=1}^{\infty} \epsilon^k V_n^{(k)}, \qquad (5.33)$$

where $V_n^{(0)} = (u_n^{(0)})^2 - 1$, $V_n^{(1)} = 2u_n^{(0)} u_n^{(1)}$, $V_n^{(2)} = 2u_n^{(0)} u_n^{(2)} + (u_n^{(1)})^2$ and so on, due to analytic dependence of the solution u_n of the nonlinear stationary problem on ϵ. For example, in the case of the inter-site-centered mode, the profile u_n, can be expanded in a series in ϵ, with the leading order correction $u_n^{(1)}$ satisfying $u_0^{(1)} = 1$, $u_1^{(1)} = -1$, and $u_n^{(1)} = 0$ for all $n \in \mathbb{Z} \setminus \{0, 1\}$. The potential $V = \epsilon V^{(1)} + O(\epsilon^2)$ of the discrete Schrödinger equation (5.33) is negative at the leading order and it traps a unique negative eigenvalue with the symmetric eigenfunction $\psi_n = \psi_{-n+1}$, $n \in \mathbb{N}$. Using the parametrization

$$\gamma = \epsilon \left(2 - e^\kappa - e^{-\kappa} \right) \qquad (5.34)$$

5.1 Dark Solitary Waves

and solving the eigenvalue problem for the eigenvector $\psi_1 = 1$, $\psi_n = Ce^{-\kappa(n-2)}$ for $n \geq 2$, we obtain $C = e^{-\kappa}$ and $e^\kappa = 3$ at the leading order of $O(\epsilon)$, which gives $\gamma = -(4/3)\epsilon + O(\epsilon^2)$. Therefore, we conclude that the pair of real eigenvalues of the stability problem is given by $\lambda = \pm\sqrt{2\gamma} = \pm\sqrt{8\epsilon/3}(1 + O(\epsilon))$. As we will see in the next section this approach yields a more accurate prediction for the inter-site dark soliton eigenvalue, than the above presented methods. Furthermore, as shown in [11], this approach can be systematically used to give predictions for multihole states in very good agreement with the numerical results.

We now turn to the numerical examination of the above results for the case of the dark solitons.

5.1.2 Numerical Results

As usual, we consider the numerical solutions of interest starting from the AC limit of $\epsilon = 0$ (fixing $\Lambda = 1$). Figure 5.1 shows an on-site dark solitary wave (top left) and its linear stability (bottom left), as well as an inter-site such solution (top right) and its linear stability (bottom right), confirming the qualitative theoretical predictions given above.

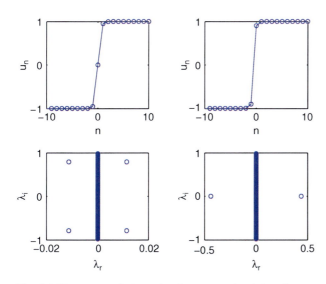

Fig. 5.1 The *top panel* shows the discrete on-site dark solitary wave (*left*) and inter-site dark solitary wave (*right*) for $\epsilon = 0.1$. The *bottom panels* show the corresponding linear stability eigenvalues, illustrating the weak oscillatory instability of the former (for $\epsilon > 0.076$) and strong exponential instability of the latter (due to a real eigenvalue pair). There is only a single pair of eigenvalues at the origin, due to the phase invariance; in the continuum case there would be two such pairs, an additional one arising from the unstable eigenmodes as translational invariance is restored in that limit. Note also that the continuous spectrum has an upper bound. This upper bound disappears in the continuum limit, where the continuous spectrum consists of the entire imaginary axis

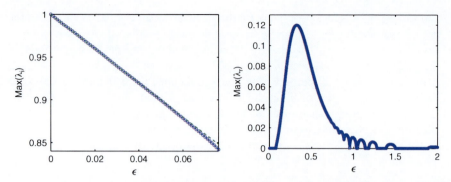

Fig. 5.2 The figure shows stability results for the on-site dark soliton from [6]. The *left panels* show the numerical dependence (*circles*) of the imaginary eigenvalue on ϵ, starting at $i\Lambda$ (the maximal imaginary eigenvalue) until the point of collision ϵ_{cr} with the upper band edge of the continuous spectrum. The *dashed line* shows the theoretical predictions of Eq. (5.23). The *right panel* shows the real part of the relevant eigenvalue, which is zero before – and, typically, non-zero after – the relevant collision due to the ensuing Hamiltonian–Hopf bifurcation leading to the emergence of an eigenvalue quartet

We now turn to a quantitative comparison of the stability results for the on-site configurations in Fig. 5.2 and the inter-site configurations in Fig. 5.3. In the case of the on-site configurations, the relevant imaginary eigenvalue (initially at $i\Lambda$, for $\epsilon = 0$) can be observed to move linearly (in a decreasing way) along the imaginary axis as ϵ is increased, in accordance with the prediction by both the Gerschgorin estimate, and also remarkably accurately (see the circles joined by the solid line in Fig. 5.2) by the anti-symmetric approximation of (5.23). Eventually, this eigenvalue collides with the band edge of the continuous spectrum at $\epsilon \approx 0.077$; this numerical result also agrees very well with the theoretical predictions of 0.07735 from Eq. (5.24), as was originally observed in [5] (see also [6] which tackles the case of a saturable nonlinearity, relevant to the experimental work of [2] in a similar

Fig. 5.3 The figure shows by means of a *solid line* the result of numerical linear stability analysis for the dependence on ϵ of the real eigenvalue of the inter-site dark discrete soliton. The thin *dashed line* shows the prediction $\lambda = \sqrt{8\epsilon/3}$ of [11], while the thick *dashed line* shows the curve $\lambda = \sqrt{2\epsilon}$ of [6]. Reprinted with permission from [11]

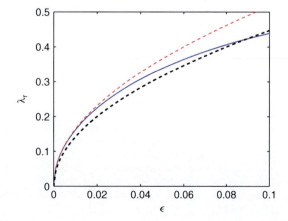

5.1 Dark Solitary Waves

way). Upon collision, a complex quartet of eigenvalues arises in the spectral plane. This eigenvalue approaches the spectral plane origin of $\lambda = 0$, as the continuum limit of $\epsilon \to \infty$ is approached. One of the important points highlighted in connection with this eigenvalue in [5] was the nature of its dependence on the finite domain size of the computations. The results presented herein are for $N = 250$ lattice sites and seem to illustrate the presence of restabilization windows where the eigenvalue "sneaks into" the imaginary axis as ϵ grows. However, this feature is a direct by-product of the finite computational size of the lattice, which results in a quantization of the relevant wave numbers k and, as a result, to the presence of gaps in the continuous spectral band of Eq. (5.19). While such features, highlighted, e.g., in Fig. 2 of [5] through computations for different domain sizes, would disappear in the infinite lattice size limit, the reader should be cautioned that they may be relevant to experimental situations such as the one of [2] where propagation over 250 channels was reported.

The relevant stability results for the inter-site dark solitary wave case are shown in Fig. 5.3. The theoretical prediction of Eq. (5.21) is also illustrated by a thick line and provides a fair approximation of the relevant real eigenvalue, especially for small ϵ (i.e., for $\epsilon < 0.2$). On the other hand, the more recent prediction of $\lambda = \sqrt{8\epsilon/3}$ of [11] appears to be more accurate for very small values of ϵ. It should be noted here that the inter-site configuration, remains *always* unstable up to the continuum limit; the corresponding eigenvalue asymptotically approaches the spectral plane origin as $\epsilon \to \infty$. Note that contrary to the complex quartet discussed above this real eigenvalue is not significantly affected by the domain size.

Lastly, a more complex multihole configuration is illustrated in Fig. 5.4. The three zero crossings, in the absence of an intermediate site of vanishing amplitude, lead to three pairs of real eigenvalues in this case. These can be accurately evaluated

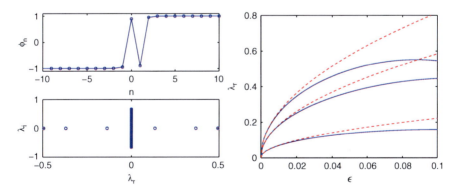

Fig. 5.4 A typical example of a more complex, multihole configuration which can be thought of as a bound state of multiple dark solitons. The *top left panel* shows a typical solution profile, while the *bottom left panel* shows the corresponding stability indicating the presence of three real eigenvalue pairs; the dependence of the latter on the coupling strength ϵ is shown on the *right*; the *solid line* indicates the numerical result, while the *dashed line* is the theoretical prediction of [11]. Reprinted with permission from [11]

for small values of ϵ via the techniques developed in [11], as can be observed from the comparison with the corresponding numerical results in Fig. 5.4.

5.2 Vortex States on a Non-Zero Background

In two-dimensional settings, the prototypical nonlinear wave solution of the continuum NLS equation is the vortex [12]. A natural question then is whether vortex states also persist for the DNLS model, a question that we now briefly address following the exposition of the recent work of [13].

To compute vortex solutions in the discrete setting, we implement a Newton method and a continuation with respect to the coupling parameter ϵ. The path-following can be initiated either near the continuum limit (for ϵ large) or at the AC limit $\epsilon = 0$, as in both cases one is able to construct a suitable initial guess for the Newton method.

For relatively high ϵ, a suitable initial condition for a vortex with topological charge S is obtained through a Padé approximation developed for the continuum limit in [14]. We set $\phi_{n,m} = \rho_{n,m} e^{iS\alpha_{n,m}}$, where

$$\rho_{n,m} = \sqrt{\frac{r_{n,m}^{2S}(a_1 + a_2 r_{n,m}^2)}{1 + b_1 r_{n,m}^2 + a_2 r_{n,m}^{2S+2}}}, \quad r_{n,m} = \sqrt{n^2 + m^2} \qquad (5.35)$$

($a_1 = 11/32$, $a_2 = a_1/12$, $b_1 = 1/3$, see [14]),

$$\alpha_{n,m} = \begin{cases} \arctan(m/n) + \dfrac{3\pi}{2} & \text{for } n \geq 1, \\ \arctan(m/n) + \dfrac{\pi}{2} & \text{for } n \leq -1, \\ \dfrac{\pi}{2}(1 - \text{sign}(m)) & \text{for } n = 0. \end{cases}$$

Once a vortex is found for a given ϵ, the solution can be continued by increasing or decreasing ϵ. Although this method was found to be efficient, it remains limited to single vortex solutions having explicit continuum approximations. Moreover, when the Newton method is applied to continue these solutions near $\epsilon = 0$, the Jacobian matrix becomes ill-conditioned (and non-invertible for $\epsilon = 0$) and the iteration no longer converges.

In [13] a different method was introduced with a wider applicability and for which the above-mentioned singularity is removed. Considering a finite $N \times N$ lattice with $(n, m) \in \Gamma = \{-M, \ldots, M\}^2$ ($N = 2M + 1$), equipped with fixed-end boundary conditions given below, we set $u_{n,m} = R_{n,m} e^{i\theta_{n,m}}$ and note $R = (R_{n,m})_{n,m}$, $\theta = (\theta_{n,m})_{n,m}$. One obtains the equivalent problem

$$R_{n,m}(1 - R_{n,m}^2) + \epsilon f(R, \theta)_{n,m} = 0, \qquad (5.36)$$

$$\epsilon g(R, \theta)_{n,m} = 0, \qquad (5.37)$$

5.2 Vortex States on a Non-Zero Background

where $f(R, \theta) = \text{Re}\,[\,e^{-i\theta}\,\Delta(R\,e^{i\theta})\,]$ and $g(R, \theta) = \text{Im}\,[\,e^{-i\theta}\,\Delta(R\,e^{i\theta})\,]$ can be rewritten

$$f(R, \theta)_{n,m} = R_{n+1,m} \cos(\theta_{n+1,m} - \theta_{n,m}) + R_{n-1,m} \cos(\theta_{n,m} - \theta_{n-1,m}) - 4 R_{n,m} + R_{n,m+1} \cos(\theta_{n,m+1} - \theta_{n,m}) + R_{n,m-1} \cos(\theta_{n,m} - \theta_{n,m-1}),$$

$$g(R, \theta)_{n,m} = R_{n+1,m} \sin(\theta_{n+1,m} - \theta_{n,m}) - R_{n-1,m} \sin(\theta_{n,m} - \theta_{n-1,m}) + R_{n,m+1} \sin(\theta_{n,m+1} - \theta_{n,m}) - R_{n,m-1} \sin(\theta_{n,m} - \theta_{n,m-1}).$$

Now we divide equation (5.37) by ϵ (this eliminates the above-mentioned degeneracy at $\epsilon = 0$) and consider Eq. (5.36) coupled to

$$g(R, \theta)_{n,m} = 0. \tag{5.38}$$

System (5.36), (5.38) is supplemented by the boundary conditions

$$R_{n,m} = 1 \quad \text{for } \text{Max}(|n|, |m|) = M, \tag{5.39}$$

$$\theta_{n,m} = \theta_{n,m}^{\infty} \quad \text{for } \text{Max}(|n|, |m|) = M. \tag{5.40}$$

The prescribed value $\theta_{n,m}^{\infty}$ of the angles on the boundary will depend on the type of vortex solution we look for, more precisely on the vortex distribution and their topological charge S. In particular, we use the boundary conditions $\theta_{n,m}^{\infty} = S\alpha_{n,m}$ for a single vortex with topological charge S centered at $(n, m) = (0, 0)$.

In this way, vortices of topological charge $S = 1, 2$ (but also higher) can be constructed [13]. Typical examples for $S = 1$ and 2 are shown in Fig. 5.5.

Figures 5.6 and 5.7 show, respectively, the real and imaginary parts of eigenfrequencies (or equivalently the imaginary and real parts of the eigenvalues) pertaining to the discrete vortices, thus detailing their linear stability properties. The vortices with $S = 1$ and 2 are, respectively, stable for $C < C_{cr} \approx 0.0395$ and $C < C_{cr} \approx 0.0425$. This instability, highlighted in the case of the $S = 1$ vortex in Fig. 5.6 can be rationalized by analogy with the corresponding stability calculations in the case of dark solitons [5]. In particular, the relevant linearization problem can be written in the form

$$i\lambda \begin{pmatrix} p_{n,m} \\ q_{n,m}^{\star} \end{pmatrix} = \begin{pmatrix} 2|\phi_{n,m}|^2 - 1 - \epsilon\Delta & \phi_{n,m}^2 \\ -(\phi_{n,m}^2)^{\star} & 1 - 2|\phi_{n,m}|^2 + \epsilon\Delta \end{pmatrix} \begin{pmatrix} p_{n,m} \\ q_{n,m}^{\star} \end{pmatrix}. \tag{5.41}$$

However, by analogy to the corresponding one-dimensional problem, the symmetry of the configuration renders it a good approximation to write for the relevant perturbations that $\Delta p_{n,m} \approx -4 p_{n,m}$ (and similarly for q), by virtue of which it can be extracted that the relevant eigenfrequency is $i\lambda \approx \pm(1 - 4\epsilon)$. On the other hand, by analogy to the one-dimensional calculation, it is straightforward to compute the dispersion relation characterizing the eigenfrequencies of the continuous

130   5 The Defocusing Case

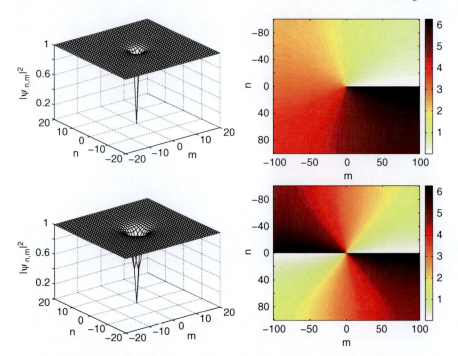

Fig. 5.5 Vortex soliton with $S = 1$ and $\epsilon = 0.2$. *Left panel*: density Profile; *right panel*: angular dependence. The *bottom panels* show the same features but for an $S = 2$ vortex

spectrum (using $\{p_{n,m}, q_{n,m}\} \propto \exp(i(k_n n + k_m m)))$ as extending through the interval $\lambda \in i[-\sqrt{64\epsilon^2 + 16\epsilon}, \sqrt{64\epsilon^2 + 16\epsilon}]$. Therefore, the collision of the point spectrum eigenvalue with the band edge of the continuous spectrum yields a prediction for the critical point of $\epsilon_{cr} \approx (2\sqrt{3} - 3)/12 \approx 0.0387$ in good agreement with the corresponding numerical result above. At $\epsilon = \epsilon_{cr}$ the system experiences

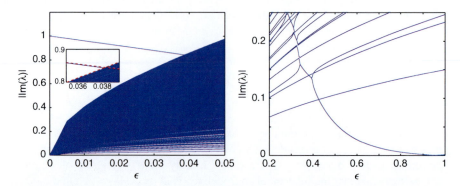

Fig. 5.6 Imaginary part of the stability eigenvalues for $S = 1$. The panels show zooms of two different regions

5.3 Gap States

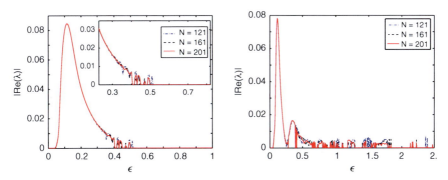

Fig. 5.7 Real part of the stability eigenvalues for $S = 1$ (*left panel*) and $S = 2$ (*right panel*), as a function of the coupling strength ε. This corresponds to the growth rate of the corresponding instability. The different lines correspond to computations with an increasing number of nodes

a Hopf bifurcation. In consequence, for larger ϵ, there exists an eigenvalue quartet $\{\lambda, \lambda^*, -\lambda, -\lambda^*\}$. As ϵ increases, a cascade of Hopf bifurcations takes place due to the interaction of a localized mode with extended modes, as it was observed in one-dimensional dark solitons [5]. This cascade implies the existence of stability windows between inverse Hopf bifurcations and direct Hopf bifurcations. For $S = 1$ vortices, each one of the bifurcations takes place for decreasing $|\text{Re}(\lambda)|$ when ϵ grows, and, in consequence, the bifurcations cease at a given value of ϵ, as $\text{Re}(\lambda)$ of the localized mode is smaller than that of the lowest extended mode frequency (however, in the infinite domain limit, this eventual restabilization would not take place but for the limit of $\epsilon \to \infty$). This fact is illustrated in Fig. 5.6. When the lattice size tends to infinity ($N \to \infty$), the linear modes band extends for zero to infinity and becomes dense; thus, these stabilization windows should be expected to disappear at this limit. To illustrate this point, we have considered lattices of up to 201×201 sites for the $S = 1$ and 2 vortices and have shown the growth rate of the corresponding instabilities in Fig. 5.7. With the increase of lattice size N, we can observe that the size of the windows decreases. The maximum growth rate (i.e., the largest imaginary part of the stability eigenfrequencies) takes place at $C \approx 0.23$ for $S = 1$ and 2 with $\text{Re}(\lambda) \approx 0.0845$ (0.0782) for $S = 1$ ($S = 2$).

5.3 Gap States

As we indicated in the previous sections, gap states of the defocusing model can be addressed rather naturally in the context of the staggering transformation, in terms of their qualitative properties of stability (in connection to the corresponding focusing states studied in Chaps. 2, 3, and 4). Nevertheless, for reasons of completeness and in order to illustrate and discuss the relevant patterns and their stability ranges, we examine the case of such states in the two-dimensional setting following the exposition of [15].

Firstly, it should be noted that the Lyapunov–Schmidt methodology developed in Chaps. 2, 3, and 4 is still applicable in the defocusing case; the only thing that changes is that the bifurcation conditions should be defined near the AC limit as

$$-2ig_\mathbf{n}(\theta, \epsilon) \equiv -\epsilon e^{-i\theta_\mathbf{n}}\Delta_2\phi_\mathbf{n} + \epsilon e^{i\theta_\mathbf{n}}\Delta_2\bar{\phi}_\mathbf{n} = 0, \qquad (5.42)$$

i.e., with an additional $-$ sign in comparison to their focusing counterpart. Then, once again the corresponding Jacobian should be computed (for waveforms satisfying these conditions) as $\mathcal{M}_{ij} = \partial g_i/\partial \theta_j$ and its eigenvalues γ are related to the eigenvalues of the full problem, to leading order, according to $\lambda = \pm\sqrt{2\gamma}$. We now consider various configurations and study their detailed stability.

5.3.1 General Terminology

We start by giving some general terminology for the configurations to be examined. An IP (in-phase) designation will be used for two sites with 0 relative phase difference, while OOP will be used for out-of-phase configurations with π phase difference. Also, OS (on-site) will mean that the center of the configuration is on an empty lattice site (between the excited ones), while IS (inter-site) will signify that the center is located between the excited lattice sites (and no empty site exists between them). For all modes, in the figures below, we show their power $P = \sum |u_\mathbf{n}|^2$ as a function of the coupling ϵ, as well as the real and imaginary parts of the key eigenvalues (the ones determining the stability of the configuration). We consider, more specifically, dipole (two-site) configurations, quadrupole (four-site) configurations, and complex-valued vortex waveforms. In all the cases, we show typical examples of the mode profiles and stability for select values of the coupling. Note that in this case the continuous spectrum band extends through the interval $\lambda_i \in \pm[\Lambda - 8\epsilon, \Lambda]$. This latter trait affects directly the stability intervals of the structures in comparison with their focusing counterparts as we will see also below (since configurations may be stable for small ϵ, but not for larger values thereof).

In each pair of the figures that follow, we show two types of configurations (one in the left column and one in the right column). The first figure of each pair will have five panels showing P as a function of ϵ, the principal real eigenvalues (second panel), and imaginary eigenvalues (third panel). In these plots, the numerical results are shown by the solid line, while the analytical results by the dashed line. The fourth and fifth panels show typical examples of the relevant configuration and its linear stability eigenvalues (shown through the spectral plane (λ_r, λ_i) for the eigenvalues $\lambda = \lambda_r + i\lambda_i$).

An overview of the results encompassing the main findings reported below is summarized in Table 5.1. The table summarizes the configurations considered, their linear stability and the outcome of their dynamical evolution for appropriate initial conditions in the instability regime. Note that if the solutions are unstable for all ϵ, they are denoted as such, while if they are partially stable for a range of coupling

5.3 Gap States

Table 5.1 Summary of the stability results for all the configurations presented below. For partially stable (near the anti-continuum limit) solutions their interval of stability (for $\Lambda = 1$) is given

Type	On-site stability	Inter-site stability
In-phase dipole	Unstable	$\epsilon < 0.064$
Out-of-phase dipole	$\epsilon < 0.092$	Unstable
In-phase quadrupole	Unstable	$\epsilon < 0.047$
Out-of-phase quadrupole	$\epsilon < 0.08$	Unstable
Vortex	$\epsilon < 0.095$	$\epsilon < 0.095$

strengths, their interval of stability is explicitly mentioned. Details of our analytical results and their connection/comparison with the numerical findings are discussed in the rest of this section. It is also important to mention in passing that, based on the predictions discussed below, very recent experiments in [3] and [16] have illustrated the experimental realizability of gap multipole and gap vortex structures, respectively, in photorefractive media with the saturable nonlinearity.

5.3.2 Dipole Configurations

5.3.2.1 Inter-Site, In-Phase Mode

Figure 5.8 summarizes our results for the two types of IP dipole solutions (i.e., initialized at the AC limit with two IP excited sites). The IS–IP mode of the left panels is theoretically found to possess 1 imaginary eigenvalue pair (and, hence, is stable for small C)

$$\lambda \approx \pm 2\sqrt{\epsilon} i. \tag{5.43}$$

The collision with the band edge of the continuous spectrum described above causes the mode to become unstable for sufficiently large coupling strength; the theoretically predicted instability threshold (obtained by equating the eigenvalue of Eq. (5.43) with the lower edge of the phonon band located at $\Lambda - 8\epsilon$) is $\epsilon = 0.0625$, in close agreement with the numerically found one is $\epsilon \approx 0.064$. Additional instability may ensue when the monotonicity of the P versus ϵ curve changes (we have found this to be a general feature of the defocusing branches). The fourth and fifth panels show the mode and its spectral plane for $\epsilon = 0.08$ and 0.116.

5.3.2.2 On-Site, In-Phase Mode

The OS–IP mode of the right panels of Fig. 5.8 is always unstable due to a real pair, theoretically found to be

$$\lambda \approx \pm 2\epsilon \tag{5.44}$$

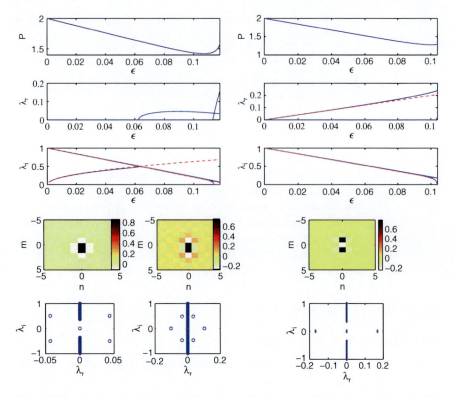

Fig. 5.8 The *first line* of panels shows the power P versus coupling ϵ for the inter-site (IS), in-phase (IP) mode (*left*) and on-site (OS), IP mode (*right*). The *second lines* show their maximal real eigenvalues and the third their first few imaginary eigenvalues. The *solid lines* illustrate the numerical results, while the *dashed lines* the analytical ones. The *fourth* and *fifth panels* show the contour plot of the mode profile (*fourth row*) and the corresponding spectral plane of eigenvalues $\lambda = \lambda_r + i\lambda_i$ (*fifth panel*); The *left two panels* are for the IS-IP mode for $\epsilon = 0.08$ and 0.116, respectively. The *right panel* shows the OS-IP mode for $\epsilon = 0.08$. Reprinted from [15]

for small ϵ. Note once again the remarkable accuracy of this theoretical prediction, in comparison with the numerically obtained eigenvalue. The fourth and fifth right panels of Fig. 5.8 show the mode and its stability for $\epsilon = 0.08$.

5.3.2.3 Inter-Site, Out-of-phase Mode

Figure 5.9 illustrates the two OOP dipole modes. As before, the left panel corresponds to the IS–OOP mode; this one is also immediately unstable (as one departs from the anti-continuum limit), due to a real pair which is

$$\lambda \approx 2\sqrt{\epsilon} \tag{5.45}$$

5.3 Gap States

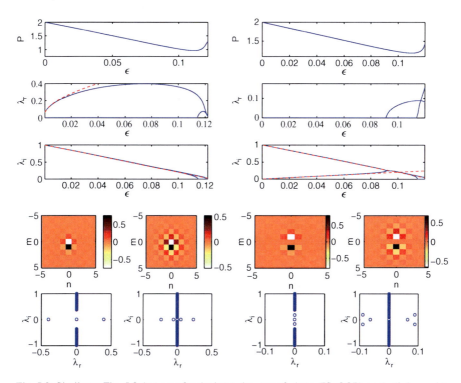

Fig. 5.9 Similar to Fig. 5.8, but now for the inter-site, out-of-phase (IS–OOP) mode (*left panels*) and for the on-site, out-of-phase mode (OS–OOP). The *fourth* and *fifth rows* of panels are for $\epsilon = 0.08$ and for $\epsilon = 0.116$ in both cases. Reprinted from [15]

for small ϵ. The fourth and fifth panels of Fig. 5.9 show the relevant mode for $\epsilon = 0.08$ and 0.116, showing its 1 and 2 unstable real eigenvalue pairs, respectively.

5.3.2.4 On-Site, Out-of-phase Mode

The right panels of the Fig. 5.9 show the OS–OOP mode. The stability analysis of this waveform shows that it possesses an imaginary eigenvalue

$$\lambda \approx 2\epsilon i. \tag{5.46}$$

This leads to an instability upon collision (theoretically, this occurs for $\epsilon = 0.1$, numerically it arises for $\epsilon \approx 0.092$) with the lower edge (located at $\Lambda - 8\epsilon$) of the continuous band of phonon modes. The mode is shown for $\epsilon = 0.08$ and 0.116 in the right panels of Fig. 5.9.

5.3.3 Quadrupole Configurations

5.3.3.1 Inter-Site, In-Phase Mode

Figure 5.10 shows the quadrupolar mode with four IP participating sites in the case where it is centered between lattice sites (left panels). The structure is theoretically predicted to have two imaginary (for small ϵ) eigenvalue pairs with

$$\lambda \approx 2\sqrt{\epsilon}i \qquad (5.47)$$

and one imaginary pair with

$$\lambda \approx \sqrt{8\epsilon}i. \qquad (5.48)$$

As a result, this mode (shown in the fourth row panels of Fig. 5.10 for $\epsilon = 0.05$ and 0.1) becomes unstable due to the collision of the above eigenvalues with the

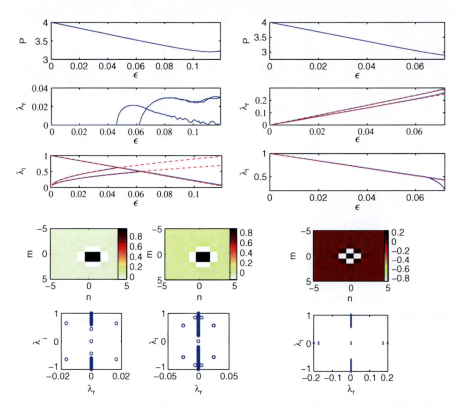

Fig. 5.10 The *first three rows* show the same features as the corresponding ones of figure 5.8 but now for the quadrupole IS–IP mode (*left*) and the quadrupole OS–IP mode (*right*). The contour plot of the real part of the modes and the spectral plane of their linearization eigenvalues are shown in the *fourth* and *fifth rows* for $\epsilon = 0.05$ and 0.1 in the case of the former mode, while the latter is only shown for $\epsilon = 0.05$. Reprinted from [15]

5.3 Gap States

band edge of the continuous spectrum occurring theoretically for $\epsilon \approx 0.0477$, while in the numerical computations it happens for $\epsilon \approx 0.047$.

5.3.3.2 On-Site, In-Phase Mode

The right panels of the Fig. 5.10 show the case of the OS–IP mode. The latter is found to be always unstable due to a real eigenvalue pair of

$$\lambda \approx \pm 4\epsilon \tag{5.49}$$

and a double, real eigenvalue pair of

$$\lambda \pm \sqrt{12\epsilon}. \tag{5.50}$$

This can also be clearly observed in the fourth and fifth rows of Fig. 5.10, showing the mode and its stability for $\epsilon = 0.05$.

5.3.3.3 Inter-Site, Out-of-Phase Mode

We next consider the case of the IS–OOP mode in Fig. 5.11. Our analytical results for this mode show that for small values of ϵ, we should expect to find it to be immediately unstable due to three real pairs of eigenvalues, namely a single one with

$$\lambda \approx \pm\sqrt{8\epsilon} \tag{5.51}$$

and a double one with

$$\lambda \approx \pm 2\sqrt{\epsilon}. \tag{5.52}$$

This expectation is once again confirmed by the numerical results of the left panel of Fig. 5.11. The fourth and fifth rows show the mode and the spectral plane of its linearization for the particular cases of $\epsilon = 0.08$ and 0.116.

5.3.3.4 On-Site, Out-of-Phase Mode

We complete our consideration of the quadrupolar modes by examining the OS–OOP mode, shown in the right panels of Fig. 5.11. Our theoretical analysis predicts that this mode should have a double imaginary eigenvalue pair of

$$\lambda \approx \pm 2\epsilon i \tag{5.53}$$

and a single imaginary pair of

$$\lambda \approx 4\epsilon i. \tag{5.54}$$

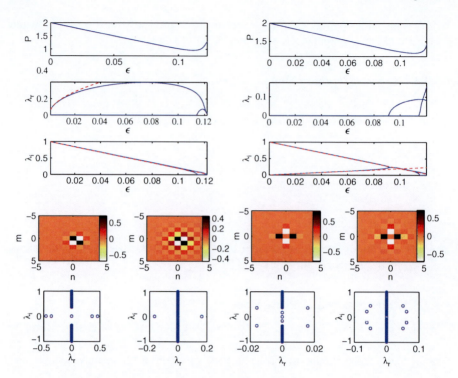

Fig. 5.11 Similar to Fig. 5.8 but for the quadrupole IS–OOP mode (*left panels*) and the quadrupole OS–OOP mode (*right panels*). The *fourth* and *fifth rows* show the modes and their stability for $C = 0.08$ and 0.116 in each case. Reprinted from [15]

Based on these predictions, we expect the mode to be stable for small ϵ (a result confirmed by numerical computations); however, it becomes destabilized upon collision of the larger one among these eigenvalues with the continuous spectral band. This is numerically found to occur for $\epsilon \approx 0.08$, while it is theoretically predicted, based on the above eigenvalue estimates, to take place for $\epsilon = 0.083$. The mode's stability analysis is shown in the fourth and fifth rows of Fig. 5.11 for $\epsilon = 0.08$ and 0.116.

5.3.4 Vortex Configuration

5.3.4.1 Inter-Site Vortices

Lastly, Fig. 5.12 shows similar features, but now for the IS (left panels) and OS (right panels) vortex solutions, discussed in the two-dimensional context in Chap. 3. The former has a theoretically predicted double pair of eigenvalues

$$\lambda \approx \pm 2\epsilon i \qquad (5.55)$$

5.3 Gap States

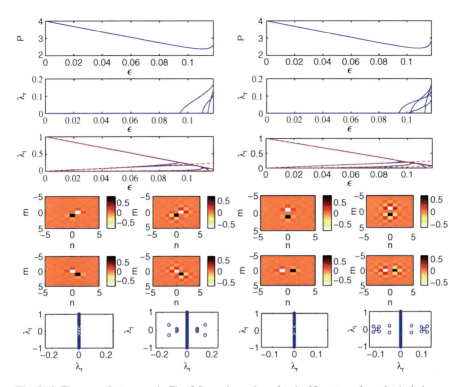

Fig. 5.12 The same features as in Fig. 5.8 are shown here for the IS vortex of topological charge $S = 1$ (*left*) and the OS vortex of $S = 1$ (*right*). In this case, both the real (*fourth row*) and imaginary (*fifth row*) parts of the solution are shown (and their stability in the *sixth row*) for $\epsilon = 0.08$ and 0.116. Reprinted from [15]

leading to an instability upon collision with the continuum band for $\epsilon \geq 0.095$ ($\epsilon \geq 0.1$ theoretically). Additionally, there is also an eigenvalue of higher order

$$\lambda \approx \pm 4\epsilon^2 i. \tag{5.56}$$

which obviously depends more weakly on ϵ. The fourth and fifth rows of Fig. 5.12 show the real and the imaginary parts of the vortex configuration for $\epsilon = 0.08$ and 0.116 and the sixth row shows the corresponding spectral planes for the corresponding (one stable and one unstable) cases.

5.3.4.2 On-Site Vortices

The OS vortices are shown in the right panels of the Fig. 5.12. In this case, we theoretically find that the vortex, for small ϵ, should have a double pair of eigenvalues

$$\lambda \approx 2\epsilon i \tag{5.57}$$

and a single, higher order pair of eigenvalues

$$\lambda \approx \pm\sqrt{32}\epsilon^3 i. \tag{5.58}$$

The first one among these, upon collision with the continuous spectrum, leads to an instability, theoretically predicted to occur at $\epsilon = 0.1$ and numerically found to happen for $\epsilon \approx 0.095$. The OS mode (and its stability) is shown in the fourth to sixth right panels of Fig. 5.12 for $\epsilon = 0.08$ and 0.116.

5.3.5 General Principles Derived from Stability Considerations

A general conclusion that it is relevant to mention is that the stability intervals of the defocusing structures are different from those of their focusing counterparts (especially when they are stable close to the AC limit) because of the collisions with the continuous spectrum band edge; the latter is at $\lambda = \Lambda$ in the focusing case, while it is at $\lambda = \Lambda - 8\epsilon$ (i.e., it is coupling dependent) in the defocusing setting. Another similarly general note is an immediate inference on whether the structures are stable or not; this can be made based on the knowledge of whether their focusing counterparts are stable or not and the conversion from the former to the latter through the staggering transformation $u_{n,m} = (-1)^{n+m} w_{n,m}$. For instance, IP two-site configurations (both OS and IS) are known to be generically unstable in the focusing regime (see Chap. 2); through the staggering transformation, OS–IP of the focusing case remains OS–IP in the defocusing, while IS-IP of the focusing becomes IS–OOP in the defocusing. Hence, these two should be expected to be always unstable, while the remaining two (OS–OOP in both focusing and defocusing and IS–OOP of the focusing, which becomes IS–IP in the defocusing) should similarly be expected to be linearly stable close to the AC limit. This is in accordance with our numerical observations in all the relevant cases. Note that, interestingly enough, for the vortex states the staggering transformation indicates that the stability is not modified between the focusing and defocusing cases. This is because for an IS vortex, it transforms an $S = 1$ state into an $S = -1$ state (which is equivalent to the former, in terms of stability properties), while the OS vortex remains unchanged by the transformation. However, as mentioned above, these considerations are not sufficient to compute the instability thresholds for initially stable modes, among other things. They do, nonetheless, provide a guiding principle for inferring the near-AC limit stability of the defocusing staggered states, based on their focusing counterparts, that we examined previously in this volume in Chaps. 2, 3, and 4.

References

1. Morandotti, R., Eisenberg, H.S., Silberberg, Y., Sorel, M., Aitchison, J.S.: Phys. Rev. Lett. **86**, 3296 (2001)
2. Smirnov, E., Rüter, C.E., Stepić, M., Kip, D., Shandarov, V.: Phys. Rev. E **74**, 065601 (2006)

References

3. Tang, L., Lou, C., Wang, X., Song, D., Chen, X., Xu, J., Chen, Z., Susanto, H., Law, K., Kevrekidis, P.G.: Opt. Lett. **32**, 3011 (2007)
4. Kivshar, Yu.S., Królikowski, W., Chubykalo, O.A.: Phys. Rev. E **50**, 5020 (1994)
5. Johansson, M., Kivshar, Yu.S.: Phys. Rev. Lett. **82**, 85 (1999)
6. Fitrakis, E.P., Kevrekidis, P.G., Susanto, H., Frantzeskakis, D.J.: Phys. Rev. E **75**, 066608 (2007)
7. Atkinson, K.: An Introduction to Numerical Analysis. Wiley, New York, p. 588 (1989)
8. Kapitula, T., Kevrekidis, P.G., Sandstede, B.: Physica D **195**, 263 (2004)
9. Kapitula, T., Kevrekidis, P.G., Sandstede, B.: Physica D **201**, 199 (2005)
10. Susanto, H., Johansson, M.: Phys. Rev. E **72**, 016605 (2005)
11. Pelinovsky, D.E., Kevrekidis, P.G.: J. Phys. A Math. Theor. **41**, 185206 (2008)
12. Pismen, L.M.: Vortices in Nonlinear Fields. Oxford University Press, Oxford (1999)
13. Cuevas, J., James, G., Kevrekidis, P.G., Law, K.J.H.: Physica D inpress, doi: 10.1016/j. physd. 2008.10.001, arXiv:0803.3379
14. Berloff, N.G.: J. Phys. A: Math. Gen. **37**, 1617(2004)
15. Kevrekidis, P.G., Susanto, H., Chen, Z.: Phys. Rev. E **74**, 066606 (2006)
16. Song, D., Lou, C., Tang, L., Wang, X., Li, W., Chen, X., Law, K.J.H., Susanto, H., Kevrekidis, P.G., Xu, J., Chen, Z.: Opt. Express **16**, 10110 (2008)

Chapter 6
Extended Solutions and Modulational Instability

In our considerations up to now, we have focused on localized solutions of the DNLS equation. In this chapter, we consider a different, yet important, class of solutions of the DNLS, namely the plane waves. Plane waves are spatially uniform (in the modulus) solutions, characterized by a wave number of the spatial modulation of their real and imaginary part, and an associated frequency (of temporal oscillation). They exist both in the continuum and in the discrete form of the NLS equation and one of the fundamental elements of their importance in this dispersive wave setting is that they are unstable, under appropriate conditions, to modulations, through a mechanism known as the modulational instability (MI). MI has a time-honored history in a number of fields traditionally associated with the NLS equation such as fluid dynamics [1, 2], nonlinear optics [3–5], and plasma physics [6, 7]. Despite its 40 years of history, the MI is still today an active field of investigation in this class of systems; for instance, it has been recently used in the dynamics of BECs as a method to produce bright solitons [8] in attractive condensates or as a scheme permitting the generation of Faraday waves in repulsive condensates [9].

Here, we will start by giving an explanation of the MI mechanism in the continuum case, and will then move on to examine the same mechanism in the discrete case, in order to illustrate the similarities but also the important differences between the two. Finally, we will consider some recent applications of this mechanism to BEC and nonlinear optics.

6.1 Continuum Modulational Instability

Starting with the continuum version of the NLS equation of the form

$$i u_t = -u_{xx} + g|u|^2 u, \tag{6.1}$$

it is straightforward to see that plane wave solutions of the form

$$u(x, t) = A \exp[i(qx - \omega t)] \tag{6.2}$$

exist provided that the nonlinear "dispersion relation"

$$\omega = q^2 + gA^2 \tag{6.3}$$

is satisfied. It is then natural to examine the linear stability of these solutions by virtue of the ansatz

$$u(x,t) = (A + \delta b(x,t))\exp[i((qx - \omega t) + \delta \psi(x,t))]. \tag{6.4}$$

Deriving the leading order ($O(\delta)$) equations for the evolution of the amplitude perturbation $b(x,t)$ and the phase perturbation $\psi(x,t)$, we realize that the ensuing linear PDEs can be solved by Fourier decomposition of the form $b(x,t) = b_0 \exp(i(Qx - \Omega t))$, and $\psi(x,t) = \psi_0 \exp(i(Qx - \Omega t))$. The resulting homogeneous linear system for b_0 and ψ_0 yields the solvability condition (which is also the modulational stability condition)

$$(-\Omega + 2qQ)^2 = Q^2(Q^2 + 2gA^2). \tag{6.5}$$

The key observation now is that

- If $g > 0$, then the right-hand side of Eq. (6.5) is positive and hence Ω is real, hence the perturbations will only lead to benign oscillations (i.e., the defocusing continuum plane wave solutions are stable under modulational perturbations).
- On the other hand, if $g < 0$, then the right-hand side of Eq. (6.5) *can* become negative, provided that the $Q^2 < Q_{cr}^2 = 2(-g)A^2$. Therefore, in this case, independently of their amplitude, there will *always* exist (i.e., for any amplitude A or equivalently negative nonlinearity strength g) wave numbers small enough such that the plane wave will be *unstable* under such modulational perturbations.

This phenomenology is illustrated in Fig. 6.1. In the case shown, $Q_{cr} = \sqrt{2}$. In the top row, the original perturbation has $Q = 1$, while in the bottom row the perturbation has $Q = 2$. In the latter, the small perturbation (of size ≈ 0.1) remains bounded in the course of the evolution and the Fourier space evolution reveals the presence of the originally excited wavenumber. On the contrary, in the top row, the small amplitude perturbation is exponentially amplified (until it saturates due to the presence of the nonlinearity). The amplification of the perturbation in real space is accompanied by the formation of large amplitude structures (the solitary waves or localized solutions that we have considered previously – hence, MI offers a "bridge" between the robust localized waveforms, and the unstable extended ones in the focusing NLS setting). On the other hand, in Fourier space, we clearly detect that sidebands of the original wave number are excited, leading to the emergence of harmonics of the original $Q = 1$.

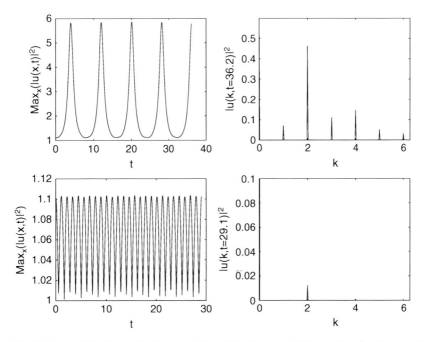

Fig. 6.1 From [10]: the result of an evolution of the focusing NLS equation for the case of excitation of a plane wave with $Q = 1$ (*top row*; unstable) and $Q = 2$ (*bottom row*; stable). The *left panels* in each case show the results of the evolution of the maximum squared modulus of the solution under the presence of the small perturbation. The *right panels* show the Fourier spectrum of the solution at a given time during the evolution (for a more detailed explanation, see text)

6.2 Discrete Modulational Instability

We now turn to the corresponding discrete case, where this type of analysis was first carried out, to the best of our knowledge, already in [11] (see also the detailed exposition of [12]). Writing the DNLS equation as

$$i\dot{u}_n = -\epsilon (u_{n+1} + u_{n-1} - 2u_n) + g|u_n|^2 u_n, \qquad (6.6)$$

we can again find a plane wave solution of the form $u_n = A \exp(i(kn - \omega t))$, which in this case satisfies

$$\omega = 4\epsilon \sin^2\left(\frac{k}{2}\right) + gA^2. \qquad (6.7)$$

Subsequently following the same steps (i.e., using linear stability analysis and decomposing the perturbation into Fourier modes), we obtain the modulational stability condition of the form

$$(\Omega - 2\epsilon \sin(k) \sin(Q))^2 = 8\epsilon \cos(k) \sin^2\left(\frac{Q}{2}\right) \left[2\epsilon \cos(k) \sin^2\left(\frac{Q}{2}\right) + gA^2\right]. \tag{6.8}$$

Firstly, it should be observed in connection to Eq. (6.6) that in the long wavelength limit of $k \ll 1$ and $Q \ll 1$, the condition degenerates into the continuum limit condition of Eq. (6.5). Despite this connection in the appropriate limit, Eq. (6.8) contains elements of new physics that are fundamentally discrete; more specifically, in this case MI arises for $g \cos(k) < 0$ (which naturally for $k \ll 1$ results in the purely focusing nonlinearity condition of the continuum limit). This signifies that instability here can emerge *even in the defocusing* setting, provided that $k > \pi/2$ (if the relevant threshold condition making the bracket expression of the right-hand side of Eq. (6.8) negative is satisfied).

This phenomenon is illustrated in Fig. 6.2 which displays a focusing case example with $k = \pi/3$ in the left panel (the top panel shows the space–time evolution of the plane wave, and the bottom panel shows the – normalized – space–time evolution of the Fourier transform of the solution's amplitude). On the other hand, the right panel shows a defocusing case example with $k = 2\pi/3$. It can clearly be seen that both cases become unstable, in accordance with the theoretical analysis. In fact, it is interesting to point out that wave numbers in the vicinity of $\pi/2$ are among the

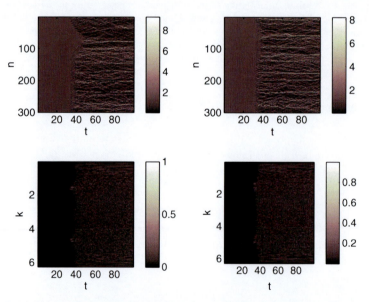

Fig. 6.2 Evolution of plane waves of $k = \pi/3$ for $g = -1$ (*left panels*) and of $k = 2\pi/3$ for $g = 1$ (*right panels*). The *top panel* show the spatiotemporal evolution of the squared modulus of the solution, while the *bottom panels* show the spatiotemporal evolution of the (normalized to its maximum value) Fourier transform of the squared modulus of the solution. Note in both cases the emergence of the modulational instability for $t < 40$, which leads to the formation of filamentary structures and the excitation of different wave numbers in Fourier space

6.2 Discrete Modulational Instability

ones to become destabilized first. Clearly, the result of the instability is to induce the emergence of filamentary structures of solitary wave type.

As a generalization of the DNLS model, and as a preamble to the following chapter, we also consider here the case of the two-component DNLS model of the form

$$i\frac{\partial u_{1,n}}{\partial t} = -\epsilon_1 \left(u_{1,n+1} + u_{1,n-1} - 2u_{1,n}\right) + \left(g_{11}|u_{1,n}|^2 + g_{12}|u_{2,n}|^2\right) u_{1,n} + c u_{2,n},$$

$$i\frac{\partial u_{2,n}}{\partial t} = -\epsilon_2 \left(u_{2,n+1} + u_{2,n-1} - 2u_{2,n}\right) + \left(g_{12}|u_{1,n}|^2 + g_{22}|u_{2,n}|^2\right) u_{2,n} + c u_{1,n}.$$
(6.9)

This is a model that is relevant both to the optics in the case of propagation of two different polarizations or of two different wavelengths within a waveguide array [13, 14], as well as in BECs in the case of other multicomponent BECs of different species (or even of different hyperfine states of the same species); see, e.g., the recent exposition of [15] and references therein. More details on the physical relevance of the model will be given in the next chapter.

We now follow the discussion of MI in this setting given in [16]. Starting first with the case where $c \neq 0$, we can again consider plane waves of the form

$$u_{jn} = A_j \exp\left[i(q_j n - \omega_j t)\right], \quad j = 1, 2, \tag{6.10}$$

where the linear coupling imposes $q_1 = q_2 \equiv q$ and $\omega_1 = \omega_2 \equiv \omega$. The resulting dispersion relations are

$$\omega A_1 = -2\epsilon_1 (\cos q - 1) A_1 + \left(g_{11} A_1^2 + g_{12} A_2^2\right) A_1 + c A_2,$$

$$\omega A_2 = -2\epsilon_2 (\cos q - 1) A_2 + \left(g_{12} A_1^2 + g_{22} A_2^2\right) A_2 + c A_1. \tag{6.11}$$

The solution of these algebraic equations yields the possible amplitudes of the plane waves as a function of their wave number and frequency.

Now, to examine the stability of the uniform solutions by imposing

$$u_{jn}(x,t) = \left[A_j + B_{jn}(x,t)\right] \exp[i(qn - \omega t)] \tag{6.12}$$

into the original equations to obtain a system of two coupled linearized equations for the perturbations $B_j(x,t)$. As before, we use a Fourier decomposition of the form

$$B_{jn} = \alpha_j \cos(Qn - \Omega t) + i\beta_j \sin(Qn - \Omega t), \tag{6.13}$$

where Q and Ω are the wave number and frequency of perturbation. This leads to a set of four homogeneous equations for α_1, β_1, α_2, and β_2. The latter, have a nontrivial solution if Q and Ω satisfy the dispersion relation

$$\left[(\Omega - 2\epsilon_1 \sin Q \sin q)^2 - \left(2\epsilon_1 r + c\frac{A_2}{A_1}\right)\left(2\epsilon_1 r + c\frac{A_2}{A_1} - 2g_{11}A_1^2\right)\right]$$

$$\times \left[(\Omega - 2\epsilon_2 \sin Q \sin q)^2 - \left(2\epsilon_2 r + c\frac{A_1}{A_2}\right)\left(2\epsilon_2 r + c\frac{A_1}{A_2} - 2g_{22}A_2^2\right)\right]$$

$$-2c(2g_{12}A_1A_2 + c)(\Omega - 2\epsilon_1 \sin Q \sin q)(\Omega - 2\epsilon_2 \sin Q \sin q)$$

$$-c^2\left(2\epsilon_1 r + c\frac{A_2}{A_1} - 2g_{11}A_1^2\right)\left(2\epsilon_2 r + c\frac{A_1}{A_2} - 2g_{22}A_2^2\right)$$

$$-(2g_{12}A_1A_2 + c)^2\left[\left(2\epsilon_1 r + c\frac{A_2}{A_1}\right)\left(2\epsilon_2 r + c\frac{A_1}{A_2}\right) - c^2\right] = 0, \quad (6.14)$$

where, for simplicity of notation, we define $r \equiv \cos q(\cos Q - 1)$.

In the absence of coupling, e.g., for $c = g_{12} = 0$, we obtain the same expression as in the one-component problem, i.e., Eq. (6.8) above.

Focusing for definiteness on the more tractable case of $\epsilon_1 = \epsilon_2 \equiv \epsilon$, we can rewrite the above equation as

$$(\Omega - 2\epsilon \sin Q \sin q)^4 - (K_1 + K_2 + K_3)(\Omega - 2\epsilon \sin Q \sin q)^2 + K_1K_2 - K_4 = 0, \quad (6.15)$$

where

$$K_1 = \left(2\epsilon r + c\frac{A_2}{A_1}\right)\left(2dr + c\frac{A_2}{A_1} - 2g_{11}A_1^2\right),$$

$$K_2 = \left(2\epsilon r + c\frac{A_1}{A_2}\right)\left(2\epsilon r + c\frac{A_1}{A_2} - 2g_{22}A_2^2\right),$$

$$K_3 = 2c(2g_{12}A_1A_2 + c),$$

$$K_4 = c^2\left(2\epsilon r + c\frac{A_2}{A_1} - 2g_{11}A_1^2\right)\left(2\epsilon r + c\frac{A_1}{A_2} - 2g_{22}A_2^2\right)$$

$$+(2g_{12}A_1A_2 + c)^2\left[\left(2\epsilon r + c\frac{A_2}{A_1}\right)\left(2\epsilon r + c\frac{A_1}{A_2}\right) - c^2\right]. \quad (6.16)$$

As before, to avoid MI, both solutions for $(\Omega - 2\epsilon \sin Q \sin q)_{1,2}^2$ should be positive. Taking into account the bi-quadratic nature of the equation, it is concluded that the spatially homogeneous solution is unstable if either the sum $\Sigma = K_1 + K_2 + K_3$ or the product $\Pi = K_1K_2 - K_4$ of the solutions is negative:

$$K_1 + K_2 + K_3 < 0, \quad (6.17)$$

$$K_1K_2 - K_4 < 0. \quad (6.18)$$

To illustrate the results, we can proceed to fix the value of the perturbation wave number, investigating the parameter region in which it would give rise to the MI.

6.2 Discrete Modulational Instability

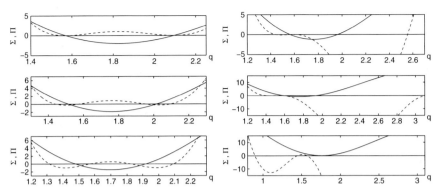

Fig. 6.3 The *left panels* of the figure show, for $g_{12} = 0$, the cases of $c = 0$ (*top panel*; unstable for $\pi/2 < q < 2.0945$), $c = 0.25$ (*middle panel*; unstable for $1.4455 < q < 2.0945$), and $c = 0.5$ (*bottom panel*; unstable for $1.318 < q < 2.0945$). The *solid line* shows the sum Σ and the *dashed line* the product Π of the solutions of the bi-quadratic equation. The instability takes place in intervals of the wave number q of the unperturbed plane-wave solution where either Σ or Π (or both) are negative. The *right panels* show the same features but fixing $c = 0.25$ and varying g_{12}. The figure shows the cases of $g_{12} = 2/3$ (*top panel*; unstable for $1.4455 < q < 2.556$), $g_{12} = 1$ (*middle panel*; unstable for $\pi/2 < q < \pi$), and $g_{12} = 2$ (*bottom panel*; unstable for $0.8955 < q < 1.4455$ and $\pi/2 < q < \pi$). Reprinted from [16] with permission

This is shown in Fig. 6.3 in which we fix $Q = \pi$ and $g_{11} = g_{22} = A_1 = A_2 = d_1 = \epsilon = 1$ and vary c and g_{12} (the coefficients of the linear and nonlinear coupling, respectively), to examine their effect on the stability interval. For these values of the parameters the modulationally unstable region is $\pi/2 < q < 2\pi/3 = 2.0945$. It can be inferred from the figure that c may widen the MI interval by decreasing its lower edge. On the other hand, g_{12} has a more complex effect: while making the instability interval larger by increasing its upper edge (until it reaches π), it may also open MI bands within the initially modulationally stable region.

Similar calculations can be performed in the case with purely nonlinear coupling. Although in the latter case, $c = 0$ and the expressions are simplified in that regard, on the other hand, it is now possible to have, in principle, $q_1 \neq q_2$, and $\omega_1 \neq \omega_2$ which complicates the logistics of the relevant calculation (although the approach is the same). If we let $\epsilon_1 = \epsilon_2 = \epsilon$ and $q_1 = q_2 = q$, then the relevant condition becomes

$$(\Omega - 2\epsilon \sin Q \sin q)^4 - 2K_5(\Omega - 2\epsilon \sin Q \sin q)^2 + K_6 = 0, \quad (6.19)$$

where

$$K_5 = 2\epsilon r \left(2\epsilon r - (g_{11}A_1^2 + g_{22}A_2^2)\right),$$
$$K_6 = (2\epsilon r)^2 \left((2\epsilon r)^2 - 2(g_{11}A_1^2 + g_{22}A_2^2)2\epsilon r + 4A_1^2 A_2^2 (g_{11}g_{22} - g_{12}^2)\right).$$
$$(6.20)$$

In this case, the MI conditions become

$$K_5 < 0 \quad \text{or} \quad K_6 < 0.$$

The latter can be rewritten, respectively, as

$$-\frac{1}{2}(g_{11}A_1^2 + g_{22}A_2^2) < 2\epsilon \cos(q) \sin^2\left(\frac{Q}{2}\right) < 0, \quad (6.21)$$

$$K_- < -4\epsilon \cos(q) \sin^2\left(\frac{Q}{2}\right) < K_+ \quad (6.22)$$

with $K_\pm \equiv g_{11}A_1^2 + g_{22}A_2^2 \pm \sqrt{(g_{11}A_1^2 + g_{22}A_2^2)^2 - 4A_1^2 A_2^2(g_{11}g_{22} - g_{12}^2)}$. From the formulas, as well as graphically from Fig. 6.4, we can observe that in this case as well, the coupling between the two components tends to augment the band of the modulationally unstable wave numbers with respect to the single-component case. Similarly to the one-component case, both for the linear and for the nonlinear coupling between the components, the dynamical manifestation of MI leads to filamentation and the formation of localized structures in both components, as is shown in [25].

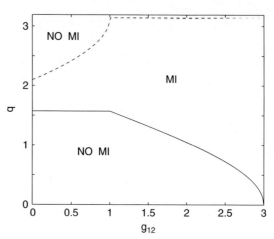

Fig. 6.4 The figure shows the threshold wave numbers q for MI as a function of g_{12} in the case of purely nonlinear coupling and for $A = \epsilon = g_{11} = g_{22} = 1$. Reprinted from [16] with permission

6.3 Some Case Examples

Finally, we briefly touch upon some case examples of recent applications of the modulational instability of the DNLS equation in BEC and optics.

In the work of [17, 18], an effective interpretation of the modulational instability was devised in the form of a dynamical superfluid to insulator transition for a BEC trapped in both a magnetic trap and an optical lattice. What was realized in these works was that if the (parabolic) magnetic trap holding the condensate (see e.g., the right panel of Fig. 6.5) is displaced, then the ground state of the system (which is

6.3 Some Case Examples   151

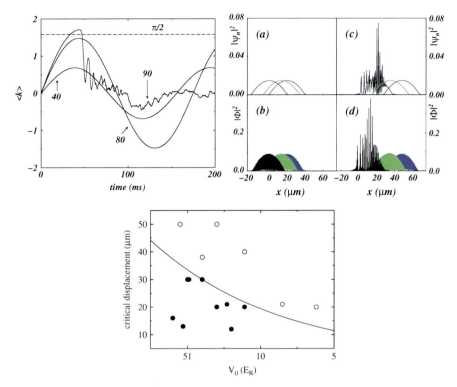

Fig. 6.5 Reprinted from [17, 18]: the *top left* shows the discrete quasi-momentum for two sub-critical (40 and 80) and one supercritical (90 sites) evolution; the latter is clearly unstable. The *top right* shows the discrete model (*top row*) and the continuum model with a periodic potential (*bottom row*) for a modulationally stable (*left*) and a modulationally unstable (*right*) case. The *bottom plot* is taken from the experimental work of the Florence group, showing coherent motion in the case of *filled circles* and *pinned motion* in the case of *empty circles*. The *solid line* separating the two corresponds to the theoretical prediction of [17, 18]

the effective analog of a plane wave for the defocusing dynamics of a ^{87}Rb BEC) will acquire a quasi-momentum. If this quasi-momentum (which is associated with the wave number of the state) becomes larger than $\pi/2$, then in accordance with the modulational instability criterion of Eq. (6.8), the system shall become unstable, and the instability will give rise to localization. Obviously, the more that the parabolic trap is initially displaced, the higher the initial potential energy of the state, hence the larger the kinetic energy that it will acquire. Hence, there will be a critical displacement, which was evaluated in [17, 18] as $\xi_{cr} = \sqrt{2\epsilon/\Omega}$ (where Ω was the parabolic trap frequency), beyond which, the induced momentum will exceed the MI threshold and will therefore induce a transition from a "superfluid" to an "insulator." This is illustrated in the top row of Fig. 6.5, which features in the left panel different initial displacements and the corresponding quasi-momentum evolution; the right panel shows both the discrete (top) and the continuum with a periodic potential (bottom) behavior of the model, for subcritical and supercritical cases. In the bottom plot, the theoretical prediction (solid line) is compared to experiments illustrating phase coherence (filled circles), or lack thereof (open circles). Clearly, the theoretical prediction is in good agreement with the experimental observations.

On the other hand, more recently, in the work of [26], an optical waveguide array manifestation of the discrete modulational instability was illustrated. The latter experiment was in the setting of a focusing Kerr nonlinearity. As a result here, the instability should arise for sufficiently small wave numbers (and sufficiently high optical power), as predicted by Eq. (6.8) and experimentally observed in [19]; for a relevant example in the case of high-input power, see Fig. 3 of [19]. In the figure one can observe clear filamentation of the optical output intensity distribution for small wave numbers (a direct consequence of MI), while for high values of the wave number, this phenomenon is absent.

References

1. Witham, G.B.: Proc. R. Soc. London **283**, 238 (1965)
2. Benjamin, T.B., Feir, J.E.: J. Fluid. Mech. **27**, 417 (1967)
3. Bespalov, V.I., Talanov, V.I.: JETP Lett. **3**, 307 (1966)
4. Karpman, V.I.: JETP Lett. **6**, 227 (1967)
5. Ostrovskii, L.A.: Sov. Phys. JETP **24**, 797 (1967)
6. Taniuti, T., Washimi, H.: Phys. Rev. Lett. **21**, 209 (1968)
7. Hasegawa, A.: Phys. Rev. Lett. **24**, 1165 (1970)
8. Strecker, K.E., Partridge, G.B., Truscott, A.G., Hulet, R.G.: Nature **417**, 150 (2002)
9. Nicolin, A.I., Carretero-González, R., Kevrekidis, P.G.: Phys. Rev. A **76**, 063609 (2007)
10. Theocharis, G., Rapti, Z., Kevrekidis, P.G., Frantzeskakis, D.J., Konotop, V.V.: Phys. Rev. A **67**, 063610 (2003)
11. Christodoulides, D.N., Joseph, R.I.: Opt. Lett. **13**, 794 (1988)
12. Kivshar, Yu.S., Peyrard, M.: Phys. Rev. A **46**, 3198 (1992)
13. Hasegawa, A.: Solitons in Optical Communications. Clarendon Press, Oxford, NY (1995)
14. Agrawal, G.P.: Nonlinear Fiber Optics. Academic Press, San Diego, CA (1995)
15. Mertes, K.M., Merrill, J.W., Carretero-González, R., Frantzeskakis, D.J., Kevrekidis, P.G., Hall, D.S.: Phys. Rev. Lett. **99**, 190402 (2007)
16. Rapti, Z., Trombettoni, A., Kevrekidis, P.G., Frantzeskakis, D.J., Malomed, B.A., Bishop, A.R.: Phys. Lett. A **330**, 95 (2004)
17. Smerzi, A., Trombettoni, A., Kevrekidis, P.G., Bishop, A.R.: Phys. Rev. Lett. **89**, 170402 (2002)
18. Cataliotti, F.S., Fallani, L., Ferlaino, F., Fort, C., Maddaloni, P., Inguscio, M.: New J. Phys. **5**, 71 (2003)
19. Meier, J., Stegeman, G.I., Christodoulides, D.N., Silberberg, Y., Morandotti, R., Yang, H., Salamo, G., Sorel, M., Aitchison, J.S.: Phys. Rev. Lett. **92**, 163902 (2004)

Chapter 7
Multicomponent DNLS Equations

One of the most interesting extensions of the DNLS equation is in the study of multi-component versions of the model. Such models are relevant both in nonlinear optics, e.g., when propagating multiple frequencies or polarizations of light; prototypical examples of these types have been discussed, e.g., in [1, 2] from the theoretical point of view and in [3] from the experimental point of view (see also references therein). On the other hand, similar models are quite relevant to two-component [4] or even one-component BECs [5] in the presence of optical lattice potentials. The coupling between the different components can be linear or nonlinear (or both) [1]. In this chapter, we show some case examples of interesting dynamics that can arise in the linear coupling case (symmetry breaking), as well as ones that can arise in the nonlinear coupling case (dynamical instabilities).

7.1 Linearly Coupled

In the context of optics, systems of linearly coupled DNLS equations are relevant to various applications: linear coupling may occur among two polarization modes inside each waveguide of a waveguide array, being induced by a twist of the core (for linear polarizations), or by the birefringence (for circular polarizations). Linear coupling between two modes also takes place in arrays of dual-core waveguides [1]. On the other hand, in BECs, the linear coupling may be imposed by an external microwave or radio frequency field, which can drive Rabi [6, 7] or Josephson [8, 9] oscillations between populations of two different states.

In both optical and atomic media, the basic linearly coupled DNLS model takes the following form:

$$\begin{cases} iU_t = K\epsilon\Delta_2 U + KV + |U|^2 U, \\ iV_t = K\epsilon\Delta_2 V + KU + |V|^2 V, \end{cases} \quad (7.1)$$

where $U = U(x, t)$ and $V = V(x, t)$ are wave functions of the two species in BEC, or electric field envelopes of the two coupled modes in optics (x is realized as a discrete vectorial coordinate), K is the strength of the linear coupling between fields U and V, and ϵ determines the couplings between adjacent sites of the lattice. For convenience, the full lattice-coupling constant is defined as $K\epsilon$ (this will allow us to scale out K from the analysis presented below).

Following the analysis of [10], we seek stationary solutions to the equations in the form

$$\begin{cases} U(x,t) = \sqrt{K}\, u(x) \exp[-iK(\mu - 2D\epsilon)t], \\ V(x,t) = \sqrt{K}\, v(x) \exp[-iK(\mu - 2D\epsilon)t], \end{cases} \quad (7.2)$$

where $u(x)$ and $v(x)$ are real-valued functions, and μ is an appropriately shifted chemical potential. Then the steady-state equations become

$$\begin{cases} \mu u_n = \epsilon \overline{\Delta}_1 u_n + v_n + u_n^3, \\ \mu v_n = \epsilon \overline{\Delta}_1 v_n + u_n + v_n^3, \end{cases} \quad (7.3)$$

with $\overline{\Delta}_1 w_n \equiv w_{n+1} + w_{n-1}$. In the two-dimensional case, the stationary equations are

$$\begin{cases} \mu u_{n,m} = \epsilon \overline{\Delta}_2 u_{n,m} + v_{n,m} + u_{n,m}^3, \\ \mu v_{n,m} = \epsilon \overline{\Delta}_2 v_{n,m} + u_{n,m} + v_{n,m}^3, \end{cases} \quad (7.4)$$

where $\overline{\Delta}_2 w_{n,m} \equiv w_{n+1,m} + w_{n-1,m} + w_{n,m+1} + w_{n,m-1}$.

In [10], both symmetric (with $u = v$) and symmetry-broken (with $u \neq v$) states were constructed as solutions of Eqs. (7.3) and (7.4). Since we are interested here in the properties of the fundamental single-site states, we will use as a reasonably accurate method to obtain an analytical handle on the waveforms the variational approximation (comparing it with the full numerical results). We start by noting that Eqs. (7.3) and (7.4) can be derived from the following Lagrangians:

$$L_{1D} = \sum_{n=-\infty}^{\infty} \left[-\frac{\mu}{2}(u_n^2 + v_n^2) + \frac{1}{4}(u_n^4 + v_n^4) + u_n v_n + \epsilon(u_{n+1} u_n + v_{n+1} v_n) \right] \quad (7.5)$$

$$L_{2D} = \sum_{m,n=-\infty}^{\infty} \left[-\frac{\mu}{2}(u_{n,m}^2 + v_{n,m}^2) + \frac{1}{4}(u_{n,m}^4 + v_{n,m}^4) + u_{n,m} v_{n,m} \right.$$
$$\left. + \epsilon(u_{n+1,m} u_{n,m} + u_{n,m+1} u_{n,m} + v_{n+1,m} v_{n,m} + v_{n,m+1} v_{n,m}) \right]. \quad (7.6)$$

Then, the discrete soliton *ansätze*, $\{u_n, v_n\} = \{A, B\} e^{-\lambda|n|}$ and $\{u_{n,m}, v_{n,m}\} = \{A, B\} e^{-\lambda|n|} e^{-\lambda|m|}$, with free constants A, B, and $\lambda > 0$, are used in the one- and two-dimensional cases, respectively, as in our earlier analysis in Chaps. 2 and 3. It should be noted in comparison to the standard one-component ansatz that by

7.1 Linearly Coupled

introducing different amplitudes A and B, we admit a possibility of *asymmetric solitons*, within the framework of the variational approximation.

The resulting expressions for the effective Lagrangians are

$$L_{1D} = \left[AB - \frac{\mu}{2}\left(A^2 + B^2\right)\right] \coth \lambda + \frac{1}{4}\left(A^4 + B^4\right) \coth(2\lambda)$$
$$+ \epsilon \left(A^2 + B^2\right) \text{cosech} \lambda,$$

$$L_{2D} = \left[AB - \frac{\mu}{2}\left(A^2 + B^2\right)\right] \coth^2 \lambda + \frac{1}{4}\left(A^4 + B^4\right) \coth^2 2\lambda$$
$$+ 2\epsilon \left(A^2 + B^2\right)(\text{cosech} \lambda) \coth \lambda.$$

Then, deriving the static version of the Euler–Lagrange equations $\partial L_{1D,2D}/\partial(\lambda, A, B) = 0$ yields

$$\frac{\mu}{2}\left(A^2 + B^2\right) \text{cosech}^2 \lambda - \frac{1}{2}\left(A^4 + B^4\right) \text{cosech}^2 2\lambda$$
$$- AB \text{cosech}^2 \lambda - \epsilon \left(A^2 + B^2\right) \text{cosech} \lambda \coth \lambda = 0,$$

$$-\mu A \coth \lambda + A^3 \coth 2\lambda + B \coth \lambda + 2\epsilon A \text{cosech} \lambda = 0,$$

$$-\mu B \coth \lambda + B^3 \coth 2\lambda + A \coth \lambda + 2\epsilon B \text{cosech} \lambda = 0$$

for the one-dimensional case, and

$$\mu \left(A^2 + B^2\right) \coth \lambda \text{cosech}^2 \lambda - \left(A^4 + B^4\right) \coth 2\lambda \text{cosech}^2 2\lambda$$
$$-2AB \coth \lambda \text{cosech}^2 \lambda - 2\epsilon \left(A^2 + B^2\right)\left(\text{cosech} \lambda \coth^2 \lambda + \text{cosech}^3 \lambda\right) = 0,$$

$$-\mu A \coth^2 \lambda + A^3 \coth^2 2\lambda + B \coth^2 \lambda + 4\epsilon A \text{cosech} \lambda \coth \lambda = 0,$$

$$-\mu B \coth^2 \lambda + B^3 \coth^2 2\lambda + A \coth^2 \lambda + 4\epsilon B \text{cosech} \lambda \coth \lambda = 0$$

for the two-dimensional fundamental waves.

An interesting observation consists of the analytically tractable AC limit of $\epsilon = 0$. For the symmetric branch, we then have $u_n = v_n = 0$ or $u_n = v_n = \sqrt{\mu - 1}$, while for the asymmetric branch, a system of algebraic equations has to be solved $\mu u_n = v_n + u_n^3$, $\mu + 1 = u_n^2 + u_n v_n + v_n^2$. The solution is shown in Fig. 7.1, which displays the symmetry-breaking bifurcation in the AC limit, by means of a plot of the *asymmetry measure*, $r \equiv (E_1 - E_2)/(E_1 + E_2)$, versus half the total norm, $E = (E_1 + E_2)/2$, where $\{E_1, E_2\} = \sum_n \{u_n^2, v_n^2\}$ are the norms of the two components of the solution. It is particularly interesting to point out that in the case of $\epsilon = 0$, the observed pitchfork bifurcation is *supercritical* (cf. with the finite ϵ case below).

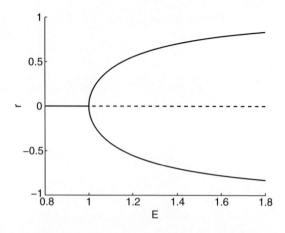

Fig. 7.1 From [10]: the bifurcation diagram for the discrete solitons in the anti-continuum limit, $\epsilon = 0$; r and E are the asymmetry parameter and the half of the total squared norm, respectively. The *solid* and *dashed* lines represent stable and unstable solutions, respectively

These results can be compared with those of direct numerical continuation of the corresponding branches from the AC limit. In [10] the relevant branches were obtained and their numerical linear stability was also examined using the perturbed solution ansatz

$$\begin{cases} U(\bm{x},t) = e^{-i\mu t}\left[u(\bm{x}) + a(\bm{x})\,e^{\lambda t} + b^*(\bm{x})\,e^{\lambda^* t}\right], \\ V(\bm{x},t) = e^{-i\mu t}\left[v(\bm{x}) + c(\bm{x})\,e^{\lambda t} + d^*(\bm{x})\,e^{\lambda^* t}\right] \end{cases} \tag{7.7}$$

in Eqs. (7.1) and solving the resulting linearized equations for the perturbation eigenmodes a, b, c, d and the eigenvalues λ associated with them.

Typical results for particular values of ϵ are shown in Figs. 7.2 and 7.3, for the one- and two-dimensional cases, respectively. In both cases, it is remarkable to observe that the relevant bifurcation is observed to be *subcritical* (instead of supercritical as in the AC limit) pitchfork due to the collision of the fundamental

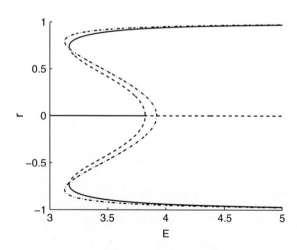

Fig. 7.2 From [10]: the bifurcation diagram is shown for $\epsilon = 1.6$ in the one-dimensional model. The *dash–dotted line* indicates solutions found through the variational approximation, while *solid* and *dashed* lines show, respectively, numerically found stable and unstable steady-state solutions

7.1 Linearly Coupled

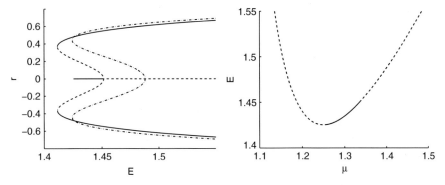

Fig. 7.3 From [10]: the *left panel* shows the bifurcation diagram in the two-dimensional model for $\epsilon = 0.25$, in the same way (i.e., with the same meaning of the different curves) as the one-dimensional diagram is shown in Fig. 7.2. The *right panel* displays the dependence of the solution's squared norm, E, upon the chemical potential, μ, for the symmetric solutions. Unlike the one-dimensional case, there are now two different symmetric solutions, resulting in both stable (*solid line*) and unstable (*dashed lines*) solutions for norms below the value at which the symmetric and asymmetric solution branches intersect

symmetric branch with two unstable asymmetric branches. The latter ones emerge through a saddle node bifurcation also generating a stable asymmetric branch. Interestingly, between the two critical points, both the symmetric branch and the outer asymmetric one are stable, hence there exists a region of bistability. We also observe that, in these typical comparisons, the results obtained from the variational approximation are quite close to the fully numerical results. This is more so in the one-dimensional case than in the two-dimensional case, since, as we have seen before (e.g., in Chap. 3), since the inaccuracy of the variational ansatz tends to accumulate the error in higher dimensions.

Typical examples of the existence and stability results obtained numerically, and how the former compare with the variational predictions are shown in Fig. 7.4 for the one-dimensional case and in Figs. 7.5 and 7.6 for the two-dimensional case. We note that in general the VA provides a fairly accurate description of the profile, although in some cases, it may yield slower decay rates (and slightly different amplitudes) than the full numerical results. It should also be reminded to the reader that in the two-dimensional case, as discussed in Chap. 3, there are typically two (symmetric) solutions corresponding to the same norm, namely a stable and an unstable one, as is shown, e.g., in Fig. 7.6 (see also the right panel of Fig. 7.3).

There are a couple of physically relevant observations to be made here. On the one hand, a conclusion following from the comparison of Figs. 7.1 and 7.2 is that the bifurcation found in the AC limit (see Fig. 7.1) is supercritical, unlike the weakly subcritical one in Fig. 7.2. This indicates that the character of the pitchfork bifurcation should change from subcritical to supercritical with the increase of discreteness, i.e., decrease of ϵ, which, in turn, should eliminate the unstable asymmetric branches. In accordance with this expectation, it was found in [10] that the unstable asymmetric solutions exist only for $\epsilon > 0.35$, in the one-dimensional case.

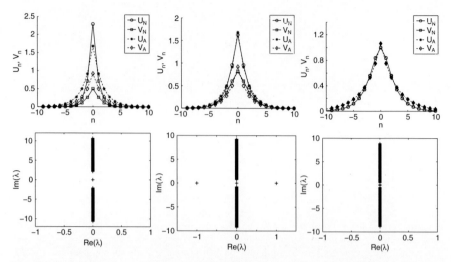

Fig. 7.4 From [10]: plots of solutions belonging to different branches in Fig. 7.2, at $E = 3.4$. The *top row* figures show the solution profiles found by means of the numerical (U_N, V_N) and variational ("analytical", U_A, V_A) methods. The *bottom row* plots illustrate the linear stability eigenvalues for the numerical solution. The *first column* presents a stable stationary asymmetric solution belonging to the outer (upper) branch in Fig. 7.2, the *second column* is an unstable asymmetric solution, and the last column shows a stable solution of the symmetric family

On the other hand, in the two-dimensional case, the bifurcation diagram has no continuum analog due to the occurrence of collapse, contrary to what is the case in one-dimension. In the two-dimensional case also, due to the existence of a minimum norm threshold below which the symmetric branch does not exist, as discussed in Chap. 3 [11–14], it is possible that the asymmetric solution (as in Fig. 7.3) will exist for powers below the symmetric solution's excitation threshold. This will enable the system to access lower norm states than in its one-component incarnation. Finally, it should be pointed out that the bistability arising from Figs. 7.2 and 7.3 above has been used in [10] to successfully "steer" the unstable asymmetric solitons dynamically toward either the stable symmetric or the stable asymmetric branch, depending on the type of the original "kick" (i.e., perturbation) to the unstable solution.

7.2 Nonlinearly Coupled

In the context of nonlinearly coupled DNLS equations, numerous studies have been present at the theoretical level discussing the properties of solitary waves, both in one dimension [15–17], and in two-dimensions [1, 15, 18–20], as well as even in three dimensions [21]. However, experimental results in this system materialized, to the best of our knowledge, only very recently in [3] (see also the longer exposition of [22]). These experimental realizations resulted in further theoretical work addressing various aspects of nonlinearly coupled multicomponent models including

7.2 Nonlinearly Coupled

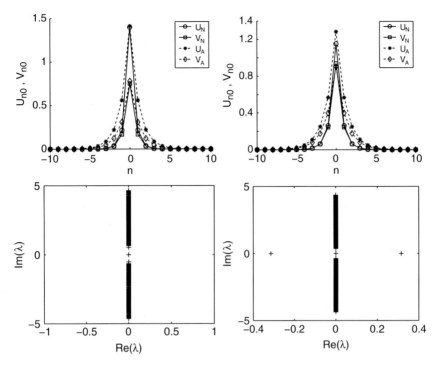

Fig. 7.5 From [10]: cross section plots of the asymmetric solutions belonging to different branches in Fig. 7.3, at $E = 1.435$. The *top row* figures show the solutions found by means of the numerical (U_N, V_N) and variational ("analytical," U_A, V_A) methods, and the *bottom row* plots display linear stability eigenvalues for the numerical solution. The *first* and *second* columns represent, respectively, stable and unstable solutions belonging to the asymmetric branches of the bifurcation diagram, respectively

switching and instability-induced amplification, modulational instability, PN barriers, and stability of localized modes among others [2, 23–25]. Here, we restrict ourselves to the study of the fundamental modes of the system in one dimension and a small sampler of the interesting possibilities that arise in higher dimensions (including multivortex structures, etc.). We refer the interested reader to the above literature for further details.

7.2.1 One Dimension

The theoretical model put forth in [3] to analyze the experimental results was of the form

$$i\dot{a}_n = -a_n - \epsilon(a_{n+1} + a_{n-1}) - \left(|a_n|^2 + A|b_n|^2\right)a_n - Bb_n^2 a_n^\star, \quad (7.8)$$

$$i\dot{b}_n = b_n - \epsilon(b_{n+1} + b_{n-1}) - \left(|b_n|^2 + A|a_n|^2\right)b_n - Ba_n^2 b_n^\star. \quad (7.9)$$

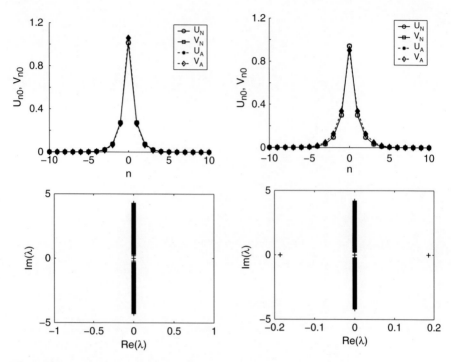

Fig. 7.6 Same as Fig. 7.5 for two symmetric solutions found at $E = 1.435$

In the experimental context, a_n and b_n are the appropriately normalized, slowly varying, complex field envelopes for the transverse electric (TE) and transverse magnetic (TM) polarized waves, respectively. The constants A and B are respectively associated with the cross-phase modulation (XPM) and four-wave mixing (FWM) and were evaluated in the experiments of [3] to be approximately equal to $A \simeq 1$ and $B \simeq 1/2$. It is interesting to note that in this case, due to the FWM term, only the total power (instead of the individual powers of each component, as would be the case if $B = 0$)

$$P = \sum_n \left(|a_n|^2 + |b_n|^2 \right) \tag{7.10}$$

is conserved and based on the analysis of [3], it is connected to the dimensional power P_d (measured in watts) through $P_d \simeq 56.4 P$. While in the analysis of [2], which we will follow here, the dimensionless coupling ϵ was considered a free parameter, in the experimental results reported in [3], it was $\epsilon \approx 0.921$.

Seeking stationary solutions in the form $a_n = \tilde{a}_n e^{iqz}$ and $b_n = \tilde{b}_n e^{iqz}$ and subsequently dropping the tildes, results in the following stationary equations:

7.2 Nonlinearly Coupled

$$(q-1)a_n - \epsilon(a_{n+1} + a_{n-1}) - \left(|a_n|^2 + A|b_n|^2\right)a_n - Bb_n^2 a_n^* = 0, \quad (7.11)$$

$$(q+1)b_n - \epsilon(b_{n+1} + b_{n-1}) - \left(|b_n|^2 + A|a_n|^2\right)b_n - Ba_n^2 b_n^* = 0. \quad (7.12)$$

The dimensionless propagation constant q is then an additional (to the dimensionless coupling ϵ) free parameter and it is in the (ϵ, q) two-parameter plane that the results presented herein are given.

The linear stability of a given stationary solution (a_n^0, b_n^0) of the stationary equations (7.11) and (7.12) can be obtained through the usual perturbation ansatz

$$a_n = a_n^0 + \delta\left(c_n e^{-i\omega z} + d_n e^{i\omega^* z}\right), \quad (7.13)$$

$$b_n = b_n^0 + \delta\left(f_n e^{-i\omega z} + g_n e^{i\omega^* z}\right). \quad (7.14)$$

Then the matrix eigenvalue problem yielding the eigenfrequency ω reads

$$\omega \begin{pmatrix} c_n \\ d_n^* \\ f_n \\ g_n^* \end{pmatrix} = \mathbf{L} \cdot \begin{pmatrix} c_n \\ d_n^* \\ f_n \\ g_n^* \end{pmatrix},$$

where

$$\mathbf{L} = \begin{pmatrix} L_{11} & L_{12} & L_{13} & L_{14} \\ L_{21} & L_{22} & L_{23} & L_{24} \\ L_{31} & L_{32} & L_{33} & L_{34} \\ L_{41} & L_{42} & L_{43} & L_{44} \end{pmatrix}.$$

The $N \times N$ (where N is the size of the lattice) blocks of the linearization matrix are given by

$$L_{11} = (q-1) - \epsilon(\Delta_2 + 2) - 2\left|a_n^0\right|^2 - A\left|b_n^0\right|^2, \quad (7.15)$$

$$L_{12} = -\left(a_n^0\right)^2 - B\left(b_n^0\right)^2, \quad (7.16)$$

$$L_{13} = -Aa_n^0 \left(b_n^0\right)^* - 2B\left(a_n^0\right)^* b_n^0, \quad (7.17)$$

$$L_{14} = -Aa_n^0 b_n^0, \quad (7.18)$$

$$L_{21} = -L_{12}^*, \quad (7.19)$$

$$L_{22} = -L_{11}, \quad (7.20)$$

$$L_{23} = -L_{14}^*, \quad (7.21)$$

$$L_{24} = -L_{13}^*, \quad (7.22)$$

$$L_{31} = L_{13}^\star, \tag{7.23}$$

$$L_{32} = L_{14}, \tag{7.24}$$

$$L_{33} = (q+1) - \epsilon(\Delta_2 + 2) - 2\left|b_n^0\right|^2 - A\left|a_n^0\right|^2, \tag{7.25}$$

$$L_{34} = -\left(b_n^0\right)^2 - B\left(a_n^0\right)^2, \tag{7.26}$$

$$L_{41} = -L_{14}^\star, \tag{7.27}$$

$$L_{42} = -L_{13}, \tag{7.28}$$

$$L_{43} = -L_{34}^\star, \tag{7.29}$$

$$L_{44} = -L_{33}. \tag{7.30}$$

In the above, we use the shorthand notation $(\Delta_2 + 2)z_n = z_{n+1} + z_{n-1}$. In [2], this eigenvalue problem was solved fully in the AC limit of $\epsilon = 0$ for the fundamental branches and subsequent numerical continuation was used to determine the stability of the branches for finite values of ϵ.

We first examine the AC limit of individual sites whose complex fields we decompose as $a_n = r_n e^{i\theta_n}$ and $b_n = s_n e^{i\phi_n}$, obtaining from Eqs. (7.11) and (7.12)

$$(q-1) - \left(r_n^2 + As_n^2\right) - Bs_n^2 e^{2i(\phi_n - \theta_n)} = 0, \tag{7.31}$$

$$(q+1) - \left(s_n^2 + Ar_n^2\right) - Br_n^2 e^{-2i(\phi_n - \theta_n)} = 0. \tag{7.32}$$

From these equations, we obtain

$$\theta_n - \phi_n = k\frac{\pi}{2} \tag{7.33}$$

with $k \in \mathcal{Z}$. The simplest possible solutions are the ones that involve only one of the two branches and were hence termed TE and TM modes, respectively, in [26]. The TE solution of Eqs. (7.31) and (7.32) has the form (in the present limit)

$$r_n = \pm\sqrt{q-1}, \quad s_n = 0 \tag{7.34}$$

and exists only for $q > 1$. On the other hand, the TM mode features

$$r_n = 0, \quad s_n = \pm\sqrt{q+1} \tag{7.35}$$

and is only present for $q > -1$.

In addition to these, there are two possible mixed mode solutions allowed by Eq. (7.33). The first one ($e^{2i(\theta_n - \phi_n)} = 1$) was characterized as a linearly polarized (LP) branch in [3], involving in-phase contributions from both the TE and TM components. In this case, the linear system of Eqs. (7.31) and (7.32) has the general solution

7.2 Nonlinearly Coupled

$$r_n = \pm\sqrt{\frac{(A+B)(q+1) - (q-1)}{(A+B)^2 - 1}}, \tag{7.36}$$

$$s_n = \pm\sqrt{\frac{(A+B)(q-1) - (q+1)}{(A+B)^2 - 1}}. \tag{7.37}$$

If $(A+B)^2 > 1$ (as was the case in the experiment of [3]), this branch only exists for $(A+B)(q+1) > (q-1)$ and $(A+B)(q-1) > (q+1)$ (the sign of the two above inequalities should be reversed for existence conditions in the case of $(A+B)^2 < 1$). Among the two conditions, in the present setting, the second one is the most "stringent" for the case $A = 2B = 1$, which yields the constraint $q \geq 5$ (while the first condition requires for the same parameters $q \geq -5$). Finally, the second mixed mode possibility with $e^{2i(\theta_n - \phi_n)} = -1$ represents the so-called elliptically polarized mode (EP) with amplitudes

$$r_n = \pm\sqrt{\frac{(A-B)(q+1) - (q-1)}{(A-B)^2 - 1}}, \tag{7.38}$$

$$s_n = \pm\sqrt{\frac{(A-B)(q-1) - (q+1)}{(A-B)^2 - 1}}. \tag{7.39}$$

If $(A-B)^2 < 1$ (as is experimentally the case), the EP branch will exist if $q - 1 \geq (A-B)(q+1)$ and $q + 1 \geq (A-B)(q-1)$ (once again the signs should be reversed if $(A-B)^2 > 1$). Here, the first condition is more constraining than the second, imposing for $A = 2B = 1$ that $q \geq 3$ (while the second condition only requires $q \geq -3$).

We now turn to the analysis of the stability of the various single-site branches (TE, TM, LP, and EP) that can be constructed at the AC limit with one excited site, while all others are inert. It is straightforward to see [2] from direct inspection of the stability matrix that the inert sites yield a pair of eigenfrequencies at $\pm(q-1)$, as well as one at $\pm(q+1)$. On the other hand, the excited site will yield a non-trivial 4×4 block in the stability matrix. One pair of the eigenvalues of that block will be at $\omega = 0$ due to the gauge invariance of the solution, associated with the conservation of the total power P. The other pair in the case of the TE mode will be

$$\omega_{TE} = \pm\sqrt{(q+1)^2 + (A^2 - B^2)(q-1)^2 - 2A(q^2 - 1)}. \tag{7.40}$$

Examining our model for the experimental case of $A = 1$, $B = 1/2$, and $q > 0$, we find that this eigenfrequency is real for $q < 5$, while it is imaginary for $q > 5$ (hence implying the presence of an instability, due to the corresponding eigenvalue $\lambda = i\omega$ becoming real). The mode is stable, on the other hand, for $1 < q < 5$. Similar considerations for the TM mode yield

$$\omega_{TM} = \pm\sqrt{(q-1)^2 + (A^2 - B^2)(q+1)^2 - 2A(q^2-1)}, \qquad (7.41)$$

leading to stability for $-1 < q < 3$, and instability for $q > 3$. Finally, for the LP and EP modes, following [2], we only give the results for $A \approx 2B \approx 1$ as

$$\omega_{LP} = \pm \frac{2\sqrt{2}}{5}\sqrt{q^2 - 25} \qquad (7.42)$$

and

$$\omega_{EP} \pm \frac{2\sqrt{2}}{3}\sqrt{q^2 - 9}, \qquad (7.43)$$

suggesting stability for $q \geq 5$ and $q \geq 3$, respectively.

Based on the above observations, one can reconstruct the full picture at the AC limit (and, to be specific, for $A \approx 2B \approx 1$, although it is possible to do for any value of A and of B). In particular, the TE branch exists for $q \geq 1$ and is stable for $1 \leq q \leq 5$. For $q > 5$, the branch is destabilized as a new branch emerges, namely the LP branch, through a pitchfork bifurcation; note that the TM component of this branch, per Eq. (7.37) is exactly zero at $q = 5$, hence it directly bifurcates from the TE branch. The two branches of this supercritical pitchfork correspond to the two signs of s_n in Eq. (7.37). The bifurcating branch "inherits" the stability of the TE branch for all larger values of q, while the latter branch remains unstable thereafter. Similarly, the TM branch exists for $q \geq -1$ and is stable in the interval $-1 \leq q \leq 3$. However, at $q = 3$, a new branch (in fact, a pair thereof) bifurcates with non-zero r_n, beyond the bifurcation point, as in Eq. (7.38). This is accompanied by the destabilization of the TM branch (due to a real eigenvalue) and the *apparent* stability of the ensuing EP branch for all values of $q > 3$.

In the presence of finite coupling, it is firstly important to determine the nature of the continuous spectrum, by using $a_n \sim e^{i(kn - \omega z)}$ and $b_n \sim e^{i(kn - \omega z)}$. The resulting dispersion relations then read

$$\omega = (q - 1) - 2\epsilon \cos(k), \qquad (7.44)$$

$$\omega = (q + 1) - 2\epsilon \cos(k). \qquad (7.45)$$

Hence the continuous spectrum extends through the frequency intervals $[q - 1 - 2\epsilon, q - 1 + 2\epsilon]$ and $[q + 1 - 2\epsilon, q + 1 + 2\epsilon]$ (and their opposites), which for $\epsilon = 0$ degenerate to the isolated points $q - 1$ and $q + 1$ (obtained previously). Also, importantly, for $q - 1 < 2\epsilon$ (equivalently for $\epsilon > (q-1)/2$), the continuous spectrum branch will be crossing the origin, leading to the collision of the eigenvalues with their mirror symmetric opposites, that will, in turn, lead to instabilities. For this reason, we need *only* consider couplings in the interval $\epsilon \in [0, (q-1)/2]$. Using the above pieces of information and two-parameter numerical continuation

7.2 Nonlinearly Coupled

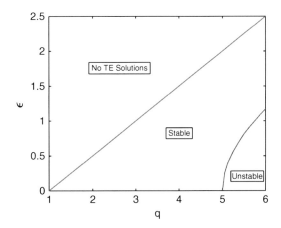

Fig. 7.7 TE branch from [2]: the panel shows the two parameter bifurcation diagram of the coupling ϵ as a function of q. All the relevant existence and stability regimes have been accordingly labeled

in [2], the numerical bifurcation diagrams of the different single-site branches were constructed, which we reproduce below.

The continuation of the TE branch is detailed in Fig. 7.7. For this branch, solutions cannot be obtained for $\epsilon > (q-1)/2$, i.e., the branch terminates at that point with its amplitude tending to zero at this point. Within its region of existence, the branch has a domain of stability and one of instability. The point of separation between the two in the AC limit, studied previously, was the critical point of $q = 5$. For $\epsilon \neq 0$, the separatrix curve is shown in Fig. 7.7 and can be well approximated numerically by the curve $\epsilon_{TE}^c \approx (4\sqrt{2}/5)\sqrt{q-5}$. Hence for $q \leq 5$, the solution is stable for all values of ϵ in its range of existence ($0 < \epsilon < (q-1)/2$), while for $q \geq 5$, the solution is only stable for $\epsilon_{TE}^c < \epsilon < (q-1)/2$ and unstable (due to a real eigenvalue pair) for $0 < \epsilon < \epsilon_{TE}^c$.

The TM branch is somewhat more complicated than the TE one. Firstly, it does not disappear beyond the critical $\epsilon = (q-1)/2$; however, it does become unstable as predicted previously, hence we will, once again, restrict ourselves to this parameter range. Also, similarly to the TE branch case, there is an ϵ_{TM}^c below which the branch is *always* unstable, whereas for $\epsilon > \epsilon_{TM}^c$, the branch *may* be stable. At the AC limit, the critical point for the instability is $q = 3$, as discussed previously; for $q > 3$, the critical point is obtained numerically in Fig. 7.8. It can be well approximated numerically (close to $q = 3$) by $\epsilon_{TM}^c \approx (9/10)\sqrt{q-3}$.

However, within the range of *potential* stability ($0 \leq \epsilon \leq (q-1)/2$ for $q \leq 3$, and $\epsilon_{TM}^c \leq \epsilon \leq (q-1)/2$ for $q \geq 3$), we observe an additional large region of instability in the two-parameter bifurcation diagram of Fig. 7.8, due to a complex quartet of eigenvalues. This instability appears to stem from the point with $(q, \epsilon) = (2.2, 0)$ in the AC limit, and to linearly expand its range as ϵ increases. At $q = 2.2$ in the AC limit, the point spectrum eigenfrequency of Eq. (7.41) "collides" with the continuous spectrum point of concentration, corresponding to $\omega = q - 1$. However, this eigenvalue is associated with a negative Krein signature, i.e., opposite to that of the continuous band at $\omega = q - 1$ (see Chap. 2 for a detailed discussion of the Krein signature). The resulting collision therefore leads to the formation of a quartet

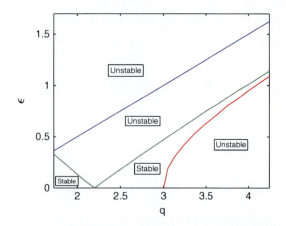

Fig. 7.8 Similarly to Fig. 7.7, the bifurcation diagram shows the two-parameter plane of stability of the TM branch

of eigenvalues emerging in the complex plane and, in turn, implying the instability of the TM configuration. As ϵ grows, the continuous spectrum band grows linearly in ϵ, hence the corresponding interval of q's, where this instability is present also grows at the same rate. Along the same vein, it is worth pointing out that the line of this instability threshold and that of $\epsilon = (q - 1)/2$ are parallel.

The LP branch has in-phase contributions of the TE and TM modes and exists for $q > 5$. It emerges through a supercritical pitchfork as q is varied for fixed ϵ. Since this branch stems from the TE one, it only arises for $0 < \epsilon < \epsilon_{TE}^c$ and $q > 5$ and it is stable throughout its interval of existence, which is exactly the interval of Fig. 7.7 where the TE branch is found to be unstable.

Finally, the EP branch is shown in Fig. 7.9 and its description is somewhat analogous to that of the LP one. In particular, for fixed q close to (and larger than) 3 and varying ϵ, the branch exists and is *stable* for $0 < \epsilon < \epsilon_{TM}^c$, since it emerges from the TM branch through a supercritical pitchfork (as q is increased). Interestingly, for $q > 3.62$, this phenomenology appears to change and an expanding

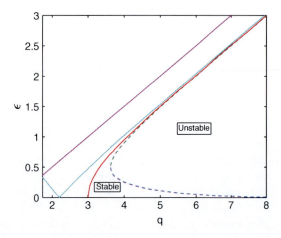

Fig. 7.9 The panel shows the two-parameter diagram for the EP mode from [2]. The region for $q > 3$ and $0 < \epsilon < \epsilon_{TM}^c$, where this modes exists is separated by the dashed line into a stable and an unstable regime. For comparison the extension of the regions of stability/instability of the TM mode from Fig. 7.8 are also included

(for increasing q) interval of oscillatory instability within the range of existence of the EP branch appears to arise. Returning to the AC limit, we note that the EP branch has a point spectrum eigenfrequency given by Eq. (7.43) with a negative Krein signature which upon collision with the continuous spectrum band of eigenfrequencies leads to instability. Setting the frequency of Eq. (7.43) equal to $q - 1$, we obtain that this collision occurs at $q = 9$. For lower values of q, this "collision" will occur for a finite (non-zero) interval of values of ϵ, which is the source of the oscillatory instability of the EP mode shown by the dashed line in Fig. 7.9.

7.2.2 Higher Dimensions

In addition to the one-dimensional incarnation of the above nonlinearly coupled mode, a number of studies have considered ground states [1] or excited states [19, 20] in two-dimensional DNLS lattices and even three-dimensional examples thereof [21]. Here we give some representative examples of these results and refer the interested reader to the corresponding references for more details.

In particular, in two-dimensional settings, among the most interesting solutions that it is possible to construct are the vortex pairs that were considered in [19]. Such pairs can have the form of the so-called double-charge configuration (S, S), where S is the topological charge of the structure, or the so-called hidden charge configuration $(S, -S)$, where the pair denotes the vorticity of each of the components. It was shown even in the continuum analog of NLS-type models that these distinct possibilities have *different* stability windows in terms of the model parameters [27, 28]. In particular, in the setting of [19], a so-called vortex cross configuration of charge $S = 1$ was considered as the single-component building block consisting of a vortex on the four sites: $(-1, 0)$, $(1, 0)$, $(0, 1)$, and $(0, -1)$. In fact, such a vortex cross inspired by a prototypical configuration of the form $u_{n,m} \propto \exp(i\phi) = \cos(\phi) + i \sin(\phi)$ was the original motivation in [29] for suggesting the existence of a discrete vortex in the context of the two-dimensional DNLS model. For this discrete vortex cross, the technique of the Lyapunov–Schmidt reductions as developed in Chap. 3 yields two pairs of imaginary eigenvalues $\lambda = \pm 2i\epsilon$, while a higher order calculation yields for the remaining pair of nonzero eigenvalues (since the fourth pair is at the origin due to the $U(1)$ invariance) the approximation $\lambda = \pm 4i\epsilon^2$. The comparison of these predictions with the full numerical results is shown in Fig. 7.10, indicating good agreement with the analytical predictions for coupling strengths up to $\epsilon \approx 0.1$. The instability of this mode arises for $\epsilon \approx 0.395$.

Subsequently, the case of coupled vortices of the double charge and of the hidden charge variety were examined in the model

$$\left[i\frac{d}{dt} + C\Delta_2 + \begin{pmatrix} |\phi_{m,n}|^2 & \beta|\psi_{m,n}|^2 \\ \beta|\phi_{m,n}|^2 & |\psi_{m,n}|^2 \end{pmatrix}\right] \begin{pmatrix} \phi_{m,n} \\ \psi_{m,n} \end{pmatrix} = 0, \qquad (7.46)$$

where β is used to denote the strength of the XPM (and equal SPMs [self-phase modulations] are assumed). In that setting, it was found by appropriately extending

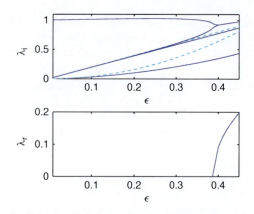

Fig. 7.10 Eigenvalues of the scalar vortex cross versus ϵ from [19]. The *top panel* shows the imaginary part of the relevant eigenvalues, while the *bottom panel* shows the real part. The *solid lines* display the numerical results, while the *dashed* ones correspond to the asymptotic approximations given in the text

the LS technique to the multicomponent setting (see [19] for details) that it is possible to compute the relevant eigenvalues as a function of β. Two leading order pairs of these eigenvalues preserve the form of the one-component problem ($\lambda = \pm 2i\epsilon$), but then there exists a pair which is intrinsically dependent on β at the leading order, namely $\lambda = \pm 2i\epsilon\sqrt{(1-\beta)/(1+\beta)}$ (which is shared by both double and hidden charge configurations). More importantly, at the next order, the eigenvalues of the $(1, 1)$ configuration *differ* from those of the $(1, -1)$ configuration. In particular, for the former we have a pair $\lambda = \pm 4i\epsilon^2$ and one which is $\lambda = \pm 4i|(1-\beta)/(1+\beta)|\epsilon^2$, while for the latter there is a double pair $\lambda = \pm 4i\sqrt{(1-\beta)/(1+\beta)}\epsilon^2$. Interestingly, these differences in eigenvalues are evident also in the numerical results illustrated in Fig. 7.11; note, in particular, the marked differences between the $(1, 1)$ and $(1, -1)$ eigenvalues, and the good agreement of both with the corresponding theoretical result for small $\epsilon < 0.1$. Along the same vein, it should be pointed out that the double charge branch $(1, 1)$ becomes unstable for $\epsilon > 0.395$, while the $(1, -1)$ branch becomes unstable only for $\epsilon > 0.495$, i.e., has a wider stability interval. We have found this to be generally true for the cases with $\beta < 1$. On the other hand, for values of $\beta > 1$, both branches are always unstable (i.e., $\forall \epsilon$). Lastly, the most delicate case is that of $\beta = 1$, whereby there is an additional homotopic symmetry between the two components, as both the transformations $\phi_{n,m} = \cos(\delta)\Phi_{n,m}$ and $\psi = \sin(\delta)\Phi_{n,m}$ (pertaining to the $(1, 1)$ solution for $\delta = \pi/4$) and $\phi_{n,m} = \cos(\delta)\Phi_{n,m}$ and $\psi = \sin(\delta)\Phi^*_{n,m}$ (pertaining to the $(1, -1)$ solution for $\delta = \pi/4$), yield a one-component equation (and δ is a free parameter). Hence, these cases need to be treated specially, as was done in [19], with the, perhaps somewhat unexpected, result that this special case leads to stability for small ϵ in the $(1, 1)$ case, while it results in *immediate* instability for the $(1, -1)$ case due to a double real eigenvalue pair $\lambda = \pm 2\sqrt{3}\epsilon^2$ (note the marked difference between this special case and the cases with $\beta < 1$).

In addition to these solutions associated with $S = 1$, it is also possible to obtain similar results for the different vortex pairs of $S = 3$, in the form of both double charge $(3, 3)$ and hidden charge $(3, -3)$ vortices, as is illustrated in Fig. 7.12 for $\epsilon = 0.25$ and $\beta = 2/3$. In this particular case, for both types of configuration,

7.2 Nonlinearly Coupled

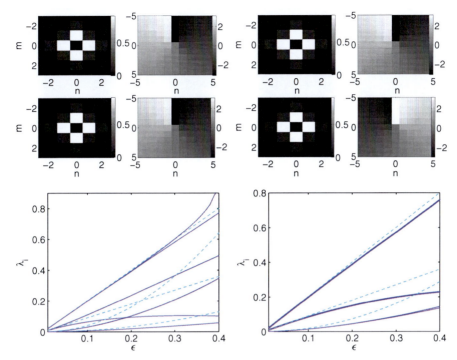

Fig. 7.11 From [19]: the contour plots of the *top two rows* show the amplitude and phase (*left and right panels*, respectively) of the two components (*top* and *bottom*, respectively) for a $(1, 1)$ (*left four subplots*) and a $(1, -1)$ (*right four subplots*) vortex configuration, in the case of $\beta = 2/3$, and $\epsilon = 0.1$. The *bottom two rows* show for the case of $\beta = 2/3$ the eigenvalues of the vector vortex cross as a function of ϵ. Left: $(1, 1)$. Right: $(1, -1)$. The *solid lines* show the numerical results, while the *dashed lines* show the asymptotic approximations. Bold curves correspond to double eigenvalues

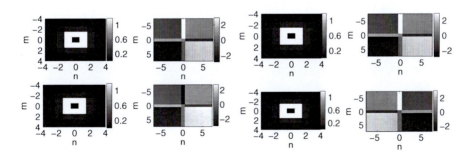

Fig. 7.12 The *left panels* show the case of a double charge with $S = 3$ (i.e., a $(3, 3)$ two-component vortex, similarly to the *top two rows* of Fig. 7.11 above). The *right panels* illustrate the hidden charge case of a $(3, -3)$ two-component vortex. Both are for $\epsilon = 0.25$ and $\beta = 2/3$

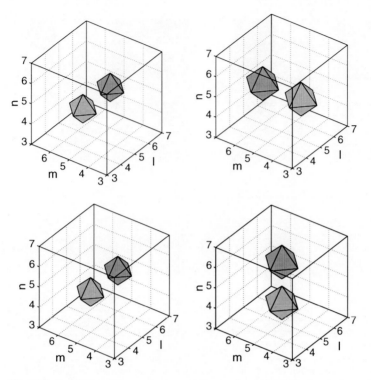

Fig. 7.13 A complex of two-component orthogonal vortices with $S = 1$ from [21] in the two-component three-dimensional system is shown for $\epsilon = 0.01$ and $\beta = 0.5$. The *top panels* correspond to the first component, while the bottom ones to the second component; the *left panels* show the respective real parts, while the right ones the corresponding imaginary parts

the instability occurs for $\epsilon \geq 0.1$; however, again the differences in the relevant eigenvalues and instability growth rates are evident. For instance, for $\epsilon = 0.4$, the growth rate of the most unstable eigenvalue for the $(3, 3)$ configuration is 0.3077, while for the $(3, -3)$ one it is 0.2097.

Finally, as an interesting example of a possibility that arises in such multicomponent systems in three spatial dimensions, we illustrate the result of Fig. 7.13, whereby a stable vortex complex has been constructed in which one component has a vortex in the (l, m) plane, while the other has a vortex in the (m, n) plane (perpendicular to the first one). Such configurations were indicated in [21] as being stable for sufficiently small ϵ, again in the case where $\beta < 1$ (while they should be expected to be unstable for $\beta > 1$).

References

1. Hudock, J., Kevrekidis, P.G., Malomed, B.A., Christodoulides, D.N.: Phys. Rev. E **67**, 056618 (2003)
2. Horne, R., Kevrekidis, P.G., Whitaker, N.: Phys. Rev. E **73**, 066601 (2006)

References

3. Meier, J., Hudock, J., Christodoulides, D.N., Stegeman, G., Silberberg, Y., Morandotti, R., Aitchison, J.S.: Phys. Rev. Lett. **91**, 143907 (2003)
4. Jin, G.-R., Kim, C.-K., Nahm, K.: Phys. Rev. A **72**, 045601 (2005)
5. Alfimov, G.L., Kevrekidis, P.G., Konotop, V.V., Salerno, M.: Phys. Rev. E **66**, 046608 (2002)
6. Ballagh, R.J., Burnett, K., Scott, T.F.: Phys. Rev. Lett. **78**, 1607 (1997)
7. Deconinck, B., Kevrekidis, P.G., Nistazakis, H.E., Frantzeskakis, D.J.: Phys. Rev. A **70**, 063605 (2004)
8. Ohberg, P., Stenholm, S.: Phys. Rev. A **59**, 3890 (1999)
9. Williams, J., Walser, R., Cooper, J., Cornell, E., Holland, M.: Phys. Rev. A **59**, R31 (1999)
10. Herring, G., Kevrekidis, P.G., Malomed, B.A., Carretero-González, R., Frantzeskakis, D.J.: Phys. Rev. E **76**, 066606 (2007)
11. Flach, S., Kladko, K., MacKay, R.S.: Phys. Rev. Lett. **78**, 1207 (1997)
12. Weinstein, M.I.: Nonlinearity **12**, 673 (1999)
13. Kastner, M.: Phys. Rev. Lett. **93**, 150601 (2004)
14. Kevrekidis, P.G., Rasmussen, K.Ø., Bishop, A.R.: Phys. Rev. E **61**, 2006 (2000)
15. Lederer, F., Darmanyan, S., Kobyakov, A.: In: Spatial Optical Solitons. Trillo, S., Torruellas, W.E. (eds.) Springer-Verlag, New York, (2001)
16. Darmanyan, S., Kobyakov, A., Schmidt, E., Lederer, F.: Phys. Rev. E **57**, 3520 (1998)
17. Kevrekidis, P.G., Nistazakis, H.E., Frantzeskakis, D.J., Malomed, B.A., Carretero-González, R.: Eur. Phys. J. D **28**, 181 (2004)
18. Ablowitz, M.J., Musslimani, Z.H.: Phys. Rev. E **65**, 056618 (2002)
19. Kevrekidis, P.G., Pelinovsky, D.E.: Proc. R. Soc. A **462**, 2671 (2006)
20. Kevrekidis, P.G., Malomed, B.A., Frantzeskakis, D.J., Carretero-González, R.: In: Focus on Soliton Research. Chen, L.V. (ed.) pp. 139–166. Nova Science Publishers, New York, (2006)
21. Kevrekidis, P.G., Malomed, B.A., Frantzeskakis, D.J., Carretero-González, R.: Phys. Rev. Lett. **93**, 080403 (2004)
22. Meier, J., Hudock, J., Christodoulides, D.N., Stegeman, G.I., Yang, H.Y., Salamo, G., Morandotti, R., Aitchison, J.S., Silberberg, Y.: J. Opt. Soc. Am. B **22**, 1432 (2005)
23. Vicencio, R.A., Molina, M.I., Kivshar, Yu.S.: Opt. Lett. **29**, 2905 (2004)
24. Vicencio, R.A., Molina, M.I., Kivshar, Yu.S.: Phys. Rev. E **71**, 056613 (2005)
25. Rapti, Z., Trombettoni, A., Kevrekidis, P.G., Frantzeskakis, D.J., Malomed, B.A., Bishop, A.R.: Phys. Lett. A **330**, 95 (2004)
26. Meier, J., Stegeman, G.I., Christodoulides, D.N., Silberberg, Y., Morandotti, R., Yang, H., Salamo, G., Sorel, M., Aitchison, J.S.: Phys. Rev. Lett. **92**, 163902 (2004)
27. Desyatnikov, A.S., Mihalache, D., Mazilu, D., Malomed, B.A., Denz, C., Lederer, F.: Phys. Rev. E **71**, 026615 (2005)
28. Ye, F., Wang, J., Dong, L., Li, Y.P.: Opt. Commun. **230**, 219 (2004)
29. Malomed, B.A., Kevrekidis, P.G.: Phys. Rev. E **64**, 026601 (2001)

Part II
Special Topics

Chapter 8
Experimental Results Related to DNLS Equations

Mason A. Porter

8.1 Introduction

Discrete nonlinear Schrödinger (DNLS) equations can be used to model numerous phenomena in atomic, molecular, and optical physics. The general feature of these various settings that leads to the relevance of DNLS models is a competition between nonlinearity, dispersion, and spatial discreteness (which can be periodic, quasiperiodic, or random). In three dimensions (3D), the DNLS with cubic nonlinearity is written in normalized form as

$$i\dot{u}_{l,m,n} = -\epsilon \Delta u_{l,m,n} \pm |u_{l,m,n}|^2 u_{l,m,n} + V_{l,m,n}(t) u_{l,m,n}, \tag{8.1}$$

where Δ is the discrete Laplacian, ϵ is a coupling constant, $u_{l,m,n}$ is the value of the field at site (l, m, n), and $V_{l,m,n}(t)$ is the value of the external potential at that site. In Eq. (8.1), a $+$ sign represents the defocusing case and a $-$ sign represents the focusing one. Both of these situations have been discussed extensively throughout this book.

The study of DNLS equations dates back to theoretical work on biophysics in the early 1970s [1]. In the late 1980s, this early research motivated extensive analysis of such equations for the purpose of modeling the dynamics of pulses in optical waveguide arrays [2]. One decade later, experiments using fabricated aluminum gallium arsenide (AlGaAs) waveguide arrays [3] stimulated a huge amount of subsequent research, including experimental investigations of phenomena such as discrete diffraction, Peierls barriers, diffraction management [4, 5], gap solitons [6], and more [7]. As was first suggested theoretically in [8] and realized experimentally in [9–12], DNLS equations also accurately predict the existence and stability properties of nonlinear localized waves in optically induced lattices in photorefractive media such as strontium barium niobate (SBN). Because of this success, research in this

---

M.A. Porter (✉)
Oxford Centre for Industrial and Applied Mathematics, Mathematical Institute, University of Oxford, Oxford, England, UK
e-mail: porterm@maths.ox.ac.uk

arena has exploded; structures such as dipoles [13], quadrupoles [14], multiphase patterns (including soliton necklaces and stripes) [15, 16], discrete vortices [17, 18], and rotary solitons [19] have now been theoretically predicted and experimentally obtained in lattices induced with a self-focusing nonlinearity. As discussed in [20] (and references therein), self-defocusing realizations have also been obtained. These allow the construction of dipole-like gap solitons, etc.

DNLS equations have also been prominent in investigations of Bose–Einstein condensates (BECs) in optical lattice (OL) potentials, which can be produced by counterpropagating laser beams along one, two, or three directions [21]. This field has also experienced enormous growth in the last 10 years; major experimental results that have been studied using DNLS equations include modulational ("dynamical") instabilities [22, 23], gap soliton dynamics [24], Bloch oscillations and Landau–Zener tunneling [25], and the production of period-doubled solutions [26].

8.2 Optics

For many decades, optics has provided one of the traditional testbeds for investigations of nonlinear wave propagation [27]. For example, the (continuous) nonlinear Schrödinger (NLS) equation provides a dispersive envelope wave model for describing the electric field in optical fibers. In the presence of a spatially discrete external potential (such as a periodic potential), one can often reduce the continuous NLS to the discrete NLS. In this section, we will consider some appropriate situations that arise in optical waveguide arrays and photorefractive crystals. Many of these optical phenomena have direct analogs in both solid state and atomic physics [28].

8.2.1 Optical Waveguide Arrays

An optical waveguide is a physical structure that guides electromagnetic waves in the optical spectrum. Early proposals in nonlinear optics suggested that light beams can trap themselves by creating their own waveguide through the nonlinear Kerr effect [29]. Waveguides confine the diffraction, allowing spatial solitons to exist. In the late 1990s, Eisenberg et al. showed experimentally that a similar phenomenon (namely, discrete spatial solitons) can occur in a coupled array of identical waveguides [3]. One injects low-intensity light into one waveguide (or a small number of neighboring ones); this causes an ever increasing number of waveguides to couple as it propagates, analogously to what occurs in continuous media. If the light has high intensity, the Kerr effect changes the refractive index of the input waveguides, effectively decoupling them from the rest of the array. That is, certain light distributions propagate with a fixed spatial profile in a limited number of waveguides; see Figs. 8.1 and 8.2.

The standard theoretical approach used to derive a DNLS equation in the context of one-dimensional (1D) waveguide arrays is to decompose the total field

8 Experimental Results Related to DNLS 177

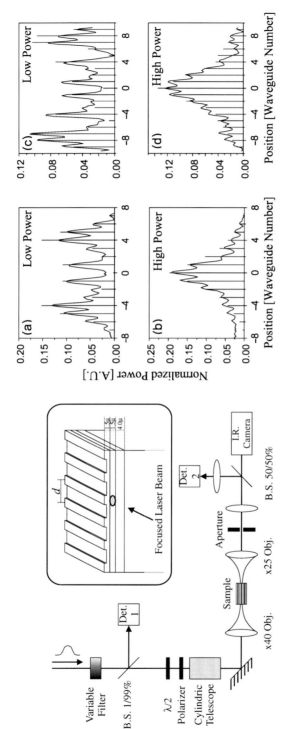

Fig. 8.1 (*Left*) Experimental setup for the waveguide array experiments reported in [3]. (*Right*) Low-power (diffraction) versus high-power experiments. The latter result in discrete spatial solitons. Reprinted from Figs. 2, 4, and 5 with permission from [3]. Copyright 1998 by the American Physical Society

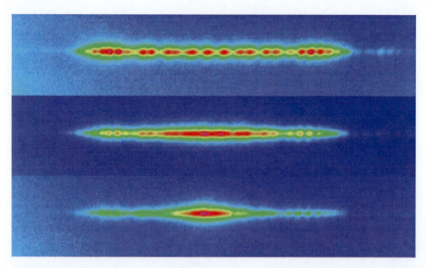

Fig. 8.2 Experimental observation of a soliton (*bottom panel*) at the output facet of a waveguide array. The peak powers are 70 W (*top*), 320 W (*center*), and 500 W (*bottom*). Reprinted from Fig. 3 with permission from [3]. Copyright 1998 by the American Physical Society

(describing the envelope amplitude) into a sum of weakly coupled fundamental modes that are excited in each individual waveguide [2, 30]. If one supposes that each waveguide is only coupled to its nearest neighbors, this approach amounts to the *tight-binding approximation* of solid-state physics. For lossless waveguides with a Kerr (cubic) nonlinearity, one obtains the DNLS

$$i\frac{dE_n}{dz} + \beta E_n + c(E_{n+1} + E_{n-1}) + \lambda |E_n|^2 E_n + \mu(|E_{n+1}|^2 + |E_{n-1}|^2) E_n = 0, \quad (8.2)$$

where E_n is the mode amplitude of the nth waveguide, z is the propagation direction, β is the field propagation constant of each waveguide, c is a coupling coefficient, and λ and μ are positive constants that, respectively, determine the strengths of the self-phase and cross-phase modulations experienced by each waveguide. The quantity λ is proportional to the optical angular frequency and Kerr nonlinearity coefficient, and is inversely proportional to the effective area of the waveguide modes [30]. In most situations, the self-phase modulation dominates the cross-phase modulation (which arises from the nonlinear overlap of adjacent modes), so that $\mu \ll \lambda$. This allows one to set $\mu = 0$ and yields the model that Davydov employed for α-spiral protein molecules [1]. The transformation $E_n(z) = \Phi_n(z) \exp[i(2c + \beta)z]$ and a rescaling then gives a 1D version of Eq. (8.1).

The impact of [3] was immediate and powerful, as numerous subsequent experiments reported very interesting phenomena. For example, this setting provided the first experimental demonstration of the Peierls–Nabarro (PN) potential in a macroscopic system [4], thereby explaining the strong localization observed for high-intensity light in the original experiments [3]. That is, the PN potential describes the energy barrier between the (stable) solitons that are centered on a waveguide and propagate along the waveguide direction and the (unstable) ones that are centered symmetrically between two waveguides and tend to shift away from the waveguide direction. Eisenberg et al. [5] have also exploited diffraction management (which is analogous to the dispersion management ubiquitously employed in the study of temporal solitons [31]) to produce structures with designed (reduced, canceled, or reversed) diffraction properties. The ability to engineer the diffraction properties has paved the way for new possibilities (not accessible in bulk media) for controlling light flow. For example, using two-dimensional (2D) waveguide networks, discrete solitons can travel along essentially arbitrarily curves and be routed to any destination [32]. This may prove extremely helpful in the construction of photonic switching architectures.

A waveguide array with linearly increasing effective refractive index, which can be induced using electro- or thermo-optical effects, has also been used to demonstrate Bloch oscillations (periodic recurrences) in which the initial distribution is recovered after one oscillation period [33]. A single-waveguide excitation spreads over the entire array before refocusing into the initial guide. More recently, Morandotti et al. investigated the interactions of discrete solitons with structural defects produced by modifying the spacing of one pair of waveguides in an otherwise uniform array [34]. This can be used to adjust the PN potential. It has also been demonstrated experimentally that even a binary array is sufficient to generate discrete gap solitons, which can then be steered via inter-band momentum exchange [35]. From a nonlinear dynamics perspective, an especially exciting result is the experimental observation of discrete modulational instabilities [36]. Using an AlGaAs waveguide array with a self-focusing Kerr nonlinearity, Meier et al. found that such an instability occurs when the initial spatial Bloch momentum vector is within the normal diffraction region of the Brillouin zone. (It is absent even at very high-power levels in the anomalous diffraction regime.) More recent experimental observations include discrete spatial gap [37] and dark [38] solitons in photovoltaic lithium niobate ($LiNbO_3$) waveguide arrays, evidence for the spontaneous formation of discrete X waves in AlGaAs waveguide arrays [39], and an analog of Anderson localization (which occurs in solid-state physics when an electron in a crystal becomes immobile in a disordered lattice) [40].

8.2.2 Photorefractive Crystals

Photorefractive crystals can be used to construct 2D periodic lattices via plane wave interference by employing a technique known as optical induction. This method, which was developed theoretically in [8] and subsequently demonstrated experimentally for 1D discrete solitons in [10] and bright 2D solitons in [9], has become a very important playground for investigations of nonlinear waves in optics [41]. One obtains a periodic lattice in real time through the interference of two or more plane waves in a photosensitive material (see Fig. 8.3). One then launches a probe beam,

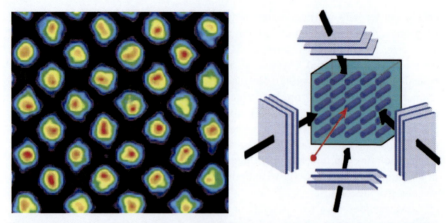

Fig. 8.3 (*Right*) Diagram of experimental setup for the creation of a photorefractive crystal lattice with electro-optic anisotropy. (*Left*) Typical observation of the lattice at the terminal face of the crystal. Each waveguide has a diameter of about 7 μm and is about 11 μm away from its nearest neighbors. Reprinted with permission from Fig. 1 in [9]. Copyright 2003 by the Nature Publishing Group

which experiences discrete diffraction (the optical equivalent of quantum tunnelling in a periodic potential) and can form a discrete soliton provided the nonlinearity is sufficiently large. The model for photorefractive crystals is a continuous NLS equation with saturable nonlinearity [9],

$$iU_z + U_{xx} + U_{yy} - \frac{E_0}{1 + I_l + |U|^2} U = 0, \tag{8.3}$$

where z is the propagation distance, (x, y) are transverse coordinates, U is the slowly varying amplitude of the probe beam (normalized by the dark irradiance of the crystal), and E_0 is the applied dc field, and I_l is a lattice intensity function. For a square lattice, $I_l = I_0 \sin^2\{(x+y)/\sqrt{2}\} \sin^2\{(x-y)/\sqrt{2}\}$, where I_0 is the lattice's peak intensity. DNLS equations have been enormously insightful in providing corroborations between theoretical predictions and experimental observations (see, in particular, the investigations of discrete vortices in [17, 18, 42]), although they do not provide a prototypical model in this setting the way they do with waveguide arrays.

For optical induction to work, it is essential that the interfering waves are unaffected by the nonlinearity (to ensure that the "waveguides" are as uniform as possible) but that the probe (soliton-forming) beam experiences a significant nonlinearity. This can be achieved by using a photorefractive material with a strong electro-optic anisotropy. In such materials, coherent rays interfere with each other and form a spatially varying pattern of illumination (because the local index of refraction is modified, via the electro-optic effect, by spatial variations of the light intensity). This causes ordinary polarized plane waves to propagate almost linearly (i.e., with practically no diffraction) and extraordinary polarized waves to propagate

8 Experimental Results Related to DNLS

in a highly nonlinear fashion. The material of choice in the initial experiments of [9] was the (extremely anisotropic) SBN:75 crystal.

The theoretical prediction and subsequent experimental demonstration of 2D discrete optical solitons has led to the construction and analysis of entirely new families of discrete solitons [8, 9]. As has been discussed throughout this book, the extra dimension allows much more intricate nonlinear dynamics to occur than is possible in the 1D waveguides discussed above. Early experiments demonstrated novel self-trapping effects such as the excitation of odd and even nonlinear localized states [11]. They also showed that photorefractive crystals can be used to produce index gratings that are more controllable than those in fabricated waveguide arrays.

Various researchers have since exploited the flexibility of photorefractive crystals to create interesting, robust 2D structures that have the potential to be used as carriers and/or conduits for data transmission and processing in the setting of all-optical communication schemes (see [41, 42] and references therein). In the future, 1D arrays in 2D environments might be used for multidimensional waveguide junctions, which has the potential to yield discrete soliton routing and network applications. For example, Martin et al. observed soliton-induced dislocations and deformations in photonic lattices created by partially incoherent light [12]. By exploiting the photorefractive nonlinearity's anisotropy, they were able to create optical structures analogous to polarons[1] from solid-state physics [43]. An ever-larger array of structures has been predicted and experimentally obtained in lattices induced with a self-focusing nonlinearity. For example, Yang et al. demonstrated discrete dipole (two-hump) [13] and quadrupole (four-hump) solitons [14] both experimentally and theoretically. They also showed that both dipole and quadrupole solitons are stable in a large region of parameter space when their humps are out of phase with each other. (The stable quadrupole solitons were square-shaped; adjacent humps had a phase difference of π, so that diagonal humps had the same phase.) Structures that have been observed in experiments in the self-defocusing case include dipole-like gap solitons [20] and gap-soliton vortices [44].

More complicated soliton structures have also been observed experimentally. For example, appropriately launching a high-order vortex beam (with, say, topological charge $m = 4$) into a photonic lattice can produce a stationary necklace of solitons [15]. Stripes of bright [16] and gap [45] solitons have also been created [16], providing an interesting connection with several other pattern-forming systems [46]. Another very fruitful area has been the construction of both off-site and on-site discrete vortices [17, 18, 42]. Recent experimental observations in this direction have included self-trapping and charge-flipping of double-charged optical vortices (which lead to the formation of rotating quasivortex solitons) [47]. Discrete rotary solitons [19] and discrete random-phase solitons [48] have also been observed. In fact, the results of [48] are reminiscent of the Fermi–Pasta–Ulam (FPU)

---

[1] A polaron is a quasiparticle composed of a conducting electron and an induced polarization field that moves with the electron.

Fig. 8.4 Experimental image of the decagonal field-intensity pattern in an optically induced nonlinear photonic quasicrystal. Adapted with permission from Fig. 1 in [50]. Copyright 2006 by the Nature Publishing Group

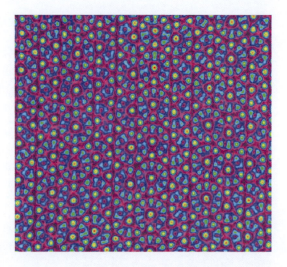

numerical experiments [49], as an initially homogeneous distribution in momentum space evolved into a steady-state multihumped soliton power spectrum.

The investigation of photorefractive crystals continues to produce experimental breakthroughs, offering ever more connections to solid-state physics. One particularly exciting experiment was the observation of dispersive shock waves [51]. Another fascinating result was the observation of an analog of Anderson localization in disordered 2D photonic lattices [52]. In this context, the transverse localization of light is caused by random fluctuations. Wave, defect, and phason dynamics (including discrete diffraction and discrete solitons) have recently been investigated experimentally in optically induced nonlinear photonic *quasicrystals* [50, 53] (see Fig. 8.4), whose theoretical investigation provides one of the outstanding challenges for DNLS models.

8.3 Bose–Einstein Condensation

At sufficiently low temperature, bosonic particles in a dilute 3D gas occupy the same quantum (ground) state, forming a BEC [54–57]. Seventy years after they were first predicted theoretically, dilute (i.e., weakly interacting) BECs were finally observed experimentally in 1995 in vapors of rubidium and sodium [58, 59]. In these experiments, atoms were loaded into magnetic traps and evaporatively cooled to temperatures well below a microkelvin. To record the properties of the BEC, the confining trap was then switched off, and the expanding gas was optically imaged [55]. A sharp peak in the velocity distribution was observed below a critical temperature T_c, indicating that condensation had occurred.

If the temperature is well below T_c, then considering only two-body, mean-field interactions, the BEC dynamics is modeled using the 3D Gross–Pitaevskii (GP) equation (i.e., the continuous cubic NLS equation),

8 Experimental Results Related to DNLS 183

$$i\hbar\Psi_t = \left(-\frac{\hbar^2\nabla^2}{2m} + g_0|\Psi|^2 + \mathcal{V}(r)\right)\Psi, \tag{8.4}$$

where $\Psi = \Psi(r, t)$ is the condensate wave function (order parameter) normalized to the number of atoms, $\mathcal{V}(r)$ is the external potential, and the effective self-interaction parameter is $\tilde{g} = [4\pi\hbar^2 a/m][1 + O(\zeta^2)]$, where a is the two-body scattering length and $\zeta \equiv \sqrt{|\Psi|^2|a|^3}$ is the dilute gas parameter [55, 60, 61]. The cubic nonlinearity arises from the nearly perfect contact (delta function) interaction between particles.

In a quasi-1D ("cigar-shaped") BEC, the transverse dimensions are about equal to the healing length, and the longitudinal dimension is much larger than the transverse ones. One can then average (8.4) in the transverse plane to obtain the 1D GP equation [55, 62],

$$i\hbar u_t = -\left[\frac{\hbar^2}{2m}\right]u_{xx} + g|u|^2 u + V(x)u, \tag{8.5}$$

where u, g, and V are, respectively, the rescaled 1D wave function, interaction parameter, and external trapping potential. The interatomic interactions in BECs are determined by the sign of g: they are repulsive (producing a defocusing nonlinearity) when $g > 0$ and attractive (producing a focusing nonlinearity) when $g < 0$.

BECs can be loaded into OL potentials (or superlattices, which are small-scale lattices subjected to a large-scale modulation), which are created experimentally as interference patterns of laser beams. Consider two identical laser beams with parallel polarization and equal peak intensities, and counterpropagate them as in Fig. 8.5a so that their cross sections overlap completely. The two beams create an interference pattern with period $d = \lambda_L/2$ (half of the optical wavelength) equal to the distance between consecutive maxima of the resulting light intensity. The potential experienced by atoms in the BEC is then [21]

$$V(x) = V_0 \cos^2\left(\frac{\pi x}{d}\right), \tag{8.6}$$

where V_0 is the lattice depth. See [21] for numerous additional details.

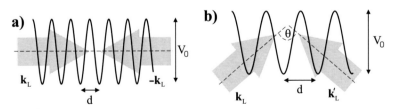

Fig. 8.5 Diagram of the creation of a 1D optical lattice potential using (**a**) counterpropagating laser beams and (**b**) beams intersecting at an angle θ inducing a spacing $d = \lambda_L \cos(\theta/2)/2$. The quantities $\mathbf{k}_L$ and $\mathbf{k}'_L$ denote the wave vectors of the beams. The lattice period is given by the distance d between consecutive maxima of light intensity in the interference pattern. Reprinted with permission from Fig. 1 in [21]. Copyright 2006 by the American Physical Society

BECs were first successfully placed in OLs in 1998 [25], and numerous labs worldwide now have the capability to do so. Experimental and theoretical investigations of BECs in OLs (and related potentials) have developed into one of the most important subdisciplines of BEC investigations [21, 54, 63]. We focus here on results that can be modeled using a DNLS framework; see the reviews [21, 64] for discussions of and references to many other outstanding experiments. The first big experimental result was the observation of Bloch oscillations in a repulsive BEC by Anderson and Kasevich [25]. They used a trapping potential with both an OL and a linear (gravitational) component to create a sloping periodic ("washboard") potential. When the slope was small, wave packets remained confined in a single band and oscillated coherently at the Bloch frequency. This effect is closely related to the ac Josephson effect in superconducting electronic systems. For larger slopes, wave packets were able to escape their original band and transition to higher states. Trombettoni and Smerzi analyzed the results of these experiments using a DNLS equation that they derived from the GP equation with the appropriate (OL plus gravitational) potential using the tight-binding approximation valid for moderate amplitude potentials in which Bloch waves are strongly localized in potential wells [65]. In this paper, they also showed that discrete breathers (specifically, bright gap solitons) can exist in BECs with repulsive potentials (i.e., defocusing nonlinearities).

Other fundamental experimental work on BECs in OLs has concerned superfluid properties. In 2001, Burger et al. treated this setting as a homogeneous superfluid with density-dependent critical velocity [66]. Cataliotti et al. then built on this research to examine a classical transition between superfluid and Mott insulator behavior in BECs loaded into an OL superimposed on a harmonic potential [23]. (The better-known quantum transition was first shown in [67].) The BEC exhibits coherent oscillations in the "superfluid" regime and localization in the harmonic trap in the "insulator" regime, in which each site has many atoms of its own and is effectively its own BEC. The transition from superfluidity to the insulating state occurs when the condensate wave packet's initial displacement is larger than some critical value or, equivalently, when the velocity of its center of mass is larger than a critical velocity that depends on the tunneling rate between adjacent OL sites. These experiments confirmed the predictions of [22], which used a DNLS approach to predict the onset of this superfluid–insulator transition via a discrete modulational ("dynamical") instability and to derive an analytical expression for the critical velocity at which it occurs.

Subsequent theoretical work with DNLS equations predicted that modulational instabilities could lead to "period-doubled" solutions in which the BEC wave function's periodicity is twice that of the underlying OL [68]. (Period-doubled wave functions were simultaneously constructed using a GP approach [69].) The modulational instability mechanism was exploited experimentally the next year to construct these solutions by parametrically exciting a BEC via periodic translations (shaking) of the OL potential [26]. Parametric excitation of BECs promises to lead to many more interesting insights in the future.

By balancing the spatial periodicity of the OL with the nonlinearity in the DNLS, one can also construct intrinsic localized modes (discrete breathers) known as bright gap solitons, which resemble those supported by Bragg gratings in nonlinear optical systems. In BECs, such breathers have been predicted in two situations:

1. The small amplitude limit in which the value of chemical potential is close to forbidden zones ("gaps") of the underlying linear Schrödinger equation with a periodic potential [70].
2. In the tight-binding approximation, for which the continuous NLS equation with a periodic potential can be reduced to the DNLS equation [22]. (As mentioned earlier, this corresponds to the standard manifestation of the DNLS in waveguide arrays.)

Recent experiments [24] have confirmed the first prediction (see Fig. 8.6).

Another important development was the experimental construction of 2D and 3D OLs [21], which as in optical systems leads to much more intricate nonlinear localized structures. One can obtain higher dimensional OL potentials by using additional pairs of laser beams. The simplest way to do this is to have pairs of counter-propagating laser beams along each of two or three mutually orthogonal axes. The interference pattern obtained with this many laser beams depends sensitively on their polarizations, relative phases, and orientations. This allows experimentalists to construct a large variety of OL geometries in 2D and 3D. In 2001, Greiner et al. showed that BECs can be efficiently transferred into 2D lattice potentials by adiabatically increasing the depth of the lattice [71, 72]. They confined atoms to an array of narrow potential tubes, each of which was filled with a 1D quantum gas. Around the same time, Burger et al. confined quasi-2D BECs into the lattice sites of a 1D OL potential [73]. By adding more laser beams and/or controlling their polarizations and relative phases, experimentalists can in principle create even more complicated potentials (such as quasiperiodic or Kagomé lattices) [74]. Very recently, there has been also been a great deal of theoretical and experimental interest in rotating optical lattices [75], which can likely be modeled using an appropriate DNLS framework to study interesting vortex dynamics. Such experiments and (more generally) investigations of BECs with effective fields obtained by this and other [76] means promise to yield considerable insights into quantum hall physics.

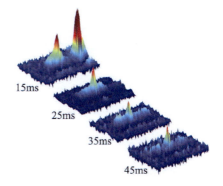

Fig. 8.6 Experimental demonstration of a bright gap soliton (the shading shows the atomic density). A small, stable peak forms after about 25 ms. Reprinted with permission from Fig. 1 in [24]. Copyright 2004 by the American Physical Society

In addition to the fascinating insights into basic physics discussed above, two other major consequences of investigations of BECs in OLs and related potentials have been to bring quantum computation one (small) step closer to reality and to help bridge the gap between condensed matter physics and atomic/molecular physics. One of the key proposed systems for constructing a quantum computer is a BEC in optical lattice and related potentials [77, 78]. This was the motivation for the experimental implementation of optical superlattice potentials [79] and its 2D egg-carton descendent, which consists of an optical lattice potential in one cardinal direction and a double-well potential in another [80]. This has led very recently to the experimental realization of a two-qubit quantum gate [81]. The second front has been advanced experimentally in the OL context by investigations of Fermi condensates in OL potentials [82, 83], Bose–Fermi mixtures in OL potentials [84], and more. Finally, in parallel with the recent insights in optics discussed above, Anderson localization has been observed recently both for a BEC placed in a 1D waveguide with controlled disorder [85] and for a BEC in a 1D quasiperiodic OL [86].

8.4 Summary and Outlook

In this review, I have discussed applications in nonlinear optics and Bose–Einstein condensation in which DNLS equations have been used to explain fundamental and striking experimental results. In optics, DNLS equations provide a prototypical model for the dynamics of discrete solitons in waveguide arrays. The same is true for BECs in optical lattice (and related) potentials. DNLS equations have also been used successfully to predict robust experimental features in photorefractive crystals, although they do not provide a prototypical model in this setting.

DNLS equations arise in a number of other contexts as envelope models for several types of nonlinear lattice equations (such as ones of Klein–Gordon type). Related experiments have revealed the existence of intrinsic localized modes in these systems [87]. Relevant settings include quasi-1D antiferromagnets [88], micromechanical oscillator arrays [87], and electric transmission lines [89].

DNLS equations have also been used in a variety of other settings to make interesting predictions that have not yet been verified experimentally. For example, in composite metamaterials, Shadrivov et al. have analyzed the modulational instability of different nonlinear states and demonstrated that nonlinear metamaterials support the propagation of domain walls (kinks) that connect regions of positive and negative magnetization [90]. Very recently, dissipative discrete breathers were constructed in a model of rf superconducting quantum interference device (SQUID) arrays [91]. (Similar discrete breathers have also recently been studied theoretically in both 1D and 2D in the setting of metamaterials [92].) This model is reminiscent of a DNLS with dissipation, except that the nonlinearity was sinusoidal rather than cubic. A bit farther afield, 1D chains of granular materials (sometimes called "phononic crystals") have been given increasing attention from both experimentalists and theorists in recent years [93, 94]. When given an initial pre-compression, they can exhibit optical modes that are expected to be describable as gap solitons in a nonlinear lattice model reminiscent of FPU chains.

In conclusion, DNLS equations and related models have been incredibly successful in the description of numerous experiments in nonlinear optics and Bose–Einstein condensation. They also show considerable promise in a number of other settings, and related nonlinear lattice models are also pervasive in a huge number of applications. To borrow a phrase from the defunct rock band Timbuk3, the future's so bright that we've got to wear shades.

Acknowledgments I would like to thank Panos Kevrekidis for the invitation to write this article. I would also like to acknowledge Martin Centurion, Panos Kevrekidis, Alex Nicolin, Yaron Silberberg, and Ian Spielman for reading and providing critical comments on drafts of this manuscript. Finally, I also thank Markus Oberthaler, Moti Segev, and Yaron Silberberg for permission to use their figures.

References

1. Davydov, A.S.: J. Theor. Bio. **38**, 559 (1973)
2. Christodoulides, D.N., Joseph, R.I.: Opt. Lett. **13**, 794 (1988)
3. Eisenberg, H.S., Silberberg, Y., Morandotti, R., Boyd, A.R., Aitchison, J.S.: Phys. Rev. Lett. **81**, 3383 (1998)
4. Morandotti, R., Peschel, U., Aitchison, J.S., Eisenberg, H.S., Silberberg, Y.: Phys. Rev. Lett. **83**, 2726 (1999)
5. Eisenberg, H.S., Silberberg, Y., Morandotti, R., Aitchison, J.S.: Phys. Rev. Lett. **85**, 1863 (2000)
6. Mandelik, D., Morandotti, R., Aitchison, J.S., Silberberg, Y.: Phys. Rev. Lett. **92**, 093904 (2004)
7. Morandotti, R., Eisenberg, H.S., Silberberg, Y., Sorel, M., Aitchison, J.S.: Phys. Rev. Lett. **86**, 3296 (2001)
8. Efremidis, N.K., Sears, S., Christodoulides, D.N., Fleischer, J.W., Segev, M.: Phys. Rev. E **66**, 046602 (2002)
9. Fleischer, J.W., Segev, M., Efremidis, N.K., Christodoulides, D.N.: Nature **422**, 147 (2003)
10. Fleischer, J.W., Carmon, T., Segev, M., Efremidis, N.K., Christodoulides, D.N.: Phys. Rev. Lett. **90**, 023902 (2003)
11. Neshev, D., Ostrovskaya, E., Kivshar, Yu.S., Krolikowski, W.: Opt. Lett. **28**, 710 (2003)
12. Martin, H., Eugenieva, E.D., Chen, Z., Christodoulides, D.N.: Phys. Rev. Lett. **92**, 123902 (2004)
13. Yang, J., Makasyuk, I., Bezryadina, A., Chen, Z.: Opt. Lett. **29**, 1662 (2004)
14. Yang, J., Makasyuk, I., Bezryadina, A., Chen, Z.: Stud. App. Math. **113**, 389 (2004)
15. Yang, J., Makasyuk, I., Kevrekidis, P.G., Martin, H., Malomed, B.A., Frantzeskakis, D.J., Chen, Z.: Phys. Rev. Lett. **94**, 113902 (2005)
16. Neshev, D., Kivshar, Yu.S., Martin, H., Chen, Z.: Opt. Lett. **29**, 486 (2004)
17. Neshev, D.N., Alexander, T.J., Ostrovskaya, E.A., Kivshar, Yu.S., Martin, H., Makasyuk, I., Chen, Z.: Phys. Rev. Lett. **92**, 123903 (2004)
18. Fleischer, J.W., Bartal, G., Cohen, O., Manela, O., Segev, M., Hudock, J., Christodoulides, D.N.: Phys. Rev. Lett. **92**, 123904 (2004)
19. Wang, X., Chen, Z., Kevrekidis, P.G.: Phys. Rev. Lett. **96**, 083904 (2006)
20. Tang, L., Lou, C., Wang, X., Song, D., Chen, X., Xu, J., Chen, Z., Susanto, H., Law, K., Kevrekidis, P.G.: Opt. Lett. **32**, 3011 (2007)

21. Morsch, O., Oberthaler, M.: Rev. Mod. Phys. **78**, 179 (2006)
22. Smerzi, A., Trombettoni, A., Kevrekidis, P.G., Bishop, A.R.: Phys. Rev. Lett. **89**, 170402 (2002)
23. Cataliotti, F.S., Fallani, L., Ferlaino, F., Fort, C., Maddaloni, P., Inguscio, M.: New J. Phys. **5**, 71 (2003)
24. Eiermann, B., Anker, Th., Albiez, M., Taglieber, M., Treutlein, P., Marzlin, K.-P., Oberthaler, M.K.: Phys. Rev. Lett. **92**, 230401 (2004)
25. Anderson, B.P., Kasevich, M.A.: Science **282** 1686 (1998)
26. Gemelke, N., Sarajlic, E., Bidel, Y., Hong, S., Chu, S.: Phys. Rev. Lett. **95**, 170404 (2005)
27. Kivshar, Yu.S., Agrawal, G.P.: Optical Solitons: From Fibers to Photonic Crystals. Academic Press, San Diego (2003)
28. Scott, A. (ed.): Encyclopedia of Nonlinear Science. Lectures in Applied Mathematics, Routledge, Taylor & Francis Group, New York (2005)
29. Agrawal, G.P.: Nonlinear Fiber Optics. Academic Press, San Diego, (1995)
30. Sukhorukov, A.A., Kivshar, Y.S., Eisenberg, H.S., Silberberg, Y.: IEEE J. Quant. Elect. **39**, 31 (2003)
31. Malomed, B.A.: Soliton Management in Periodic Systems. Springer, New York (2005)
32. Christodoulides, D.N., Lederer, F., Silberberg, Y.: Nature **424**, 817 (2003)
33. Morandotti, R., Peschel, U., Aitchison, J.S., Eisenberg, H.S., Silberberg, Y.: Phys. Rev. Lett. **83**, 4756 (1999)
34. Morandotti, R., Eisenberg, H.S., Mandelik, D., Silberberg, Y., Modotto, D., Sorel, M., Stanley, C.R., Aitchison, J.S.: Opt. Lett. **28**, 834 (2003)
35. Morandotti, R., Mandelik, D., Silberberg, Y., Aitchison, J.S., Sorel, M., Christodoulides, D.N., Sukhorukov, A.A., Kivshar, Yu.S.: Opt. Lett. **29**, 2890 (2004)
36. Meier, J., Stegeman, G.I., Christodoulides, D.N., Silberberg, Y., Morandotti, R., Yang, H., Salamo, G., Sorel, M., Aitchison, J.S.: Phys. Rev. Lett. **92**, 163902 (2004)
37. Chen, F., Stepic, M., Ruter, C.E., Runde, D., Kip, D., Shandarov, V., Menela, O., Segev, M.: Opt. Expr. **13**, 4314 (2005)
38. Smirnov, E., Rüter, C.E., Stepić, M., Kip, D., Shandarov, V.: Phys. Rev. E **74**, 065601 (2006)
39. Lahini, Y., Frumker, E., Silberberg, Y., Droulias, S., Hizanidis, K., Morandotti, R., Christodoulides, D.N.: Phys. Rev. Lett. **98**, 023901 (2007)
40. Lahini, Y., Avidan, A., Pozzi, F., Sorel, M., Morandotti, R., Christodoulides, D.N., Silberberg, Y.: Phys. Rev. Lett. **100**, 013906 (2008)
41. Fleischer, J., Bartal, G., Cohen, O., Schwartz, T., Manela, O., Freedman, B., Segev, M., Buljan, H., Efremidis, N.: Opt. Expr. **13**, 1780 (2005)
42. Chen, Z., Martin, H., Bezryadina, A., Neshev, D., Kivshar, Yu.S., Christodoulides, D.N.: J. Opt. Soc. Am. B **22**, 1395 (2005)
43. Ashcroft, N.W., Mermin, N.D.: Solid State Physics. Brooks/Cole, Australia (1976)
44. Song, D., Lou, C., Tang, L., Wang, X., Li, W., Chen, X., Law, K.J.H., Susanto, H., Kevrekidis, P.G., Xu, J., Chen, Z.: Opt. Express **16**, 10110 (2008) Personal communication
45. Wang, X., Chen, Z., Wang, J., Yang, J.: Phys. Rev. Lett. **99**, 243901 (2007)
46. Cross, M.C., Hohenberg, P.C.: Rev. Mod. Phys. **65**, 851 (1993)
47. Bezryadina, A., Eugenieva, E., Chen, Z.G.: Opt. Lett. **31**, 2456 (2006)
48. Cohen, O., Bartal, G., Buljian, H., Carmon, T., Fleischer, J.W., Segev, M., Christodoulides, D.N.: Nature **433**, 500 (2005)
49. Campbell, D.K., Rosenau, P., Zaslavsky, G.: Chaos **15**, 015101 (2005)
50. Freedman, B., Bartal, G., Segev, M., Lifshitz, R., Christodoulides, D.N., Fleischer, J.W.: Nature **440**, 1166 (2006)
51. Wan, W., Jia, S., Fleischer, J.W.: Nature Physics **3**, 46 (2007)
52. Schwartz, T., Bartak, G., Fishman, S., Segev, M.: Nature **446**, 52 (2007)
53. Freedman, B., Lifshitz, R., Fleischer, J.W., Segev, M.: Nature Materials **6**, 776 (2007)
54. Pethick, C.J., Smith, H.: Bose-Einstein Condensation in Dilute Gases, 2nd edn. Cambridge University Press, Cambridge, UK (2008)
55. Dalfovo, F., Giorgini, S., Pitaevskii, L.P., Stringari, S.: Rev. Mod. Phys. **71**, 463 (1999)

56. Ketterle, W.: Phys. Today, **52**, 30 (Dec. 1999)
57. Burnett, K., Edwards, M., Clark, C.W.: Phys. Today **52**, 37 (Dec. 1999)
58. Anderson, M.H., Ensher, J.R., Matthews, M.R., Wieman, C.E., Cornell, E.A.: Science **269**, 198 (1995)
59. Davis, K.B., Mewes, M.-O., Andrews, M.R., van Druten, N.J., Durfee, D.S., Kurn, D.M., Ketterle, W.: Phys. Rev. Lett. **75**, 3969 (1995)
60. Köhler, T.: Phys. Rev. Lett. **89**, 210404 (2002)
61. Baizakov, B.B., Konotop, V.V., Salerno, M.: J. Phys. B: At. Mol. Opt. Phys. **35**, 5105 (2002)
62. Salasnich, L., Parola, A., Reatto, L.: J. Phys. B: At. Mol. Opt. Phys **35**, 3205 (2002)
63. Porter, M.A., Carretero-González, R., Kevrekidis, P.G., Malomed, B.A.: Chaos **15**, 015115 (2005)
64. Fallani, L., Fort, C., Inguscio, M.: Adv. At. Mol. Opt. Phys. **56**, 119 (2008)
65. Trombettoni, A., Smerzi, A.: Phys. Rev. Lett. **86**, 2353 (2001)
66. Burger, S., Cataliotti, F.S., Fort, C., Minardi, F., Inguscio, M.: Phys. Rev. Lett. **86**, 4447 (2001)
67. Greiner, M., Mandel, O., Esslinger, T., Hänsch, T., Bloch, I.: Nature **415**, 39 (2002)
68. Machholm, M., Nicolin, A., Pethick, C.J., Smith, H.: Phys. Rev. A **69**, 043604 (2004)
69. Porter, M.A., Cvitanović, P.: Phys. Rev. E 2 **69**, 047201 (2004)
70. Konotop, V.V., Salerno, M.: Phys. Rev. A **65**, 021602(R) (2002)
71. Greiner, M., Bloch, I., Mandel, O., Hänsch, T.W., Esslinger, T.: Appl. Phys. B **73**, 769 (2001)
72. Greiner, M., Bloch, I., Mandel, O., Hänsch, T.W., Esslinger, T.: Phys. Rev. Lett. **87**, 160405 (2001)
73. Burger, S., Cataliotti, F.S., Fort, C., Maddaloni, P., Minardi, F., Inguscio, M.: Europhys. Lett. **57**, 1 (2002)
74. Santos, L., Baranov, M.A., Cirac, J.I., H.-Everts, U., Fehrmann, H., Lewenstein, M.: Phys. Rev. Lett. **93**, 030601 (2004)
75. Tung, S., Schweikhard, V., Cornell, E.A.: Phys. Rev. Lett. **97**, 240402 (2006)
76. Juzeliūnas, G., Ruseckas, J., Öhberg, P., Fleischhauer, M.: Phys. Rev. A **73**, 025602 (2006)
77. Vollbrecht, K.G.H., Solano, E., Cirac, J.L.: Phys. Rev. Lett. **93**, 220502 (2004)
78. Cirac, I.J., Zoller, P.: Phys. Today **57**, 38 (Mar 2004)
79. Peil, S., Porto, J.V., Laburthe Tolra, B., Obrecht, J.M., King, B.E., Subbotin, M., Rolston, S.L., Phillips, W.D.: Phys. Rev. A **67**, 051603(R) (2003)
80. Anderlini, M., Sebby-Strabley, J., Kruse, J., Porto, J.V., Phillips, W.D.: J. Phys. B: At. Mol. Opt. Phys. **39**, S199 (2006)
81. Anderlini, M., Lee, P.J., Brown, B.L., Sebby-Strabley, J., Phillips, W.D., Porto, J.V.: Nature **448**, 452 (2007)
82. Modugno, G., Ferlaino, F., Heidemann, R., Roati, G., Inguscio, M.: Phys. Rev. A **68**, 011601 (2003)
83. Köhl, M., Moritz, H., Stöferle, T., Günter, K., Esslinger, T.: Phys. Rev. Lett. **94**, 080403 (2005)
84. Ott, H., de Mirandes, E., Ferlaino, F., Roati, G., Modugno, G., Inguscio, M.: Phys. Rev. Lett. **92**, 160601 (2004)
85. Billy, J., Josse, V., Zuo, Z., Bernard, A., Hambrecht, B., Lugan, P., Clément, D., Sanchez-Palencia, L., Bouyer, P., Aspect, A.: Nature **453**, 891 (2008)
86. Roati, G., D'Errico, C., Fallani, L., Fattori, M., Fort, C., Zaccanti, M., Modugno, G., Modugno, M., Inguscio, M.: Nature **453**, 895 (2008)
87. Sato, M., Hubbard, B.E., Sievers, A.J.: Rev. Mod. Phys. **78**, 137 (2006)
88. Wrubel, J.P., Sato, M., Sievers, A.J.: Phys. Rev. Lett. **95**, 264101 (2005)
89. Sato, M., Yasui, S., Kimura, M., Hikihara, T., Sievers, A.J.: Europhys. Lett. **80**, 30002 (2007)
90. Shadrivov, I.V., Zharov, A.A., Zharova, N.A., Kivshar, Yu.S.: Photon. and Nanostruct. – Fund. App. **4**, 69 (2006)
91. Lazarides, N., Tsironis, G.P., Eleftheriou, M.: Nonlinear Phenom. Complex Systems **11**, 250 (2008)
92. Eleftheriou, M., Lazarides, N., Tsironis, G.P.: Phys. Rev. E **77**, 036608 (2008)
93. Nesterenko, V.F.: Dynamics of Heterogeneous Materials. Springer-Verlag, New York, (2001)
94. Daraio, C., Nesterenko, V.F., Herbold, E.B., Jin, S.: Phys. Rev. E **72**, 016603 (2005)

Chapter 9
Numerical Methods for DNLS

Kody J.H. Law and Panayotis G. Kevrekidis

9.1 Introduction

In this section, we briefly discuss the numerical methods that have been used extensively throughout this book to obtain the numerical solutions discussed herein, as well as to analyze their linear stability and to propagate them in time (e.g., to examine their dynamical instability, or to confirm their numerical stability).

Our tool of preference, regarding the numerical identification of solutions consists of the so-called Newton–Raphson (or simply Newton) method. We choose the Newton method because of its quadratic convergence, upon the provision of a suitably good initial guess [1]. It should be clearly indicated here that different groups use different methods to obtain stationary solutions. For instance, methods based on rewriting the standing wave problems of interest in Fourier space and applying Petviashvili's iteration scheme have been proposed [2, 3] and shown to converge for nonlinear Schrödinger-type problems under suitable conditions [4]. Also, methods based on imaginary time integration have been proposed and suitably accelerated [5]; finally, also methods based on constraint minimization of appropriate (e.g., energy) functionals have been developed [6]. However, for the standing wave discrete nonlinear Schrödinger (DNLS) problem, given the existence of the anti-continuum limit of zero coupling, and its analytical tractability (which provides an excellent initial guess and a starting point for parametric continuations), the Newton method posed on the lattice works extremely efficiently. Note that although the Newton method will be presented herein in a simple parametric continuation format, with respect to the coupling parameter ϵ, it is straightforward to combine it also with pseudo-arclength ideas such as the ones discussed in [7], in order to be able to continue the solution past fold points (and to detect relevant saddle-node bifurcations). Although for one-dimensional problems (and even for two-dimensional discrete problems), it is straightforward to use the direct Newton algorithm with full matrices and then perform the linear stability analysis with full eigensolvers,

K.J.H. Law (✉)
University of Massachusetts, Amherst, MA, 01003, USA
e-mail: law@math.umass.edu

in three dimensions (or e.g., in two-dimensional multicomponent systems), the relevant computations become rather intensive. To bypass this problem, we offer a possibility to perform the Newton method (and the subsequent eigenvalue computations) using sparse iterative solvers and sparse matrix eigensolvers that considerably accelerate the computation, based on the work of Kelley [8].

As concerns the direct numerical integration of the DNLS model, our tool of choice for the time stepping herein will be the fourth-order Runge–Kutta method [1]. Although both lower order methods (such as the efficient split-step Fourier method [9]), as well as higher order methods (including even the eighth-order Runge–Kutta method [10]) have been presented and used in the literature, our use of the fourth-order method, we feel, represents a good balance between a relatively high-order local truncation error (accuracy) and stability properties that allow a relatively high value of the time step ($dt = 10^{-3}$ or higher for most cases of interest here) without violating stability conditions.

All of the above methods (existence, linear stability and direct integration) will be presented by means of Matlab [11] scripts in what follows. The scripts will be vectorized to the extent possible to allow for efficient numerical computation and will also be set up to provide "on the fly" visualization of the relevant parametric continuations (for our bifurcation calculations) and the time-stepping evolution (for our direct integrations). The codes presented below can be found at the website [12].

9.2 Numerical Computations Using Full Matrices

We start with a numerical implementation of the one-dimensional Newton algorithm. We recall that the algorithm assumes the simple form $x^{m+1} = x^m - f(x^m)/f'(x^m)$ for approximating the solution x^s such that $f(x^s) = 0$ (m here denotes the algorithm iteration index). The N-dimensional vector generalization for a lattice of N sites in our one-dimensional problem reads

$$\mathbf{J} \cdot \left(\mathbf{x}^{m+1} - \mathbf{x}^m\right) = -\mathbf{F}(\mathbf{x}^m), \tag{9.1}$$

where $\mathbf{J}$ is the Jacobian of the (vector of) N equations $\mathbf{F}$ with respect to the (vector of) N unknowns $\mathbf{x}$, i.e., $J_{ij} = \partial F_i/\partial x_j$. We are writing Eq. (9.1) as indicated above for a reason, namely to highlight that it is far less expensive to perform the Newton algorithm iteration step as a solution of a linear system (for the vector $\mathbf{x}^{m+1}$), rather than through the inversion of the Jacobian. The vector $\mathbf{x}$ in the computations below consists of the lattice field variables u_n, satisfying the vector of equations

$$F_n = \epsilon \Delta_2 u_n + u_n^3 - u_n = 0, \tag{9.2}$$

where we have taken advantage of the real nature of the one-dimensional solutions (although the algorithm can be straightforwardly generalized to complex solutions as needed in higher dimensions). Once the solutions of Eq. (9.2) are identified to a prescribed accuracy (set below to 10^{-8}), linear stability analysis is performed

9 Numerical Methods for DNLS

around the solution as is explained in Chap. 2. The code detailing these bifurcation computations, along with relevant commenting of each step is given below.

```
clear; format long;
% number of sites;
n=100;
% initial coupling; typically at AC-limit eps=0;
eps=0;
% propagation constant; typically set to 1.
l=1;
% field initialization
u1=zeros(1,n);
u1(n/2)=sqrt(l);
u1(n/2+1)=sqrt(l);
% iteration index
it=1;
% lattice index
x=linspace(1,n,n)-n/2;
% continuation in coupling epsilon
while (eps<0.101)
u=zeros(1,n);
while (norm(u-u1)>1e-08)
u=u1;
% evaluation of second difference with free boundaries
sd2=diff(u,2); sd1=u1(2)-u1(1); sdn=u1(n-1)-u1(n);
sd=[sd1,sd2,sdn];
% equation that we are trying to solve
f=-l*u+eps*sd+(u1.^2).*u1;
% auxiliary vectors in Jacobian
ee = eps*ones(1,n);
ee0=ee; ee0(1)=ee(1)/2; ee0(n)=ee(n)/2;
ee1=(-2*ee0-l*ones(1,n)+3*(u1.^2));
%tridiagonal Jacobian
jj1 = spdiags([ee' ee1' ee'], -1:1, n, n);
% Newton correction step
cor=( jj1 \ f' )'; u1=u-cor;
% convergence indicator: should converge quadratically
norm(cor)
end;
% auxiliary vector for stability
ee1=(-2*ee0-l*ones(1,n)+2*abs(u1.^2));
% construction of stability matrix
jj2=spdiags([ee' ee1' ee'], -1:1, n, n);
jj3=diag(u1.^2);
jj4=[ jj2, jj3;
-conj(jj3) -conj(jj2)];
% eigenvalues d and eigenvectors v of stability matrix
[v,d]=eig(full(jj4));
d1=diag(d);
% store solution and stability
u_store(:,it)=u;
d_store(:,it)=d1;
e_store(it)=eps;
% visualize the continuation profiles and stability on the fly
```

```
subplot(2,1,1)
plot(x,u,'-o')
drawnow;
subplot(2,1,2)
plot(imag(d1),real(d1),'o')
drawnow;
% increment indices and epsilon
it=it+1;
eps=eps+0.001;
end;
% save the profiles and other data
save('sol_eps_1d.mat','u_store','d_store','e_store')
```

A prototypical result of the continuation of the above code has been given below (the code addresses the unstable case with two excited sites – the inter-site mode) using the command

```
imagesc(0.001*linspace(0,100,101),linspace(1,100,100)-50,u_store)
```

to spatially visualize the branch for different values of ϵ. Also the dominant stability eigenvalues of this unstable branch are shown (more specifically, λ^2) via the commands

```
d2=sort(real(d_store.^2));
plot(e_store,d2(1,:),e_store,d4(3,:),'b--',e_store,d4(5,:),'b-.')
```

The plots for the solution continuation and its stability are depicted in Fig. 9.1. We now turn to the numerical integration of one of the unstable solutions of the above branch (namely, of the solution for $\epsilon = 0.1$) that we saved at the end of the

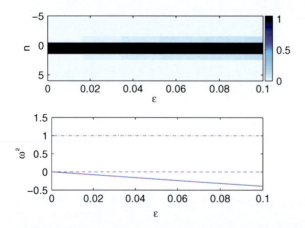

Fig. 9.1 The *top panel* shows the result of continuation as a function of ϵ of the solution profile (shown in contour plot). The *bottom panel* shows the result of the linear stability analysis, indicating the instability of this inter-site mode, through the presence of a negative squared eigenfrequency (*solid line*). The *dashed* pair of eigenvalues at the origin is due to the phase invariance, while the *dash–dotted* pair at 1 (due to the choice of propagation constant $\Lambda = 1$) indicates the lower edge of the continuous spectrum

9 Numerical Methods for DNLS

previous bifurcation code. As indicated above, we use the fourth-order Runge–Kutta method whose four steps and subsequent integration step, we now remind, for the solution of the vector of ordinary differential equations $\mathbf{x}' = \mathbf{f}(t, \mathbf{x})$ with initial condition $\mathbf{x}(t_0) = \mathbf{x}_0$:

$$\mathbf{k}^{(1)} = dt\, \mathbf{f}\left(t^m, \mathbf{x}^m\right), \tag{9.3}$$

$$\mathbf{k}^{(2)} = dt\, \mathbf{f}\left(t^m + \frac{dt}{2}, \mathbf{x}^m + \frac{dt}{2}\mathbf{k}^{(1)}\right), \tag{9.4}$$

$$\mathbf{k}^{(3)} = dt\, \mathbf{f}\left(t^m + \frac{dt}{2}, \mathbf{x}^m + \frac{dt}{2}\mathbf{k}^{(2)}\right), \tag{9.5}$$

$$\mathbf{k}^{(4)} = dt\, \mathbf{f}\left(t^m + dt, \mathbf{x}^m + dt\,\mathbf{k}^{(3)}\right), \tag{9.6}$$

$$\mathbf{x}^{m+1} = \mathbf{x}^m + \frac{1}{6}\left(\mathbf{k}^{(1)} + 2\mathbf{k}^{(2)} + 2\mathbf{k}^{(3)} + \mathbf{k}^{(4)}\right). \tag{9.7}$$

The commented version of the Matlab script that implements this algorithm for the DNLS equation is given below.

```
% parameters
n=100; eps=0.1;
% load solutions from Newton
load sol_eps_1d.mat
u_num=10
u=u_store(:,u_num)'+1e-04*rand(1,n);
x=real(u); y=imag(u);
% spatial lattice index
sp=linspace(1,n,n)-n/2;
% iteration indices and time step
it=1;
dt=0.001;
it1=1;
it2=1;
% integration up to t=100
while ((it-1)*dt<100)
% computation of second differences and 1st RK integration step
d2y=diff(y,2); ad1y=(y(2)-y(1)); ad3y=(y(n-1)-y(n));
d2x=diff(x,2); ad1x=(x(2)-x(1)); ad3x=(x(n-1)-x(n));
p1=[ad1y,d2y,ad3y]; p2=[ad1x,d2x,ad3x];
k1x=dt*(-eps*p1-y.*(x.^2+y.^2));
k1y=dt*(eps*p2+x.*(x.^2+y.^2));
a=x+k1x/2;
b=y+k1y/2;
% computation of second differences and 2nd RK integration step
d2y=diff(b,2); ad1y=(b(2)-b(1)); ad3y=(b(n-1)-b(n));
d2x=diff(a,2); ad1x=(a(2)-a(1)); ad3x=(a(n-1)-a(n));
p1=[ad1y,d2y,ad3y]; p2=[ad1x,d2x,ad3x];
k2x=dt*(-eps*p1-b.*(a.^2+b.^2));
k2y=dt*(eps*p2+a.*(a.^2+b.^2));
```

```
a=x+k2x/2;
b=y+k2y/2;
% computation of second differences and 3rd RK integration step
d2y=diff(b,2); ad1y=(b(2)-b(1)); ad3y=(b(n-1)-b(n));
d2x=diff(a,2); ad1x=(a(2)-a(1)); ad3x=(a(n-1)-a(n));
p1=[ad1y,d2y,ad3y]; p2=[ad1x,d2x,ad3x];
k3x=dt*(-eps*p1-b.*(a.^2+b.^2));
k3y=dt*(eps*p2+a.*(a.^2+b.^2));
a=x+k3x;
b=y+k3y;
% computation of second differences and 4th RK integration step
d2y=diff(b,2); ad1y=(b(2)-b(1)); ad3y=(b(n-1)-b(n));
d2x=diff(a,2); ad1x=(a(2)-a(1)); ad3x=(a(n-1)-a(n));
p1=[ad1y,d2y,ad3y]; p2=[ad1x,d2x,ad3x];
k4x=dt*(-eps*p1-b.*(a.^2+b.^2));
k4y=dt*(eps*p2+a.*(a.^2+b.^2));
% completion of integration from t -> t+dt
x1=x+(k1x+2*k2x+2*k3x+k4x)/6;
y1=y+(k1y+2*k2y+2*k3y+k4y)/6;
% square modulus profile
uu=x1.^2+y1.^2;
% evaluate energy and (l^2 norm)^2 & visualize the solution every
few steps
if (mod(it,100)==0)
% time counter
tim(it1)=(it-1)*dt;
% square l^2 norm; should be conserved
l2(it1)=sum(uu);
% energy calculation; energy should also be conserved
% although to lower accuracy than l2
gr=[diff(x1,1),0]; gr1=[diff(y1,1),0];
ener(it1)=sum(eps*(gr.^2+gr1.^2)-uu.^2/2);
% plot the solution and its energy and l2 norm
subplot(2,1,1)
plot(sp,uu,'-')
drawnow
subplot(2,1,2)
plot(tim,l2,tim,ener,'--')
drawnow
if (mod(it,1000)==0)
u_store(:,it2)=x1+sqrt(-1)*y1;
it2=it2+1;
end;
it1=it1+1;
end;
it=it+1;
x=x1;
y=y1;
end;
```

The result of the integration for the unstable evolution of the solution is shown in Fig. 9.2 (note that in the initial condition the exact solution was perturbed by a random uniformly distributed noise field of amplitude 10^{-4} in order to seed the

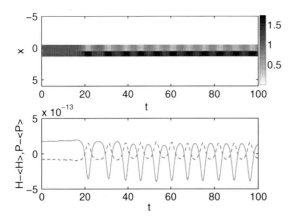

Fig. 9.2 The *top panel* shows the result of the direct integration of the unstable inter-site-centered solution for $\epsilon = 0.1$, with the instability being seeded by a random (uniformly distributed) perturbation of amplitude 10^{-4}. The space–time contour plot of the solution shows how it results into a breathing mode oscillating between the initial condition and a single-site-centered mode. The bottom panel of the figure shows the deviation from the energy H (*solid line*) and the squared l^2 norm P (*dashed line*) conservation. The average energy during the simulation is ≈ -1.007, while the mean of the squared l^2 norm is ≈ 2.199. In both cases, we can see that the deviations from this conservation law are of $O(10^{-13})$

instability). It can be seen how the two-site solution transforms itself into a breathing mode oscillating between a two-site and a single-site solution. On the other hand, the bottom panel of the figure shows the deviation from the relevant (for the DNLS) conservation laws of the energy and the squared l^2 norm. Both of these deviations are of $O(10^{-13})$ as can be seen in the panel while the means of these quantities are of $O(1)$ for the presented simulation. This confirms the good preservation by the proposed scheme of the important conservation laws of the underlying physical model. The above panels are created in Matlab through the use of the commands:

```
subplot(2,1,1)
imagesc(linspace(1,100,100),sp,abs(u_store.^2))
subplot(2,1,2)
plot(tim,l2-mean(l2),'--',tim,ener-mean(ener))
```

Extending the above type of numerical considerations to higher dimensions is conceptually straightforward, although both computationally tedious and obviously far more numerically intensive. We briefly indicate how the above considerations would generalize in two-dimensions (three-dimensional generalizations would naturally extend along the same vein); however, we focus in the next section on how to render these computations more efficient in higher dimensional settings by means of the use of sparse matrix computations and iterative linear solvers.

In the two-dimensional case, the Newton iteration has to extend over a grid of $N \times N$ points, hence the relevant vectors have N^2 elements (more generally N^d for d-dimensional computations). The key realization concerning the performance of operations such as those of Eq. (9.1) with such vectors stemming from higher dimensional grids is that not all points should be treated on the same footing. Taking perhaps the simplest case of vanishing Dirichlet boundary conditions at the edges of our two-dimensional domain, it should be appreciated that while the "inner" $(N-2) \times (N-2)$ nodes of the domain are "regular" points possessing all 4 of their neighbors, there exist an additional $4 \times (N-2)$ "edge" points with only 3 neighbors, while the 4 corner points only have 2 neighbors. That is to say, these points should be treated separately regarding both the equation they satisfy and the nature of their corresponding Jacobian elements. Upon this realization, one can treat the two-dimensional grid as a one-dimensional vector whose elements $(1, 1), \ldots, (1, N)$ become elements $1, \ldots, N$, elements $(2, 1), \ldots, (2, N)$ becomes $N+1, \ldots, 2N$, and so on. Then, when constructing the full Jacobian, one should test whether the vector index i running from 1 to N^2 lies at the corners $(1, N, N^2 - N + 1$ and $N^2)$, or at the edges $1 < i < N$, $N^2 - N + 1 < i < N^2$, $\mathrm{mod}(i, N) = 1$ or $\mathrm{mod}(i, N) = 0$). This testing can be constructed through appropriate if statements, or equivalently by more clever vector manipulations particularly well suited for Matlab. If none of the above happens, then one has all four neighbors (which for the element i are $i+1, i-1, i+N$, and $i-N$ in the quasi-one-dimensional vector implementation of the grid). Using these considerations one can construct the corresponding Jacobian and perform the same bifurcation computations as above.

As regards the two-dimensional Runge–Kutta simulations, things are in fact a bit simpler, as no Jacobian evaluations are needed. Then assuming that the field (and its second differences) are vanishing at the boundaries, which is a reasonable assumption, for the vast majority of the configurations considered in this book, we can construct the second difference operators in a simple vectorized manner as follows:

```
e=zeros(n,1);
Dxmm = diff([e';x;e'],2,1);
Dxnn = diff([e,x,e],2,2);
Dymm = diff([e';y;e'],2,1);
Dynn = diff([e,y,e],2,2);
```

Using those second differences along the two lattice directions, it is straightforward to again perform the steps used to obtain the intermediate integration vectors $\mathbf{k}^{(j)}$ $j = 1, \ldots, 4$, e.g., as follows:

```
k1x=dt*(-eps*(Dymm+Dynn)-y.*(x.^2+y.^2));
k1y=dt*(eps*(Dxmm+Dxnn)+x.*(x.^2+y.^2));
```

Modulo this small modification, the one-dimensional Runge–Kutta realization given above can be essentially immediately transferred into a two-dimensional integrator, upon suitable provision $N \times N$ vector initial conditions $\mathbf{x}_0, \mathbf{y}_0$. The same type of considerations can be immediately extended to three-dimensional computations with Dxmm, Dxnn, Dxll computed similarly using the diff command.

9.3 Numerical Computations Using Sparse Matrices/Iterative Solvers

We discuss two among the many methods for efficiently computing the solution of the Newton fixed point in two dimensions using finite difference derivatives. The more straightforward method is to utilize the sparse banded structure of the Jacobian and the efficiency of the Matlab command "backslash," which will automatically recognize the banded structure and use a banded solver. If memory is not the main consideration, this method is faster. However, one can save a fraction of the memory at the cost of a slightly slower computation utilizing a Newton–Krylov GMRES scheme as implemented by the Matlab script `nsoli` [8]. The memory is minimized by using an Arnoldi iterative algorithm to solve the linear system at each iteration of the Newton method and approximating the Jacobian only in the direction of the Krylov subspace. We note that beyond the standard case outlined here this has far-reaching benefits, particularly when the explicit form of the Jacobian is unknown, or when employing a pseudo-arclength method, for instance, which spoils the banded structure of the Jacobian.

First, we will outline the analogous two-dimensional standard Newton solver (without the tedium of `if` statements) utilizing the structure and sparsity of the Jacobian.

```
clear; format long;
% number of sites;
n=100;
% initial coupling; typically at AC-limit eps=0;
eps=0;
% propagation constant; typically set to 1.
l=1;
% field initialization
ult=zeros(n,n);
ult(n/2,n/2)=sqrt(l);
ult(n/2+1,n/2)=sqrt(l);
% reshape the field into a column vector with real and
% imaginary parts separated for solving
u1=[reshape(real(ult),n*n,1);reshape(imag(ult),n*n,1)];
% iteration index
it=1;
% lattice indices as vectors
x=linspace(1,n,n)-n/2;
y=linspace(1,n,n)-n/2;
% lattice indices as matrices
[X,Y]=meshgrid(x,y);
% continuation in coupling epsilon
eps=0;
% define increment
inc = .001;
while (eps<0.101)
eps=eps+inc;
u=zeros(2*n*n,1);
```

```
% BEGIN - see alternative method below
%
while (norm(u-u1)>1e-08)
u=u1;
% fill in the real and imaginary parts of the original 2d field
uur = u(1:n*n);
ur = reshape(uur,n,n);
uui = u(n*n+1:2*n*n);
ui = reshape(uui,n,n);
% evaluation of second difference with zero fixed boundaries
e=zeros(n,1);
Durmm = diff([e';ur;e'],2,1);
Durnn = diff([e,ur,e],2,2);
Duimm = diff([e';ui;e'],2,1);
Duinn = diff([e,ui,e],2,2);
Dur = Durmm + Durnn;
Dui = Duimm + Duinn;
% Other boundary conditions can be implemented similarly,
% for instance use the following for free boundaries
% diff([ur(1,:);ur;ur(n,:)],2,1); ... etc.
% equation that we are trying to solve (vectorized)
f=[reshape(-1*ur+eps*Dur+(ur.^2+ui.^2).*ur,n*n,1); ...
    reshape(-1*ui+eps*Dui+(ur.^2+ui.^2).*ui,n*n,1)];
% auxiliary vectors for linear part of the Jacobian
ee = eps*ones(n,1);
ee0 = ee; ee0(1) = ee(1)/2; ee0(n) = ee(n)/2;
ee1 = -2*ee0;
% tridiagonal linear component for 1d
jj1 = spdiags([ee ee1 ee], -1:1, n, n);
% quick conversion of the 1d n × n tridiagonal
% into the 2d n² × n² quintidiagonal via the Kronecker product.
% [ note that the linear component of the Jacobian only needs to
% be constructed once at the beginning of the code, but is left
% here for clarity. ]
jj22 = kron(speye(n),jj1) + kron(jj1,speye(n))-1*speye(n*n,n*n);
% nonlinear component of the Jacobian
n1 = sparse(n*n,n*n);
n2 = sparse(n*n,n*n);
n3 = sparse(n*n,n*n);
n1 = spdiags(uui.^2 + 3*uur.^2,0,n1);
n2 = spdiags(uur.^2 + 3*uui.^2,0,n2);
n3 = spdiags(2*uur.*uui,0,n3);
nn = [n1 n3;n3 n2];
clear n1 n2 n3
% completion of the Jacobian
jj2 = nn + [ jj22, sparse(n*n,n*n);sparse(n*n,n*n), jj22];
clear nn
% Newton correction step
cor = jj2 \ f ;
clear jj2
u1 = u-cor;
% convergence indicator: should converge quadratically
norm(cor)
```

```
end;
% END - see alternative method below
%
% convert back into a complex valued vector
uu1 = u1(1:n*n)+sqrt(-1)*u1(n*n+1:2*n*n);
% and it's 2d representation
uu2 = reshape(uu1,n,n);
% construct nonlinear part of stability matrix
n1 = sparse(n*n,n*n);
n2 = sparse(n*n,n*n);
n1 = spdiags(uu1.^2,0,n1);
n2 = spdiags(2*abs(uu1).^2,0,n2);
jj2 = jj22 + n2;
jj3=[ jj2, n1;
-conj(n1) -conj(jj2)];
% smallest magnitude ('SM') 100 eigenvalues d and
% eigenvectors v of stability matrix
% (utilizing a shift in order to improve stability of the
% algorithm)
d1=eigs(jj3+sqrt(-1)*3*speye(2*n*n),100,'SM')-sqrt(-1)*3;
% store solution and stability
u_store(:,it)=uu1;
d_store(:,it)=d1;
e_store(it)=eps;
% visualize the continuation profiles and stability on the fly
subplot(2,2,1)
imagesc(x,y,abs(uu2).^2)
drawnow;
subplot(2,2,2)
imagesc(x,y,angle(uu2))
drawnow;
subplot(2,2,3)
imagesc(x,y,imag(uu2))
drawnow;
subplot(2,2,4)
plot(imag(d1),real(d1),'o')
drawnow;
% increment indices and epsilon
it=it+1;
eps=eps+0.001;
end;
% save the data
save('sol_eps_2d.mat','u_store','d_store',e_store');
```

We should make a few comments here about the code given above. We have illustrated in the existence portion (the Newton method) and the stability section the respective formulations of the complex-valued problem in terms of the real and imaginary components of the field and the field and its complex conjugate (rotation of the former), respectively. The formulations can be interchanged, and it is unnecessary to represent both components in the existence section for real-valued solutions (as with one-dimensional solutions) or equivalently purely imaginary ones, while it is always necessary in the linearization problem because we must consider complex valued perturbations even of real solutions. There are many alternative options in eigs other than "SM," which the interested readers should perhaps explore. On the other hand, the latter is particularly efficient here because the relevant eigenvalues bifurcate from the origin in the anticontinuum limit. The reader should be aware, however, that eigs uses an Arnoldi iterative method to calculate the eigenvalues and the (unshifted) linearization system is singular due to phase invariance. Therefore, some care has to be taken, by a shift or otherwise.

Now, we briefly outline the alternative method using the Matlab script nsoli [8] for the existence portion of the above routine. The reader should note that in addition to the benefits mentioned above, it may be attractive as a black box since it slims down the code. Consult the help file or [8] for more details about it.

```
% setup the parameters for nsoli
max_iter=200;
max_iter_linear=100;
etamax=0.9;
lmeth=1;
restart_limit=20;
sol_parms=[max_iter,max_iter_linear,etamax,lmeth,restart_limit];
error_flag=0;
tolerance=1e-8*[1,1]
% solve with nsoli the equation F(u_)
[u1,iter_hist,error_flag]=nsoli(u1,@(u_)F(u_,eps,n,l),
tolerance,sol_parms);
if error_flag==0
else
disp('there was an error')
iter_hist
break
end
% Now define the function F as a new script F.m in the
% same directory. (Functions cannot be defined within scripts,
% but can be defined within functions, so if the whole code is
% made into a function, then this routine can be embedded.)
function f = F(u,eps,n,l)
ur = reshape(u(1:n*n),n,n);
ui = reshape(u(n*n+1:2*n*n),n,n);
% evaluation of second difference with zero fixed boundaries
e=zeros(n,1);
Durmm = diff([e';ur;e'],2,1);
Durnn = diff([e,ur,e],2,2);
Duimm = diff([e';ui;e'],2,1);
Duinn = diff([e,ui,e],2,2);
Dur = Durmm + Durnn;
Dui = Duimm + Duinn;
% equation that we are trying to solve (vectorized)
f(1:n*n)=reshape(-l*ur+eps*Dur+(ur.^2+ui.^2).*ur,n*n,1);
f(n*n+1:2*n*n)=reshape(-l*ui+eps*Dui+(ur.^2+ui.^2).*ui,n*n,1);
```

9.4 Conclusions

In this section, we have presented an overview of numerical methods used in order to perform both bifurcation, as well as time-stepping computations with the discrete nonlinear Schrödinger equation. We have partitioned our presentation into roughly two broad classes of such methods, namely the ones that use full matrices and direct linear solvers, and ones that use sparse matrices and iterative linear solvers. Our tool of choice for bifurcation calculations has been the Newton method, especially due to the existence of the so-called anti-continuum limit, whereby configurations of interest can be constructed in an explicit form and subsequently continued via either parametric or pseudo-arclength continuation techniques. On the other hand, for the direct dynamical evolution aspects of the problem, we advocated that the use of Runge–Kutta methods affords us the possibility to use relatively large time steps, while at the same time preserving to a satisfactory degree the conservation laws (such as energy and l^2 norm) associated with the underlying Hamiltonian system. The codes given above can be found at the website [12].

References

1. Atkinson, K.: An Introduction to Numerical Analysis. John Wiley, New York (1989)
2. Ablowitz, M.J., Musslimani, Z.H.: Opt. Lett. **30**, 2140 (2005)
3. Lakoba, T.I., Yang, J.: J. Comp. Phys. **226**, 1668 (2007)
4. Pelinovsky, D.E., Stepanyants, Yu.A.: SIAM J. Numer. Anal. **42**, 1110 (2004)
5. Yang, J., Lakoba, T.I.: Stud. Appl. Math. **120**, 265 (2008)
6. Garcia-Ripoll, J.J., Perez-Garcia, V.M.: SIAM J. Sci. Comput. **23**, 1315 (2001)
7. http://en.wikipedia.org/wiki/Numerical_continuation
8. Kelley, C.T.: Iterative methods for linear and nonlinear equations. Frontiers in Applied Mathematics, vol. 16, SIAM, Philadelphia (1995)
9. Taha, T.R., Ablowitz, M.J.: J. Comp. Phys. **55**, 203 (1984)
10. Hairer, E., Norsett, S.P., Wanner, G.: Solving ordinary differential equations. I: Nonstiff Problems. Springer-Verlag, Berlin (1987)
11. http://www.mathworks.com
12. http://www.math.umass.edu/~law/Research/DNLS_codes/

Chapter 10
The Dynamics of Unstable Waves

Kody J.H. Law and Q. Enam Hoq

10.1 Introduction

Discretized equations in which the evolution variable is continuous while the spatial variables are confined to points on the lattice, have had a significant presence and impact across multiple disciplines [1–6] (cf. Chap. 8 for a review on experiments related to the DLNS equation). This should not be surprising considering that many natural processes and phenomena exhibit discrete structure [7–11]. As particular examples, we see that in physics the discrete nonlinear Schrödinger equation was used to model periodic optical structures [12], while in biology the Davydov equations model energy transfer in proteins [13]. These diverse phenomena are testament to the ubiquitousness of discrete regimes across diverse settings, and hence validate the need to study them. The study of discretized equations can be traced at least back to the work of Frenkel and Kontorova on crystal dislocations [14] and the Fermi–Pasta–Ulam problem [15]. The literature has since grown significantly to include novel and engaging ideas (e.g. [16–25]), and one of the points of interest is the dynamical behavior of solutions [26–28].

In this section, we investigate the dynamics of unstable wave solutions to the cubic discrete nonlinear Schrödinger (DNLS) in one, two, and three spatial dimensions which were described in the previous chapters. While these earlier chapters focused on the analysis of the existence and linear stability/instability of the relevant solutions, the present section sheds light into typical dynamical evolution examples to illustrate the outcome of the previously identified instabilities.

The equation of interest is

$$i\dot{u}_\mathbf{n} = -\epsilon \Delta_d u_\mathbf{n} + g|u_\mathbf{n}|^2 u_\mathbf{n} \tag{10.1}$$

where $u_\mathbf{n}$ is the complex lattice field with $\mathbf{n}$ being the vectorial lattice index, Δ_d is the standard d-dimensional discrete Laplacian extrapolated from a three-point stencil, ϵ is the inter-site (IS) coupling, and $g = -1$ in Sect. 10.2 while $g = 1$ in Sect. 10.3.2.

K.J.H. Law (✉)
University of Massachusetts, Amherst, MA, 01003, USA
e-mail: law@math.umass.edu

The overdot represents the derivative with respect to the evolution variable (which, for example, could be z in the case of optical arrays, or t in BEC models). Also, we write $\mathbf{u} = (u_{\mathbf{n}_1}, \ldots, u_{\mathbf{n}_{N^d}})$, to denote the complex lattice field of N sites in each of d dimensions.

As before, we are interested in stationary solutions of the form

$$u_{\mathbf{n}} = \exp(i\Lambda t) v_{\mathbf{n}} \tag{10.2}$$

for all $\mathbf{n}$. It can easily be seen that Eq. (10.1) admits such solutions in the anticontinuum (AC) limit ($\epsilon = 0$) with the additional structure $v_{\mathbf{n}} = e^{i\theta_{\mathbf{n}}}$ for $\theta_{\mathbf{n}} \in [0, 2\pi)$. This leads to the persistence as well as stability criteria detailed in the previous chapters (see also [29, 30] for further details). We look at solutions (from the previous chapters) for coupling values for which they have been predicted to be unstable. There exist parameter values such that an unstable solution of each configuration family eventually settles into a single site structure, and also that the time it takes for the original structure to break up is dependent on the magnitude of the real part of the linearization, the magnitude of the coupling between sites, and the initial perturbation.

The organization will be as follows. Section 10.2 will be devoted to 1(+1)-dimensional (10.2.1), 2(+1)-dimensional (10.2.2), and 3(+1)-dimensional (10.2.3) solutions of the standard focusing DNLS (Eq.(10.1) with $g = -1$). In Sect. 10.3, we discuss the more exotic settings of a grid with hexagonal geometry (10.3.1) and then solutions in the case of a defocusing nonlinearity ($g = +1$) in 1(+1) and 2(+1) dimensions (10.3.2). All the dynamical evolutions confirm stability predictions given theoretically and numerically in the preceding chapters, though some of the dynamical behavior is interesting and not a priori predictable.

10.2 Standard Scenario

We begin by discussing the space–time evolution of unstable solutions of the standard focusing DNLS (Eq. (10.1) with $g = -1$).

10.2.1 1(+1)-Dimensional Solutions

In this first section we examine the evolution of $\{|u_{\mathbf{n}}|^2\}$ (it is understood that the set is taken over all indices $\mathbf{n}$ in the d-dimensional lattice, where here $d = 1$ and the boldface is unnecessary, but we will use this notation throughout for consistency) for four one-dimensional configurations (see Fig. 10.1). Each is placed in the bulk of a lattice with 201 sites. Denoting a positive excited node by "+" and a negative one by "−", the configurations we consider are that of two in-phase (IP) adjacent nodes (i) ++, and the following with three adjacent nodes (ii) +++, (iii) ++−, and (iv) +−+. These can be found in Chap. 2. To expedite the onset of instability, in each

10 The Dynamics of Unstable Waves

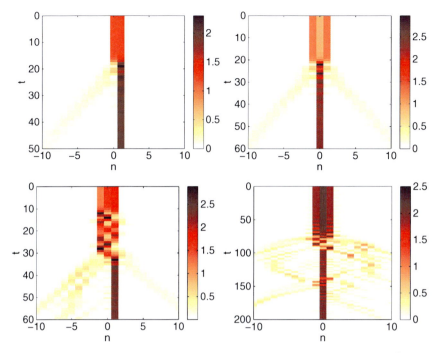

Fig. 10.1 These panels exhibit the space–time evolution of the density field $\{|u_\mathbf{n}(t)|^2\}$ of four modes to the one-dimensional DNLS equation. The *top left panel* shows the (i) $++$ configuration, the *top right* is (ii) $+++$, the *bottom left* is (iii) $++-$, and the bottom right is (iv) $+-+$

case the coupling parameter is $\epsilon = 0.3$, which is beyond the instability threshold of each configuration. We present the intensity of the field $\{|u_\mathbf{n}(t)|^2\}$, where $\mathbf{u} = \mathbf{u}_s + \mathbf{u}_r$, $\mathbf{u}_s$ being the lattice field with the respective stationary solution and $\mathbf{u}_r$ is a uniformly distributed random noise field in the interval $(0, a)$. For the discrete solitons (i)–(iv) above, The amplitude of the perturbation is taken to be (i) 10^{-8}, (ii) 10^{-7}, (iii) 10^{-5}, and (iv) 10^{-3}.

In all panels we clearly see a single surviving site that persists for long times. We see that the two IP modes dissolve from their original forms via a short turbulent stage into a (stable) single site structure. This is not surprising since, as was discussed in Chapter 2, (adjacent) IP excitations are found to be unstable for any $\epsilon \neq 0$ due to a positive real eigenvalue. The structure $++-$ also has adjacent IP excitations and is unstable, for any $\epsilon \neq 0$, due to a positive eigenvalue, with also a pair of bifurcating imaginary eigenvalues with negative Krein signature which eventually collide with the continuous spectrum and become complex (see Chap. 2 and [29–32]). The presence of the out-of-phase (OP) site complicates the dynamics pattern as seen in the lower left panel. The lower right panel shows the dynamics for the OP mode, $+-+$, which is stable for small couplings, but has two pairs of imaginary eigenvalues which (for larger coupling) become complex as a result of two Hamiltonian–Hopf bifurcations. This last configuration is clearly the most robust of the four ($\epsilon = 0.3$ for all configurations) since it is perturbed the most and yet persists for the longest time before turbulence sets in.

10.2.2 2(+1)-Dimensional Solutions

The panels depicted in this section exhibit the dynamics for the field intensity $\{|u_\mathbf{n}|^2\}$ for several two-dimensional configurations found in Chap. 3. As before, the value of the coupling is always beyond the threshold of instability and the configuration is perturbed by a random noise. All the dynamical evolutions are performed in a 21 × 21 grid. The dynamics in larger grid sizes (i.e., 31 × 31) were examined in a few of the cases and there was no qualitative difference found for the timescales considered herein. Each exhibited the same outcome in that one site remained for long times. A characteristic density isosurface $D_k = \{(\mathbf{n}, t) \mid |u_\mathbf{n}(t)|^2 = k\}$ is used here to represent the space–time evolution of the fields. The coupling and perturbation in each case were adjusted to exhibit complete destruction of the initial configuration to a single site in a reasonable time span. The details are supplied below each figure.

As already detailed in Chap. 3, the first five-site configuration, the symmetric vortices with $L = 1, M = 2$, and $L = M = 2$, as well as the asymmetric vortices with $L = 1, M = 2$, and $L = 3, M = 2$ have purely real eigenvalues. All other solutions break up due to oscillatory instability arising from Hamiltonian–Hopf bifurcations of the linearized problem when the pure imaginary eigenvalues collide with the phonon band (or continuous spectrum). The dynamical evolutions depicted in the following images qualitatively corroborate this earlier analysis. Figure 10.2 depicts the dynamical instability for the three-site configuration of Chap. 3. Next, we depict an example of the five-site configuration (lower left panel, Fig. 10.3) which exhibits complete breakdown from the initial state earlier than the other modes even

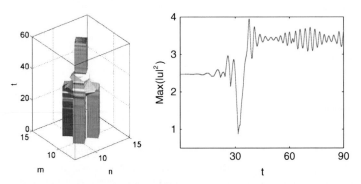

Fig. 10.2 The *left panel* shows the dynamics in time of $\{|u_\mathbf{n}|^2\}$ for the three-site configuration from Chap. 3 (see Eq. (3.88)). Here $\epsilon = 0.4$, the maximum of the perturbation is of amplitude $a = 10^{-4}$, and the isosurface is taken at $k = 0.2$. The *right panel*, which exhibits the maximum of the amplitude of the field intensity for this configuration up to time $t = 90$, clearly shows that the resulting single site has an oscillating amplitude ("breathes"). This can also be seen as the undulations in the isosurface picture on the *left*. Similar diagnostics confirmed this for the other cases

10 The Dynamics of Unstable Waves

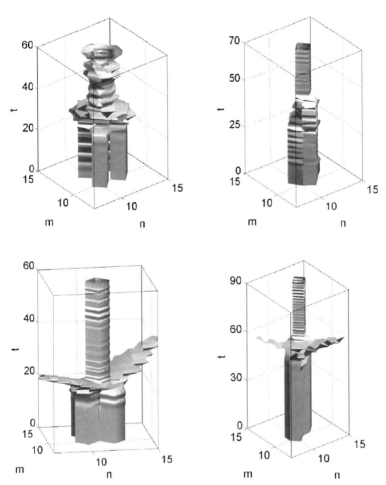

Fig. 10.3 The *two top panels* show the dynamics for the first (*top left panel*) and second (*top right panel*) four-site configurations of Eq. (3.89) in Chap. 3. The *bottom two panels* are the dynamics for the first (*bottom left panel*) and second (*bottom right panel*) five-site configurations, respectively, of the form of Eq. (3.90), Chap. 3. In each case $\epsilon = 0.4$ and the perturbation is of amplitude $a = 10^{-4}$. The isosurface is taken as $k = 0.2$ for all figures except the one at the *bottom right*, for which $k = 0.4$

though the coupling is the same and they have all been perturbed by noise with the same amplitude. This illustrates the stronger instability from purely real eigenvalues. We see a similar situation in Fig. 10.4 where the oscillatory instabilities manifest themselves in the dynamics at later times. Of the configurations in Fig. 10.4, the longest surviving one is the one with purely oscillatory instability (the symmetric vortex with $L = M = 1$ (ii)). Note that this solution is perturbed by a random noise with $a = 10^{-4}$ while the amplitude of the perturbation in (iii) is $a = 10^{-6}$. Still the solution (ii) has greater longevity.

An interesting observation is that the number of sites that remain can depend on the coupling strength and also on the magnitude of the perturbation. Take for

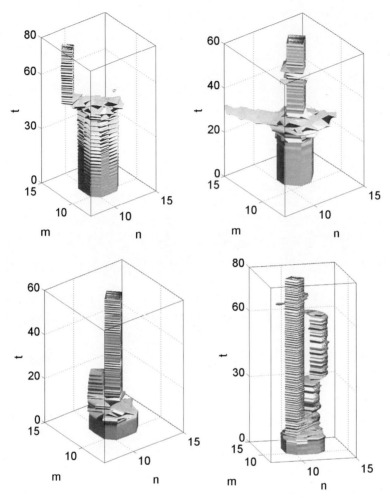

Fig. 10.4 These panels show the dynamics for the (i) nine-site configuration (*top left panel* with $a = 10^{-2}, k = 1.2$), and the three symmetric vortices with (ii) $L = M = 1$ (*top right panel* with $a = 10^{-4}, k = 0.2$), (iii) $L = 1$ and $M = 2$ (*bottom left panel* with $a = 10^{-6}, k = 0.7$), and (iv) $L = M = 2$ (*bottom right panel* with $a = 10^{-2}, k = 0.7$). In each case, $\epsilon = 0.6$

example the second five-site configuration shown in the bottom right panel of Fig. 10.3. In the figure shown, $\epsilon = 0.4$, $a = 10^{-4}$, and $k = 0.4$, and a single site remains, while for a weaker perturbation ($a = 10^{-5}$) two sites actually remain for long times. However, for a larger coupling value of $\epsilon = 0.5$ one site remains for perturbations with amplitude as small as $a = 10^{-8}$. Finally, in Fig. 10.5, we depict the dynamical instabilities for the different vortex configurations discussed in Chap. 3. Further study is needed to build on existing knowledge of discrete breathers [32, 33], in order to explain this curious interplay between the various components, and also to elucidate the underlying mechanisms responsible for these (as well as other) observations.

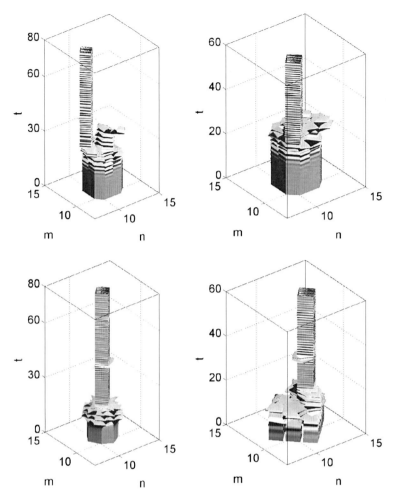

Fig. 10.5 These panels show the dynamics for the (i) symmetric vortex $L = 3$, $M = 2$ (*top left panel* with $\epsilon = 0.6$, $a = 10^{-4}$, $k = 1.2$), (ii) asymmetric vortex with $L = 1$ $M = 2$ (*top right panel* with $\epsilon = 0.6$, $a = 10^{-4}$, $k = 0.9$), (iii) asymmetric vortex with $L = 3$, $M = 2$ (*bottom left panel* with $\epsilon = 0.5$, $a = 10^{-4}$, $k = 0.7$), and (iv) vortex cross of $L = 2$ (*bottom right panel* with $\epsilon = 0.6$, $a = 10^{-2}$, $k = 0.7$). As before, these are from configurations laid out in Chap. 3

10.2.3 3(+1)-Dimensional Solutions

This section shows the dynamics of the field intensity, $\{|u_\mathbf{n}|^2\}$, for the three-dimensional configurations, the (a) diamond ($S_{\pm 1} = \{\pi/2, 3\pi/2\}$) [see Fig. 10.6], (b) octupole ($\theta_0 = \pi$, $s_0 = 1$, i.e., $S_1 = \{\pi, 3\pi/2, 0, \pi/2\}$) [see Fig. 10.7], and (c) double-cross ($\theta_0 = \pi$, $s_0 = 1$, i.e., $S_1 = \{\pi, 3\pi/2, 0, \pi/2\}$) [see Fig. 10.8]. All three of these structures were shown previously (Chap. 4) to persist. For each, we

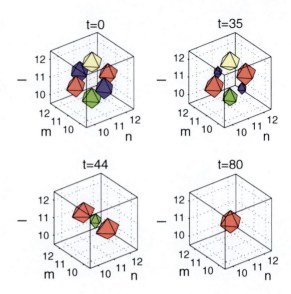

Fig. 10.6 These panels show four times in the evolution of the three-dimensional diamond structure. The coupling is taken to be $\epsilon = 0.3$, the perturbation amplitude is $a = 10^{-2}$ and and all iso-contours are taken at $\text{Re}(u_\mathbf{n}) = \pm 0.75$ and $\text{Im}(u_\mathbf{n}) = \pm 0.75$. The *dark gray* and *gray* are real iso-contours, while the *light* and *very light gray* are the imaginary contours

show the evolution of the instability with characteristic density isosurfaces of the three-dimensional field at four times beginning with $t = 0$.

As before, in each case a perturbation is applied at a value of the coupling well past the threshold of instability. As with the lower dimensional configurations, the coupling parameter and perturbation play a role in the evolution of the instability. For appropriate values of each, a single site will remain at the end of the time frame considered here. In the case of the octupole, it is seen that when we take $\epsilon = 0.3$

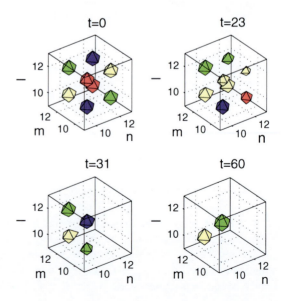

Fig. 10.7 These panels show four times in the evolution of the octupole. The coupling is $\epsilon = 0.3$, the perturbation amplitude is $a = 10^{-2}$ and and all iso-contours are taken at $\text{Re}(u_\mathbf{n}) = \pm 0.75$ and $\text{Im}(u_\mathbf{n}) = \pm 0.75$. The *dark gray* and *gray* are real iso-contours, while the *light* and *very light gray* are the imaginary contours

10 The Dynamics of Unstable Waves

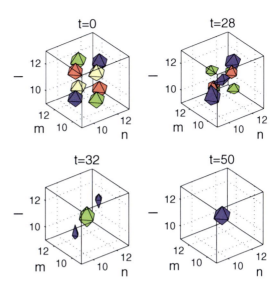

Fig. 10.8 These panels show four times in the evolution of the double cross. The coupling is $\epsilon = 0.4$, the perturbation amplitude is $a = 10^{-4}$ and and all iso-contours are taken at $\text{Re}(u_\mathbf{n}) = \pm 0.75$ and $\text{Im}(u_\mathbf{n}) = \pm 0.75$. The *dark gray* and *gray* are real iso-contours, while the *light* and *very light gray* are the imaginary contours

with a perturbation of amplitude $a = 10^{-2}$, two sites remain as seen in Fig. 10.7, while for $\epsilon = 0.6$ with a perturbation of amplitude $a = 10^{-4}$, a single site remains. A similar phenomenon is observed for the double cross where for $\epsilon = 0.3$ with a perturbation of amplitude $a = 10^{-2}$, two sites remain as seen in Fig. 10.8, while for $\epsilon = 0.4$ with a perturbation of amplitude $a = 10^{-4}$, a single site remains.

In all cases, the grid size is $21 \times 21 \times 21$. It should be noted that for the diamond configuration, larger grid sizes (i.e., $25 \times 25 \times 25$, $27 \times 27 \times 27$) were examined, with no qualitative change in behavior witnessed. In each case, the end result for a given coupling and perturbation was always the same number of surviving sites.

10.3 Non-Standard Scenario

We will now consider a few more exotic settings. First, we look at the vortex solutions with a six neighbor hexagonal geometry as seen in the end of Chap. 3. Then, we will look at the same DNLS equation (10.1), except with a defocusing nonlinearity ($g = 1$) in 1(+1) and 2(+1) dimensions from Chap. 5.

10.3.1 Hexagonal Lattice

In this section we consider the variation of Eq. (10.1) in which the terms $\Delta_d u_\mathbf{n}$ are replaced by the non-standard extrapolation of the two-dimensional five-point stencil, in which each site has four neighbors as in the previous section and following subsection, to the natural variation for the six-neighbor lattice, $\sum_{\langle \mathbf{n}' \rangle} u_{\mathbf{n}'} - 6 u_\mathbf{n}$, where $\langle \mathbf{n}' \rangle$ is the set of nearest neighbors to the node indexed by $\mathbf{n}$.

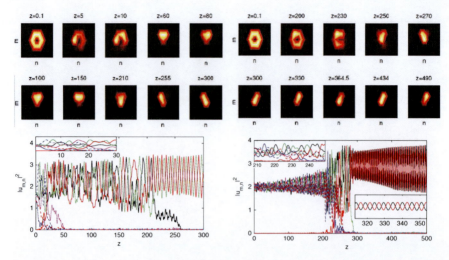

Fig. 10.9 Space–time evolution of $\{|u_\mathbf{n}(t)|^2\}$, where the lattice geometry is hexagonal, $\mathbf{u} = \mathbf{u}_s(1 + \max_\mathbf{n}\{|u_{\mathbf{n},s}(0)|^2\}\mathbf{u}_r)$, $\mathbf{u}_r$ is a uniform random variable in $(-0.05, 0.05)$ random variable, and $\mathbf{u}_s$ is a single charged vortex on the *left* ($\epsilon = 0.1$) and a double charged vortex on the *right* ($\epsilon = 0.125$). The *top panels* are snapshots and the *bottom* are amplitudes of the individual excited sites. Note the almost harmonic oscillations depicted in the inset of the *bottom right*

The vortex solutions in this geometry are displayed at the end of Chap. 3. The single charged vortex is actually more unstable than the double charged one (the eigenvalues bifurcating from the origin in the AC limit are real for the former and imaginary for the latter). The dynamics of these solutions given in Fig. 10.9 confirm this theoretical prediction. The single charged vortex ($\epsilon = 0.1$) breaks up very rapidly before $z = 20$ and subsequently degenerates into a lopsided dipole-type configuration, while the double charged vortex ($\epsilon = 0.125$) persists until well past $z = 200$. Each configuration ultimately becomes a two-site breather for long distances, with one site being the initially unpopulated center site.

10.3.2 Defocusing Nonlinearity

We now study the dynamics of typical 1(+1)- and 2(+1)-dimensional solutions with defocusing nonlinearity ($g = +1$ in Eq. (10.1)).

10.3.2.1 1(+1)-Dimensional Solutions

We begin by examining the dynamics of the unstable one-dimensional configurations. Two one-dimensional dark soliton configurations with defocusing nonlinearity are principally considered in this book; in both cases the absolute value squared of the background is one and there is a π phase jump, which can either occur between two sites (IS) or between three sites (on-site, OS), where there exists a node with zero amplitude in the middle of the latter. As shown in Chap. 5, the OS dark soliton is stable for small coupling, and subject only to oscillatory instability (as the coupling increases) due to complex quartets of eigenvalues which emerge when the null eigenvalues from the AC limit with negative Krein signature reaches the continuous spectrum, at which point Hamiltonian–Hopf bifurcations occur. On the other hand, the pair which bifurcates from the spectral plane origin (at the AC limit) in the case of the IS configuration becomes real and therefore this configuration is subject to a strong (exponential) instability.

Figure 10.10 shows the space–time evolution of the dark solitons which confirm the theoretical and numerical predictions. The left panel shows the solution $\mathbf{u} = \mathbf{u}_s + \mathbf{u}_r$ where $\mathbf{u}_s$ is the IS dark soliton and $\mathbf{u}_r$ is random noise field uniformly distributed in the interval $(-5, 5) \times 10^{-4}$. Note even with such a mild perturbation from the stationary state, this configuration disintegrates after $t = 20$. On the other hand, in the right panel $\mathbf{u}_s$ is the OS configuration with $\mathbf{u}_r$ uniformly distributed in the interval $(-5, 5) \times 10^{-2}$, and yet the original configuration persists until $t = 500$. This not only confirms, but really highlights the accuracy of the theoretical stability calculations from Chap. 5, Sect. 5.1.2.

10.3.2.2 2(+1)-Dimensional Solutions

Next, we will consider the two-dimensional configurations from Chap. 5, Sect. 5.2. As in the one-dimensional case described above, the configurations $\mathbf{u}_s$ are perturbed by a field $\mathbf{u}_r$ randomly distributed in the interval $(-5, 5) \times 10^{-2}$. The same field $\mathbf{u}_r$ is added to each solution for consistency so they may be more easily compared (and it is so large because the least stable among these configurations took a considerable time to degenerate even with this perturbation). We present the results of the dynamical evolution of the two-site, four-site, and $S = 1$ vortices organized with

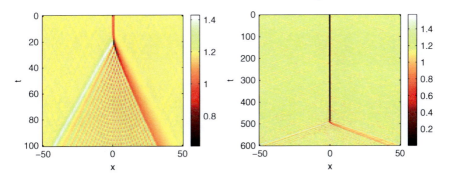

Fig. 10.10 Dynamics of the inter-site (*left*) and on-site (*right*) dark soliton configurations are represented by space–time plots of the intensity of the field $\{|u_\mathbf{n}(t)|^2\}$ as defined in the previous section with $\epsilon = 0.1$. These solutions correspond to those in Chap. 5, where the inter-site one is perturbed by a random noise of only $\pm 5 \times 10^{-4}$, and is visibly distorted by $t = 20$, while the on-site configuration is perturbed by a noise amplitude of $\pm 5 \times 10^{-2}$ and yet the original configuration persists until $t = 500$

the IS solutions in the left column, the OS solutions in the right column, and, for the former two, the top are IP, and the bottom are OP. Again as in Sect. 10.2.2, we choose density isosurfaces, $D_k = \{(\mathbf{n}, t) \mid |u_\mathbf{n}(t)|^2 = k\}$ as our visualization tool. The magnitude of the density isosurface k is chosen as half the maximum of the initial density field $k = (1/2)|u_\mathbf{n}(0)|^2$ in most cases, except when a smaller magnitude was necessary to visualize the relevant dynamics. All solutions degenerate into a single site configuration for long times, although it is worth mentioning here that, as in the focusing case, for smaller coupling values than those chosen here (but still significantly far from the AC limit), even unstable solutions may only undulate and not actually break up at all.

The two-site configurations are given in Fig. 10.11. Results confirm the stability analysis in Chap. 5, Sect. 5.3.2. In particular, note that the mild instability of the ISIP

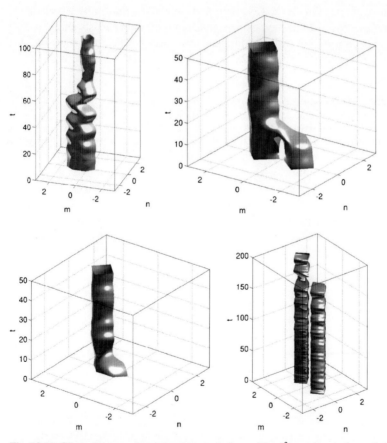

Fig. 10.11 Given above are the space–time evolutions of $\{|u_\mathbf{n}|^2\}$ as given before where $\mathbf{u}_s$ are the two-dimensional dipole configurations for defocusing nonlinearity. The *top row* is in-phase, the *bottom row* is out-of-phase, the *left column* is inter-site, and the *right column* is on-site. These solutions correspond to those found in Chap. 5, Sect. 5.2, for the coupling values (clockwise from *top left*) $\epsilon = 0.116, 0.08, 0.116$, and 0.08

solution for $\epsilon = 0.116$ given in the upper left panel takes a considerable amount of time to break up the initial configuration as compared to the strong instability of the ISOP for $\epsilon = 0.08$ in the bottom left, which leads to degeneration almost immediately into a single site upon evolution. In the right column we can see the fast degeneration of the OSIP for $\epsilon = 0.08$ as a manifestation of the strong instability on the top row and the much slower degeneration of the OSOP for $\epsilon = 0.116$ depicting the oscillatory Hamiltonian–Hopf instability on the bottom.

Figure 10.12 depicts the quadrupole solutions, which again confirm the theoretical and numerical findings of Chap. 5, Sect. 5.3.3. The more stable ISIP ($\epsilon = 0.1$)

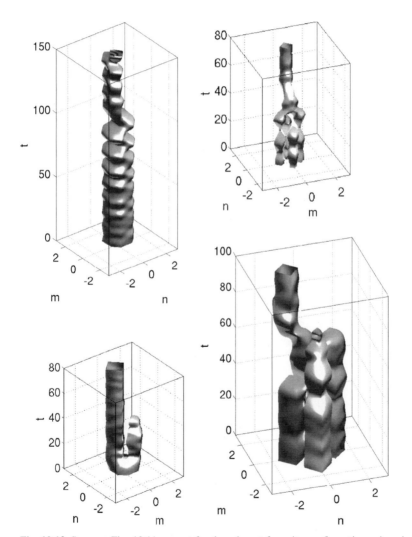

Fig. 10.12 Same as Fig. 10.11, except for the relevant four-site configurations given in Chap. 5, Sect. 5.3 and the coupling value is $\epsilon = 0.1$ for all panels

in the upper left panel and OSOP ($\epsilon = 0.1$) in the lower right panel confirm the predictions. So do the more unstable OSIP and ISOP (both for $\epsilon = 0.1$) in the top right and bottom left panels, respectively. For this coupling value, the ultimate single site configuration of the ISOP is robust to perturbation (down to 10^{-8}), as was the second five-site configuration presented in Sect. 10.2.2. Again for a slightly smaller coupling of $\epsilon = 0.08$, a two-site configuration remains, which breathes up to at least $t = 900$ (not shown), and again there is a larger perturbation which spoils the breathing two-site structure for this coupling. Additionally, on investigating the intermediate coupling value of $\epsilon = 0.9$, one finds that the perturbation necessary to break the two-site structure becomes smaller, suggesting that the necessary perturbation to eliminate the breather is inversely proportional to the coupling prior to the lower bound of the region in coupling space for which a single site invariably survives. Also, for $\epsilon = 0.05$ the OSIP remains a breathing four-site structure even for longer times despite the strong instability of the linearization and almost instantaneous breathing behavior. This again supports the same hypothesis mentioned above. Finally, Fig. 10.13 shows the single charge vortex solutions from Chap. 5, Sect. 5.3.4. Both of these solutions are approximately equivalently unstable. Each has both Hamiltonian–Hopf quartets and real pairs of eigenvalues.

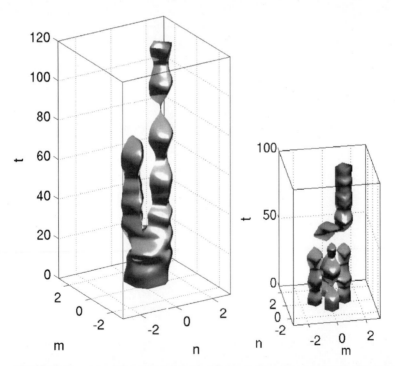

Fig. 10.13 Dynamics of the two-dimensional charge 1, four-site gap vortices with defocusing nonlinearity. The inter-site version is on the *left*, while the on-site version is on the *right* and $\epsilon = 0.1$ in both cases

10.4 Conclusion and Future Challenges

We have numerically examined the dynamics of discrete solitons for the DNLS equation in one, two, and three spatial dimensions with the standard focusing nonlinearity. We also investigated the more exotic hexagonal lattice and the case of defocusing nonlinearity, the latter of which has been observed in the experimental setting of nonlinear optics [34, 35]. It is found that the numerical dynamics aligns itself with the theoretical predictions for linear stability established in the previous chapters. It is noted that the dynamics within the timescales considered here depend not only on the linear stability, but also sensitively on the coupling parameter and (to a lesser degree) on the perturbation. For each case, there exists some coupling and perturbation such that the eventual result is a single robust site (as displayed in the images).

A major challenge for the future would be to devise and further develop a theoretical framework to understand [33] the instability process. The single site is an attractive equilibrium of the system. Since it is stable and has low energy, it is no surprise that an unstable system would tend to this state. But, it would be interesting to attempt to develop some more precise theory relating these aspects of the nonlinear evolution and perhaps elucidate general characteristics and features that may not be visible from the numerics. It would also be of interest to examine more complex systems, such as multicomponent and higher dimensional systems, and to determine whether some general features persist.

Acknowledgments K.J.H.L. would like to extend his grateful appreciation to P.G. Kevrekidis, first and foremost, not only for the endless stream of advising during the revision process of the current work, but also for the financial support that enabled its completion. He would also like to acknowledge UMass for the office and mathematical support. Q.E.H. is grateful for the opportunity to work on this project.

References

1. Dodd, R.K., Eilbeck, J.C., Gibbon, J.D., Morris, H.C.: Solitons and Nonlinear Wave Equations. Academic Press, London (1982)
2. Krumhansl, J.A.: The intersection of nonlinear science, molecular biology, and condensed matter physics. Viewpoints. In: Nonlinear Excitations in Biomolecules. Peyrard, M. (ed.), Springer-Verlag, Berlin and les Editions de Physique, Les Ulis, 1–9, (1995).
3. Remoissenet, M.: Waves Called Solitons: Concepts and Experiments. Springer, New York (1999)
4. Kevrekidis, P.G., Rasmussen, K.Ø., Bishop, A.R.: Int. J. Mod. Phys. B **15**, 2833–2900 (2001)
5. Scott, A.: Nonlinear Science: Emergence and Dynamics of Coherent Structures. Oxford Texts in Applied and Engineering Mathematics, London (2003)
6. Dauxois, T., Peyrard, M.: Physics of Solitons. Cambridge University Press, New York (2006)
7. Careri, G., Buontempo, U., Galluzzi, F., Scott, A.C., Gratton, E., Shyamsunder, E.: Phys. Rev. B **30**, 4689–4702 (1984)
8. Feddersen, H.: Phys. Lett. A **154**, 391–395 (1991)
9. Eisenberg, H.S., Silberberg, Y., Morandotti, R., Boyd, A.R., Aitchison, J.S.: Phys. Rev. Lett. **81**, 3383–3386 (1998)

10. Binder, P., Abraimov, D., Ustinov, A.V.: Phys. Rev. E **62**, 2858–2862 (2000)
11. Trombettoni, A., Smerzi, A.: Phys. Rev. Lett. **86**, 2353–2356 (2001)
12. Christodoulides, D.N., Joseph, R.I.: Opt. Lett. **13**, 794–796 (1988)
13. Davydov, A.S.: J. Theor. Biol. **38**, 559–569 (1973)
14. Frenkel, J., Kontorova, T.: J. Phys. (USSR) **1**, 137–149 (1939)
15. Fermi, E., Pasta, J.R., Ulam, S.M.: Studies of Nonlinear Problems. Los Alamos Sci. Lab. Rep. LA-1940, (1955), published later in Collected Papers of Enrico Fermi, Serge, E. (ed.), University of Chicago Press, NewYork (1965)
16. Toda, M.: Theory of Nonlinear Lattices. Springer-Verlag, Berlin (1981)
17. Carr, J., Eilbeck, J.C., Phys. Lett. A **109**, 201–204 (1985)
18. Mackay, R.S., Aubry, S.: Nonlinearity **7**, 1623–1643 (1994)
19. Efremidis, N.K., Sears, S., Christodoulides, D.N., Fleischer, J.W., Segev, M.: Phys. Rev. E **66**, 046602 (2002)
20. Kivshar, Y.S., Agrawal, G.P., Optical Solitons: From Fibers to Photonic Crystals. Academic Press, NewYork (2003)
21. Ablowitz, M.J., Prinari, B., Trubatch, A.D.: Discrete and Continuous Nonlinear Schrödinger Systems. London Mathematical Society Lecture Note Series 302, Cambridge University Press, NewYork (2004)
22. Makris, K.G., Suntsov, S., Christodoulides, D.N., Stegeman, G.I., Heche, A., Opt. Lett. **30**, 2466 (2005)
23. Bludov, Yu.V., Konotop, V.V.: Phys. Rev. E **76**, 046604 (2007)
24. Susanto, H., Kevrekidis, P.G., Carretero-González, R., Malomed, B.A., Frantzeskakis, D.J.: Phys. Rev. Lett. **99**, 214103 (2007)
25. Kevrekidis, P.G., Frantzeskakis, D.J., Carretero-González, R.: Emergent Nonlinear Phenomena in Bose-Einstein Condensates: Theory and Experiment. Springer Series on Atomic, Optical, and Plasma Physics, vol. 45, (2008)
26. Hoq, Q.E., Gagnon, J., Kevrekidis, P.G., Malomed, B.A., Frantzeskakis, D.J., Carretero-González, R.: Extended Nonlinear Waves in Multidimensional Dynamical Lattices. Math. Comput. Simulat. 2009
27. Hoq, Q.E., Gagnon, J., Kevrekidis, P.G., Malomed, B.A., Frantzeskakis, D.J., Carretero-González, R.: http://www-rohan.sdsu.edu/~rcarrete/
28. Hoq, Q.E., Carretero-González, R., Kevrekidis, P.G., Malomed, B.A., Frantzeskakis, D.J., Bludov, Yu.V., Konotop, V.V.: Phys. Rev. E **78**, 036605 (2008)
29. Pelinovsky, D.E., Kevrekidis, P.G., Frantzeskakis, D.J.: Physica D **212**, 1–19 (2005)
30. Pelinovsky, D.E., Kevrekidis, P.G., Frantzeskakis, D.J.: Physica D **212**, 20–53 (2005)
31. Kapitula, T., Kevrekidis, P.G., Sandstede, B.: Physica D **195**, 263–282 (2004)
32. Aubry, S.: Physica D **103**, 201–250 (1997)
33. Alfimov, G.L., Brazhnyi, V.A., Konotop, V.V.: Physica D **194**, 127–150 (2004)
34. Tang, L., Lou, C., Wang, X., Chen, Z., Susanto, H., Law, K.J.H., Kevrekidis, P.G.: Opt. Lett. **32**, 3011–3013 (2007)
35. Song, D., Tang, L., Lou, C., Wang, X., Xu, J., Chen, Z., Susanto, H., Law, K.J.H., Kevrekidis, P.G.: Opt. Exp. **16**, 10110–10116 (2008)

Chapter 11
A Map Approach to Stationary Solutions of the DNLS Equation

Ricardo Carretero-González

11.1 Introduction

In this chapter we discuss the well-established map approach for obtaining stationary solutions to the one-dimensional (1D) discrete nonlinear Schrödinger (DNLS) equation. The method relies on casting the ensuing stationary problem in the form of a recurrence relationship that can in turn be cast into a two-dimensional (2D) map [1–5]. Within this description, any orbit for this 2D map will correspond to a steady state solution of the original DNLS equation.

The map approach is extremely useful in finding localized solutions such as bright and dark solitons. As we will see in what follows, this method allows for a global understanding of the types of solutions that are present in the system and their respective bifurcations.

This chapter is structured as follows. In Sect. 11.2 we introduce the map approach to describe steady states for general 1D nonlinear lattices with nearest-neighbor coupling. In Sect. 11.3 we present some of the basic properties of the 2D map generated by the 1D DNLS lattice and how these properties, in turn, translate into properties for the steady-state solutions to the DNLS. We also give an exhaustive account of the possible orbits that can be generated using the map approach. Specifically, we describe in detail the families of extended steady-state solutions (homogeneous, periodic, quasi-periodic, and spatially chaotic) as well as spatially localized steady states (bright and dark solitons and multibreather solutions). In Sect. 11.4 we study the limiting cases of small and large couplings. We briefly describe the bifurcation process that is responsible for the mutual annihilation of localized solutions through a series of bifurcations. For a more detailed account of the bifurcation scenaria for the DNLS using the map approach, see [3].

R. Carretero-González (✉)
Nonlinear Dynamical Systems Group, Computational Science Research Center, and Department of Mathematics and Statistics, San Diego State University, San Diego CA, 92182-7720, USA
e-mail: carreter@sciences.sdsu.edu

11.2 The 2D Map Approach for 1D Nonlinear Lattices

The 2D map approach that we present can be used in general for any 1D nonlinear lattice as long as the coupling between lattice sites is restricted to nearest neighbors. The most common form of such coupling scheme is the discrete Laplacian $\Delta u_n = u_{n-1} - 2u_n + u_{n+1}$. In order to describe the map approach in its more general form, let us consider a generic nonlinear lattice of the form

$$\dot{u}_n = G(u_{n-1}, u_n, u_{n+1}) + F(u_n), \tag{11.1}$$

where G is the nearest-neighbor coupling function and F corresponds to the on-site nonlinearity. The case of the DNLS with the standard cubic nonlinearity is obtained by choosing $G = (\epsilon/i)\Delta$ (Δ will be used to denote the discrete Laplacian) and $F(u) = (\beta/i)|u|^2 u$, where $\epsilon \geq 0$ is the coupling constant and $\beta = \pm 1$ corresponds to defocusing and focusing nonlinearities, respectively. For the map approach to be directly applicable we need to rewrite the steady-state solution of Eq. (11.1) as a recurrence relationship. Therefore, the only requirement for the map approach to work in the general case of the system (11.1) is that the coupling function needs to be invertible with respect to u_{n+1} such that $G(u_{n-1}, u_n, u_{n+1}) = G_0$ can be *explicitly* rewritten as $u_{n+1} = G^{-1}(u_{n-1}, u_n, G_0)$. In particular, this is the case for any coupling function defined as a linear combination of nearest neighbors (which is the case of the discrete Laplacian). For the sake of definitiveness, let us concentrate on the DNLS with cubic nonlinearity but keeping in mind that the technique can be applied in more general scenaria (for example, in [4] and [6] unstaggered and staggered solutions of the cubic-*quintic* DNLS are studied in detail).

Let us then start with the 1D DNLS with cubic on-site nonlinear term

$$i\dot{u}_n = -\epsilon \Delta u_n + \beta |u_n|^2 u_n. \tag{11.2}$$

It can be shown [7] that any steady-state solution of Eq. (11.2) must be obtained by separating space and time as $u_n = \exp(i\Lambda t)v_n$, where Λ is the frequency of the solution, which yields the steady-state equation for the real amplitudes v_n:

$$\Lambda v_n = \epsilon(v_{n-1} - 2v_n + v_{n+1}) - \beta v_n^3. \tag{11.3}$$

It is worth noting at this point that in the 1D case the stationary state is determined, without loss of generality, by the *real* amplitude v_n. In higher dimensions, for topologically charged solutions such as discrete vortices and supervortices in 2D [8–12], discrete diamonds and vortices in 3D [13, 14], and discrete skyrmion-type solutions [15], it is necessary to consider a *complex* steady-state amplitude v_n. Nonetheless, it is crucial to stress that the 2D map approach is only applicable for 1D lattices since the steady-state problem for higher dimensional dynamical lattices cannot be reduced to a recurrence relationship as it is the case (see below) for the 1D lattice.

The steady-state equation described by Eq. (11.3) can now be rewritten as the recurrence relationship

$$v_{n+1} = R(v_n, v_{n-1}) \equiv \frac{1}{\epsilon}\left[(\Lambda + 2\epsilon)v_n - \epsilon v_{n-1} + \beta v_n^3\right], \tag{11.4}$$

which in turn can be cast as the 2D map

$$\begin{pmatrix} v_{n+1} \\ w_{n+1} \end{pmatrix} = M \begin{pmatrix} v_n \\ w_n \end{pmatrix}, \qquad M : \begin{cases} v_{n+1} = R(v_n, w_n) \\ w_{n+1} = v_n \end{cases}, \tag{11.5}$$

where the second equation defines the intermediate variable $w_n \equiv v_{n-1}$. It is important to stress that, by construction, any orbit of the 2D map (11.5) will correspond to a steady-state solution of the DNLS (11.2). In particular, any given initial condition $P_0 = (v_0, w_0)^T$ for the 2D map will generate the orbit described by the *doubly* infinite sequence of points $(\ldots, P_{-2}, P_{-1}, P_0, P_1, P_2, \ldots)$ where $P_{n+1} = M(P_n)$ and negative subindexes correspond to backward iterates of the 2D map $[P_{n-1} = M^{-1}(P_n)]$. This 2D orbit will in turn correspond to the steady state $\{\ldots, w_{-2}, w_{-1}, w_0, w_1, w_2, \ldots\}$, where $w_n = [M^n(P_0)]_y$ is the y-coordinate (projection) of the nth iterate of P_0 through M. Alternatively, one could also obtain the steady state as $\{v_n\}_{n=-\infty}^{\infty}$, where $v_n = [M^{n+1}(P_0)]_x$ is the x-coordinate (projection) of the $(n+1)$th iterate of P_0.

It is also important to mention that the 2D map approach, although helpful in describing/finding steady-state solutions of the associated nonlinear lattice, does not give any information about the *stability* of the steady states themselves. This is a consequence of separating time from the steady state where one loses all the temporal information (including stability properties). Nonetheless, the 2D map approach does indicate the *genericity* or parametric/structural stability of certain types of orbits. Specifically, if the type of steady state that is been considered corresponds to a 2D map orbit (including fixed points, periodic orbits, and quasi-periodic orbits) that is isolated (i.e., away in physical and parameter space) from a bifurcation point, then this orbit will still exist in the presence of, small, generic parametric *and* external perturbations. This genericity property might be useful in realistic applications where the presence of (a) small errors in the determination of the parameters of the system and (b) external noise is ubiquitous. Note, however, that if the steady state is unstable to start with, the parametric perturbation will not modify its existence but it will remain unstable.

11.3 Orbit Properties and Diversity in the DNLS

Now that we have established the equivalence between a steady state of the DNLS (11.2) and orbits of the 2D map (11.5), let us discuss the different types of orbits that can be generated using the 2D map approach, their bifurcations and some of their basic properties.

11.3.1 Symmetries and Properties of the Cubic DNLS Steady States

All symmetries and properties inherent to the 2D map (11.5) generate respective symmetries and properties for the steady-state solutions to the DNLS. In particular, for the cubic DNLS (cf. Eq. (11.4)), we have the following symmetries and properties:

(a) The inverse map M^{-1}: $M^{-1}(v_n, w_n)^T = (v_{n-1}, w_{n-1})^T$ is identical to M after exchanging $v \leftrightarrow w$. Therefore any forward orbit of the 2D map will have a symmetric backward orbit that is symmetric with respect to the identity line.
(b) Exchanging $v_n \rightarrow (-1)^n v_n$ and $w_n \rightarrow (-1)^n w_n$ transforms the 2D map M onto $(-1)^n M$ with $\Lambda \rightarrow -\Lambda - 4\epsilon$ and $\beta \rightarrow -\beta$. This corresponds to the so-called staggering transformation where every solution to the *focusing* ($\beta = -1$) cubic DNLS has a corresponding solution to the *defocusing* ($\beta = +1$) cubic DNLS with adjacent sites alternating signs (and after a rescaling of the frequency).
(c) The 2D map is area preserving and, as a consequence, the steady-state solutions to the DNLS have the following properties. (i) Linear centers of the 2D map are also nonlinear centers and thus there will be periodic and quasi-periodic orbits around (linearly) neutrally stable fixed points. These 2D map orbits correspond, respectively, to spatially periodic and quasi-periodic steady state solutions to the DNLS (see below). (ii) Saddle fixed points of the 2D map will have stable and unstable manifold with the same exponential rates of convergence. Thus, localized steady state solutions of the DNLS will have symmetric tails at $n \rightarrow \pm\infty$.

11.3.2 Homogeneous, Periodic, Modulated, and Spatially Chaotic Steady States

In this section we concentrate on describing steady states that are spatially extended (i.e., not localized in space). These correspond to (a) fixed points, (b) periodic orbits, (c) quasi-periodic orbits, and (d) chaotic orbits of the 2D map M.

11.3.2.1 Homogeneous Steady States

The most straightforward orbit that can be described by the 2D map approach is a fixed point. Suppose that $P^* = (v^*, w^*)^T$ is a fixed point of M, namely $M(P^*) = P^*$. This trivial orbit generates the *homogeneous* steady solution $v_n = v^*$. Note that, by construction, all fixed points of M must satisfy $v^* = w^*$. For the DNLS case under consideration, the 2D map fixed point equation $(\Lambda + \beta v^2)v = 0$ has three fixed points $v^* = \{0, \pm\sqrt{-\Lambda/\beta}\}$, that in turn correspond to the two spatially homogeneous solutions $u_n(t) = 0$ and $u_n(t) = \sqrt{-\Lambda/\beta} \exp(i\Lambda t)$.

11.3.2.2 Periodic Steady States

Let us now consider a periodic orbit of the 2D map. Suppose that $\{P_0, P_1, ..., P_{p-1}\}$ is a period-p orbit of M (i.e., $M(P_{p-1}) = P_0$). This periodic orbit for M will generate a *spatially periodic* steady-state solution for the DNLS, where $v_n = [P_{n \bmod(p)}]_y$. A particular case of this spatially periodic steady state stems from period-2 orbits $\{T_0, T_1\}$. There are at most three such period-2 solutions depending on the $(\epsilon, \Lambda, \beta)$-parameter values. One of these solutions has the form $T_0 = -T_1 = (+a, -a)^T$ where $a = \sqrt{-(\Lambda + 4\epsilon)/\beta}$. This symmetric period-2 orbit is a consequence of the symmetry of the 2D map under consideration where the transformations $v \leftrightarrow -v$ and $w \leftrightarrow -w$ leave the equations invariant. This symmetric period-2 orbit generates an oscillatory steady-state profile of the form $v_n = (..., -a, +a, -a, +a, ...)$. In general a period-$p$ orbit of the 2D map generates an spatially periodic steady state with spatial wavelength (period) of p.

11.3.2.3 Quasi-Periodic Steady States

An interesting steady-state solution is generated when one considers quasi-periodic solutions of the 2D map. For example, the origin is a nonlinear center for $-4\epsilon < \Lambda < 0$ in both the focusing and defocusing case. Around this center point the 2D map exhibits an infinite family of quasi-periodic solutions rotating about the origin (cf. Fig. 11.1). These 2D map orbits correspond to steady-state modulated waves about the fixed point (in this case the origin) for the DNLS. An example of such an orbit is depicted in Fig. 11.1. In the left panel of the figure we depict with circles the quasi-periodic orbit around the origin, while in the right panel we depict (also with circles) its corresponding steady-state solution to the DNLS. The spatial periodicity of these modulated waves is approximately determined by the argument of the eigenvalues of the Jacobian at the fixed point.

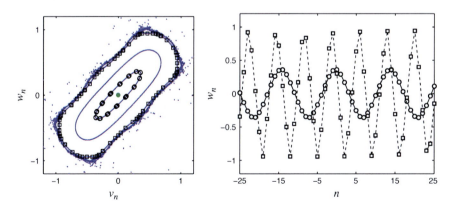

Fig. 11.1 Periodic, quasi-periodic and chaotic orbits of the 2D map (*left*). The *right panel* depicts the corresponding steady-state solutions to the DNLS. Circles (*squares*) correspond to a quasi-periodic (chaotic) orbit. Parameter values correspond to: $\Lambda = -0.1$, $\beta = -1$, and $\epsilon = 1$

11.3.2.4 Spatially Chaotic Steady States

As a last example of a non-localized steady state let us consider the next level of complexity for a 2D orbit: a chaotic orbit. Chaotic orbits will be a common occurrence in nonlinear maps. For the case under consideration, the 2D map induced by the DNLS becomes chaotic close to the separatrix between higher periodic orbits. In Fig. 11.1 we depict such a chaotic region around the separatrix of a pair of period-7 orbits (see outer orbits). Such a chaotic orbit naturally generates a steady-state solution (see squares in the right panel) that resembles a period-7 orbit that is chaotically modulated. It is important to mention that, typically, these chaotic orbits exhibit "stickiness" close to the separatrix (see [16] and references therein for more details on chaotic transport) and thus will stay close to a periodic orbit for some time. However, the chaotic orbit is eventually expelled (both in forward and backward time) and therefore the steady state becomes unbounded at $n \to \pm\infty$. See [17] for a discussion of the relationship between these chaotic orbits and the transmission properties in nonlinear Schrödinger-type lattices.

11.3.3 Spatially Localized Solutions: Solitons and Multibreathers

Undoubtedly, the most interesting steady-state solutions are generated by homoclinic and heteroclinic orbits of the 2D map. These orbits correspond, respectively, to *bright* and *dark* solitons of the DNLS.

11.3.3.1 Homoclinic Orbits

Let us concentrate our attention on homoclinic orbits emanating from the origin. A homoclinic orbit corresponds to an orbit that connects, in forward and backward time, a fixed point with itself. In turn, this corresponds to a non-trivial steady-state solution that decays to the fixed point for $n \to \pm\infty$. This is the so-called *bright soliton* solution. A sufficient condition for the existence of a homoclinic orbit for a 2D map is that the stable (W^s) and unstable (W^u) manifolds of the fixed point intersect. Thus, a necessary condition for the existence of these manifolds is that the fixed point must be a saddle. This latter condition, in turn, translates into a necessary (but not sufficient) condition on the parameters of the system. For example, in the $\Lambda < 0$ case, one needs a coupling constants $\epsilon < -\Lambda/4$ to ensure the origin is a saddle point (for $\Lambda > 0$ the origin is always a saddle point). It is important to stress that the existence of a saddle does not guarantee the existence of a homoclinic connection since the stable and unstable manifolds might not intersect at all. It is possible to formally establish the existence of homoclinic orbits of nearly integrable 2D maps through the Mel'nikov approach [18]. This method has been successfully applied to the single DNLS chain [1] as well as to systems of coupled DNLS equations [19] by means of a higher dimensional Mel'nikov approach [20]. Another approach to establish the existence of the homoclinic orbit is to expand them in a power series using a center manifold reduction [1, 7, 21, 22]. This has the advantage that one is able to extract an approximation for the homoclinic orbits and thus be able to approximate their bifurcations [23]. See [21] for a comprehensive list of different techniques to approximate the homoclinic connections arising from the DNLS system.

Any intersection between the stable and unstable manifolds (a so-called homoclinic point) will generate a localized steady-state solution for the DNLS. Generically, the stable and unstable manifolds cross transversally giving rise to a so-called homoclinic tangle (see left panel of Fig. 11.2 for a typical example). The transversality of the intersection of the manifolds establishes the *parametric* stability for the existence of homoclinic points and thus localized solutions. This property is extremely important for applications since it guarantees that, despite inaccuracies in the model parameters and external perturbations, localized solutions will still survive. This, for example, allows for *approximate* dynamical reductions to the interactions of continuous chains of bright solitons to be able to perform localized oscillations [24]. Two examples of soliton solutions generated by a homoclinic point of the focusing ($\beta = -1$) 2D map are depicted in Fig. 11.2 and they correspond to *bond-centered* (circles) and *site-centered* (squares) solutions. These two families are generated by the odd and even crossing of the stable and unstable manifolds starting at the points labeled by the points Q_0 and P_0 in the left panel. In general, the 2D map approach not only establishes the existence of bright soliton solutions (as well as dark soliton solutions, see below) but also determines their decay rate. Specifically, the eigenvalues $\lambda_\pm$ ($\lambda_- < 1 < \lambda_+$) for the saddle fixed point supporting the homoclinic orbit (the origin in the case under consideration) determine the *exponential* decay $\lambda_-^{|n|} = \lambda_+^{-|n|}$ for $n \to \pm\infty$ ($\lambda_- = \lambda_+^{-1}$ is a consequence of the

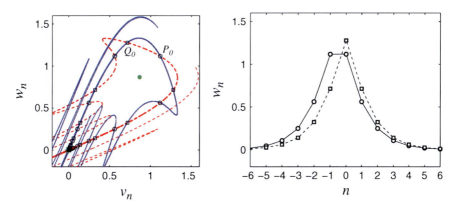

Fig. 11.2 Homoclinic connection of the 2D map (*left*). Stable and unstable manifolds are depicted by *solid* and *dashed lines*, respectively. The *right panel* depicts the corresponding *bright soliton* steady-state solutions to the DNLS. *Circles* (*squares*) correspond to a bond (site) centered bright soliton solution generated by the initial condition depicted in the *left panel* by P_0 (Q_0). Parameter values correspond to: $\Lambda = 0.75$, $\beta = -1$, and $\epsilon = 1$

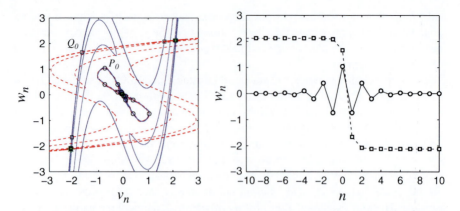

Fig. 11.3 Homoclinic and heteroclinic connections of the 2D map (*left*). Stable and unstable manifolds are depicted by *solid* and *dashed lines*, respectively. The *right panel* depicts the corresponding *dark soliton* (*squares*) and staggered *bright soliton* (*circles*) steady state solutions to the DNLS generated by the initial conditions depicted in the *left panel* by Q_0 and P_0, respectively. Parameter values correspond to: $\Lambda = -4.5$, $\beta = 1$, and $\epsilon = 1$

properties described in Sect. 11.3.1). In our case the eigenvalues at the origin are given by $2\epsilon\lambda_\pm = \Lambda + 2\epsilon \pm \sqrt{\Lambda(\Lambda + 4\epsilon)}$.

The staggering transformation generated by the symmetry described in Sect. 11.3.1.(b) establishes the existence of a staggered companion to the above described bright soliton. In Fig. 11.3 we depict with circles such a staggered bright soliton emanating from the initial condition labeled with P_0 in the left panel. The decaying properties for the staggered bright soliton are the same as for its unstaggered sibling.

11.3.3.2 Heteroclinic Orbits

Instead of considering connections involving a single fixed point, consider the stable manifold $W^s(x_1^*)$ emanating from the fixed point x_1^* and the unstable manifold $W^u(x_2^*)$ emanating from the fixed point x_2^* ($x_1^* \neq x_2^*$). If these manifolds intersect then it is possible to induce an orbit that connects, in forward time, x_1^* with, in backward time, x_2^*. This is a so-called *heteroclinic* connection and it corresponds to a steady state that connects to distinct homogeneous steady states (x_1^* and x_2^*), namely a *dark soliton* (or front).

Two examples of dark solitons generated by heteroclinic orbits of the 2D map are depicted in Figs. 11.4 and 11.3. Figure 11.4 depicts a dark soliton in the focusing case which has staggered tails, while Fig. 11.3 depicts (see orbit depicted with squares emanating from the initial condition labeled by Q_0) a standard dark soliton for the defocusing case.

The decaying properties for the tails of the dark soliton can be obtained, as in the case of the bright soliton, by the appropriate eigenvalues of the fixed points supporting the solution.

11 A Map Approach to Stationary Solutions of the DNLS Equation

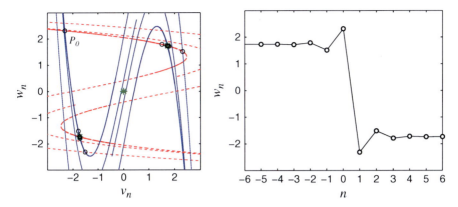

Fig. 11.4 Heteroclinic connection of the 2D map (*left*). Stable and unstable manifolds are depicted by *solid* and *dashed lines*, respectively. The *right panel* depicts the corresponding staggered *dark soliton* steady-state solution to the DNLS generated by the initial condition depicted in the *left panel* by P_0. Parameter values correspond to: $\Lambda = 3$, $\beta = -1$, and $\epsilon = 1$

11.3.3.3 Multibreathers

By following higher order intersections of the homoclinic connections it is possible to construct localized solutions with more than one localized hump [2, 3]. These solutions are usually referred to as *multibreathers*. In Fig. 11.5 we depict three examples of *bright* multibreathers for the same parameters but starting at different intersections on the homoclinic tangle. For a detailed classification of these multibreather solutions see [2] and [3]. Naturally, multibreather solutions are also possible in the defocusing case in the form of dark multisolitons (several contiguous

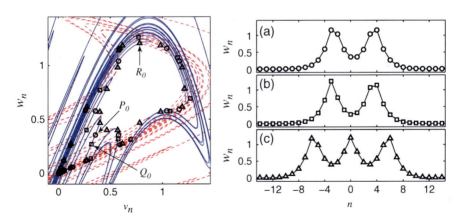

Fig. 11.5 Higher order homoclinic connections corresponding to multibreather solutions. The three multibreather solutions correspond to: (**a**) symmetric two-hump multibreather generated by the initial condition P_0 (see *circles*), (**b**) asymmetric two-hump multibreather generated by the initial condition Q_0 (see *squares*), and (**c**) three-hump multibreather generated by the initial condition R_0 (see *triangles*). Parameter values correspond to: $\Lambda = 0.75$, $\beta = -1$, and $\epsilon = 1$

troughs asymptotic to the constant homogeneous steady-state background) [25, 26]. It is worth mentioning that all the multibreather structures described herein are genuinely discrete solutions and are not related to the multisoliton solutions of the continuous nonlinear Schrödinger equation that can be generated from the single soliton solution by using the inverse scattering theory [27].

11.4 Bifurcations: The Road from the Anti-Continuous to the Continuous Limit

One of the most appealing aspects of the map approach to study steady states of nonlinear lattices is not only the elucidation of the extremely rich variety of structures that can be described but, perhaps more importantly, its usefulness in fully characterizing their bifurcations. The idea is to start at the so-called anti-continuous [28] (uncoupled) limit, $\epsilon = 0$, where any solution $v_n \in \{0, \pm\sqrt{-\Lambda/\beta}\}$ is valid. It is known that *all* possible solutions for $\epsilon = 0$ can be continued to *finite* coupling $\epsilon^* > 0$ [28]. In fact, several works have been devoted to finding bounds for ϵ^* (threshold for coupling below which *any* solution can be found) and they range from $\epsilon^* > 1/(10 + 4\sqrt{2}) \approx 0.0639$ to $\epsilon^* > (3\sqrt{3} - 1)/52 = 0.0807$ [3, 7, 29]. In terms of the 2D map description, the existence of any solution $v_n \in \{0, \pm\sqrt{-\Lambda/\beta}\}$ is a consequence of the fractal structure of the homoclinic tangle for small coupling. In fact, for small ϵ the homoclinic tangles tend to accumulate close to the basic nine points (x, y) with $x, y \in \{0, \pm\sqrt{-\Lambda/\beta}\}$ allowing orbits consisting of any combination of states $v_n \approx \{0, \pm\sqrt{-\Lambda/\beta}\}$ to be possible [3]. This effect can be clearly seen in panel (a) of Fig. 11.6 that corresponds to a very weak coupling $\epsilon = 0.05$ that is below the critical coupling ϵ^* and thus any orbit connecting any possible combination of neighboring basic points is valid.

As the coupling parameter ϵ is increased from the anti-continuous limit, solutions start to dissappear through mutual collisions in saddle node and pitchfork bifurcations. A detailed description of this scenario pertaining to the DNLS can be found in [3]. This work was in turn inspired by a similar analysis performed on the Hénon map [30]. In both [30] and [3] it is conjectured (the so-called no-bubbles-conjecture)

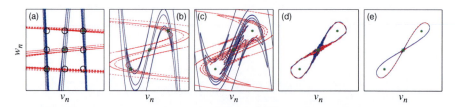

Fig. 11.6 Homoclinic tangle progression as the coupling parameter is increased from the anti-continuous limit toward the continuous limit. The coupling for each panel corresponds, from *left to right*, to $\epsilon = 0.05, 0.2, 0.6, 1$, and 1.5. In panel (**a**) the nine *black circles* correspond to the areas of the 2D map points giving rise to any possible combination $v_n \approx \{0, \pm\sqrt{-\Lambda/\beta}\}$ close to the anti-continuous limit. The other parameter values correspond to: $\Lambda = 0.75$ and $\beta = -1$

that, as the coupling ϵ grows, *only* annihilation of solutions (through saddle node and pitchfork bifurcations) occurs and that no new solutions emerge. In Fig. 11.6 we show the progression of the homoclinic tangle of the origin as the coupling parameter is increased from the anti-continuous limit toward the continuous limit. As it is clearly suggested by the figure, the amount of crossings between the different stable/unstable manifolds is greatly reduced as ϵ is increased. The disappearance of these crossings is accounted by a series of saddle node and pitchfork bifurcations – the saddle node being the most common one. By following the number of different possible homoclinic connections as the coupling is increased one would obtain a Devil (fractal) staircase [31–33] as it is evidenced in Fig. 14 of [3].

In the limit $\epsilon \to \infty$ (the continuous limit), the homoclinic tangle of the origin gets thinner and asymptotically tends towards a *simple* homoclinic connection where the stable and unstable manifolds coincide exactly and correspond to a simple loop as it can be observed from panel (e) of Fig. 11.6. In this continuous limit both, the bond-centered and the site-centered solutions, coalesce into the bright soliton solution to the standard *continuum* nonlinear Schrödinger equation.

11.5 Summary and Future Challenges

In this chapter we presented a review of the so-called map approach whereby the steady-state problem for general nonlinear dynamical lattices, with nearest-neighbor coupling, can be cast as a second-order recurrence equation that, in turn, generates a 2D map. Within this framework, *any* orbit of this 2D map generates a corresponding steady-state solution for the nonlinear lattice. Concentrating on the 2D map generated by the DNLS equation, we first described some generic properties of the steady solutions that are straightforward consequences of the underlying symmetries of the 2D map. Then, we comprehensively studied the diversity of DNLS steady-state solutions that can be generated using this map approach. We partition the possible solutions into spatially extended and localized steady states. Spatially extended states correspond to homogeneous, periodic, modulated, and spatially chaotic steady states of the DNLS and are generated, respectively, by fixed points, periodic orbits, quasi-periodic orbits, and chaotic orbits of the 2D map. The more interesting case of spatially localized steady states is generated by homoclinic or heteroclinic connections of the 2D map that in turn generate, respectively, bright and dark soliton steady-state solutions of the DNLS. We also elaborated on the staggered (oscillating) and multibreather variants thereof. We also briefly described the bifurcation road whereby the extremely rich diversity of solutions generated at the anti-continuum limit (zero coupling) is reduced through a series of saddle node and pitchfork bifurcations to a single solution (the standard bright or dark soliton) at the continuum limit.

Some future challenges related to the map approach would include the corroboration of the the so-called no-bubbles-conjecture, originally put forward by Sterling et al. for the Hénon map [30] and then re-stated for the DNLS 2D map by Alfimov et al. [3], whereby it is noted that as the coupling parameter is increased only *annihilation* of solutions occurs (through saddle node and pitchfork bifurcations) and thus no "birth" of new solutions may occur. Another topic that has not yet been fully explored is the use of the Mel'nikov approach for higher dimensional maps [20, 34–36] for more complex coupling schemes. In particular, this higher dimensional Mel'nikov approach has been successfully applied to a 1D nonlinear double Ablowitz–Ladik chain [37, 38] (see also Sect. 2.1 in Chap. 2) in [39]. It would be interesting to explore this higher dimensional approach in 1D lattices with higher order neighboring couplings (i.e., not only nearest neighbors) that will naturally generate higher order recurrence relationships between successive lattice sites and therefore higher dimensional maps. Finally, the direct application of the map approach for higher dimensional lattices is not possible because the recurrence relationship equivalent to Eq. (11.4) would involve two and three indices for the 2D and 3D cases, respectively. Nonetheless, it should be in principle possible to treat, for example, a 2D lattice chain as an infinite array of 1D coupled chains and apply the higher dimensional Mel'nikov approach mentioned above for the double Ablowitz–Ladik chain [39]. However, such a scheme is anticipated to be extremely cumbersome and involve complicated numerical methods to evaluate the Mel'nikov approach in high dimensions.

References

1. Hennig, D., Rasmussen, K.Ø., Gabriel, H., Bülow, A.: Phys. Rev. E **54**, 5788 (1996)
2. Bountis, T., Capel, H.W., Kollmann, M., Ross, J.C., Bergamin, J.M., van der Weele, J.P.: Phys. Lett. A **268**, 50 (2000)
3. Alfimov, G.L., Brazhnyi, V.A., Konotop, V.V.: Physica D **194**, 127 (2004)
4. Carretero-Gonzáles, R., Talley, J.D., Chong, C., Malomed, B.A.: Physica D **216**, 77 (2006)
5. Qin, W.X., Xiao, X.: Nonlinearity **20** 2305 (2007)
6. Maluckov, A., Hadžievski, L., Malomed, B.A.: Phys. Rev. E **77**, 036604 (2008)
7. Hennig, D., Tsironis, G.P.: Phys. Rep. **307** 333 (1999)
8. Malomed, B.A., Kevrekidis, P.G.: Phys. Rev. E **64**, 026601 (2001)
9. Kevrekidis, P.G., Malomed, B.A., Chen, Z., Frantzeskakis, D.J.: Phys. Rev. E **70**, 056612 (2004)
10. Sakaguchi, H., Malomed, B.A.: Europhys. Lett. **72**, 698 (2005)
11. Neshev, D.N., Alexander, T.J., Ostrovskaya, E.A., Kivshar, Yu.S., Martin, H., Makasyuk, I., Chen, Z.: Phys. Rev. Lett. **92**, 123903 (2004)
12. Fleischer, J.W., Bartal, G., Cohen, O., Manela, O., Segev, M., Hudock, J., Christodoulides, D.N.: Phys. Rev. Lett. **92**, 123904 (2004)
13. Kevrekidis, P.G., Malomed, B.A., Frantzeskakis, D.J., Carretero-González, R.: Phys. Rev. Lett. **93**, 080403 (2004)
14. Carretero-González, R., Kevrekidis, P.G., Malomed, B.A., Frantzeskakis, D.J.: Phys. Rev. Lett. **94**, 203901 (2005)
15. Kevrekidis, P.G., Carretero-González, R., Frantzeskakis, D.J., Malomed, B.A., Diakonos, F.K.: Phys. Rev. E **75**, 026603 (2007)
16. Balasuriya, S.: Physica D, **202**, 155 (2005)
17. Hennig, D., Sun, N.G., Gabriel, H., Tsironis, P.: Phys. Rev. E **52**, 255 (1995)
18. Glasser, M.L., Papageorgiou, V.G., Bountis, T.C.: SIAM J. Appl. Math. **49**, 692 (1989)
19. Kollmann, M., Bountis, T.: Physica D **113**, 397 (1998)

20. Bountis, T., Goriely, A., Kollmann, M.: Phys. Lett. A **206**, 38 (1995)
21. James, G., Sánchez-Rey, B., Cuevas, J.: Rev. Math. Phys. **21**, 1 (2009)
22. Cuevas, J., James, G., Malomed, B.A., Kevrekidis, P.G., Sánchez-Rey, B.: J. Nonlinear Math. Phys. **15**, 134 (2008)
23. Palmero, F., Carretero-González, R., Cuevas, J., Kevrekidis, P.G., Królikowski, W.: Phys. Rev. E **77** 036614 (2008)
24. Carretero-González, R., Promislow, K.: Phys. Rev. A **66**, 033610 (2002).
25. Susanto, H., Johansson, M.: Phys. Rev. E **72**, 016605 (2005)
26. Pelinovsky, D.E., Kevrekidis, P.G.: J. Phys. A: Math. Theor. **41** 185206 (2008)
27. Debnath, L.: Nonlinear Partial Differential Equations for Scientists and Engineers. Birkhäuser Boston, (2005)
28. MacKay, R.S., Aubry, S.: Nonlinearity **7**, 1623 (1994)
29. Dullin, H.R., Meiss, J.D.: Physica D **143**, 265 (2000)
30. Sterling, D., Dullin, H.R., Meiss, J.D.: Physica D **134**, 153 (1999)
31. Bak, P.: Phys. Today **12** 38 (1986)
32. Schroeder, M.: Fractals, Chaos, Power Laws. W.H. Freeman and Company, New York 1991. Chap. 7
33. Carretero-González, R., Arrowsmith, D.K., Vivaldi, F.: Physica D **103** 381 (1997)
34. Palmer, J.K.: J. Diff. Eqns. **55**, 225 (1984)
35. Chow, S.N., Hale, J.K., Mallet-Paret, J.: J. Diff. Eqns. **37**, 351 (1980)
36. Chow, S.N., Yamashita, M.: Nonlinear Equations in the Applied Sciences. Academic, New York, (1992)
37. Ablowitz, M.J., Ladik, J.F.: J. Math. Phys. **16**, 598 (1975)
38. Ablowitz, M.J., Ladik, J.F.: J. Math. Phys. **17**, 1011 (1976)
39. Bülow, A., Hennig, D., Gabriel, H.: Phys. Rev. E **59**, 2380 (1999)

Chapter 12
Formation of Localized Modes in DNLS

Panayotis G. Kevrekidis

12.1 Introduction

For the most part in this volume, we have analyzed solitary wave coherent structures that constitute the prototypical nonlinear wave solutions of the ubiquitous discrete nonlinear Schrödinger (DNLS) model. A natural question, however, that arises is how do these structures emerge from general initial data. One possible answer to that question stems from the modulational instability mechanism that we addressed in some detail in both focusing and defocusing DNLS equations in Chap. 6. Another possibility that we will address in this section is the formation of such nonlinear excitations from localized initial data. In fact, the latter approach was experimentally pioneered in [1], where an injected beam of light was introduced into one waveguide of a waveguide array. It was observed that when the beam had low intensity, then it dispersed through quasi-linear propagation. This is natural in our cubic nonlinearity setting, as for sufficiently small amplitudes the nonlinear term becomes irrelevant. On the other hand, in the same work, experiments with high intensity of the input beam led to the first example of formation of discrete solitary waves in waveguide arrays. A very similar "crossover" from linear to nonlinear behavior was observed also in arrays of waveguides with the defocusing nonlinearity [2]. The common feature of both works is that they used the DNLS equation as the supporting model to illustrate this behavior at a theoretical/numerical level. However, this crossover phenomenon is certainly not purely discrete in nature. Even the *integrable* continuum NLS equation possesses this feature. More specifically, it is well-known that, e.g., in the case of a square barrier of initial conditions of amplitude V_0 and width L, the product $V_0 L$ determines the nature of the resulting soliton, and if it is sufficiently small the initial condition disperses without the formation of a solitonic structure [3]. It should also be noted that the existence of the threshold is not a purely one-dimensional feature either. For instance, experiments on the formation of solitary waves in two-dimensional photorefractive crystals show that low intensities lead to diffraction, whereas higher intensities induce localization [4–6]. Similar

---

P.G. Kevrekidis (✉)
University of Massachusetts, Amherst, MA, 01003, USA
e-mail: kevrekid@math.umass.edu

phenomena were observed even in the formation of higher order excited structures such as vortices (as can be inferred by carefully inspecting the results of [7, 8]).

An interesting question, examined recently in the work of [9] which we will follow herein, concerns the identification of the above-mentioned crossover between linear and nonlinear behavior. In the work of [9], this was addressed in, arguably, the simplest context of a localized initial condition (which was nevertheless to experiments such as the ones reported in [1, 2]), namely a Kronecker-δ initial condition parametrized by its amplitude for the DNLS equation. For earlier work on the same general theme, see also [10, 11]. It was thus found that there exists a well-defined value of the initial state amplitude such that initial states with higher amplitude always give rise to localized modes. The condition may be determined by comparing the energy of the initial state with the energy of the localized excitations that the model supports. This sufficient, but not necessary, condition for the formation of localized solitary waves provides an intuitively and physically appealing interpretation of the dynamics that is in very good agreement with our numerical observations. It is interesting to compare/contrast this dynamics with the corresponding one of the continuum NLS model and also the discrete integrable Ablowitz–Ladik (AL)-NLS model (such a comparison is presented below). In addition to the one-dimensional DNLS lattice, we also consider the two-dimensional case where the role of both energy and beam power (mathematically the squared l^2-norm) become apparent. We should note here that our tool of choice for visualizing the "relaxational process" (albeit in a Hamiltonian system) of the initial condition will be energy–power diagrams. Such diagrams have proven very helpful in visualizing the dynamics of initial conditions in a diverse host of nonlinear wave equations. In particular, they have been used in the nonlinear homogeneous systems such as birefringent media and nonlinear couplers as is discussed in Chaps. 7 and 8 of [12], as well as for general nonlinearities in continuum dispersive wave equations in [13], while they have been used to examine the migration of localized excitations in DNLS equations in [14].

We start this discussion by presenting the theory of the integrable continuum and discrete models. We then examine how the DNLS differs from the former models in its one-dimensional version, how the relevant results generalize in higher dimensions and lastly pose some interesting questions for future study. As indicated above, our exposition chiefly follows that of [9], although some interesting new results and associated questions are posed in the last part of the relevant discussion.

12.2 Threshold Conditions for the Integrable NLS Models

12.2.1 The Continuum NLS Model

For the focusing NLS equation [3]

$$iu_t = -\frac{1}{2}u_{xx} - |u|^2 u \tag{12.1}$$

with squared barrier initial data

$$u(x, 0) = \begin{cases} V_0 & \text{if } -L \leq x \leq L, \\ 0 & \text{otherwise} \end{cases} \qquad (12.2)$$

(the inverse of) the transmission coefficient, $S_{11}(E)$, which is the first entry of the scattering matrix, is given by

$$S_{11}(E) = v(E)\cos(2v(E)L) - iE\sin(2v(E)L) \qquad (12.3)$$

with $v(E) = \sqrt{E^2 + V_0^2}$ where E is the spectral parameter and V_0 the amplitude of the barrier. It is well-known that the number of zeros of this coefficient represents the number of solitons produced by the square barrier initial condition [3]. It can be proved that the roots of this equation are purely imaginary. (This initial condition satisfies the single-lobe conditions of Klaus–Shaw potentials, from which it follows that the eigenvalues are purely imaginary [15]). Let us define $\eta \geq 0$ and use $E = i\eta$. Then, Eq. (12.3) becomes

$$\sqrt{1-\eta^2}\cos\left(2V_0\sqrt{1-\eta^2}L\right) + \eta\sin\left(2V_0\sqrt{1-\eta^2}L\right) = 0. \qquad (12.4)$$

We can verify that Eq. (12.4) does not have any roots (i.e., leads to no solitons in Eq. (12.1)) if

$$V_0 < \frac{\pi}{2}. \qquad (12.5)$$

Furthermore, the condition to generate n solitons, i.e., so that Eq. (12.4) has n roots is

$$(2n-1)\frac{\pi}{2} < 2V_0 L < (2n+1)\frac{\pi}{2} \qquad (12.6)$$

and the corresponding count of eigenvalues is given by

$$\frac{2}{\pi}V_0 L - \frac{1}{2} < n < \frac{2}{\pi}V_0 L + \frac{1}{2}. \qquad (12.7)$$

The limit $V_0 \to \infty$ together with $L \to 0$ can be reached if we impose $2V_0 L = $ const. In this instance, the number of eigenvalues stays the same.

12.2.2 The Ablowitz–Ladik Model

We now turn to the integrable discretization of Eq. (12.1) and examine its dynamics. For the AL-NLS of the form

$$i\dot{u}_n = -\frac{1}{2}(u_{n+1} + u_{n-1} - 2u_n) - \frac{1}{2}|u_n|^2(u_{n+1} + u_{n-1}), \qquad (12.8)$$

there exists a Lax pair of linear operators

$$\mathcal{L}_n = Z + M_n, \qquad (12.9)$$

$$\mathcal{B}_n = \left(\frac{z - z^{-1}}{2}\right)^2 D + \frac{1}{2}(ZM_n - Z^{-1}M_{n-1}) - \frac{1}{2}DM_n M_{n-1} \qquad (12.10)$$

with the definitions for the matrices

$$Z = \begin{pmatrix} z & 0 \\ 0 & z^{-1} \end{pmatrix}, \quad D = \begin{pmatrix} -1 & 0 \\ 0 & 1 \end{pmatrix}, \quad \text{and} \quad M_n = \begin{pmatrix} 0 & U_n \\ -U_n^\star & 0 \end{pmatrix}, \qquad (12.11)$$

where z is the spectral parameter and $U_n = U_n(t)$ is a solution of the equation. These two operators (12.9) and (12.10) define the system of differential–difference equations

$$\Psi_{n+1} = \mathcal{L}_n \Psi_n, \qquad (12.12)$$

$$i\frac{d}{d\tau}\Psi_n = \mathcal{B}_n \Psi_n \qquad (12.13)$$

for a complex matrix function Ψ_n, of which the compatibility condition

$$i\frac{d}{d\tau}\Psi_{n+1} = i\left(\frac{d}{d\tau}\Psi_m\right)\bigg|_{m=n+1}$$

is the AL-NLS model. $U_n = U_n(t)$ is referred to as the potential of the AL-NLS eigenvalue problem.

For U_n decaying rapidly at $\pm\infty$, and for $n \to \pm\infty$, from Eq. (12.12) we have

$$\Psi_{n+1} \sim Z\Psi_n.$$

We normalize this type of solutions as follows: let Ψ_n denote the solution of Eq. (12.12) such that

$$\Psi_n \sim Z^n \quad \text{as} \quad n \to +\infty,$$

and let Φ_n be the solution of Eq. (12.12) such that

$$\Phi_n \sim Z^n \quad \text{as} \quad n \to -\infty.$$

Ψ_n and Φ_n are known as the Jost functions. Each of these forms a system of linearly independent solutions of the AL eigenvalue problem (12.12). These sets of solutions are interrelated by the scattering matrix $S(z)$,

$$\Phi_n = \Psi_n S(z). \tag{12.14}$$

The first column of this equation is given by

$$(\Phi_1)_n = S_{11}(z)(\Psi_1)_n + S_{21}(z)(\Psi_2)_n, \tag{12.15}$$

where $(\Phi_1)_n$ denotes the first column of Φ_n. Similar definitions apply to $(\Psi_1)_n$ and $(\Psi_2)_n$. Since $\Phi_n \sim Z^n$ as $n \to -\infty$, then

$$(\Phi_1)_n \sim z^n \begin{pmatrix} 1 \\ 0 \end{pmatrix}.$$

To obtain decay, $(\Phi_1)_n \to 0$ when $n \to -\infty$, we require

$$|z| > 1. \tag{12.16}$$

On the other hand,

$$\Psi_n = ((\Psi_1)_n, (\Psi_2)_n) \sim \begin{pmatrix} z^n & 0 \\ 0 & z^{-n} \end{pmatrix} \quad \text{as} \quad n \to \infty.$$

Therefore, if $|z| > 1$ then $(\Psi_1)_n \to \infty$ and $(\Psi_2)_n \to 0$, as $n \to \infty$. Now, from Eq. (12.15) it follows that $(\Phi_1)_n \to \infty$ as $n \to \infty$, unless $S_{11}(z) = 0$.

We therefore seek solutions $z_1, z_2, \ldots, z_N$ of the equation

$$S_{11}(z_k) = 0, \qquad k = 1, 2, \ldots, N, \tag{12.17}$$

such that $|z_k| > 1$. Then,

$$(\Phi_1)_n(z_k) = S_{21}(z_k)(\Psi_2)_n(z_k), \qquad k = 1, 2, \ldots, N.$$

From this, it follows that $(\Phi_1)_n(z_k)$ decays at $\pm\infty$:

$$(\Phi_1)_n(z_k) \to 0 \qquad \text{as} \qquad n \to \pm\infty.$$

It is then said that $(\Phi_1)_n(z_k)$ is an eigenfunction ($k = 1, 2, \ldots, N$), with corresponding eigenvalue z_k.

In the case of $U_n(t=0) = U_0 \delta_{n,n_0}$, the Jost function is

$$\Phi_n = Z^{n-1}(Z + M_0)\Phi_0, \qquad \text{for} \quad n \geq 1 \tag{12.18}$$

with

$$M_0 = \begin{pmatrix} 0 & U_0 \\ -U_0 & 0 \end{pmatrix}. \tag{12.19}$$

Furthermore, $\Psi_n = Z^n$ for $n \geq 1$ and Φ_0 is the identity matrix: $\Phi_0 = I$. Hence, Eq. (12.18) reads

$$\Phi_n = \Psi_n Z^{-1}(Z + M_0), \quad \text{for} \quad n \geq 1.$$

Its comparison with Eq. (12.14) leads to a scattering matrix

$$S(z) = Z^{-1}(Z + M_0).$$

We thus obtain that the transmission coefficient

$$S_{11}(z) = 1,$$

which never vanishes. This means that the one-site potential (i.e., a single-site initial condition) does *not* admit solitonic solutions, *independent* of the amplitude U_0 of initial excitation. This theoretical result has also been confirmed by numerical simulations for different values of U_0, always leading to dispersion of the solution.

We should note in passing that the problem of few-site initial conditions in the AL-NLS problem has been recently considered in [16]. There, it was found that the lowest number of sites needed in order to create a soliton is, in fact, two, in which case, there is a threshold initial amplitude above which a single soliton will be generated. The same conclusion is valid for a three-site initial condition, while for four or more sites, the dynamical behavior can become more complex. For more details, we refer the interested reader to [16].

12.3 Threshold Conditions for the Non-Integrable DNLS Model

We now turn to the DNLS

$$i\dot{u}_n = -\epsilon \Delta_2 u_n - |u_n|^{2\sigma} u_n, \tag{12.20}$$

where we rescale $\epsilon = 1$, through $t \to \epsilon t$ and $u_n \to u_n/\sqrt{\epsilon}$. For completeness, we recall the Hamiltonian

$$H = \sum_n |u_{n+1} - u_n|^2 - \frac{1}{\sigma + 1} |u_n|^{2\sigma+2}, \tag{12.21}$$

which will be relevant in our considerations below, being one of the two main conserved quantities of the DNLS (the other one being the squared l^2 norm $P = \sum_n |u_n|^2$).

In examining the role of localized initial conditions in the formation of solitary waves, we start with, arguably, the simplest possibility, namely a "compactum" of mass $u_n = A\delta_{n,0}$. The question of interest concerns what is the critical value of A that is necessary for this single-site initial condition to excite a localized mode.

12 Formation of Localized Modes in DNLS

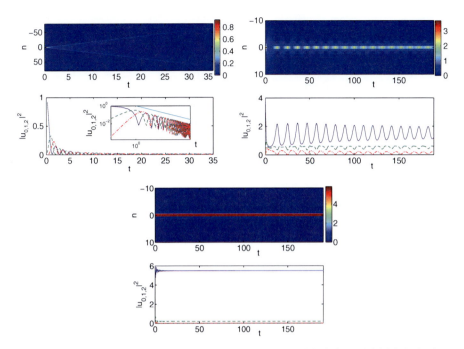

Fig. 12.1 Subcritical case (*top left*), critical (*top right*), and supercritical (*bottom*) initial single-site excitations on the lattice. In all cases, the initial condition is $u_n = A\delta_{n,0}$, with $A = 1$ in the *left*, 2 in the *middle*, and 2.5 in the *right* panels. In each case the *top panel* shows the space–time contour plot of the evolution. The *bottom panel* shows the dynamical evolution of $|u_n|^2$ for sites $n = 0, 1, 2$ (*solid*, *dashed*, and *dash–dotted lines*, respectively). For the first case the inset shows the same evolution in a log–log plot and a t^{-1} decay for comparison. This indicates that the damped oscillation of the field modulus has an envelope of $t^{-1/2}$. Reprinted from [9] with permission

That such a threshold definitely exists is illustrated in Fig. 12.1. The case of the top left panel is subcritical, leading to the discrete dispersion of the initial datum. This follows the well-known $t^{-1/2}$ amplitude decay which is implied by the solution of the (infinite lattice) problem in the absence of the nonlinearity

$$u_n(t) = Ai^n J_n(2t), \qquad (12.22)$$

where J_n is a Bessel function of order n. The bottom panel, on the other hand, shows a nonlinearity-dominated regime with the rapid formation of a solitary wave strongly localized around $n = 0$. In the intermediate case of the top right panel, the system exhibits a long oscillatory transient reminiscent of a separatrix between the basins of attraction of the two different regimes.

It was argued in [9] that this separatrix can be identified through the examination of the stationary (localized) states of the model. As we saw in Chap. 2, such standing wave solutions of the form $u_n = \exp(i\Lambda t)v_n$, can be found for arbitrary frequency Λ (and arbitrary power P in one spatial dimension). On the other hand, the single-site initial condition has an energy of

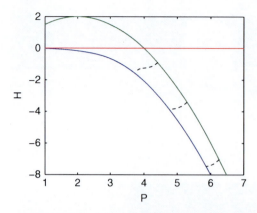

Fig. 12.2 The lower *solid line* shows the energy H versus the power P of the discrete solitary wave solutions. The upper *solid line* shows the energy versus power of the initial condition, obtained from Eq. (12.23) and $P_{ss} = A^2$. The *horizontal line* denotes $H = 0$ and its intersection with the initial condition curve defines the single-site, initial amplitude A^*. The dynamical evolution of three different supercritical initial states with $A = 2.1$, $A = 2.3$, and $A = 2.5$ is shown by *dashed lines*. Reprinted from [9] with permission

$$H_{ss} = 2A^2 - \frac{1}{\sigma + 1} A^{2\sigma+2}, \tag{12.23}$$

where the subscript denotes single site. Figure 12.2 summarizes succinctly the power dependence of the energy for these two cases for $\sigma = 1$. Both the energy of the stationary solutions as a function of their power and the single-site energy as a function of single-site power ($P_{ss} = A^2$), are shown. Note that the two curves $H_{ss}(P_{ss})$ and $H(P)$ do not intersect (except at the trivial point $H = P = 0$) since for $\epsilon \neq 0$ single-site states are not stationary ones.

The examination of the energy–power diagram of Fig. 12.2 provides information on the existence of a sufficient condition for the formation of a localized mode and on the dynamics of the single-site initial condition. Figure 12.2 shows that localized solutions exist for arbitrarily small values of the input power P and that the energy of the localized states is negative. This implies that the crucial quantity to determine the fate of the process is the energy H and not the power P. The role of the power will become more evident in the two-dimensional setting. Moreover, if the system starts at a given point on the curve defined by $H_{ss} = 2P_{ss} - P_{ss}^{\sigma+1}/(\sigma + 1)$, due to conservation of total H and total P, it can only end up in a stationary state in the quadrant $H < H_{ss}$ and $P < P_{ss}$. That is to say, some of the initial energy and power are typically "shed off" in the form of radiation (i.e., converted to other degrees of freedom which is the only way that "effective dissipation" can arise in a purely Hamiltonian system), so that the initial condition can "relax" to the pertinent final configuration. As mentioned above, a localized solution with the same power as that of the initial condition exists for arbitrary A. *However*, emergence of a localized mode occurs only for those initial conditions whose core energy (i.e., the energy of a region around the initially excited state) is negative, after the profile is "reshaped" by radiating away both energy and power.

Therefore, if $H_{ss} < 0$, then the compactum of initial data will always yield a localized excitation: this inequality provides the sufficient condition for the excitation of solitary waves. The condition on the energy, in turn, provides a condition on the single-site amplitude that leads to the formation of solitary waves, namely solitary waves always form if $A > A^*$ with

$$A^* = [2(\sigma + 1)]^{\frac{1}{2\sigma}} . \qquad (12.24)$$

For the case of $\sigma = 1$ considered in Figs. 12.1 and 12.2, this amplitude value is $A^* = 2$ in agreement with our numerical observations of Fig. 12.1 and the earlier numerical results of [10]. Whether an initial state with $H_{ss} > 0$ yields a localized state depends on the explicit system dynamics, corroborating the observation that the previous energetic condition is a sufficient, but not necessary, condition.

It is relevant to make a few important observations here. Firstly, we note the significant differences between the non-integrable discrete model and both of its integrable (continuum and discrete) counterparts. In the (singular) continuum limit, it is possible to excite a single soliton or a multisoliton depending on the barrier height and width. On the other hand, in the integrable discretization, a single-site excitation *never* leads to solitary wave formation, contrary to what is the case here where either none or one solitary wave may arise, depending on the amplitude of the initial one-site excitation.

Focusing on the DNLS model, we observe that even though a localized solution with the same power as that of the initial condition exists for arbitrary A, formation of a localized mode will always occur only for $A > A^*$. Secondly, the answer that the formation always occurs for $A > A^*$ generates two interconnected questions: what is the threshold for the formation of the localized mode, and given an initial $A > A^*$, which one among the monoparametric family of solutions will the dynamics of the model select as the end state of the system? (It should be noted in connection to the latter question that single-site initial conditions have *always* been found in our numerical simulations to give rise to *at most* a single-site-centered solution, i.e., multisoliton states cannot be produced by this process, contrary to what is the case, e.g., in the continuum NLS equation.) Some examples of this dynamical process are illustrated in Fig. 12.2, where the energy and power of a few sites (typically 20–40) around the originally excited one are measured as a function of time and are parametrically plotted in the H–P plane. As it should, the relevant curve starts from the H_{ss}–P_{ss} curve, and asymptotically approaches, as a result of the dynamical evolution, the H–P curve of the stationary states of the system. However, the relaxation process happens *neither at fixed energy, nor at fixed power*. Instead, it proceeds through a more complex, dynamically selected pathway of loss of both H and P to relax eventually to one of the relevant stationary states. This is shown for three different values of supercritical amplitude in Fig. 12.2 ($A = 2.1$, 2.3, and 2.5). We have noted (numerically) that the loss of energy and power, at least in the initial stages of the evolution happens at roughly the same rate, resulting in

$dH/dP \approx$ const. However, the later stages of relaxation are more complex and no longer preserve this constant slope. Therefore, the formation of a localized mode is neither equi-energetic, nor does it occur at fixed power; similar features have been previously observed in the continuum version of the system in [13].

In the H–P space, the dynamics of the system can be qualitatively understood by considering the frequencies Λ associated with the *instantaneous* H, P values along the system trajectories. At each instantaneous energy H a frequency Λ_H may be defined as the frequency of a single-site stationary state with energy H. Such a frequency is unique as the stationary state energy is a monotonically decreasing function of the frequency [17]. Similarly, a frequency Λ_P may be uniquely associated with the instantaneous power P, i.e., Λ_P is the frequency of a stationary state with power P. Note, however, that the stationary state power is a monotonically increasing function of Λ. At the final stationary state where the system relaxes $\Lambda_H = \Lambda_P = -dH/dP$. Hence, the system trajectories are such that Λ_H increases (consequently the energy decreases) and Λ_P decreases (the power decreases). The final stationary state is reached when the two frequencies become equal, the point where they meet depending on their rate of change along the trajectory, i.e., on their corresponding "speeds" along the trajectory. Nevertheless, the precise mechanism of selection of the particular end state (i.e., of the particular "equilibrium Λ") that a given initial state will result in remains a formidable outstanding question that would be especially interesting to address in the future. This is perhaps one of the fundamental remaining open questions in connection to the DNLS equation (see also the relevant discussion at the end of this special section).

Turning now to the two-dimensional DNLS equation of (12.20) with $\Delta_2 u_{n,m} = (u_{n+1,m} + u_{n-1,m} + u_{n,m+1} + u_{n,m-1} - 4u_{n,m})$, with an the initial condition $u_{n,m} = A\delta_{n,0}\delta_{m,0}$, we note that similar considerations apply. In particular, in this case the H_{ss}–P_{ss} curve is given by

$$H_{ss} = 4P_{ss} - \frac{P_{ss}^{\sigma+1}}{\sigma+1}. \tag{12.25}$$

In d dimensions the first term would be $2dP_{ss}$. While the standing wave branch of solutions is also well known [18] (see also Chap. 3), we recall that there are some important differences from the one-dimensional case. Firstly, a stable and an unstable branch of solutions exists; in the wedge-like curve indicating the standing waves, only the lower energy branch is stable. Furthermore, the maximal energy of the solutions is no longer $H = 0$, but finite and positive. Finally, solutions no longer exist for arbitrarily low powers, but they may only exist above a certain power (often referred to as the excitation threshold [19–21]). We can now appreciate the impact of these additional features in Fig. 12.3. Comparing H_{ss} with H_0 we find two solutions: one with $A \approx 0.7$ and one with $A \approx 2.73$. However, for the lower one there are no standing wave excitations with the corresponding power (this illustrates the role of the power in the higher dimensional problem). Hence, the relevant amplitude that determines the sufficient condition in the two-dimensional case is the latter. The figure presents the dynamical pathway for three supercritical values of $A =$

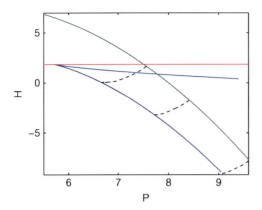

Fig. 12.3 Similarly to Fig. 12.2, we show the H–P diagram for the solution branch (wedge-like solid line), the H_{ss}–P_{ss} graph of the initial conditions (upper *solid line*) and the trajectories of three supercritical cases in H–P space for $A = 2.75$, 2.9, and 3.1 (*dashed lines*). The *horizontal line* represents the maximal energy for which solutions are found to exist, namely $H_0 \approx 1.85$. Reprinted from [9] with permission

2.75, 2.9, and 3.1. Once again, the dynamics commences on the H_{ss}–P_{ss} curve, as it should, eventually relaxing on the *stable* standing wave curve. As before, the initial dynamics follows a roughly constant dH/dP, but the relaxation becomes more complex at later phases of the evolution.

12.4 Conclusions and Future Challenges

The work of [9], as well as the earlier work of [10, 11] illustrated systematically (as was summarized above) the existence of a sharp crossover between the linear and nonlinear dynamics of the DNLS lattice. This crossover has been observed experimentally for both focusing [1] (as considered here) and defocusing nonlinearity [2]. Since the latter can be transformed into the former under the so-called staggering transformation $u_n = (-1)^n w_n$, where w_n is the field in the defocusing case, the solitary waves of the defocusing problem discussed in [2] will be "staggered" (i.e., of alternating phase between neighboring sites), yet the phenomenology discussed above will persist. The crossover was quantified on the basis of an energetic comparison of the initial state energy with the branch of corresponding stable localized solutions "available" in the model. A sufficient, but not necessary, condition for the excitation of a localized mode based on the initial state, single-site amplitude was discussed and tested in numerical simulations. Similar findings were obtained in the two-dimensional analog of the problem: a crossover behavior dictated by the energy was found, but the crossover was also affected by the power and its excitation thresholds. Furthermore, these results were fundamentally different from the case of the continuum version of the model, where depending on the strength of the excitation, also multisolitons can be obtained and from the integrable discrete

AL-NLS model where due to the nature of the nonlinearity, no single-site excitation can produce a solitary wave, independent of its excitation amplitude.

Clearly, however, there exists a number of important open questions arising in the context of the formation of solitary waves from different localized initial conditions. One of the foremost among them concerns how the dynamics "selects" among the available steady-state excitations with energy and power below that of the initial condition the one to which the dynamical evolution leads. According to our results, this evolution is neither equi-energetic, nor power-preserving, hence it would be an extremely interesting question from the point of view of understanding this Hamiltonian "relaxation to equilibrium" to identify the leading physical principle which dictates it.

A related important question concerns the stationary states that result from the excitation of multiple sites on the lattice. That is an interesting question which may exhibit sensitive dependence on the initial conditions and for which the stability issues examined in the earlier chapters (Chaps. 2, 3, and 4) may play a critical role. As a related example, we present two prototypical cases of two-site excitations in the one-dimensional DNLS in Fig. 12.4. In the top left panel, the two central

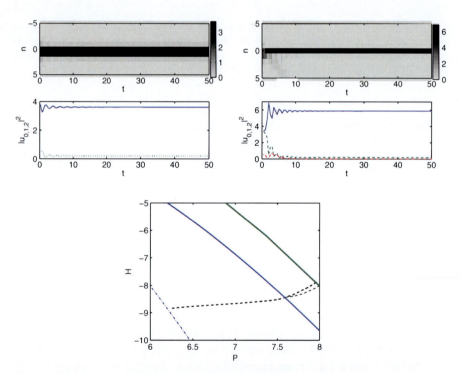

Fig. 12.4 The *top left* and *top right panels* are as in Fig. 12.1, but for two-site excitations with $A = 2$ in both sites (*left*) and $A = 2$ and 1.99 in the two sites (*right*). The *bottom panel* shows these two dynamical trajectories as *dashed lines* in the energy–power diagram featuring the initial condition of two excited sites (*thick solid line*): the two-site steady-state branch (intermediate *solid line*) and the single-site steady-state branch (*thin solid line*)

sites are both excited with amplitudes $A = 2$, while in the top right panel the right site has $A = 1.99$ (instead of $A = 2$ as the left). It can be readily observed that the dynamical evolution of the two states is dramatically different. In the weakly asymmetric initial excitation, this asymmetry seeds the instability of the inter-site state, and instead of permitting the initial condition to relax to an inter-site mode, it forces it to eventually relax to a single-site mode (contrary to what is the case in the left panel). The bottom panel illustrates that although both initial conditions are very close to being on the two site, equal excitation curve (and very close to each other in the energy–power space), they lead to significantly different dynamical evolutions. Thus, it seems that predicting the resulting stationary state for general initial data on multiple sites may be a very difficult, if not generically intractable, task.

References

1. Eisenberg, H.S., Silberberg, Y., Morandotti, R., Boyd, A.R., Aitchison, J.S.: Phys. Rev. Lett. **81**, 3383 (1998)
2. Matuszewski, M., Rosberg, C.R., Neshev, D.N., Sukhorukov, A.A., Mitchell, A., Trippenbach, M., Austin, M.W., Krolikowski, W., Kivshar, Yu.S.: Opt. Express **14**, 254 (2006)
3. Zakharov, V.E., Shabat, A.B.: Soviet Physics-JETP **34**, 62 (1972)
4. Fleischer, J.W., Carmon, T., Segev, M., Efremidis, N.K., Christodoulides, D.N.: Phys. Rev. Lett. **90**, 023902 (2003)
5. Fleischer, J.W., Segev, M., Efremidis, N.K., Christodoulides, D.N.: Nature **422**, 147 (2003)
6. Martin, H., Eugenieva, E.D., Chen, Z., Christodoulides, D.N.: Phys. Rev. Lett. **92**, 123902 (2004)
7. Neshev, D.N., Alexander, T.J., Ostrovskaya, E.A., Kivshar, Yu.S., Martin, H., Makasyuk, I., Chen, Z.: Phys. Rev. Lett. **92**, 123903 (2004)
8. Fleischer, J.W., Bartal, G., Cohen, O., Manela, O., Segev, M., Hudock, J., Christodoulides, D.N.: Phys. Rev. Lett. **92**, 123904 (2004)
9. Kevrekidis, P.G., Espinola-Rocha, J.A., Drossinos, Y., Stefanov, A.: Phys. Lett. A **372**, 2247 (2008)
10. Molina, M.I., Tsironis, G.: Physica D **65**, 267 (1993)
11. Molina, M.I., Tsironis, G.: Phys. Rev. B **47**, 15330 (1993)
12. Akhmediev, N.N., Ankiewicz, A.: Solitons: Nonlinear pulses and beams. Chapman & Hall, London (1997)
13. Akhmediev, N.N., Ankiewicz, A., Grimshaw, R.: Phys. Rev. E **59**, 6088 (1999)
14. Rumpf, B.: Phys. Rev. E **70**, 016609 (2004)
15. Klaus, M., Shaw, J.K.: Phys. Rev. E **65**, 036607 (2002)
16. Espinola-Rocha, J.A., Kevrekidis, P.G.: Thresholds for Soliton Creation in the Ablowitz-Ladik Lattice, Math. Comp. Simulat. (in press, 2009)
17. Johansson, M., Aubry, S.: Phys. Rev. E **61**, 5864 (2000)
18. Kevrekidis, P.G., Rasmussen, K.Ø., Bishop, A.R.: Phys. Rev. E **61**, 2006-2009 (2000)
19. Flach, S., Kladko, K., MacKay, R.S.: Phys. Rev. Lett. **78**, 1207 (1997)
20. Weinstein, M.I.: Nonlinearity **12**, 673 (1999)
21. Kevrekidis, P.G., Rasmussen, K.Ø., Bishop, A.R.: Phys. Rev. E **61**, 4652 (2000)

Chapter 13
Few-Lattice-Site Systems of Discrete Self-Trapping Equations

Hadi Susanto

13.1 Introduction

In this section, we will review work on few-lattice-site systems of the so-called discrete self-trapping (DST) equations where we will discuss the integrability of few-lattice-site DST systems, the presence of chaos in nonintegrable ones, their applications as well as experimental observations of the systems.

The DST equation which is a generalization of the discrete nonlinear Schrödinger (DNLS) equation was introduced by Eilbeck, Lomdahl, and Scott in [1]. The equation takes the form

$$i\frac{d}{dt}\mathbf{A} + \Gamma \mathbf{D}\left(|\mathbf{A}|^2\right)\mathbf{A} + \epsilon M \mathbf{A} = 0, \tag{13.1}$$

where $\mathbf{A} = \mathrm{col}(\mathbf{A}_1, \mathbf{A}_2, \ldots, \mathbf{A}_n)$ is a complex n-component vector and $\mathbf{D}$ is an $n \times n$ matrix denoting the nonlinearity, given by

$$\mathbf{D}\left(|\mathbf{A}|^2\right) \equiv \mathrm{diag}\left(|\mathbf{A}_1|^2, |\mathbf{A}_2|^2, \ldots, |\mathbf{A}_n|^2\right). \tag{13.2}$$

The parameter vector $\Gamma = (\gamma_1, \gamma_2, \ldots, \gamma_n)$ denotes the strength of the nonlinearity. The matrix $M = [m_{jk}]$ is a real symmetric matrix ($m_{jk} = m_{kj}$) representing the linear dispersive interactions between the jth and kth site with the constant strength ϵ. When the $n \times n$ matrix M is taken explicitly as the tridiagonal matrix

$$M = \begin{bmatrix} 0 & 1 & 0 & \ldots & 0 & 1 \\ 1 & 0 & 1 & 0 & \ldots & 0 \\ & \ddots & \ddots & \ddots & & \\ & & \ddots & \ddots & \ddots & \\ 0 & \ldots & 0 & 1 & 0 & 1 \\ 1 & 0 & \ldots & 0 & 1 & 0 \end{bmatrix}, \tag{13.3}$$

H. Susanto (✉)
School of Mathematical Sciences, University of Nottingham, University Park, Nottingham, NG7 2RD, UK
e-mail: hadi.susanto@nottingham.ac.uk

the DST equation is then nothing else but the 1D DNLS equation[1] with a periodic boundary condition (see also, e.g., [2] for a brief review of the DNLS over the last two decades). The higher dimensional DNLS equation[2] can also be written in the general form (13.1) with a properly defined matrix M.

The DST equation (13.1) can be derived from the conservation of the Hamiltonian/energy

$$H = -\sum_{j=1}^{n}\left(\frac{\gamma_j}{2}|A_j|^4 + \epsilon \sum_k m_{jk} A_j^* A_k\right) \quad (13.4)$$

with the canonical variables $q_j = A_j$ and $p_j = iA_j^*$. Equation (13.1) also conserves the norm

$$N = \sum_j |A_j|^2. \quad (13.5)$$

Historically, one of the original motivations of the formulation of the DST equation (13.1) was to investigate the self-trapping of vibrational energy in molecular crystals and proteins [1, 3]. The term "self-trapping" itself, which is also called self-localization, refers to an inhibition of the energy dispersion of a coupled nonlinear oscillators system. The concept was introduced long ago in a note by Landau [4] on the motion of a (localized) electron in a crystal lattice.

As one can consider a general choice of matrix M representing longer range couplings or different topologies of the lattice, it is also possible to extend the definition of matrix $\mathbf{D}$. In this case, one will obtain a generalized DST system. As a particular instance, for the same coupling matrix M (13.3), choosing

$$\mathbf{D}(|\mathbf{A}|^2) = \begin{bmatrix} 0 & |A_1|^2 & 0 & \cdots & 0 & |A_1|^2 \\ |A_2|^2 & 0 & |A_2|^2 & 0 & \cdots & 0 \\ & \ddots & \ddots & \ddots & & \\ & & \ddots & \ddots & \ddots & \\ 0 & \cdots & 0 & |A_{n-1}|^2 & 0 & |A_{n-1}|^2 \\ |A_n|^2 & 0 & \cdots & 0 & |A_n|^2 & 0 \end{bmatrix} \quad (13.6)$$

yields the integrable Ablowitz–Ladik (AL) model on the periodic domain.

Looking at the DST equation (13.1) as a model of the nonlinear dynamics of molecules, including small polyatomic ones such as water, ammonia, methane, acetylene, and benzene [5], it is then suggestive to consider the DST system for small n, i.e., few-lattice-site systems of DST equations. The case of $n = 1, 2, 3$, and 4 has been originally studied in detail in [1] which describes the molecular

---

[1] See Chap. 2 of this book.
[2] See Chaps. 3 and 4 of this book.

stretching vibrations in water ($n = 2$), ammonia ($n = 3$), and methane ($n = 4$) [1, 5]. Theoretically, the study of few-lattice-site systems is of interest by itself as is shown later on that the finite size of the system can bring up nontrivial properties, such as an instability to a (stable in the infinite system) coherent structure [6] and a nonstandard type of bifurcation in the study of modulational instabilities of a small size system [7].

13.2 Integrability

When $n = 1$, the DST equation (monomer) becomes an uncoupled nonlinear oscillator

$$i\frac{d}{dt}A_1 + \gamma_1|A_1|^2 A_1 = 0,$$

which is integrable and can be solved analytically. It is straightforward to see that the solution is

$$A_1 = 0, \quad \pm A_1(0)e^{i\gamma_1 A_1(0)^2 t} \tag{13.7}$$

with the norm $N = 0$, $A_1(0)^2$, respectively.

When $n = 2$ (dimer), the DST

$$\begin{aligned} i\frac{d}{dt}A_1 + \gamma_1|A_1|^2 A_1 + \epsilon A_2 &= 0, \\ i\frac{d}{dt}A_2 + \gamma_2|A_2|^2 A_2 + \epsilon A_1 &= 0 \end{aligned} \tag{13.8}$$

is also exactly integrable. It is so by the Liouville–Arnold theorem (or Liouville–Mineur–Arnold theorem) [8, 9], since the degree of freedom is equal to the number of conserved quantities, i.e., H (13.4) and N (13.5).

The dimer (13.8) is the simplest few-lattice-site DST system where one can observe the notion of a self-trapping transition, i.e., a transition from a self-trapped state to a non-self-trapped (oscillating) one. To study it, let us consider the system with a uniform nonlinearity strength $\gamma_1 = \gamma_2 = \gamma$, which is scaled to $\gamma = 1$, subject to the completely localized initial condition

$$A_1(0) = 1, \quad A_2(0) = 0. \tag{13.9}$$

The dimer (13.8) is then solved numerically for several values of the coupling constant ϵ.

When ϵ is small enough, it is natural to expect that the dynamics of $A_1(t)$ and $A_2(t)$ will resemble the case of the integrable monomer Eq. (13.7). It is indeed the case as is presented in the top left panel of Fig. 13.1 for $\epsilon = 0.1$. By defining the

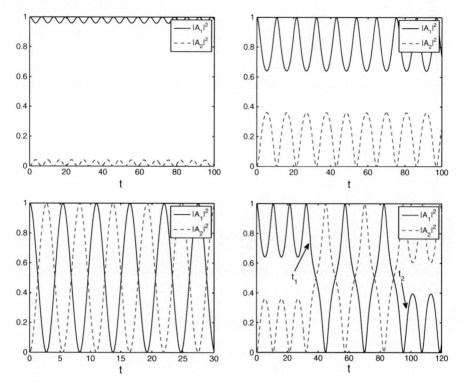

Fig. 13.1 The dynamics of the initial condition (13.9), governed by the DST dimer Eq. (13.8), for several values of the coupling constant ϵ. See the text for the details

so-called site-occupation probability difference $p = |A_1|^2 - |A_2|^2$, it can be seen that even though p is oscillating, it is sign definite. This is the state we refer to as a self-trapped state.

For a slightly larger value of ϵ, e.g., $\epsilon = 0.24$, the dynamics of the initial condition (13.9) still belongs to a self-trapped state, as is plotted in the top right panel of Fig. 13.1. It is clearly indicated that the oscillation period increases as a function of the coupling ϵ.

If ϵ is increased further, there is a critical value of the coupling constant ϵ_{cr} above which the dynamics belongs to a different state. An example is $\epsilon = 0.6 > \epsilon_{\text{cr}}$, depicted in the bottom left panel of Fig. 13.1. It can be deduced that the site-occupation probability difference p is no longer sign definite, which corresponds to a non-self-trapped state and is usually referred to as a Josephson oscillation or a Josephson tunneling state.

Regarding the dynamics of $|A_j|^2$, $j = 1, 2$, Kenkre and Campbell [10] interestingly showed that p satisfies a ϕ^4 equation. This then implies that the general solutions of (13.8) can be written explicitly in terms of the Jacobi elliptic functions. Later, it was also shown [11, 12] that by writing $\dot{q} = \gamma p$, then q will satisfy the sine-Gordon equation. Using the analytical results in [10], the above self-trapping

13 Few-Lattice-Site Systems of Discrete Self-Trapping Equations

transition can be clearly observed as one varies ϵ, e.g., that at the critical coupling ϵ_{cr} the oscillation period becomes infinite.

Later on, Khomeriki et al. [13, 14] demonstrated that the dynamics of $|A_j|^2$ can be controlled through varying the coupling constant for a short time period, such that Josephson oscillations and self-trapped states coexist. The bottom right panel of Fig. 13.1 depicts an example of this. In the panel, the coupling constant is generally set to $\epsilon = 0.24$. Yet, the state can be made to change from initially self-trapping to oscillating by instantaneously setting $\epsilon = 1$ at the point indicated as t_1 for a time period $\Delta t = 0.1$. The Josephson oscillation state has also been made to change to a self-trapped one by setting $\epsilon = 0$ at t_2 for the same period of time.

Besides the DST dimer Eq. (13.8), several generalizations of integrable dimers have been derived and studied as well. An example is the dimer studied by Scott and Christiansen [15]:

$$\begin{aligned} i\frac{d}{dt}A_1 + \gamma|A_1|^{2\sigma}A_1 + \epsilon A_2 &= 0, \\ i\frac{d}{dt}A_2 + \gamma|A_2|^{2\sigma}A_2 + \epsilon A_1 &= 0, \end{aligned} \qquad (13.10)$$

where σ is a non-negative integer. The system is integrable as it conserves the same norm (13.5) and the Hamiltonian

$$H = -\frac{\gamma}{\sigma}[|A_1|^{2+2\sigma} + |A_2|^{2+2\sigma}] - \epsilon(A_1^*A_2 + A_1A_2^*).$$

Jørgensen et al. [16] showed that when the coupling constant ϵ is allowed to be complex-valued, the dimer

$$\begin{aligned} i\frac{d}{dt}A_1 + \gamma_1|A_1|^2A_1 + \epsilon A_2 &= 0, \\ i\frac{d}{dt}A_2 + \gamma_2|A_2|^2A_2 \pm \epsilon^*A_1 &= 0 \end{aligned} \qquad (13.11)$$

is integrable.

Another notable dimer is the one derived by Jørgensen et al. [16]:

$$\begin{aligned} i\frac{d}{dt}A_1 + \gamma|A_1|^2A_2 + \beta_1|A_2|^2A_2 + \epsilon_1 A_2 + i\alpha A_1 &= 0, \\ i\frac{d}{dt}A_2 + \gamma|A_2|^2A_1 + \beta_2|A_1|^2A_1 + \epsilon_2 A_1 - i\alpha A_2 &= 0, \end{aligned} \qquad (13.12)$$

in which the general system possesses blow-up solutions that can also be written in terms of the Jacobi elliptic functions [16–18].

When $n = 3$ (trimer), the DST is not integrable, as well as the case of $n \geq 4$. Nonetheless, Hennig [19, 20] is able to obtain a generalized integrable trimer using

SU(3) notation, where SU(3) is the group of 3 × 3 unitary matrices with unit determinant. It is a system of eight real first-order ordinary differential equations.

For a general value of n, one integrable system is the well-known AL equation. The reader is also referred to [21] for another two integrable DSTs formulated in terms of Lie–Poisson algebra derived by Christiansen et al. [21], where one of the models is a Toda lattice-like system.

13.3 Chaos

Since the DST equation is not integrable when $n > 2$, it is then expected that the equation will admit Hamiltonian chaos. It is indeed the case as is demonstrated in the first study [1]. The Lyapunov exponent of the DST trimer was calculated and analyzed numerically in [22] confirming the presence of chaos in the system. Another numerical study on trimers in which the third oscillator is a linear one also shows the presence of chaotic regimes for some well-defined values of the nonlinearity and linear coupling parameters [23]. Calculations using a Melnikov method were also presented by Hennig et al. [24], where they showed analytically the presence of homoclinic chaos in the trimer. The idea was to use the integrable DST dimer as the underlying unperturbed system and to treat the additional oscillator as a small perturbation.

A similar idea as [24] was also used to show the presence of chaotic dynamics in the case of $n = 4$, where now the polymer is assumed to consist of two integrable dimers connected by a perturbative coupling [25]. A numerical computation enabled by the presence of Arnold diffusion was also presented to show the presence of chaos in the general $n = 4$ case [26].

The simplest equations describing two coupled quadratic nonlinear ($\chi^{(2)}$) systems ($n = 2$), each of which consists of a fundamental mode resonantly interacting with its second harmonic forming a four degrees of freedom system, have been studied in [27]. A gradual transition from a self-trapping solution to chaotic dynamics when going away from the near integrable limit $n = 1$ has been discussed as well [27].

Chaos can also be made to exist in an integrable model by perturbing it. Using the SU(2) representation, a study of the analytical structure of a harmonically perturbed nonlinear dimer has been done by Hennig [28]. The study shows that in this case, even though the unperturbed system is integrable, chaos is also observed and proven analytically.

13.4 Applications and Experimental Observations

As was already mentioned that one application of the DST equation was to describe experimental observations of anomalous amide resonances in molecular crystals and proteins, the DST (DNLS) equation also has applications in the study of the propagation of the complex electromagnetic field envelope in the waveguide arrays [29, 30] and the dynamics of bosons cooled to temperatures very near to absolute zero that are confined in external potentials [31].

In the context of Bose–Einstein physics [31, 32], the DST dimer has been proposed to model the tunneling between two zero-temperature Bose-Einstein condensates (BECs) confined in a double-well magnetic trap [33]. Using the so-called Wannier function-based expansion [34], the infinite-dimensional double-well potential problem [35] can be asymptotically reduced to the DST dimer [33, 36]. All the dynamics of the integrable system which have been solved analytically by Kenkre and Campbell [10] find their new interpretation in this context, including the so-called "macroscopic quantum self-trapping" [33, 36] which is nothing else but the self-trapped states observed in [1]. Following the theoretical prediction, successful experiments have been performed confirming the presence of periodic oscillations and self-trapping in the system [37, 38].

Buonsante et al. [39] have also considered trimers where chaotic solutions of the systems are now studied in the context of BECs and argued that they may correspond to macroscopic effects that can be viable to experiments. Pando and Doedel [40, 41] extended the study of few-site DST systems modeling BECs to the case of $n = 3, 5, 7$ with periodic boundary conditions in the presence of a single on-site defect. In the case of the trimer, it is found that $1/f$ noise is a robust phenomenon taking place as a result of intermittency [41]. In the case of $n = 5$ and 7, another robust effect, where chaotic synchronization of symbolic information arises in the Hamiltonian system, is then observed [40]. Related to this multisite DST system, successful experiments on BECs trapped by an optical lattice, that using again Wannier function expansion can be modeled by such DST equations, have been conducted and reported in the seminal paper [42].

In the context of waveguide arrays [29], the study of the DST dimer has actually been done a couple of years before the work [1] by Jensen in his seminal paper [43]. The dimer is used to model the coherent interaction of two optical waveguides placed in close proximity. Jensen also observed the exchange of power between the waveguides which is nothing else but the periodic solutions of Kenkre and Campbell [10]. An experimental observation has been reported as well in this context in which it is shown that in a double-trap potential system, there is a spontaneous symmetry breaking [44], i.e., the ground state of the system becomes asymmetric beyond a critical power N (cf. Eq. (13.5)).

A numerical study on a DST trimer to model a three-waveguide nonlinear directional coupler also has been conducted by Finlayson and Stegeman [45]. In the paper, they reported that transitions from quasi-periodic to chaotic behavior and back take place as the power N is varied.

A similar trimer as the one studied by [24] has been considered as well in the context of an optical coupler configuration consisting of two nonlinear waveguides coupled to a linear one [46, 47]. In relation to its physical context, it is shown that this type of coupler system can act as an optical switching device with switching properties superior to that of conventional two and three all nonlinear coupler configurations.

Experimental observations for the case of systems with a three-well potential have been performed in strontium barium niobate crystals [48]. Corroborated by an analytical study using the Lyapunov–Schmidt reduction method, it is shown that the presence of a third well causes all bifurcations of static solutions to be of saddle node type [48].

Recently, a few-lattice-site $n = 5$ DST system modeling waveguide arrays, i.e., a nonlinear trimer equation coupled to linear waveguides at the boundaries, was also studied by Khomeriki and Leon [49]. They showed that by controlling the intensity of light along the linear waveguides, the middle one can be sensitive to perturbations. The observation then demonstrates that such a system can be utilized as a weak signal detector [49].

13.5 Conclusions

To conclude, we have briefly reviewed the study of few-lattice-site systems of DST equations. It has to be admitted that the present review is far from covering and summarizing all the work that has been done on the subject. As an example, the study of quantum versions of DST equations [3] is totally omitted in this section. Nonetheless, it is expected that this review will give an idea that even in rather simple small-size systems, a lot of interesting problems and nontrivial properties can be observed and possibly technologically exploited. This then indicates that it is of interest to further study few-lattice-site systems of DST equations. One possible direction would be understanding further connections between few-site and many-site systems. As recently showed numerically by Buonsante et al. [7], even investigating the ground states of the system already reveals interesting behaviors, such as coexistence of a single-pulse and a uniform solution in a finite range of the coupling constant when $n < 6$ and the disappearance of the single-pulse mode beyond a critical parametric threshold.

References

1. Eilbeck, J.C., Lomdahl, P.S., Scott, A.C.: Physica D **16**, 318 (1985)
2. Eilbeck, J.C., Johansson, M.: The Discrete Nonlinear Schrödinger equation – 20 Years on. In: Localization and Energy Transfer in Nonlinear Systems, Vázquez, L., Mackay, R.S., Zorzano, M.P. (eds.), pp. 44–67. World Scientific, New Jersey (2003)
3. Scott, A.C.: Nonlinear Science: Emergence and Dynamics of Coherent Structures, 2nd edn. Oxford University Press, New York (2003)
4. Landau, L.D.: Phys. Zeit. Sowjetunion **3**, 664 (1933)
5. Scott, A.C., Lomdahl, P.S., Eilbeck, J.C.: Chem. Phys. Letts. **113**, 29–36 (1985)
6. Marín, J.L., Aubry, S.: Physica D **119**, 163–174 (1998)
7. Buonsante, P., Kevrekidis, P.G., Penna, V., Vezzani, A.: Phys. Rev. E **75**, 016212 (2007)
8. Arnold, V.I. (ed.): Dynamical Systems III. Springer, Berlin (1988)
9. Libermann, P., Marle, C.M.: Symplectic Geometry and Analytical Mechanics. Reidel, Dordrecht (1987)
10. Kenkre, V.M., Campbell, D.K.: Phys. Rev. B **34**, 4959 (1986)

11. Jensen, J.H., Christiansen, P.L., Elgin, J.N., Gibbon, J.D., Skovgaard, O.: Phys. Lett. A **110**, 429 (1985)
12. Cruzeiro-Hansson, L., Christiansen, P.L., Elgin, J.N.: Phys. Rev. B **37**, 7896 (1988)
13. Khomeriki, R., Leon, J., Ruffo, S.: Phys. Re. Lett. **97**, 143902 (2006)
14. Khomeriki, R., Leon, J., Ruffo, S., Wimberger, S.: Theor. Math. Phys. **152**, 1122 (2007)
15. Scott, A.C., Christiansen, P.L.: Phys. Scri. **42**, 257–262 (1990)
16. Jørgensen, M.F., Christiansen, P.L., Abou-Hayt, I.: Physica D **68**, 180–184 (1993)
17. Jørgensen, M.J., Christiansen, P.L., Chaos, Solitons & Fractals **4**, 217–225 (1994)
18. Jørgensen, M.F., Christiansen, P.L.: J. Phys. A: Math. Gen. **31**, 969–976 (1998)
19. Henning, D.: J. Phys. A: Math. Gen. **25**, 1247 (1992)
20. Hennig, D.: Physica D **64**, 121 (1993)
21. Christiansen, P.L., Jørgensen, M.F., Kuznetsov, V.B.: Lett. Math. Phys. **29**, 165-73 (1993)
22. Defilippo, S., Girard, M.F., Salerno, M.: Physica D **26**, 411 (1987)
23. Molina, M.I., Tsironis, G.P.: Phys. Rev. A **46**, 1124–1127 (1992)
24. Hennig, D., Gabriel, H., Jørgensen, M.F., Christiansen, P.L., Clausen, C.B.: Phys. Rev. E **51**, 2870 (1995)
25. Hennig, D., Gabriel, H.: J. Phys. A: Math. Gen. **28**, 3749 (1995)
26. Feddersen, H., Christiansen, P.L., Salerno, M.: Phys. Scr. **43**, 353 (1991)
27. Bang, O., Christiansen, P.L., Clausen, C.B.: Phys. Rev. E **56**, 7257–7266 (1997)
28. Hennig, D.: Phys. Scr. **46**, 14–19 (1992)
29. Christodoulides, D.N., Lederer, F., Silberberg, Y.: Nature **424**, 817 (2003)
30. Slusher, R.E., Eggleton, B.J. (eds.): Nonlinear Photonic Crystals. Springer, Berlin Heidelberg New York (2004)
31. Pitaevskii, L.P., Stringari, S.: Bose-Einstein Condensation. Oxford University Press, Oxford (2003)
32. Kevrekidis, P.G., Frantzeskakis, D.J., Carretero-González, R., (eds.): Emergent Nonlinear Phenomena in Bose-Einstein Condensates: Theory and Experiment. Springer, Berlin Heidelberg New York (2008)
33. Smerzi, A., Fantoni, S., Giovanazzi, S., Shenoy, S.R.: Phys. Rev. Lett. **79**, 4950 (1997)
34. Pelinovsky, D.E.: Asymptotic reductions of the Gross-Pitaevskii equation. In: Kevrekidis, P.G., Frantzeskakis, D.J., Carretero-González, R., (eds.): Emergent Nonlinear Phenomena in Bose-Einstein Condensates: Theory and Experiment. Springer, Berlin Heidelberg New York, pp. 377–398 (2008)
35. Theocharis, G., Kevrekidis, P.G., Frantzeskakis, D.J., Schmelcher, P.: Phys. Rev. E **74**, 056608 (2006)
36. Raghavan, S., Smerzi, A., Fantoni, S., Shenoy, S.R.: Phys. Rev. A **59**, 620 (1999)
37. Albiez, M., Gati, R., Fölling, J., Hunsmann, S., Cristiani, M., Oberthaler, M.K.: Phys. Rev. Lett. **95**, 010402 (2005)
38. Gati, R., Albiez, M., Fölling, J., Hemmerling, B., Oberthaler, M.K.: Appl. Phys. B **82**, 207 (2006)
39. Buonsante, P., Franzosi, R., Penna, V.: Phys. Rev Lett. **90**, 050404 (2003)
40. Pando, C.L., Doedel, E.J.: Phys. Rev. E **69**, 042403 (2004)
41. Pando, C.L., Doedel, E.J.: Phys. Rev. E **71**, 056201 (2007)
42. Cataliotti, F.S., Burger, S., Fort, C., Maddaloni, P., Minardi, F., Trombettoni, A., Smerzi, A., Inguscio, M.: Science **293**, 843 (2001)
43. Jensen, S.M., IEEE J. Quantum Electron. **QE-18**, 1580 (1982)
44. Kevrekidis, P.G., Chen, Z., Malomed, B.A., Frantzeskakis, D.J., Weinstein, M.I.: Phys. Lett. A **340**, 275 (2005)
45. Finlayson, N., Stegeman, G.I.: Appl. Phys. Lett. **56**, 2276 (1990)
46. Molina, M.I., Deering, W.D., Tsironis, G.P.: Physica D **66**, 135–142 (1993)
47. Tsironis, G.P., Deering, W.D., Molina, M.I.: Physica D **68**, 135–137 (1993)
48. Kapitula, T., Kevrekidis, P.G., Chen, Z.: SIAM J. Appl. Dyn. Syst. **5**, 598 (2006)
49. Khomeriki, R., Leon, J.: Phys. Rev. Lett. **94**, 243902 (2005)

Chapter 14
Surface Waves and Boundary Effects in DNLS Equations

Ying-Ji He and Boris A. Malomed

14.1 Introduction

Surface waves represent excitations which may propagate along interfaces between different media. These waves occur in diverse areas of physics, chemistry, and biology, often displaying properties that find no counterparts in bulk media [1]. The study of waves on the free surface of water and internal surfaces in stratified liquids is a classical chapter of hydrodynamics. The investigation of surface modes in solid-state physics was initiated by Tamm in 1932, who used the Kronig–Penney model to predict specific electron modes (*Tamm states*) localized at the edge of the solid [2]. This line of research was extended by Shockley in 1939 [3]. In linear optics, Kossel had predicted the existence of localized states near the boundary between homogeneous and layered media in 1966 [4], which were later observed in AlGaAs multilayer structures [5, 6]. Such waves were also shown to exist at metal–dielectric interfaces [7], as well as at interfaces between anisotropic materials [8]. In nonlinear optics, surface waves, which include transverse electric (TE), transverse magnetic (TM) and mixed polarization modes propagating at the interface between homogeneous dielectric media with different properties, were theoretically predicted in the works [9, 10] (see also review [11]).

The formation of surface solitons of the gap type (with their propagation constant falling in a bandgap of the linear spectrum generated by the respective linearized system) was predicted too [12] and observed in experiments carried out in an optical system described by such a model [13]. Surface solitons have also been predicted at an interface between two different semi-infinite waveguide arrays [14], as well as at boundaries of two-dimensional (2D) nonlinear lattices [14–18]. It has been shown that surface solitons of the vectorial [19, 20] and vortical [21] types, as well as surface kinks [22], can exist too. In addition to that, multicomponent (polychromatic) surface modes have been predicted and experimentally observed [23–25].

Y.-J. He (✉)
School of Electronics and Information, Guangdong Polytechnic Normal University, Guangzhou 510665; State Key Laboratory of Optoelectronic Materials and Technologies, Sun Yat-Sen University, Guangzhou 510275, China
e-mail: hyj8409@sxu.edu.cn

Closer to the main topic of the present book are *discrete* surface solitons. The existence of such localized lattice modes was first analyzed in one-dimensional (1D) arrays of nonlinear optical waveguides [26]. These states, predicted to exist at the edge of a semi-infinite array, feature a power threshold necessary for their formation, similar to that encountered by nonlinear surface waves at interfaces between continuous media [26, 27]. The formation and properties of discrete surface solitons have been explored theoretically in detail [26–32], and these solitons were quickly created in experiments performed in arrays of optical waveguides [33, 34]. Discrete surface solitons have been predicted in a number of other settings, such as those based on vectorial models [35, 36] and superlattices [37], as well as in a system with the quadratic nonlinearity [38].

In higher dimensions, 2D discrete surface solitons have been reported in theoretical and experimental forms [17, 39]. Recently, the creation of discrete solitons of a *corner type*, in a 2D array of optical waveguides confined by two orthogonal surfaces that form the corner, was reported in the work [18]. Finally, spatiotemporal discrete surface solitons have been predicted in the theoretical works [40, 41].

In this chapter, we present an outline of several basic theoretical and experimental results obtained for discrete surface solitons which can be supported by boundaries of various types. We first consider the simplest case of the surface solitons in 1D Kerr media (those with the cubic nonlinearity), starting with the underlying theory and then proceeding to the experimental realization. This will be further extended into 2D and 3D settings.

14.2 Discrete Nonlinear Schrödinger Equations for Surface Waves

14.2.1 The One-Dimensional Setting

In a semi-infinite nonlinear lattice (in the experiment, it represents a long array of weakly coupled nonlinear optical waveguides, as schematically depicted in Fig. 14.1), the normalized field amplitudes at lattice sites $n \geq 0$ obey a discrete nonlinear Schrödinger (DNLS) equation, which incorporates the boundary condition at the surface [26],

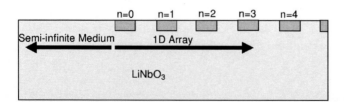

Fig. 14.1 A typical scheme of a semi-infinite array of optical waveguides, buried into a bulk medium, which gives rise to effectively one-dimensional quasi-discrete surface solitons. Reprinted from [38] with permission

$$i\frac{d}{dz}u_0 + u_1 + \beta|u_0|^2 u_0 = 0, \tag{14.1}$$

$$i\frac{d}{dz}u_n + (u_{n+1} + u_{n-1}) + \beta|u_n|^2 u_n = 0. \tag{14.2}$$

In the model of the array of optical waveguides, the evolution variable z is the distance of the propagation of electromagnetic signals along the waveguides, and β is the coefficient of the on-site nonlinearity, the self-focusing and self-defocusing nonlinearities corresponding, respectively, to $\beta > 0$ and $\beta < 0$. Unless it is said otherwise, we set $\beta \equiv 1$, by means of an obvious rescaling of the lattice field. Equation (14.1) governs the evolution of the field at the edge of the array, which corresponds to site $n = 0$, and Eq. (14.2) applies at every other site, with $n \geq 1$. The actual electric field in the optical wave is expressed in terms of scaled amplitudes u_n as follows: $E_n = \sqrt{2C\lambda_0\eta_0/(\pi n_0 \hat{n}_2)}u_n$, where C is the inter-site coupling coefficient in physical units (in Eqs. (14.1) and (14.2), normalization $C = 1$ is adopted), λ_0 is the free-space wavelength, η_0 is the free-space impedance, $\hat{n}_2$ the nonlinear Kerr coefficient, and n_0 the linear refractive index of the waveguides' material.

14.2.2 The Two-Dimensional Setting

The model of the 2D semi-infinite array of optical waveguides with a horizontal edge, whose plane is parallel to the direction of the propagation of light in individual waveguides, is based on the accordingly modified DNLS equation for the 2D set of amplitudes $u_{m,n}(z)$ of the electromagnetic waves in the guiding cores (see, e.g., [18]):

$$i\frac{d}{dz}u_{m,n} + C(u_{m+1,n} + u_{m-1,n} + u_{m,n+1} + u_{m,n-1} - 4u_{m,n}) + |u_{m,n}|^2 u_{m,n} = 0 \tag{14.3}$$

for $n \geq 0$ and all integer values of m. Unlike Eqs. (14.1) and (14.2), the constant accounting for inter-site coupling, C, is not scaled here to be 1, as it will be used in an explicit form below. Note that the corresponding coupling length in the waveguide array, which may be estimated as $z_{\text{coupling}} \sim C^{-1/2}$ in terms of Eq. (14.3), usually takes values on the order of a few millimeters, in physical units. At the surface row, which corresponds to $n = 0$ in Eq. (14.3), one should set $u_{m,-1} \equiv 0$, as there are no waveguides at $n < 0$. Equation (14.3) admits the usual Hamiltonian representation, and also conserves the total power (norm), $P = \sum_{m=-\infty}^{\infty} \sum_{n=0}^{\infty} |u_{m,n}|^2$.

14.2.3 The Three-Dimensional Setting

The equations for the slow spatiotemporal evolution of the optical signal propagating in a 2D array of linearly coupled waveguides can be cast in the following form [40, 41]:

$$\left(i\frac{\partial}{\partial z} - \gamma\frac{\partial^2}{\partial \tau^2}\right)u_{m,n} + (u_{m+1,n} + u_{m-1,n} + u_{m,n+1} + u_{m,n-1} - 4u_{m,n})u_{m,n}$$
$$+ |u_{m,n}|^2 u_{m,n} = 0, \tag{14.4}$$

where z is, as in Eq. (14.3), the propagation distance, while τ is the temporal variable and, accordingly, γ is the coefficient of the temporal dispersion in each waveguiding core. In the case of the corner configuration considered in the works [40, 41], $u_{m,n} \equiv 0$ for $m \leq -1$ and $n \leq -1$.

A different type of the 3D model is possible in the case when a planar surface borders a full 3D lattice. In that case, the basic dynamical model takes the following form:

$$i\frac{d}{dz}u_{l,m,n} + (u_{l+1,m,n} + u_{l-1,m,n} + u_{l,m+1,n} + u_{l,m-1,n} + u_{l,m,n+1} + u_{l,m,n-1}$$
$$- 6u_{l,m,n})u_{l,m,n} + |u_{l,m,n}|^2 u_{l,m,n} = 0, \tag{14.5}$$

where the 3D discrete coordinates assume the following integer values: $-\infty < l, m < +\infty$, $0 \leq n < +\infty$, and $u_{l,m,n} \equiv 0$ for $n < 0$. This equation gives rise to 3D discrete solitons of various types, and those among them which abut on the surface, or are set at a distance from it corresponding to few lattice cells, may be considered as three-dimensional surface solitons.

14.3 Theoretical Investigation of Discrete Surface Waves

14.3.1 Stable Discrete Surface Solitons in One Dimension

Here, we outline the first theoretical prediction of 1D discrete surface solitons at the interface between an array of waveguides and a continuous medium, as in [26]. Stationary surface waves in the semi-infinite lattice system correspond to the substitution $u_n = v_n \exp(i\Lambda z)$ in Eqs. (14.1) and (14.2), where Λ is the corresponding propagation constant, and all amplitudes v_n are assumed to be positive, which corresponds to an in-phase solution. In the system under consideration, solitons can be found with values of the propagation constant falling into the *semi-infinite gap*, $\Lambda \geq 2$, where localized solutions are possible in principle.

The family of soliton solutions, found numerically by means of the relaxation method, is presented in Fig. 14.2, in the form of the dependence of the respective total power (alias norm), $P = \sum_{n=-\infty}^{+\infty} |u_n|^2$, on the propagation constant Λ. In particular, in the region of $\Lambda > 3$, the 1D surface solitons are strongly localized, and may be approximated by a simple *ansatz*, $u_n = A\exp(-np + i\Lambda t)$, where the amplitude is given by $A^2 = \Lambda/2 + \sqrt{\Lambda^2/4 - 1} \approx \Lambda - 1/\Lambda$ and $p = 2\ln A$.

The dependence plotted in Fig. 14.2 demonstrates a minimum in the $P(\Lambda)$ curve at $\Lambda = 2.998$. This feature, in turn, implies that the discrete solitons exist only above a certain power threshold, which, in the present case, is $P_{\text{thr}} = 3.27$. The situation

Fig. 14.2 The total power (norm) versus propagation constant Λ for the family of in-phase 1D discrete surface solitons. Reprinted from [26] with permission

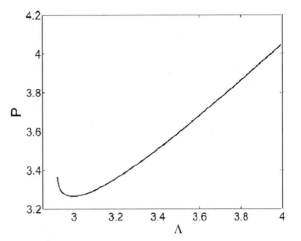

is quite similar to that found earlier in continuum models for interfaces between nonlinear dielectric media, where the well-known Vakhitov–Kolokolov (VK) stability criterion is applicable. According to it, a necessary (but, generally speaking, not sufficient) condition for the stability of the soliton family is $dP/d\Lambda > 0$ (more accurately, the VK criterion guarantees only the absence of growing eigenmodes of infinitesimal perturbations around the solitons with a purely real growth rate, but it cannot detect unstable eigenmodes corresponding to a complex growth rate).

The full stability of these solutions was also tested in direct simulations of Eq. (14.3). Figure 14.3 demonstrates that the family of the 1D surface solitons is split into stable and unstable subfamilies – in fact, in exact accordance with the VK criterion.

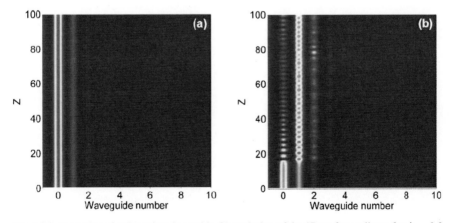

Fig. 14.3 Examples of stable (**a**) and unstable (**b**) evolution of the 1D surface solitons for $\Lambda = 3.2$ and 2.92, respectively. Reprinted from [26] with permission

14.3.2 Discrete Surface Solitons at an Interface Between Self-Defocusing and Self-Focusing Lattice Media

A specific type of interface corresponds to that between two lattices, with the self-focusing and defocusing nonlinearities, i.e., $\beta > 0$ and $\beta < 0$ in Eqs. (14.1) and (14.2). Following [42], we will present here an example with $\beta_n = -0.9$ for $n < 0$ and $\beta_n = 1.1$ for $n > 0$. Discrete solitons are looked for in the same general form as above, i.e., $u_n = v_n \exp(i\Lambda t)$, where Λ is, as before, the propagation constant, and the stationary lattice field v_n obeys the following equation:

$$\Lambda v_n - C\Delta_2 v_n - \beta_n |v_n|^2 v_n = 0. \tag{14.6}$$

In the anticontinuum (AC) limit, i.e., for $C = 0$, solutions to Eq. (14.6) can be immediately constructed in the form of one or several "excited" sites carrying a nonzero amplitude, $v_n = \pm\sqrt{\Lambda/\beta_n}$ (provided that $\Lambda\beta_n > 0$), while at all other sites the amplitude is zero. Carrying out subsequent numerical continuation of the solution to $C > 0$, this approach makes it possible to generate various species of discrete solitons, seeded at $C = 0$ by the respective "skeletons."

In [42], a number of soliton families were constructed, starting, in the AC limit, from the "skeletons" of the following types: a single excited site at $n = 0$; a pair of in-phase or out-of-phase excited sites at $n = 0$ and 1; and a triplet consisting of an excited site at $n = 0$ and ones with the opposite signs at $n = 1, 2$, and similar patterns based on four- and five-site "skeletons." Figure 14.4 displays two lowest order solution branches found in this model, viz., those seeded by the single-site configuration, and the dual-site one of the in-phase type.

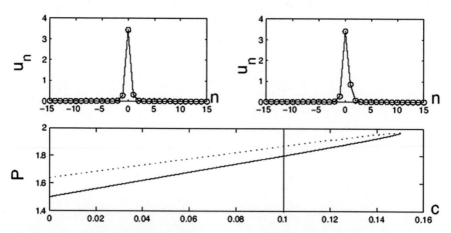

Fig. 14.4 *Top left and right panels* display, for $C = 0.1$, examples of states near the interface between self-focusing and defocusing lattices, which are engendered, respectively, by a single excited site, or a pair of excited in-phase sites, in the anticontinuum limit ($C = 0$), as in [42]. *Bottom*: families of these solutions are presented through the dependence of their power on C. As usual, *the solid and dashed lines* designate stable and unstable solutions, respectively. *The vertical line* marks the examples shown in the *top panels*

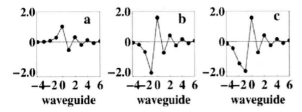

Fig. 14.5 Examples of hybrid (unstaggered–staggered) discrete solitons, attached to the interface between lattice media with the self-focusing and defocusing nonlinearities. The solitons in panels (**a**) and (**b**) are stable, while the one in (**c**) is unstable. Reprinted from [43] with permission

As stressed in [43], the interface of the same type as considered in this subsection, i.e., between lattice media with self-focusing and self-defocusing on-site nonlinearities, may give rise to *hybrid solitons*. They look as unstaggered and staggered states on the two sides of the interface, see examples in Fig. 14.5. Probably, similar hybrid states can be found in 2D and 3D models including an interface between self-focusing and self-defocusing nonlinearities.

14.3.3 Tamm Oscillations of Unstaggered and Staggered Solitons

Here we outline the theoretical prediction of *Tamm oscillations* of discrete solitons created near the edge of a 1D lattice (of optical waveguides), following [31]. These are oscillations of the position of a narrow soliton, due to the interplay of its repulsion from the lattice's edge and Bragg reflection from the bulk of the lattice (see below).

The analysis starts with the discrete equation for the propagation of light in the array, cf. Eqs. (14.1) and (14.2):

$$i\frac{d}{dz}u_n + C(u_{n+1} + u_{n-1} - 2u_n) + g|u_n|^2 u_n = 0, \quad (14.7)$$

where C is, as above, the lattice-coupling constant, while $g(|u_n|^2) = \beta|u_n|^2$ and $g(|u_n|^2) = \beta|u_n|^2/(1 + |u_n|^2)$ in cubic and saturable media, respectively, with nonlinearity coefficient $\beta > 0$ and $\beta < 0$ corresponding to the self-focusing and self-defocusing signs of the nonlinearity. The boundary conditions added to Eq. (14.7) are the same ones as considered above, see Eq. (14.1).

Stationary solutions to Eq. (14.7) are looked for in the usual (*unstaggered*) form, $u_n = v_n \exp(-i\Lambda t)$, in the case of the self-focusing, and in the *staggered* form, with alternating signs of the stationary fields at adjacent sites of the lattice, in the opposite case. The staggering substitution, $u_n = v_n(-1)^n \exp(i\Lambda z)$, makes the self-defocusing nonlinearity equivalent to its self-focusing counterpart. Staggered solitons, which may be found in various models with self-defocusing, may also be regarded as *gap solitons*, as they exist at values of the propagation constant which fall in one of finite bandgaps in the spectrum of the corresponding linearized model.

On the contrary to that, the propagation constant of ordinary – unstaggered – solitons belongs to the semi-infinite spectral gap.

Assuming that solitons generated by Eq. (14.7) are narrow (strongly localized), one can construct analytical approximations for two different types of such solitons, on-site-centered and inter-site-centered ones. The solutions of the former type have the largest value of the field at the edge, $n = 0$, with the lattice field decaying as $v_n \approx \alpha^n v_0$ at $n > 0$. Inter-site-centered solitons feature the largest local amplitude at $n = 1$, and decay as $v_n \approx \alpha^{n-1} v_1$ for $n > 1$. In either case, the localization of the soliton is determined by a small parameter, $\alpha = C/(\Lambda + 2C)$.

In direct simulations of Eq. (14.7), these narrow solitons feature swinging motions, as shown in Fig. 14.6, which are the Tamm oscillations of the narrow solitons, staggered and unstaggered ones. The oscillations are not quite persistent: in the course of its motion in the lattice, the soliton gradually loses energy due to emission of linear waves ("lattice radiation," alias "phonons"), which leads to gradual damping of the oscillations. Eventually, the soliton comes to a halt at a position at some distance from the surface (the distance may be as small as two lattice sites, see further details in [31]).

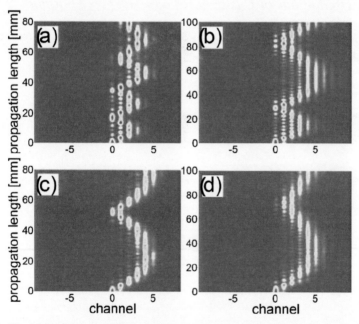

Fig. 14.6 Examples of Tamm oscillations of several types of narrow discrete solitons created near the edge of the lattice. The simulations were performed for typical values of parameters corresponding to arrays of nonlinear optical waveguides in the following models: (**a**) self-defocusing saturable, (**b**) self-defocusing cubic, (**c**) self-focusing saturable, and (**d**) self-focusing cubic. In panels (a) and (b), the soliton is staggered (otherwise, it cannot exist in the defocusing medium), while in (c) and (d) it is the usual unstaggered soliton. Reprinted from [31] with permission

For the interpretation of the oscillations, it is relevant to notice that the edge of the lattice (the surface) induces an effective repulsive potential acting on the narrow discrete soliton. Stronger inter-site coupling results in the stronger surface-induced potential, while enhanced nonlinearity suppresses it [31]. Then, Tamm oscillations of the soliton may be understood as oscillations in the effective Peierls–Nabarro (PN) potential, induced by the underlying lattice, under the action of the additional repulsive potential. In other words, the oscillations are a result of the interplay of the repulsion of the soliton from the lattice surface and Bragg reflection in the depth of the lattice. The combination of the edge-induced and PN potentials gives rise to a stable equilibrium position of the soliton at a finite distance from the edge, where the soliton eventually gets trapped.

14.3.4 Discrete Surface Solitons in Two Dimensions

The aim of this subsection is to present an outline of the theoretical prediction of 2D discrete surface solitons at the interface between a 2D lattice of optical waveguides and a substrate, following [39]. The model is based on Eq. (14.3), stationary solutions to which are looked for as $u_{m,n} = \exp(i\Lambda z)v_{m,n}$, with Λ scaled to be 1, and the stationary lattice distribution obeying the respective equation, $(1 - |v_{m,n}|^2)v_{m,n} - C(v_{m,n+1} + v_{m,n-1} + v_{m+1,n} + v_{m-1,n} - 4v_{m,n}) = 0$.

In [39], it was shown that the interaction with the edge expands the stability region for fundamental solitons, and induces a difference between dipoles (bound states of two fundamental lattice solitons with opposite signs) oriented perpendicular and parallel to the surface. A notable finding is that the edge supports a species of localized patterns which exists too but is *always unstable* in the uniform lattice, namely, a horseshoe-shaped soliton. As shown in Fig. 14.7, the "skeleton" of the horseshoe structure consists of three lattice sites.

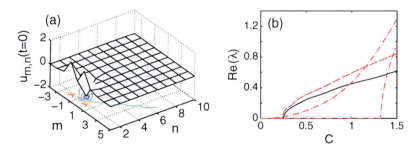

Fig. 14.7 (**a**) An example of the 2D surface soliton of the "horseshoe" type, as in [39]. *The solid curve* in panel (**b**) displays the real part of a critical instability eigenvalue for the soliton family of this type. For comparison, *the dashed–dotted lines* in (**b**) show the instability eigenvalues for the horseshoe family in the uniform lattice (without the edge). The latter family is completely (although weakly) unstable, due to a very small nonzero eigenvalue extending to $C = 0$, while the horseshoes trapped at the edge of the lattice have a well-defined stability region – in the present case, it is, approximately, $C < 0.25$

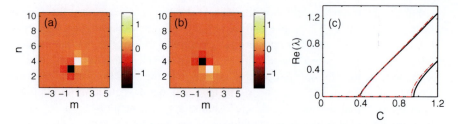

Fig. 14.8 An example of a supersymmetric vortex cell. Panels (**a**) and (**b**) show, respectively, the real and imaginary parts of the solution. In panel (**c**), *the solid lines* show instability eigenvalues of these states, as in [39]. For comparison, *dashed–dotted lines* depict the same numerically found characteristics for the supersymmetric vortex in the infinite lattice (the one without the edge)

The edge of the 2D lattice may also act in an opposite way, *impeding* the existence of localized solutions of other types. A relevant example of that is provided the so-called supersymmetric lattice vortex, i.e., one with the intrinsic vorticity ($S = 1$) equal to the size of the square (a set of four excited sites) which seeds the vortex at $C = 0$ in the above-mentioned *AC limit* (i.e., for the lattice composed of uncoupled sites), as shown in Fig. 14.8a, b. The configuration displayed in the figure is placed at the minimum separation from the edge admitting its existence, which amounts to two lattice sites. Numerically found stability eigenvalues for this structure are presented in Fig. 14.8c.

The numerical analysis of 3D equation (14.5) reveals similar effects for several species of discrete 3D solitons. In particular, three-site horseshoes are also completely unstable in the bulk 3D lattice, but are stabilized if they abut upon the lattice's surface. As for 3D vortex solitons, their properties strongly depend on the orientation with respect to the surface: the ones set parallel to the surface are essentially stabilized by it, while localized vortices with the perpendicular orientation cannot exist close to the surface.

14.3.5 Spatiotemporal Discrete Surface Solitons

The theoretical prediction of spatiotemporal discrete surface solitons at the interface between a lattice of optical waveguides and a continuous medium was reported in [40, 41], using the model based on Eq. (14.4). Stationary solutions for a spatiotemporal soliton can be looked as $u_{m,n}(z, \tau) = v_{m,n}(\tau) \exp(i \Lambda z)$, where envelopes $v_{m,n}(\tau)$ describe the temporal shape of soliton-like pulses at lattice sites (n, m). Several examples of the spatiotemporal solitons found in [40, 41] by means of numerical methods in the lattice with the corner are shown in Fig. 14.9.

14.3.6 Finite Lattices and the Method of Images

In both 1D and 2D settings, some of the results outlined above can also be obtained by means of the *method of images*. For instance, in the 1D case with the *fixed* (zero)

14 Surface Waves and Boundary Effects in DNLS Equations

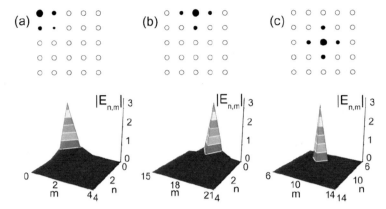

Fig. 14.9 *Top*: Panels (**a**)–(**c**) display typical examples of spatiotemporal modes localized at the lattice's corner, at the edge, and in the center of the lattice. *Bottom*: Spatial cross sections of the corresponding stable spatiotemporal solitons. Reprinted from [40, 41] with permission

boundary condition, i.e., $u_n = 0$ for $n \leq -1$ (see Eqs. (14.1) and (14.2)), the solution is equivalent to that in an infinite lattice which is subject to the anti-symmetry constraint, $u_{-(n+2)} \equiv -u_n$, $n = -1, 0, +1, +2, \ldots$ (which, obviously, includes condition $u_{-1} \equiv 0$). This way of the extension of the semi-infinite lattice into the full infinite one implies that a localized excitation created at a lattice site with number m comes together with its image, of the opposite sign, placed at site $-(m+2)$, as shown in Fig. 14.10. In [44], the image method was also elaborated in detail (but chiefly within the framework of linear models) for 2D lattices, including corner- and sector-shaped ones.

The same method may be applied to finite lattices, which are composed of a finite number of sites between two edges. An example of such a configuration was investigated in detail for discrete solitons in [45] – not in terms of the DNLS equation, but rather for solitons in the Ablowitz–Ladik (AL) model, which is based on the following discrete equation, $i du_n/dz = -(u_{n+1} + u_{n-1})(1 + |u_n|^2)$. The infinite AL lattice, as well as a finite one subject to periodic boundary conditions, are integrable systems. Under the fixed boundary conditions ($u_n \equiv 0$ for $n \geq N+1$ and $n \leq -(N+1)$, if the truncated lattice consists of $2N+1$ sites), the integrability is lost. For this case, an effective potential accounting for the interaction of a soliton (which is treated as a quasi-particle) with the edges was derived in

Fig. 14.10 (**a**) A semi-infinite lattice with local excitation at site m; (**b**) the equivalent configuration, with the negative image at site $-(m+1)$, in the respective infinite lattice. Reprinted from [44] with permission

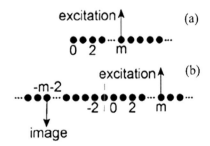

[45], and it was demonstrated that this effective potential predicts oscillations of the soliton in the finite AL lattice with a very good accuracy, if compared to direct simulations.

Finally, it is worth mentioning that the stability and instability of solitons in 1D and 2D DNLS lattices with an edge can be understood, in terms of the interaction of the soliton with its image, as manifestations of the general results for the stability and instability of bound states of two solitons with opposite or identical signs, that were reported in the works [46, 47].

14.4 Experimental Results

14.4.1 Discrete Surface Solitons in One Dimension

The first experimental observations of discrete surface solitons at the edge of a lattice of nonlinear optical waveguides were reported in [33, 34] (the material used to build the corresponding setup was AlGaAs, which is known for a very large value of the Kerr coefficient). The corresponding experimental configuration is shown in Fig. 14.11a. Its parameters were close to those previously used to observe discrete highly localized Kerr solitons in the bulk lattice (far from the edges). The array contained 101 cores and was 1 cm long, while the coupling length, determined by the linear interaction between adjacent cores, was estimated to be 2.2 mm. A set of experiments and simulations dealing with the excitation of the channel (core) at the edge of the array ($n = 0$) is depicted in Fig. 14.11.

14.4.2 Staggered Modes

The experimental observation of *staggered* discrete modes at the interface between a waveguide array, built of a copper-doped LiNb crystal, and a continuous medium was reported in [32]. The experimental sample contained 250 parallel waveguides. The width and height of the single-mode channel waveguides were 4 and 2.5 μm, respectively, while the distance between adjacent channels was 4.4 μm, the corresponding coupling length being 1.1 mm.

Optical beams of equal power, overlapping under a small angle, were coupled into the waveguide array, taking care to make the grating period of the resulting interference pattern matching the period of the array (which is 8.4 μm, according to what was said above). This input pattern had an elliptical shape whose height was adjusted to match the depth of each waveguide (approximately 2.5 μm).

In this way, a staggered input pattern was created. It consisted of a central maximum and a small number of satellites with alternating signs. The sample was placed so as to match the maximum of the staggered input and the first channel of the array. The experimental results are plotted in Fig. 14.12.

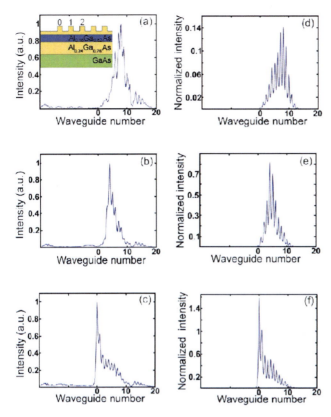

Fig. 14.11 Intensity patterns observed in the experiments and numerical simulations reported in [33, 34] (reprinted with permission) at the output of the AlGaAs waveguide array, for three different values of the power of the beam injected into channel $n = 0$. *Left-hand side*: experimental results for (**a**) $P = 450$ W, (**b**) $P = 1300$ W, and (**c**) $P = 2100$ W. *Right-hand side*: results of the numerical simulations for (**d**) $P = 280$ W, (**e**) $P = 1260$ W, and (**f**) $P = 2200$ W. The inset in panel (a) displays the experimental setup

14.4.3 Discrete Surface Solitons in Two Dimensions

Experimental observations of 2D optical discrete surface solitons were reported in [17] and [18]. Here, we summarize the results of the work [17], which used the interface between a *virtual* (photoinduced) waveguide array, created in an SBN photorefractive crystal, and a uniform medium.

The application of the positive bias voltage to a 10-mm-long sample of the crystal induced sufficiently strong self-focusing nonlinearity in the medium, and thus made it possible to create in-phase lattice surface solitons. Typical experimental results are shown in Fig. 14.13, where panel (a) displays a part of the underlying lattice pattern. The discrete diffraction in this case is stronger in the direction perpendicular to the edge than in the direction parallel to it, as seen in Fig. 14.13b. For a sufficiently high bias voltage, the self-action of the input beam provided for the formation of a

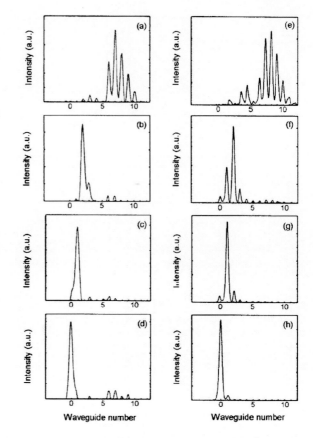

Fig. 14.12 Output patterns corresponding to the staggered input excitation of three channels, with intensity ratio 1:0.5:0.1, as reported in [32] (reprinted with permission). The results are shown for four different input powers. *Left column*: experimental results for (**a**) discrete diffraction (low-power linear regime), (**b**) power $P = 9\,\mu\text{W}$, (**c**) $P = 22.5\,\mu\text{W}$, and (**d**) $P = 225\,\mu\text{W}$. *Right column*: results of respective numerical simulations for (**e**) the linear discrete diffraction, (**f**) $P = 10\,\mu\text{W}$, (**g**) $P = 22\,\mu\text{W}$, and (**h**) $P = 230\,\mu\text{W}$

discrete surface soliton, see Fig. 14.13c, d. On the other hand, if the intensity of the beam was reduced by a factor of ≥ 8, it was not able to form a soliton, undergoing strong discrete diffraction at the surface, as seen in Fig. 14.13h. In the same work [17], 2D surface solitons were also observed at the corner of the 2D lattice, as shown in Fig. 14.13e–g.

The application of the negative bias voltage turned the crystal into a self-defocusing medium, which made it possible to create staggered 2D surface solitons, which belong to the first bandgap of the respective lattice-induced linear spectrum. With the appropriate defocusing nonlinearity, a surface *gap soliton* could be created, using a single input beam.

The difference between the in-phase surface solitons (see Fig. 14.13i, j) and staggered ones (Fig. 14.13k, l) may be clearly illustrated by the interference fringes, which break and interleave in the latter case, as shown in Fig. 14.13l. The power spectrum for the staggered solitons was also drastically different from that of the in-phase surface solitons.

Typical results of numerical simulations of the model corresponding to the experiment are presented in the right panels of Fig. 14.13, where the top one shows 3D

14 Surface Waves and Boundary Effects in DNLS Equations

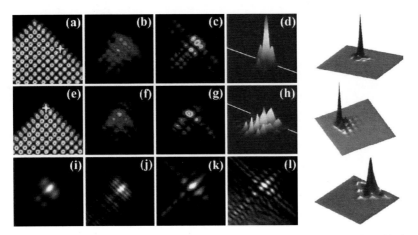

Fig. 14.13 The formation of discrete 2D surface solitons at the lattice edge (**a–d**) and lattice corner (**e–g**). Adapted from [17] with permission

plots of the 2D surface soliton at the lattice edge, corresponding to Fig. 14.13d, the middle panel shows the predicted surface soliton at the lattice corner corresponding to Fig. 14.13g, and the bottom panel presents the prediction for the surface gap soliton corresponding to Fig. 14.13k. Small dips around the central peak in the bottom figure indicate the staggered phase structure of the corresponding surface soliton.

14.5 Conclusions

In this chapter, basic theoretical and experimental results for discrete surface (and corner) solitons in lattices with the cubic on-site nonlinearity were summarized, in terms of 1D and 2D semi-infinite arrays of optical waveguides. Theoretical predictions for spatiotemporal corner solitons were presented too, as well as for 2D surface solitons with a nontrivial intrinsic structure, such as localized discrete vortices and "horseshoes." Solitons of both the in-phase (unstaggered) and staggered types have been considered (solitons of the staggered type, which are supported by the self-defocusing nonlinearity, are also called gap solitons).

As predicted theoretically and demonstrated experimentally, these solitons demonstrate various noteworthy phenomena such as power thresholds, Tamm oscillations (which are akin to Bloch oscillations), and the stabilization by the lattice edge of localized structures (such as 2D and 3D "horseshoes") which cannot be stable in bulk lattices.

Theoretical and experimental studies of surface and corner solitons can be extended in various directions. In particular, the analysis of surface states in full 3D models has started very recently. Dynamical 3D bulk lattices of the DNLS type (with the cubic on-site nonlinearity) give rise to many species of stable discrete solitons with specific arrangements and topological features, that have no counterparts in lower dimensions. These include octupoles, diagonal vortices, vortex "cubes" (stack of two quasiplanar vortices) "diamonds" (formed by two mutually orthogonal vortices) [48, 49], and discrete quasi-Skyrmions of a toroidal shape (that were found together with 2D discrete patterns of the "baby-Skyrmion" type) [50]. An obvious possibility is to investigate such 3D objects when they are placed on or close to the edge of the 3D lattice.

Virtually unexplored surfaces remain of 2D and 3D lattices whose shape is different from the simplest square and cubic types (such as triangular and hexagonal lattices in two dimensions). On the other hand, it may also be interesting to study surface states in 2D and 3D models with the quadratic (rather than cubic) on-site nonlinearity. Another challenging issue is a possibility to find 2D and, possibly, 3D discrete solitons attached to the lattice's surface that would feature *mobility* along the surface.

References

1. Zangwill, A.: Physics at Surfaces, Cambridge University Press, Cambridge (1988)
2. Tamm, I.: Phys. Z. Sowjetunion **1**, 733 (1932)
3. Shockley, W.: Phys. Rev. **56**, 317 (1939)
4. Kossel, D.: J. Opt. Soc. Am. **56**, 1434 (1966)
5. Yeh, P., Yariv, A., Cho, A.Y.: Appl. Phys. Lett. **32**, 104 (1978)
6. Ng, W., Yeh, P., Chen, P.C., Yariv, A.: Appl. Phys. Lett. **32**, 370(1978)
7. Barnes, W.L., Dereux, A., Ebbesen, T.W.: Nature **424**, 824 (2003)
8. Artigas, D., Torner, L.: Phys. Rev. Lett. **94**, 013901 (2005)
9. Seaton, C.T., Valera, J.D., Shoemaker, R.L., Stegeman, G.I., Chilwell, J.T., Smith, S.D.: IEEE J. Quantum Electron. **21**, 774 (1985)
10. Mihalache, D., Stegeman, G.I., Seaton, C.T., Wright, E.M., Zanoni, R., Boardman, A.D., Twardowski, T.: Opt. Lett. **12**, 187 (1987)
11. Mihalache, D., Bertolotti, M., Sibilia, C.: Prog. Opt. **27**, 229 (1989)
12. Kartashov, Y.V., Vysloukh, V.A., Torner, L.: Phys. Rev. Lett. **96**, 073901 (2006)
13. Rosberg, C.R., Neshev, D. N., Krolikowski, W., Mitchell, A., Vicencio, R.A., Molina, M.I., Kivshar, Y.S.: Phys. Rev. Lett. **97**, 083901 (2006)
14. Makris, K.G., Hudock, J., Christodoulides, D.N., Stegeman, G.I., Manela, O., Segev, M.: Opt. Lett. **31**, 2774 (2006)
15. Kartashov, Y.V., Vysloukh, V.A., Mihalache, D., Torner, L.: Opt. Lett. **31**, 2329 (2006)
16. Kartashov, Y.V., Torner, L.: Opt. Lett. **31**, 2172 (2006)
17. Wang, X., Bezryadina, A., Chen, Z., Makris, K.G., Christodoulides, D.N., Stegeman, G.I.: Phys. Rev. Lett. **98**, 123903 (2007)
18. Szameit, A., Kartashov, Y.V., Dreisow, F., Pertsch, T., Nolte, S., Tunnermann, A., Torner, L.: Phys. Rev. Lett. **98**, 173903 (2007)
19. Kartashov, Y.V., Ye, F., Torner, L.: Opt. Express **14**, 4808 (2006)
20. Garanovich, I.L., Sukhorukov, A.A., Kivshar, Y.S., Molina, M.: Opt. Express **14**, 4780 (2006)
21. Kartashov, Y.V., Vysloukh, V.A., Egorov, A.A., Torner, L.: Opt. Express **14**, 4049 (2006)
22. Kartashov, Y.V., Vysloukh, V.A., Torner, L.: Opt. Express **14**, 12365 (2006).
23. Motzek, K., Sukhorukov, A.A., Kivshar, Y.S.: Opt. Express **14**, 9873 (2006)
24. Motzek, K., Sukhorukov, A.A., Kivshar, Y.S.: Opt. Lett. **31**, 3125 (2006)
25. Sukhorukov, A.A., Neshev, D.N., Dreischuh, A., Fischer, R., Ha, S., Krolikowski, W., Bolger, J., Mitchell, A., Eggleton, B.J., Kivshar, Y.S.: Opt. Express **14**, 11265 (2006)

26. Makris, K.G., Suntsov, S., Christodoulides, D.N., Stegeman, G.I., Hache, A.: Opt. Lett. **30**, 2466 (2005)
27. Suntsov, S., Makris, K.G., Christodoulides, D.N., Stegeman, G.I., Morandotti, R., Yang, H., Salamo, G., Sorel, M.: Opt. Lett. **32**, 3098 (2007)
28. Molina, M.I., Garanovich, I.L., Sukhorukov, A.A., Kivshar, Y.S.: Opt. Lett. **31**, 2332 (2006)
29. Molina, M.I., Vicencio, R.A., Kivshar, Y.S.: Opt. Lett. **31**, 1693 (2006)
30. Mihalache, D., Mazilu, D., Lederer, F., Kivshar, Y.S.: Opt. Express **15**, 589 (2007)
31. Stepić, M., Smirnov, E., Rüter, C.E., Kip, D., Maluckov, A., Hadžievski, L.: Opt. Lett. **32**, 823 (2007)
32. Smirnov, E., Stepić, M., Rüter, C.E., Kip, D., Shandarov, V.: Opt. Lett. **31**, 2338 (2006)
33. Suntsov, S., Makris, K.G., Christodoulides, D.N., Stegeman, G.I., Hache, A., Morandotti, R., Yang, H., Salamo, G., Sorel, M.: Phys. Rev. Lett. **96**, 063901 (2006)
34. Suntsov, S., Makris, K.G., Siviloglou, G.A., Iwanow, R., Schiek, R., Christodoulides, D.N., Stegeman, G.I., Morandotti, R., Yang, H., Salamo, G., Volatier, M., Aimez, V., Ares, R., Sorel, M., Min, Y., Sohler, W., Wang, X.S., Bezryadina, A., Chen, Z.: J. Nonlin. Opt. Phys. Mat. **16**, 401 (2007)
35. Hudock, J., Suntsov, S., Christodoulides, D.N., Stegeman, G.I.: Opt. Express **13**, 7720 (2005)
36. Garanovich, L., Sukhorukov, A.A., Kivshar, Y.S., Molina, M.: Opt. Express **14**, 4780 (2006)
37. He, Y.J., Chen, W.H., Wang, H.Z., Malomed, B.A.: Opt. Lett. **32**, 1390 (2007)
38. Siviloglou, G.A., Makris, K.G., Iwanow, R., Schiek, R. Christodoulides, D.N., Stegeman, G.I.: Opt. Express **14**, 5508 (2006)
39. Susanto, H., Kevrekidis, P.G., Malomed, B.A., Carretero-González, R., Frantzeskakis, D.J.: Phys. Rev. E **75**, 056605 (2007)
40. Mihalache, D., Mazilu, D., Lederer, F., Kivshar, Y.S.: Opt. Lett. **32**, 3173 (2007)
41. Mihalache, D., Mazilu, D., Kivshar, Y., Lederer, F.: Opt. Express **15**, 10718 (2007)
42. Machacek, D.L., Foreman, E.A., Hoq, Q.E., Kevrekidis, P.G., Saxena, A., Frantzeskakis, D.J., Bishop, A.R.: Phys. Rev. E **74**, 036602 (2006)
43. Molina, M.I., Kivshar, Y.S.: Phys. Lett. A **362**, 280 (2007)
44. Makris, K.G., Christodoulides, D.N.: Phys. Rev. E **73**, 036616 (2006)
45. Rasmussen, K.Ø., Cai, D., Bishop, A.R., Grønbech-Jensen, N.: Phys. Rev. E **55**, 6151 (1997)
46. Kapitula, T., Kevrekidis, P.G., Malomed, B.A.: Phys. Rev. E **63**, 036604 (2001)
47. Kevrekidis, P.G., Malomed, B.A., Bishop, A.R.: J. Phys. A Math. Gen. **34**, 9615 (2001)
48. Kevrekidis, P.G., Malomed, B.A., Frantzeskakis, D.J., Carretero-González, R.: Phys. Rev. Lett. **93**, 080403 (2004)
49. Carretero-González, R., Kevrekidis, P.G., Malomed, B.A., Frantzeskakis, D.J.: Phys. Rev. Lett. **94**, 203901 (2005)
50. Kevrekidis, P.G., Carretero-González, R., Frantzeskakis, D.J., Malomed, B.A., Diakonos, F.K.: Phys. Rev. E **75**, 026603 (2007)

Chapter 15
Discrete Nonlinear Schrödinger Equations with Time-Dependent Coefficients (*Management of Lattice Solitons*)

Jesús Cuevas and Boris A. Malomed

15.1 Introduction

The general topic of the book into which this chapter is incorporated is the discrete nonlinear Schrödinger (DNLS) equation as a fundamental model of nonlinear lattice dynamics. The DNLS equation helps to study many generic features of nonintegrable dynamics in discrete media [1]. Besides being a profoundly important model in its own right, this equation has very important direct physical realizations, in terms of arrays of nonlinear optical waveguides (as was predicted long ago [2] and demonstrated in detail more recently, see [3, 4] and references therein), and arrays of droplets in Bose–Einstein condensates (BECs) trapped in a very deep optical lattice (OL), see details in the original works [5–10] and the review [11].

In all these contexts, discrete solitons are fundamental localized excitations supported by the DNLS equation. As explained in great detail in the rest of the book, the dynamics of standing solitons, which are pinned by the underlying lattice, is understood quite well, by means of numerical methods and analytical approximations (the most general approximation is based on the variational method [12, 13]). A more complex issue is posed by moving discrete solitons [14–17]. While, strictly speaking, exact solutions for moving solitons cannot exist in nonintegrable lattice models because of the radiation loss, which accompanies their motion across the lattice, direct simulations indicate that a soliton may move freely if its norm ("mass") does not exceed a certain critical value [17]. In the quasi-continuum approximation, the moving soliton may be considered, in the lowest (adiabatic) approximation, as a classical mechanical particle which moves across the effective Peierls–Nabarro (PN) potential induced by the lattice [18–21]. In this limit, the radiation loss is a very weak nonadiabatic effect, which attests to the deviation of the true soliton dynamics from that of the point-like particle.

In the case of the DNLS equation describing arrays of nearly isolated droplets of a BEC trapped in a deep OL, an interesting possibility is to apply the *Feshbach*

J. Cuevas (✉)
Grupo de Física No Lineal, Departamento de Física Aplicada I, Escuela Universitaria Politécnica, C/ Virgen de África, 7, 41011 Sevilla, Spain
e-mail: jcuevas@us.es

resonance management (FRM) to this system, as was first proposed and studied for immobile discrete solitons in [22], and later elaborated in detail, for moving solitons, in [23]. The FRM may be induced by an external low-frequency ac magnetic field, which periodically (in time) changes the sign of the nonlinearity by dint of the FRM, i.e., formation of quasi-bound states in collisions between atoms [24]. For the effectively one-dimensional BEC in the absence of the OL, the concept of the FRM was elaborated in (1D) [25], see also the book [26].

The objective of the chapter is to summarize basic results for the quiescent and moving 1D DNLS solitons subjected to the time-periodic management, following, chiefly, the lines of works [22] and [23]. For the quiescent solitons, most significant findings are FRM-induced resonances in them, and stability limits for the solitons which are affected by the resonant mechanisms. In particular, resonances with an external time-periodic modulation may stimulate self-splitting of the solitons. For the moving solitons, an essential conclusion is that the FRM may strongly facilitate their mobility, which is an essentially novel dynamical effect in discrete media.

The DNLS equation which includes the FRM mechanism can be cast in the following form:

$$i\dot{u}_n + C(u_{n+1} + u_{n-1} - 2u_n) + g(t)|u_n|^2 u_n = 0, \quad (15.1)$$

where $u_n(t)$ denotes the BEC wave function at the lattice sites, C is the strength of the linear coupling between adjacent sites of the lattice, and the real time-dependent nonlinear coefficient is

$$g(t) = g_{dc} + g_{ac} \sin(\omega t) \quad (15.2)$$

with the time-dependent term accounting for the FRM ($-g$ is proportional to the scattering length of atomic collisions, whose magnitude and sign may be directly altered by the FRM). In what follows below, we fix, by means of obvious rescaling, $C \equiv 1$. We also note that g_{dc} may always be chosen positive, as it can be transformed by means of the so-called staggering transformation, $u_n(t) \equiv (-1)^n e^{-4it} \tilde{u}_n(t)$. Equation (15.1) with the time-dependent nonlinear coefficient has a single dynamical invariant, the norm (which is proportional to the number of atoms in the BEC),

$$\mathcal{N} = \sum_{n=-\infty}^{+\infty} |u_n|^2. \quad (15.3)$$

The chapter is divided into two major parts that deal with quiescent and mobile discrete solitons (the latter one also briefly considers collisions between the moving solitons). Each part contains sections which present analytical and numerical results. In either case, the analytical approach is based on using a particular *ansatz* for the shape of the soliton, while numerical results are produced by systematic direct simulations of Eq. (15.1) with appropriate initial conditions.

15.2 Quiescent Solitons Under the Action of the "Management"

15.2.1 Semi-Analytical Approximation

In this section, we put Eqs. (15.1) and (15.2) in a slightly different form, namely,

$$i\dot{u}_n + \frac{1}{2}(u_{n+1} + u_{n-1} - 2u_n) + a(t)|u_n|^2 u_n = 0, \quad a(t) = 1 + a_1 \sin(\omega t), \quad (15.4)$$

where, as said above, we fix $C = 1$, and the dc part of $a(t)$ is also set equal to 1 by means of an additional rescaling. A (semi-)analytical approximation for soliton solutions to Eq. (15.4) is based on the fact that it can be derived from the Lagrangian,

$$L = \frac{1}{2}\sum_{n=-\infty}^{\infty}\left[i\left(u_n^*\dot{u}_n - u_n\dot{u}_n^*\right) - |u_{n+1} - u_n|^2 + a(t)|u_n|^4\right] \quad (15.5)$$

(* stands for the complex conjugate). Then, the variational approximation (VA) [13] represents the solitons by the following *ansatz*, following [12]:

$$u_n(t) = A \exp(i\phi + ib|n| - \alpha|n|), \quad (15.6)$$

where $A, \phi, b,$ and α are real functions of time. Substituting this ansatz in the Lagrangian (15.5), it is easy to perform the summations explicitly and thus arrive at the corresponding *effective Lagrangian* (an inessential term, proportional to $\dot{\phi}$, is dropped here):

$$\frac{L}{N} = -\frac{1}{\sinh(2\alpha)}\frac{db}{dt} + \frac{\cos b}{\cosh \alpha} + \frac{1}{4}\mathcal{N}a(t)\frac{\sinh \alpha}{\cosh^3 \alpha}\cosh(2\alpha), \quad (15.7)$$

where $\mathcal{N} = A^2 \coth \alpha$ is the norm of the ansatz (recall the norm is the dynamical invariant, calculated as per Eq. (15.3)). The variational equations for the soliton's chirp b and inverse width α, derived from Lagrangian (15.7) are

$$\frac{db}{dt} = 2(\cos b)\frac{\sinh^3 \alpha}{\cosh(2\alpha)} - \frac{1}{2}\mathcal{N}a(t)(\tanh^2 \alpha)\frac{2\cosh(2\alpha) - 1}{\cosh(2\alpha)}, \quad (15.8)$$

$$\frac{d\alpha}{dt} = -(\sin b)(\sinh \alpha)\tanh(2\alpha) \quad (15.9)$$

(the amplitude A was eliminated here in favor of α, due to the conservation of $\mathcal{N}$). First, in the absence of the FRM, i.e., for $a(t) = \text{const} \equiv 1$ (see Eq. (15.4)) Eqs. (15.8) and (15.9) give rise to a stationary solution (*fixed point*, FP), with $b_{FP} = 0$ and α_{FP} defined by equation $\sinh(\alpha_{FP}) = \mathcal{N}\left[1 + 3\tanh^2(\alpha_{FP})\right]/4$ [12]. The VA-predicted stationary soliton is quite close to its counterpart found from a numerical solution of Eq. (15.4) with $a(t) \equiv 1$ [22]. Furthermore, linearization of

Eqs. (15.8) and (15.9) around the FP yields a squared frequency of intrinsic oscillations for a slightly perturbed soliton,

$$\omega_0^2 = \frac{\sinh^3(\alpha_{FP})\cosh^2(\alpha_{FP})}{\cosh^3(2\alpha_{FP})} \left\{ 4\sinh(\alpha_{FP})[\cosh(2\alpha_{FP}) + 2] - \frac{\mathcal{N}}{\cosh^4(\alpha_{FP})} \left[5\cosh^2(2\alpha_{FP}) - 2\cosh(2\alpha_{FP}) - 1 \right] \right\} \quad (15.10)$$

(this expression can be shown to be always positive). Comparison of this prediction of the VA with numerically found frequencies of small oscillations of the perturbed soliton demonstrates good agreement too [22].

In the presence of the FRM, $a_1 \neq 0$, strong (resonant) response of the system is expected when the modulation frequency ω is close to eigenfrequency given by Eq. (15.10). Moreover, the dynamics may become chaotic, via the resonance-overlapping mechanism [27], if the modulation amplitude a_1 exceeds some threshold value. This was observed indeed in numerical simulations of Eqs. (15.8) and (15.9), as illustrated in Fig. 15.1 by a typical example of the Poincaré map [27] in the chaotic regime. The figure shows discrete trajectories initiated by two sets of the initial conditions, namely, $\left(b_0^{(1)}, \alpha_0^{(1)}\right) = (0, 0.789)$, that correspond to the stationary discrete soliton with amplitude $A = 1$ (cf. ansatz (15.6)) in the unperturbed system ($a_1 = 0$), and a different set, $\left(b_0^{(2)}, \alpha_0^{(2)}\right) = (0.13, 0.74)$. The respective modulation frequency, ω, is close to the eigenfrequency of small oscillations ω_0, as predicted by Eq. (15.10). For the former initial condition, the point in space (b, α) is chaotically moving away from the unperturbed FP. However, the chaotic evolution is a *transient feature*, as the discrete trajectory takes an asymptotic form with $\alpha(t) \to 0$, which implies decay (indefinite broadening) of the soliton. The second set of the initial conditions eventually leads to a *stable periodic solution* (in terms of the Poincaré map, it is represented by a new FP, which is found in a vicinity of the unperturbed one). The latter results predicts the existence of quasi-stationary discrete solitons under the action of the FRM.

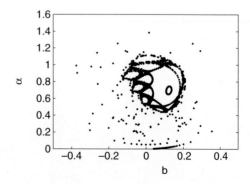

Fig. 15.1 Example of the chaotic Poincaré map generated by Eqs. (15.8) and (15.9) with $a_1 = 0.02766$, $\omega = 0.481$, and $W = 1.5202$. Reprinted from [22] with permission

Another analytically tractable case is one for the high-frequency modulation. It is then possible to perform averaging of Eq. (15.4) (without resorting to the VA). The eventual result is an effective DNLS equation for the slowly varying part of the lattice field, $q_n(t)$ [22]:

$$i\dot{q}_n + \frac{1}{2}(q_{n+1} + q_{n-1} - 2q_n) + |q_n|^2 q_n$$
$$= \left(a_1^2/8\omega^2\right)\left[3|q_n|^4(q_{n+1} + q_{n-1}) + 2|q_n|^2 q_n^2 \left(q_{n+1}^* + q_{n-1}^*\right)\right.$$
$$+ |q_{n+1}|^4 q_{n+1} + |q_{n-1}|^4 q_{n-1} - 2|q_{n+1}|^2 \left(2|q_n|^2 q_{n+1} + q_n^2 q_{n+1}^*\right)$$
$$\left. - 2|q_{n-1}|^2 \left(2|q_n|^2 q_{n-1} + q_n^2 q_{n-1}^*\right)\right]. \tag{15.11}$$

Equation (15.11) is the DNLS equation with a small inter-site quintic perturbation ($a_1^2/8\omega^2$ is a small perturbation parameter).

15.2.2 Direct Simulations

Systematic direct simulations of Eq. (15.4) demonstrate that the VA correctly predicts only an initial stage of the dynamics [22]. Emission of linear waves (lattice phonons) by the soliton, which is ignored by the VA, gives rise to an effective dissipation, that makes the resonance frequency different from the value predicted by Eq. (15.10). Actually, the soliton decouples from the resonance, as ω_0 depends on the norm $\mathcal{N}$, and the radiation loss results in a gradual decrease of the norm. Nevertheless, the general predictions of the VA turn out to be correct for $a_1 \lesssim 0.05$: under the action of the management, oscillations of the soliton's parameters are regular for very small a_1, and become chaotic at larger a_1.

Typical examples of the soliton dynamics under the action of stronger management, with $a_1 \geq 0.1$ (and $\omega = 0.5$) are displayed in Fig. 15.2. A noteworthy observation, which could not be predicted by the single-soliton ansatz, is *splitting* of the pulse, which is observed, at $a_1 = 0.2$, in Fig. 15.2, while at other values of a_1, both smaller and larger than 0.2, the soliton remains centered around $n = 0$. Note that the splitting is similar to that revealed by direct simulations of the continuum NLS equation with a term accounting for periodic modulation of the linear dispersion (*dispersion management*), which was reported in [28]. A similar phenomenon was also observed in a discrete model with the finite-difference dispersion term subject to periodic modulation [29].

Results of the systematic numerical study of the evolution of solitons in Eq. (15.4) are summarized in Fig. 15.3 (for $a_1 \gtrsim 0.2$, the pulse may split into several moving splinters). The diagram shows that the actual critical value of management amplitude a_1, past which the soliton develops the instability (via the splitting) is *much higher* than the prediction of the chaotic dynamics threshold by the VA (which also eventually leads to the decay of the soliton, as the chaotic transient is followed by the asymptotic stage of the evolution with $\alpha(t) \to 0$, see above). Thus, the VA based on Eqs. (15.8) and (15.9) *underestimates* the effective stability of the discrete

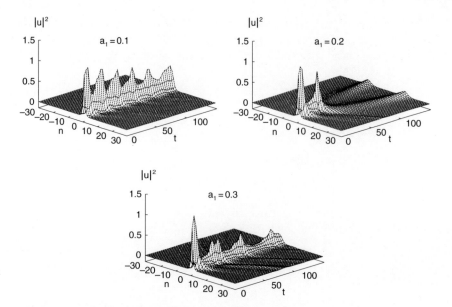

Fig. 15.2 Evolution of a discrete soliton with initial amplitude $A = 1$ in Eq. (15.4) with $\omega = 0.5$ and different values of the management amplitude, a_1. Reprinted from [22] with permission

solitons. This conclusion is explained by the fact that the radiation loss, that was ignored by the variational ansatz, plays a stabilizing role for the discrete solitons.

It is worth mentioning that the existence of a finite critical value of the modulation amplitude, past which the splitting occurs, and the fact that the actual stability area for solitons is larger than predicted by the VA are qualitatively similar to features found in the above-mentioned dispersion-management model based on the continuum NLS equation [28].

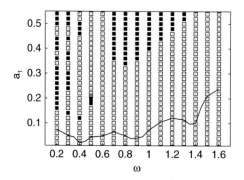

Fig. 15.3 The diagram of dynamical regimes in the plane of the FRM parameters, (ω, a_1), as produced by systematic simulations of Eq. (15.4) with $a_0 = 1$ and initial soliton's amplitude $A = 1$. *Open and solid squares* correspond to stable and splitting solitons, respectively. *The solid line* is the chaos-onset threshold predicted by the numerical solution of the variational equations (15.8) and (15.9). Reprinted from [22] with permission

15.3 Supporting Moving Solitons by Means of the "Management"

15.3.1 Analytical Approximation

In this part, we switch to the notation adopted in Eqs. (15.1) and (15.2), but again with $C \equiv 1$. The continuum limit (i.e., the ordinary NLS equation) suggests the following *ansatz* for a moving soliton [30]:

$$u(n,t) = A \exp\left[-b(n-\xi(t))^2 + i\phi(t) + \left(\frac{i}{2}\right)\dot{\xi}n - \frac{i}{4}\int (\dot{\xi}(t))^2 dt\right], \quad (15.12)$$

where $A, b, \xi(t)$, and ϕ are, respectively, the amplitude, squared inverse width, central coordinate, and phase of the soliton. Accordingly, $\dot{\xi}$ is the soliton's velocity, $\dot{\xi}/2$ simultaneously being the wave number of the wave field which carries the moving soliton. Note the difference of this ansatz, which emulates the VA for the solitons in the continuum NLS equation [30], from the above-mentioned ansatz (15.6), that was adopted for the essentially discrete model. In the framework of the continuum NLS equation, the VA for the ordinary solitons yields

$$\dot{\phi} = 3b, \quad A^2 = \frac{4\sqrt{2b}}{g} \quad (15.13)$$

(assuming that $g = \text{const} > 0$), where b is treated as an arbitrary positive constant, i.e., intrinsic parameter of the soliton family.

Then, the ansatz (15.12) with zero velocity, $\dot{\xi} = 0$, may be substituted in the Hamiltonian corresponding to the DNLS equation (15.1),

$$H = \sum_{n=-\infty}^{+\infty}\left[2|u_n|^2 - \left(u_n^* u_{n+1} + u_n u_{n+1}^*\right) - \frac{g}{2}|u_n|^4\right]. \quad (15.14)$$

In this way, an effective potential of the soliton-lattice interaction is obtained in the form of a Fourier series, $H(\xi) = \sum_{m=0}^{\infty} H_m \cos(2\pi m \xi)$. In the case of a broad soliton, for which ansatz (15.12) is relevant, it is sufficient to keep only the lowest harmonic ($m = 1$) in this expression, which yields the respective PN (Peierls–Nabarro) potential, U_{PN}. A straightforward calculation, using the Poisson summation formula, yields [23]

$$U_{\text{PN}}(\xi) = \frac{1}{2}\sqrt{\frac{\pi}{b}} A^2 \exp\left(-\frac{\pi^2}{4b}\right) \left\{4\sqrt{2}\exp\left(-\frac{\pi^2}{4b}\right)\right.$$
$$\left.(1+e^{-b/2}) - gA^2 \sqrt{\frac{\pi}{b}}\right\} \cos(2\pi\xi). \quad (15.15)$$

If the relation between b and A^2, taken as for solitons in the continuum NLS equation, i.e., as per Eq. (15.13), is substituted into Eq. (15.15), the coefficient in front of $\cos(2\pi\xi)$, i.e., the amplitude of the PN potential, never vanishes. However, it may vanish if (15.12) is considered not as a soliton, but just as a pulse with independent amplitude and width, A and $1/\sqrt{b}$; then, the PN potential in Eq. (15.15) may vanish, under the following condition:

$$1 + \exp\left(\frac{-b}{2}\right) = \left(\frac{gA^2}{4}\right)\sqrt{\frac{\pi}{2b}}\exp\left(\frac{\pi^2}{4b}\right). \tag{15.16}$$

The vanishing of the PN potential implies a possibility of unhindered motion of the soliton across the lattice.

For a broad soliton (small b), the PN potential barrier is exponentially small, hence, the soliton's kinetic energy may be much larger than the height of the potential barrier. Therefore, the velocity of the soliton moving through the potential (15.15) with period $L = 1$ contains a constant (dc) part and a small ac correction to it, with frequency $2\pi\dot{\xi}_0/L \equiv 2\pi\dot{\xi}_0$ [31]: $\dot{\xi}(t) \approx \dot{\xi}_0 + \dot{\xi}_1 \cos(2\pi\dot{\xi}_0 t)$, $\dot{\xi}_1^2 \ll \dot{\xi}_0^2$. Substituting this into condition (15.16), one can expand its left-hand side by using

$$\exp\left(\frac{-b}{2}\right)\cos\left(\frac{\dot{\xi}}{2}\right) \approx \exp\left(\frac{-b}{2}\right)\left[\cos\left(\frac{\dot{\xi}_0}{2}\right) - \frac{\dot{\xi}_1}{2}\sin\left(\frac{\dot{\xi}_0}{2}\right)\cos(2\pi\dot{\xi}_0 t)\right]. \tag{15.17}$$

Next, inserting the variable nonlinearity coefficient (15.2) into the right-hand side of Eq. (15.16), and equating the resulting expression to that (15.17), one concludes that g_{dc} and g_{ac} may be chosen so as to secure condition (15.16) to hold, provided that the average soliton's velocity takes the *resonant value*, $\dot{\xi}_0 = \omega/2\pi$. More generally, due to anharmonic effects, one may expect the existence of a spectrum of the resonant velocities,

$$\dot{\xi}_0 = (c_{\mathrm{res}})_N^{(M)} \equiv \frac{M\omega}{2\pi N} \tag{15.18}$$

with integers M and N.

Actually, an ac drive can support stable progressive motion of solitons at resonant velocities (15.18) (assuming that the spatial period is $L = 1$), even in the presence of dissipation, in a broad class of systems. This effect was first predicted for discrete systems (of the Toda lattice and Frenkel–Kontorova types) in [32–35], and demonstrated experimentally in an LC electric transmission line [36]. Later, the same effect was predicted [37] and demonstrated experimentally [38] in long Josephson junctions with a spatially periodic inhomogeneity. However, a qualitative difference of the situation considered here is that we are now dealing with *nontopological solitons*, while the above-mentioned examples involved *kinks*, i.e., discrete or continuum solitons whose topological charge directly couples to the driving field.

15.3.2 Numerical Results

Numerical simulations of Eq. (15.1) (with $C = 1$) were performed in [23], setting $g_{dc} = 1$ by means of the same rescaling which was used in the studies of the quiescent solitons. First, stationary solitons were found as solutions to the DNLS with $g_{ac} = 0$, in the form of $u_n^{(0)}(t) = v_n \exp(-i\omega_0 t)$. Then, the FRM with $g_{ac} > 0$ was switched on, and, simultaneously, the soliton was set in motion by giving it a *kick*, i.e., multiplying u_n by $\exp(iqn)$. Generic results in the plane of the ac-drive's parameters, (ω, g_{ac}), can be adequately represented by fixing $\omega_0 = -1$ and considering three values of the kick, $q = 0.25, 0.5,$ and 1.

If $g_{ac} = 0$, the kicked soliton does not start progressive motion if the thrust is relatively weak, $q \lesssim 0.7$. It remains pinned to the lattice, oscillating around an equilibrium position, which may be explained by the fact that the kinetic energy imparted to the soliton is smaller than the height of the PN potential barrier. Several types of dynamics can be observed with $g_{ac} > 0$, depending on the modulation frequency ω and kick strength q. First, the soliton may remain pinned (generally, not at the initial position, but within a few sites from it, i.e., the soliton passes a short distance and comes to a halt, as shown in Fig. 15.4a). The next generic regime is that of *irregular motion*, as illustrated in Fig. 15.4b. A characteristic feature of that regime is that the soliton randomly changes the direction of motion several times, and the velocity remains very small in comparison with regimes of persistent motion, see below. The soliton's central coordinate, the evolution of which is presented in Fig. 15.4, is defined as $X = \sum_n n|u_n|^2/\mathcal{N}$, with the norm $\mathcal{N}$ calculated as per Eq. (15.3).

Under the action of strong modulation, the soliton can sometimes split into two pulses moving in opposite directions, see Fig. 15.5. This outcome is similar to that found for quiescent solitons, cf. Fig. 15.2b, although the splitting of the kicked soliton is strongly asymmetric (unlike the nearly symmetric splitting of the quiescent

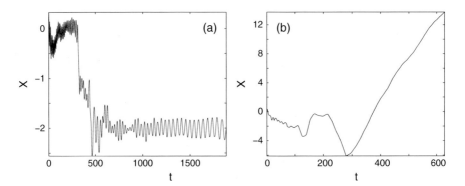

Fig. 15.4 The soliton's central position as a function of time, for typical cases in which the soliton remains pinned (**a**), at $g_{ac} = 0.03$, or develops an irregular motion (**b**), at $g_{ac} = 0.065$. In both cases, $\omega = 1$ and $q = 0.5$. Reprinted from [23] with permission

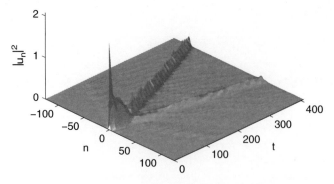

Fig. 15.5 An example of asymmetric splitting of the kicked soliton, for $q = 0.5$. The FRM parameters are $g_{ac} = 0.196$, $\omega = 0.5$. Reprinted from [23] with permission

solitons). The heavier splinter may move both forward and *backward*, relative to the initial kick.

In the case of moderately strong modulation, the moving soliton does not split. Numerical results demonstrate that, in some cases, it gradually decays into radiation, while in other cases it is completely stable, keeping all its norm, after an initial transient stage of the evolution, see examples in Figs. 15.6 and 15.7. To distinguish between the unstable and stable regimes, a particular criterion was adopted [23]. It categorizes as stable solitons those moving ones which keep $\geq 70\%$ of the initial norm in the course of indefinitely long evolution. For this purpose, very long evolution was implemented by allowing the soliton to circulate in the DNLS lattice with periodic boundary conditions.

Note that the soliton adjusting itself to the stable motion mode typically sheds off $\simeq 20\%$ of its initial norm. Although this conspicuous amount of the lattice radiation stays in the system with the periodic boundary conditions, it does not give rise to any appreciable perturbation of the established motion of the soliton. In fact, the latter observation provides for an additional essential evidence to the robustness of the moving soliton.

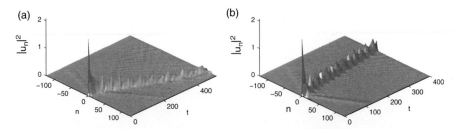

Fig. 15.6 Generic examples of the progressive motion of a decaying soliton in the straight (*forward*) direction (**a**), for $g_{ac} = 0.206$, $\omega = 0.5$, and in the reverse (*backward*) direction (**b**), for $g_{ac} = 0.170$, $\omega = 1$. In both cases, the solitons were set in motion by the application of the kick with $q = 0.5$. Reprinted from [23] with permission

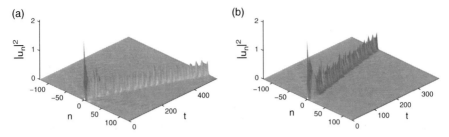

Fig. 15.7 Generic examples of the motion of propagating solitons in the straight direction (**a**), for $g_{ac} = 0.132$, and in the reverse direction (**b**), for $g_{ac} = 0.122$. In both cases, the modulation frequency is $\omega = 1$, and the kick is $q = 0.5$. Reprinted from [23] with permission

It is noteworthy too that, as seen in Figs. 15.6b and 15.7b, in both unstable and stable regimes the persistent motion of the soliton is possible in both the straight and reverse directions, relative to the initial thrust. In the latter case, the soliton starts the motion straight ahead, but very quickly bounces back. The difference from the regime of the irregular motion (cf. Fig. 15.4b) is that the direction of motion reverses only once, and the eventual velocity does not fall to very small values.

It is pertinent to compare the average velocity $\bar{c}$ of the persistent motion with the prediction given by Eq. (15.18). For example, in the cases displayed in Fig. 15.7a, b, the velocities found from the numerical data are $\bar{c}_a \approx 0.246$ and $\bar{c}_b \approx -0.155$, respectively. For $\omega = 1$, which is the corresponding modulation frequency, these values fit well to those predicted by Eq. (15.18) in the cases of, respectively, the second-order and fundamental resonance: $\bar{c}_b/(c_{res})_1^{(1)} \approx 0.974$, $\bar{c}_a/(c_{res})_3^{(2)} \approx 1.029$.

Lastly, results of systematic simulations of Eq. (15.1), which were performed, as said above, for the initial discrete soliton taken as a solution of the stationary version of the equation (for $g_{ac} = 0$ and $g_{dc} = 1$ in Eq. (15.2)) with $\omega = -1$, and for three values of the kick, $q = 0.25, 0.5$, and 1, are collected in Fig. 15.8 in the form of maps in the plane of the modulation parameters, g_{ac} and ω. The maps outline regions of the different dynamical regimes described above, as well as the distinction between regions of the straight and reverse progressive motion.

The examination of the maps shows that the increase of thrust q significantly affects the map, although quantitatively, rather than qualitatively. At all values of q, the irregular dynamics is, generally, changed by the stable progressive motion (straight or reverse) with the increase of the modulation amplitude and/or decrease of the frequency, which is quite natural. Further increase of the FRM strength, which implies the action of a strong perturbation, may lead to an instability, which indeed happens, in the form of onset of the gradual decay of the moving solitons. Finally, strong instability sets in, manifesting itself in the splitting of the soliton.

The reversal of the direction of the soliton's motion tends to happen parallel to the transition from stable moving solitons to decaying ones. For this reason, in most cases (but not always, see Fig. 15.7b) backward-moving solitons are decaying ones. Finally, a somewhat counterintuitive conclusion is that the increase of the initial

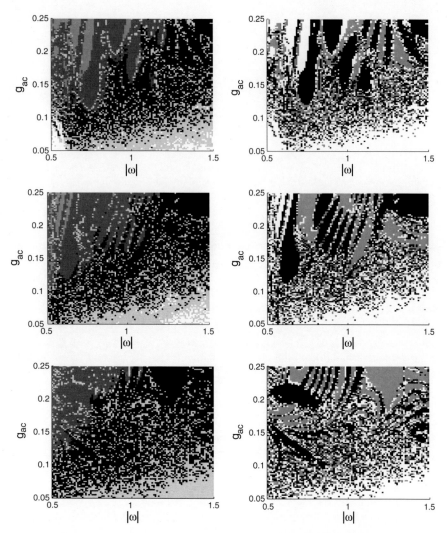

Fig. 15.8 Maps in the *left column* show areas in the plane of the FRM frequency and amplitude (ω, g_{ac}) which give rise to the following dynamical regimes. *White areas*: the soliton remains pinned; *bright gray*: irregular motion; *gray*: splitting; *dark gray*: regular motion with decay; *black*: stable motion. The maps in the *right column* additionally show the difference between the straight and reverse directions of the regular motion (marked by *dark gray and black*), relative to the direction of the initial thrust. Regular-motion regimes for both decaying and stable solitons are included in the *right-hand panels*. *Top row*: $q = 0.25$; *middle row*: $q = 0.5$; *bottom row*: $q = 1$. Reprinted from [23] with permission

thrust leads to overall *stabilization* of the soliton, making the decay and splitting zones smaller.

Collisions between solitons moving with opposite velocities (generated by thrusts $\pm q$ applied to two far separated quiescent solitons) were studied too, using the

15 DNLS with Time-Dependent Coefficients

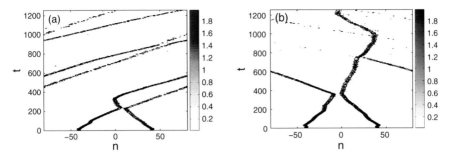

Fig. 15.9 Two generic outcomes of collisions between identical solitons moving in opposite directions, in the lattice with periodic boundary conditions. The parameters are $g_{ac} = 0.132$, $\omega = 1$, $q = 0.5$ (**a**) and $g_{ac} = 0.122$, $\omega = 1$, $q = 0.5$ (**b**). Note the presence of multiple collisions in panel (**b**). Reprinted from [23] with permission

lattice with periodic boundary conditions (which allow repeated collisions). Two different types of the interaction can be identified, see typical examples in Fig. 15.9. In the case shown in Fig. 15.9a, the solitons bounce back from each other almost elastically. Afterward, one of them spontaneously reverses the direction of motion, due to its interaction with the lattice. Eventually, a pair of virtually noninteracting solitons traveling indefinitely long in the same direction is observed.

In the other case, shown in Fig. 15.9b, the solitons also bounce after the first collision; however, in this case the collision is inelastic, resulting in transfer of mass from one soliton to the other. Repeated collisions (due to the periodicity of the lattice) lead to additional such transfer, and the weaker soliton disappears eventually. In contrast to what is known about collisions between moving solitons in the ordinary DNLS equation (with constant coefficients) [17], merger of colliding solitons into a single standing one was not observed under the action of the management (FRM).

15.4 Conclusion and Future Challenges

The use of the nonlinearity management may provide for a powerful tool for the control of the dynamics of standing and moving discrete solitons. In the present chapter, these possibilities were outlined for the 1D settings. The action of a similar "management" on two-dimensional (2D) discrete fundamental and vortex solitons, which, in terms of the BEC, may also be readily implemented by means of the FRM, has not been investigated as yet.

The stability limits of 2D solitons, against periodic modulation of the strength of the potential, in the continuum Gross–Pitaevskii equation have been recently studied. The considered model includes the self-attractive cubic nonlinearity and a quasi-1D [39] or full 2D [40] OL potential. In fact, the respective solitons may be considered as quasi-discrete ones; accordingly, in terms of discrete equation (15.1), this "lattice management" corresponds to making the lattice coupling constant a periodic function of time, $C = C(t)$.

As concerns the application of the nonlinearity management to 2D lattice solitons, challenging issues are a possibility of the generation of moving 2D lattice solitons, as well as the control of vortex solitons and bound states of fundamental solitons by means of this technique. These applications may pertain to both isotropic and anisotropic lattices.

A similar management technique may be applied to waveguide arrays or photonic crystals made of photorefractive materials. These systems can be described by DNLS equations with a saturable nonlinearity.

References

1. Kevrekidis, P.G., Rasmussen, K.Ø., Bishop, A.R.: Int. J. Mod. Phys. B **15**, 2833 (2001)
2. Christodoulides, D.N., Joseph, R.I.: Opt. Lett. **13**, 794 (1988)
3. Peschel, U., Morandotti, R., Arnold, J.M., Aitchison, J.S., Eisenberg, H.S., Silberberg, Y., Pertsch, T., Lederer, F.: J. Opt. Soc. Am. B **19**, 2637 (2002)
4. Eisenberg, H.S., Morandotti, R., Silberberg, Y., Arnold, J.M., Pennelli, G., Aitchison, J.S.: J. Opt. Soc. Am. B **19**, 2938 (2002)
5. Trombettoni, A., Smerzi, A.: Phys. Rev. Lett. **86**, 2353 (2001)
6. Abdullaev, F.Kh., Baizakov, B.B., Darmanyan, S.A., Konotop, V.V., Salerno, M.: Phys. Rev. A **64**, 043606 (2001)
7. Alfimov, G.L., Kevrekidis, P.G., Konotop, V.V., Salerno, M.: Phys. Rev. E **66**, 046608 (2002)
8. Smerzi, A., Trombettoni, A.: Phys. Rev. A **68**, 023613 (2003)
9. Smerzi, A., Trombettoni, A.: Chaos **13**, 766 (2003)
10. Efremidis, N.K., Christodoulides, D.N.: Phys. Rev. A **67**, 063608 (2003)
11. Porter, M.A., Carretero-González, R., Kevrekidis, P.G., Malomed, B.A.: Chaos **15**, 015115 (2005)
12. Malomed, B. Weinstein, M.I.: Phys. Lett. A **220**, 91 (1996)
13. Malomed, B.A.: In Progress in Optics. Wolf, E. (ed.), vol. 43, p. 71. North Holland, Amsterdam (2002)
14. Duncan, D.B., Eilbeck, J.C., Feddersen, H., Wattis, J.A.D.: Physica D **68**, 1 (1993)
15. Flach, S., Zolotaryuk, Y. Kladko, K.: Phys. Rev. E **59**, 6105 (1999)
16. Ablowitz, M.J., Musslimani, Z.H., Biondini, G.: Phys. Rev. E **65**, 026602 (2002)
17. Papacharalampous, I.E., Kevrekidis, P.G., Malomed, B.A., Frantzeskakis, D.J.: Phys. Rev. E **68**, 046604 (2003)
18. Kivshar, Yu.S., Malomed, B.A.: Rev. Mod. Phys. **61**, 763 (1989)
19. Kivshar, Y.S., Campbell, D.K.: Phys. Rev. E **48**, 3077 (1993)
20. Brizhik, L., Eremko, A., Cruzeiro-Hansson, L., Olkhovska,Y.: Phys. Rev. B **61**, 1129 (2000)
21. Kevrekidis, P.G., Kevrekidis, I.G., Bishop, A.R., Titi, E.S.: Phys. Rev. E **65**, 046613 (2002)
22. Abdullaev, F.Kh., Tsoy, E.N., Malomed, B.A., Kraenkel, R.A.: Phys. Rev. A **68**, 053606 (2003)
23. Cuevas, J., Malomed, B.A., Kevrekidis, P.G.: Phys. Rev. E **71**, 066614 (2005)
24. Donley, E.A., Claussen, N.R., Thompson, S.T., Wieman, C.E.: Nature, **417**, 529 (2002)
25. Kevrekidis, P.G., Theocharis, G., Frantzeskakis, D.J., Malomed, B.A.: Phys. Rev. Lett. **90**, 230401 (2003)
26. Malomed, B.A.: Soliton Management in Periodic Systems. Springer, New York (2006)
27. Ott, E.: Chaos in Dynamical Systems. Cambridge University Press, Cambridge (1993)
28. Grimshaw, R., He, J., Malomed, B.A.: Phys. Scripta **53**, 385 (1996)
29. Peschel, U., Lederer, F.: J. Opt. Soc. Am. B **19**, 544 (2002)
30. Anderson, D.: Phys. Rev. A **27**, 3135 (1983)
31. Peyrard, M., Kruskal, M.D.: Physica D **14**, 88 (1984)

32. Malomed, B.A.: Phys. Rev. A **45**, 4097 (1992)
33. Bonilla, L.L., Malomed, B.A.: Phys. Rev. B **43**, 11539 (1991)
34. Kuusela, T., Hietarinta, J., Malomed, B.A.: J. Phys. A **26**, L21 (1993)
35. Filatrella, G., Malomed, B.A.: J. Phys. Cond. Matt. **11**, 7103 (1999)
36. Kuusela, T.: Chaos Sol. Fract. **5**, 2419 (1995)
37. Filatrella, G., Malomed, B.A., Parmentier, R.D.: Phys. Lett. A **198**, 43 (1995)
38. Ustinov, A.V., Malomed, B.A.: Phys. Rev. B **64**, 020302 (2001)
39. Mayteevarunyoo, T., Malomed, B.A., Krairiksh, M.: Phys. Rev. A **76**, 053612 (2007)
40. Burlak, G., Malomed, B.A.: Phys. Rev. A **77**, 053606 (2008)

Chapter 16
Exceptional Discretizations of the NLS: Exact Solutions and Conservation Laws

Sergey V. Dmitriev and Avinash Khare

16.1 Introduction

Discrete nonlinear equations that admit exact solutions are interesting from the mathematical point of view and they also help us understand the properties of some physically meaningful discrete nonlinear systems. Completely integrable discrete equations, such as the Ablowitz–Ladik (AL) lattice [1, 2], constitute one class of such equations. Recently, it has been realized by many researchers that *nonintegrable* lattice equations can have subtle symmetries that allow for particular *exact* solutions that propagate with particular velocities and interact with each other inelastically, in contrast to the AL solitons that propagate with arbitrary velocity and collide elastically. For vanishing velocity, one can talk about translationally invariant (TI) stationary solutions (i.e., stationary solutions with arbitrary shift along the lattice). Nonintegrable lattice equations supporting exact moving and/or TI stationary solutions are often called exceptional discrete (ED) models and a natural question is how to identify such models. The problem is often viewed differently, namely, one can look for exceptional solution (not model) parameters when a given lattice equation is satisfied exactly. Exact stationary and moving solutions to nonintegrable discrete nonlinear Schrödinger (DNLS) equations have been constructed and analyzed in a number of recent works [3–14].

In this contribution we first review the existing literature on the exact solutions to the nonintegrable DNLS equations and closely related discrete Klein–Gordon models. Then we report on some analytical and numerical results for the DNLS equations with general cubic nonlinearity.

16.2 Review of Existing Works

The search for exact solutions to the nonintegrable DNLS systems has been carried out in two main directions: the first one is the search for DNLS models supporting

---

S.V. Dmitriev (✉)
Institute for Metals Superplasticity Problems RAS, 450001 Ufa, Khalturina 39, Russia
e-mail: dmitriev.sergey.v@gmail.com

stationary TI solutions, while the second one is the search for exact *moving* solutions to various DNLS equations. Let us summarize the results of those studies.

16.2.1 Stationary Translationally Invariant Solutions

The problem at hand can be rephrased as follows: the aim is to discretize the generalized NLS equation of the form

$$iu_t + \frac{1}{2}u_{xx} + G'(|u|^2)u = 0, \quad (16.1)$$

in a way so that the resulting DNLS equation supports the TI stationary solutions. Here $G(\xi)$ is a real function of its argument and $G'(\xi) = dG/d\xi$. It is usually assumed that the DNLS equation that one is looking for has the form

$$i\frac{du_n}{dt} = -\varepsilon \Delta_2 u_n - f(u_{n-1}, u_n, u_{n+1}), \quad (16.2)$$

where $\Delta_2 u_n \equiv u_{n-1} - 2u_n + u_{n+1}$ is the discrete Laplacian, ε is the coupling constant, the nonlinear function f in the continuum limit ($\varepsilon \to \infty$) reduces to $G'(|u|^2)u$ and possesses the property

$$f(ae^{i\omega t}, be^{i\omega t}, ce^{i\omega t}) = f(a, b, c)e^{i\omega t}. \quad (16.3)$$

Seeking stationary solutions of Eq. (16.1) in the form

$$u(x, t) = F(x)e^{i\omega t}, \quad (16.4)$$

we reduce it to an ordinary differential equation (ODE) for the real function $F(x)$,

$$D(x) \equiv \frac{d^2 F}{dx^2} - 2\omega F + 2FG'(F^2) = 0, \quad (16.5)$$

and the problem of finding the ED NLS for (16.1) is effectively reduced to finding the ED forms of the above ODE. Suppose that a discrete analog of Eq. (16.5) that supports TI solutions (with arbitrary shift x_0) is found in the form

$$D(F_{n-1}, F_n, F_{n+1}) = 0, \quad (16.6)$$

then the original problem of discretization of Eq. (16.1) can be solved by substituting in Eq. (16.6), F_n with u_n or u_n^* in such a way so as to present it in the form of Eq. (16.2), satisfying the property as given by Eq. (16.3). Usually there are many possibilities to do so.

ED models supporting TI stationary solutions with arbitrary x_0 differ from conventional discrete models supporting only discrete sets of stationary solutions, located symmetrically with respect to the lattice, typically in the on-site and inter-site configurations, one of them being stable and corresponding to an energy minimum, while another one being unstable and corresponding to an energy maximum. The energy difference between these two states defines the height of the so-called Peierls–Nabarro potential (PN). TI stationary solutions do not experience this periodic potential and can be shifted quasi-statically along the chain without any energy loss. For the non-Hamiltonian models with path-dependent forces the discussion of the PN energy relief is more complicated but in this case too, zero work is required for quasi-static shift of a TI solution along the path corresponding to continuous change of x_0.

The problem of discretization of Eq. (16.5) in the form of Eq. (16.6) has been addressed in different contexts, and will be discussed in the rest of this section.

16.2.1.1 Integrable Maps

From the theory of integrable maps [15–17] it is known that some of the second-order difference equations of the form of Eq. (16.6) can be integrated, resulting in the first-order difference equation of the form $U(F_{n-1}, F_n, K) = 0$, where K is the integration constant. Such second-order difference equations can be regarded as exactly solvable because the solution F_n can now be found iteratively, starting from any admissible initial value F_0 and solving at each step the algebraic problem. Continuous variation of the initial value F_0 results in continuous shift x_0 of the corresponding stationary solution along the lattice.

Several years ago, an integrable map was shown to be directly related to the second-order difference equations supporting the Jacobi elliptic function (JEF) solutions [15]. In [16], for the nonlinear equation $d^2 F/dx^2 + aF + bF^3 = 0$, the discrete analog of the form $F_{n-1} - 2F_n + F_{n+1} + a[c_{11} F_n + c_{12}(F_{n-1} + F_{n+1})] + b[c_{21} F_{n-1} F_n F_{n+1} + c_{22} F_n^2 (F_{n-1} + F_{n+1})] = 0$ was studied and its two-point reduction was found to be of the QRT form [15]. This type of nonlinearity was later studied in [18] and its two-point reduction was rediscovered in [19]. The QRT map appears in many other studies of discrete models, for example, in [11, 20, 21]. Recent results on the integrable maps of the non-QRT type can be found in [17].

One interesting implementation of the theory of integrable maps can be found in [13] where the methodology of [22] was employed. In this work stationary solutions to the DNLS equation with saturable nonlinearity have been analyzed through the corresponding three-point map. It was found that for some selected values of model parameters, the map generates on the plane (F_n, F_{n+1}) a set of points belonging to a line, having topological dimension equal to one. This effective reduction of dimensionality of the map means the possibility of its two-point reduction, resulting in vanishing PN potential.

16.2.1.2 Exceptional Discrete Klein–Gordon Equations

Equation (16.5) can be viewed as the static version of the Klein–Gordon equation, $F_{tt} = F_{xx} - V'(F)$, with the potential function

$$V(F) = \omega F^2 - G\left(F^2\right). \tag{16.7}$$

Thus, the ED Klein–Gordon equation can be used to write down the ED NLS models (and vice versa).

The first successful attempt in deriving the ED Klein–Gordon equation was made by Speight and Ward [23–25] using the Bogomol'nyi argument [26], and also by Kevrekidis [27]. In both cases, the authors obtained the two-point reduction of the corresponding three-point discrete models. While the Speight and Ward discretization conserves the Hamiltonian, the Kevrekidis discretization conserves the classically defined momentum. These works have inspired many other investigations in this direction [18, 19, 21, 28–36]. Later it was found that both the models can be derived using the discretized first integral (DFI) approach [21, 29].

To illustrate the DFI approach, we write down the first integral of Eq. (16.5),

$$U(x) \equiv \left(F'\right)^2 - 2\omega F^2 + 2G\left(F^2\right) + K = 0, \tag{16.8}$$

where K is the integration constant, and discretize it as

$$U(F_{n-1}, F_n, K) \equiv \frac{(F_n - F_{n-1})^2}{h^2} - 2\omega F_{n-1} F_n + 2G(F_{n-1}, F_n) + K = 0. \tag{16.9}$$

It is assumed that, in the above equation, $G(F_{n-1}, F_n)$ reduces to $G(F^2)$ in the continuum limit. On discretizing the left-hand side of the identity $(1/2)dU/dF = D(x)$, we obtain the discrete version of Eq. (16.5),

$$D(F_{n-1}, F_n, F_{n+1}) \equiv \frac{U(F_n, F_{n+1}) - U(F_{n-1}, F_n)}{F_{n+1} - F_{n-1}} = 0. \tag{16.10}$$

Clearly, solutions to the three-point problem $D(F_{n-1}, F_n, F_{n+1}) = 0$ can be found from the two-point problem $U(F_{n-1}, F_n, K) = 0$. We note that Eq. (16.10) was first proposed in [27] and it was used in [3, 10] to derive ED for Eq. (16.1) conserving norm or modified norm and momentum.

16.2.1.3 Jacobi Elliptic Function Solutions

Some DNLS equations (and discrete Klein–Gordon equations) with cubic nonlinearity support exact TI stationary [6, 7, 9, 28, 37, 38] and even moving [6, 7, 11] solutions in terms of Jacobi elliptic functions (JEF). Special cases of these solutions describe the TI stationary or moving bright and dark solitons having sech and tanh profiles, respectively. Such solutions can be derived with the help of the JEF identities reported in [39].

JEF solutions are important in their own right, and besides, they also help in establishing the integrable nonlinearities of the QRT type [11]. It is worth pointing out that, so far, no JEF solutions are known to the Kevrekidis ED model given by

Eq. (16.10), thereby indicating that the integrable map to this model is perhaps of the non-QRT type.

An inverse approach to the general problem of finding the kink or the pulse-shaped traveling solutions to the lattice equations was developed by Flach and coworkers [40]. They showed that for a given wave profile, it is possible to generate the corresponding equations of motion. In this context, also see the earlier works [41, 42]. A similar idea has also been used in other studies, see e.g., [43].

16.2.2 Exact Moving Solutions to DNLS

Exact moving solutions to the different variants of the DNLS, as was already mentioned, have been derived in terms of the JEF [6, 7, 11]. They have also been found, for DNLS with generalized cubic and saturable nonlinearities, with the help of specially tuned numerical approaches [4, 5, 8, 14]. These works suggest that the moving soliton solutions can be expected in models where, for different model parameters (or/and soliton parameters), there is a transition between stable on-site and inter-site configurations for stationary solitons. Contrary to the DNLS with saturable nonlinearity, the solitons in the classical DNLS do not show such transition and moving solutions have not been found for this system [14].

In our recent work [11] on DNLS with general cubic nonlinearity, we have derived not only moving JEF but also moving sine solutions (also given here in Sect. 16.3). Exact, extended, sinusoidal solutions of the lattice equations have been recently found by several authors [43–47]. It has been proposed that such solutions can be used to construct approximate large-amplitude *localized* solutions by truncating the sine solutions [44, 48].

16.3 Cubic Nonlinearity

Here we discuss Eq. (16.2) with the function f given by

$$\begin{aligned}
f = & \alpha_1 |u_n|^2 u_n + \alpha_2 |u_n|^2 (u_{n+1} + u_{n-1}) + \alpha_3 u_n^2 \left(u_{n+1}^\star + u_{n-1}^\star \right) \\
& + \alpha_4 u_n \left(|u_{n+1}|^2 + |u_{n-1}|^2 \right) + \alpha_5 u_n \left(u_{n+1}^\star u_{n-1} + u_{n-1}^\star u_{n+1} \right) \\
& + \alpha_6 u_n^\star \left(u_{n+1}^2 + u_{n-1}^2 \right) + \alpha_7 u_n^\star u_{n+1} u_{n-1} + \alpha_8 \left(|u_{n+1}|^2 u_{n+1} + |u_{n-1}|^2 u_{n-1} \right) \\
& + \alpha_9 \left(u_{n-1}^\star u_{n+1}^2 + u_{n+1}^\star u_{n-1}^2 \right) + \alpha_{10} \left(|u_{n+1}|^2 u_{n-1} + |u_{n-1}|^2 u_{n+1} \right) \\
& + \alpha_{11} \left(|u_{n-1} u_n| + |u_n u_{n+1}| \right) u_n + \alpha_{12} \left(u_{n+1} |u_{n+1} u_n| + u_{n-1} |u_n u_{n-1}| \right) \\
& + \alpha_{13} \left(u_{n+1} |u_{n-1} u_n| + u_{n-1} |u_n u_{n+1}| \right) \\
& + \alpha_{14} \left(u_{n+1} |u_{n-1} u_{n+1}| + u_{n-1} |u_{n-1} u_{n+1}| \right), \quad (16.11)
\end{aligned}$$

where the real-valued parameters α_i satisfy the continuity constraint

$$\alpha_1 + \alpha_7 + 2(\alpha_2 + \alpha_3 + \alpha_4 + \alpha_5 + \alpha_6 + \alpha_8$$
$$+ \alpha_9 + \alpha_{10} + \alpha_{11} + \alpha_{12} + \alpha_{13} + \alpha_{14}) = \pm 2, \tag{16.12}$$

with the upper (lower) sign corresponding to a focusing (defocusing) nonlinearity. Note that Eq. (16.11) is the most general function with cubic nonlinearity which is symmetric under $u_{n-1} \leftrightarrow u_{n+1}$.

Particular cases of the nonlinearity (16.11) have been studied in a number of works, many of them are listed in the introduction of [8]. Nonlocal cubic terms coupling the nearest neighbor lattice points naturally appear in the DNLS models approximating continuous NLS with periodic coefficients [49].

16.3.1 Conservation Laws

It is easily shown that DNLS Eqs. (16.2) and (16.11) with arbitrary α_1, α_4, α_5, α_6, α_{11}, α_{12}, with $\alpha_2 = \alpha_3 + \alpha_8$, and $\alpha_7 = \alpha_9 = \alpha_{10} = \alpha_{13} = \alpha_{14} = 0$, conserve the norm

$$N = \sum_n u_n u_n^\star. \tag{16.13}$$

On the other hand, for arbitrary α_2, α_{14}, with $\alpha_1 + \alpha_6 = \alpha_4$, $\alpha_5 = \alpha_6$, $\alpha_4 + \alpha_5 = \alpha_7$, $\alpha_8 + \alpha_9 = \alpha_{10}$, $\alpha_{12} = \alpha_{13}$ and $\alpha_3 = \alpha_{11} = 0$, the model conserves the modified norm

$$N_1 = \sum_n \left(u_n u_{n+1}^\star + u_n^\star u_{n+1} \right). \tag{16.14}$$

Instead, if only α_7 is nonzero while all other $\alpha_i = 0$, then yet another type of modified norm, given by

$$N_2 = \sum_n \left(u_n u_{n+2}^\star + u_n^\star u_{n+2} \right), \tag{16.15}$$

is conserved.

Further, for arbitrary α_2 and α_3, with $\alpha_4 + \alpha_6 = \alpha_1$, $\alpha_5 = \alpha_6$, $\alpha_5 + \alpha_7 = \alpha_4$, $\alpha_9 + \alpha_{10} = \alpha_8$, and $\alpha_{11} = \alpha_{12} = \alpha_{13} = \alpha_{14} = 0$, Eqs. (16.2) and (16.11) conserve the momentum operator

$$P_1 = i \sum_n \left(u_{n+1} u_n^\star - u_{n+1}^\star u_n \right). \tag{16.16}$$

Instead, for arbitrary α_5 and α_7 while all other $\alpha_i = 0$, another type of momentum operator, given by

$$P_2 = i \sum_n \left(u_{n+2}u_n^* - u_{n+2}^*u_n\right), \tag{16.17}$$

is conserved.

On the other hand, for arbitrary α_1, α_4, and α_6, with $2\alpha_3 = 2\alpha_8 = \alpha_2$, and $\alpha_5 = \alpha_7 = \alpha_9 = \alpha_{10} = \alpha_{11} = \alpha_{12} = \alpha_{13} = \alpha_{14} = 0$, Eq. (16.2) with f given by Eq. (16.11) can be obtained from the Hamiltonian

$$H = \sum_n \left[|u_n - u_{n+1}|^2 - \frac{\alpha_1}{2}|u_n|^4 - \frac{\alpha_6}{2}\left[(u_n^*)^2 u_{n+1}^2 + (u_{n+1}^*)^2 u_n^2\right] \right.$$
$$\left. - \alpha_4 |u_n|^2 |u_{n+1}|^2 - \frac{\alpha_2}{2} \left(|u_n|^2 + |u_{n+1}|^2\right) \left(u_{n+1}^* u_n + u_n^* u_{n+1}\right) \right], \tag{16.18}$$

by using the equation of motion

$$i\dot{u}_n = [u_n, H]_{PB}, \tag{16.19}$$

where the Poisson bracket is defined by

$$[U, V]_{PB} = \sum_n \left[\frac{dU}{du_n} \frac{dV}{du_n^*} - \frac{dU}{du_n^*} \frac{dV}{du_n} \right]. \tag{16.20}$$

Thus in this model, the energy (H) is conserved.

Finally, in case one considers a rather unconventional Poisson bracket given by

$$[U, V]_{PB1} = \sum_n \left[\frac{dU}{du_n} \frac{dV}{du_n^*} - \frac{dU}{du_n^*} \frac{dV}{du_n} \right] \left[1 + (\alpha_2 - \alpha_3)|u_n|^2 \right.$$
$$\left. + \alpha_8 \left(|u_{n+1}|^2 + |u_{n-1}|^2\right) + \alpha_7 u_n \left(u_{n+1}^* + u_{n-1}^*\right) + \alpha_7 u_n^*(u_{n+1} + u_{n-1}) \right], \tag{16.21}$$

then the DNLS Eq. (16.2) with f given by Eq. (16.11) can be obtained from the Hamiltonian

$$H_1 = \sum_n \left[|u_n - u_{n+1}|^2 - \beta |u_n|^2 \right], \tag{16.22}$$

by using the equation of motion

$$i\dot{u}_n = [u_n, H_1]_{PB1}, \tag{16.23}$$

provided

$$\alpha_7 = 2\alpha_5 = 2\alpha_6, \quad \alpha_8 = \alpha_9 = \alpha_{10}, \quad \alpha_3 = (\beta - 2)\alpha_5, \quad \alpha_4 = (\beta - 2)\alpha_8 + \alpha_5,$$
$$\alpha_1 = (\beta - 2)(\alpha_2 - \alpha_3), \quad \alpha_{11} = \alpha_{12} = \alpha_{13} = \alpha_{14} = 0. \tag{16.24}$$

Thus the energy (H_1) is conserved in this model.

A few summarizing remarks are in order here:

1. In case only α_2 is nonzero while all other $\alpha_i = 0$, we have the integrable AL lattice with infinite number of conserved quantities. Among them are, e.g., N_1, P_1, and H_1 with $\beta = 2$, but not N, N_2, P_2, and H.
2. In the case of the conventional DNLS model (i.e., only $\alpha_1 \neq 0$), N and H are conserved.
3. The model with only α_7 nonzero conserves N_2 and P_2.
4. N and P_1 are conserved in case only α_2 and α_3 are nonzero and $\alpha_2 = \alpha_3$.
5. N_1 and P_1 are conserved in case α_2 is arbitrary while $\alpha_1 = \alpha_4 = \alpha_7$, $\alpha_8 = \alpha_{10}$ while other $\alpha_i = 0$.
6. The model conserving H also conserves N.
7. The model conserving H_1 also conserves N_1 in case $\beta = 2$ and $\alpha_8 = \alpha_9 = \alpha_{10} = 0$.

16.3.2 Two-Point Maps for Stationary Solutions

With the ansatz $u_n(t) = F_n e^{-i\omega t}$, we obtain the following difference equation from the DNLS Eqs. (16.2) and (16.11)

$$\varepsilon \left[F_{n-1} - (2 - \omega/\varepsilon)F_n + F_{n+1} \right] + \alpha_1 F_n^3 + \gamma_1 F_n^2 (F_{n-1} + F_{n+1})$$
$$+ \gamma_2 F_n \left(F_{n-1}^2 + F_{n+1}^2 \right) + \gamma_3 F_{n-1} F_n F_{n+1}$$
$$+ \alpha_8 \left(F_{n-1}^3 + F_{n+1}^3 \right) + \gamma_4 F_{n-1} F_{n+1} (F_{n-1} + F_{n+1}) = 0, \quad (16.25)$$

where, for convenience, we have introduced the following notation:

$$\gamma_1 = \alpha_2 + \alpha_3 + \alpha_{11}, \quad \gamma_2 = \alpha_4 + \alpha_6 + \alpha_{12},$$
$$\gamma_3 = 2\alpha_5 + \alpha_7 + 2\alpha_{13}, \quad \gamma_4 = \alpha_9 + \alpha_{10} + \alpha_{14}. \quad (16.26)$$

In the special case of

$$\alpha_8 = \gamma_4, \quad \alpha_1 = \gamma_2 = \gamma_3, \quad 2\alpha_1 + \gamma_1 + 2\alpha_8 = 1, \quad (16.27)$$

the first integral of the second-order difference Eq. (16.25) reduces to the two-point map

$$U(F_{n-1}, F_n, K) \equiv \varepsilon \left[(F_{n-1}^2 + F_n^2) - (2 - \omega/\varepsilon) F_{n-1} F_n \right]$$
$$+ \alpha_1 \left(F_{n-1}^2 + F_n^2 \right) F_{n-1} F_n + \gamma_1 F_{n-1}^2 F_n^2 + \alpha_8 \left(F_{n-1}^4 + F_n^4 \right) + K = 0, \quad (16.28)$$

where K is an integration constant. This is so because Eq. (16.25) can be rewritten in the form

$$\frac{U(F_n, F_{n+1}) - U(F_{n-1}, F_n)}{F_{n+1} - F_{n-1}} = 0, \qquad (16.29)$$

and clearly, if $U(F_{n-1}, F_n) = 0$, then indeed Eq. (16.25) is satisfied.

On the other hand, in case only γ_1 and γ_3 are nonzero while $\alpha_1 = \alpha_8 = \gamma_2 = \gamma_4 = 0$, then the two-point map is given by

$$W(F_{n-1}, F_n, K) \equiv F_{n-1}^2 + F_n^2 - \frac{Y F_{n-1}^2 F_n^2}{(2 - \omega/\varepsilon)} - 2Z F_{n-1} F_n - \frac{KY}{(2 - \omega/\varepsilon)} = 0, \qquad (16.30)$$

which is of the QRT form [15, 16]. Here K is an integration constant while

$$Z = \frac{(2 - \omega/\varepsilon)^2 - K\gamma_3^2}{2(2 - \omega/\varepsilon) + 2K\gamma_1\gamma_3}, \quad Y = 2\gamma_1 Z + \gamma_3. \qquad (16.31)$$

This is because, in this case, Eq. (16.25) can be rewritten in the form

$$\frac{(2 - \omega/\varepsilon)}{2Z(F_{n+1} - F_{n-1})} \Big\{ W(F_n, F_{n+1}) - W(F_{n-1}, F_n)$$

$$+ \frac{\gamma_3}{(2 - \omega/\varepsilon)} \big[F_{n+1}^2 W(F_{n-1}, F_n) - F_{n-1}^2 W(F_n, F_{n+1}) \big] \Big\} = 0, \qquad (16.32)$$

and clearly, if $W(F_{n-1}, F_n, K) = 0$, then indeed Eq. (16.3.2) is satisfied. As expected, in the special case of $\gamma_3 = 0$ so that only γ_1 is nonzero, Eq. (16.3.2) reduces to Eq. (16.29).

We want to emphasize that the two-point maps, Eqs. (16.28) and (16.30), allow one to find exact solutions to Eq. (16.25) iteratively, starting from any admissible value of F_0 and solving at each step an algebraic problem. Thus, such solutions define the exact TI stationary solutions to the DNLS Eq. (16.2) with the nonlinearity function f given by Eq. (16.11).

It is worth pointing out here that some of the exact stationary TI and non-TI solutions (specially the short period and the sine solutions) can also follow from factorized two-point and reduced three-point maps. Several examples of such solutions and their relation with short-period or aperiodic stationary solutions and even with the sine solution can be found in [36]. Here we give two illustrative examples of the TI solutions which follow from factorized two-point and reduced three-point maps.

It is easy to check that Eq. (16.25) has the exact period-four solution

$$F_n = (\ldots, a, b, -a, -b, \ldots), \qquad (16.33)$$

provided

$$2\gamma_2 = \alpha_1 + \gamma_3, \quad (a^2 + b^2)\alpha_1 = 2\varepsilon - \omega. \qquad (16.34)$$

Parameter a in this solution can vary continuously resulting in the shift of the solution with respect to the lattice, which means that this is a TI solution. Now we note that in case

$$\alpha_8 = \gamma_4 = \gamma_1 = 0, \quad \alpha_1 = \gamma_2 = \gamma_3, \quad 2\alpha_1 = 1, \quad K = 2(\omega - 2\varepsilon), \quad (16.35)$$

then the map as given by Eq. (16.28) can be factorized as

$$U(F_{n-1}, F_n) = \frac{1}{2}(2\varepsilon + F_{n-1}F_n)\left(\frac{2\omega}{\varepsilon} - 4 + F_{n-1}^2 + F_n^2\right) = 0. \quad (16.36)$$

Remarkably, the second factor of this two-point map satisfies the period-four solution (16.33) with the conditions (16.34). Note that the TI solution of Eq. (16.33) is equivalent to the sine solution as given by Eq. (16.50) with $\beta = \pi/2$ and $v = k = 0$.

Our next example is for the model following from the Hamiltonian of Eq. (16.18). In this case, Eq. (16.25) assumes the form

$$\varepsilon\left[F_{n-1} - (2 - \omega/\varepsilon)F_n + F_{n+1}\right] + \alpha_1 F_n^3 + \gamma_1 F_n^2 (F_{n-1} + F_{n+1})$$
$$+ \gamma_2 F_n (F_{n-1}^2 + F_{n+1}^2) + \frac{\gamma_1}{3}\left(F_{n-1}^3 + F_{n+1}^3\right) = 0. \quad (16.37)$$

Remarkably, in case the following two-point equation holds

$$F_{n-1}^2 + \frac{4\gamma_1}{3\alpha_1} F_{n-1} F_n + F_n^2 = B, \quad (16.38)$$

then the (stationary) difference Eq. (16.37) can be rewritten as

$$(B\gamma_1 + 3\varepsilon)(F_{n+1} + F_{n-1}) + 3(\omega - 2\varepsilon + B\alpha_1)F_n = 0, \quad (16.39)$$

provided

$$\gamma_2 = \frac{\alpha_1}{2} + \frac{4\gamma_1^2}{9\alpha_1}, \quad B = \frac{\omega - 2\varepsilon - \frac{4\gamma_1\varepsilon}{3\alpha_1}}{\frac{4\gamma_1^2}{9\alpha_1} - \alpha_1}. \quad (16.40)$$

One can now show that for the Hamiltonian model (16.18), the TI stationary sine solutions of Eq. (16.50) with $v = k = 0$ and with $v = 0, k = \pi$ also follow from the two-point map (16.38) provided

$$\cos(\beta) = -\frac{2\gamma_1}{3\alpha_1}. \quad (16.41)$$

Furthermore, in this case the three-point Eq. (16.39) is also automatically satisfied.

16.3.3 Moving Pulse, Kink, and Sine Solutions

The DNLS model given by Eqs. (16.2) and (16.11) supports exact *moving* JEF solutions, e.g., cn, dn, sn, in case

$$\alpha_1 = \alpha_8 = 0. \tag{16.42}$$

In the limit $m = 1$, where m is the JEF modulus, one obtains the hyperbolic, moving pulse and kink solutions. For $\alpha_{11} = \alpha_{12} = \alpha_{13} = \alpha_{14} = 0$, JEF solutions were given in [11] and below we give the hyperbolic and sine solutions including these terms.

In particular, the DNLS model given by Eqs. (16.2) and (16.11) supports the moving pulse (bright soliton) solution,

$$u_n = A \exp[-i(\omega t - kn + \delta)] \operatorname{sech}[\beta(n - vt + \delta_1)], \tag{16.43}$$

provided the parameters satisfy

$$v\beta = 2\varepsilon s_1 S, \quad \omega = 2\varepsilon(1 - c_1 C),$$
$$2\xi_6 C + \xi_5 = 0, \quad \left[S^2 + (\alpha_3 - \alpha_2)A^2\right]s_1 = 0,$$
$$2\xi_2 C + \xi_4 = 0, \quad A^2(\xi_1 C - \xi_2 + \xi_3/2) = \varepsilon S^2 C c_1. \tag{16.44}$$

Here δ and δ_1 are arbitrary constants, A, ω, k, β, and v denote the amplitude, frequency, wavenumber, inverse width, and velocity of the moving pulse, respectively, and the following compact notation has been used to describe the relations between the parameters of the exact moving solutions:

$$S = \sinh(\beta), \quad C = \cosh(\beta), \quad T = \tanh(\beta),$$
$$s_1 = \sin(k), \quad s_2 = \sin(2k), \quad s_3 = \sin(3k),$$
$$c_1 = \cos(k), \quad c_2 = \cos(2k), \quad c_3 = \cos(3k).$$
$$\xi_1 = (\alpha_2 + \alpha_3)c_1 + \alpha_{11}, \quad \xi_2 = \alpha_4 + \alpha_6 c_2 + \alpha_{12} c_1,$$
$$\xi_3 = 2\alpha_5 c_2 + \alpha_7 + 2\alpha_{13} c_1, \quad \xi_4 = \alpha_9 c_3 + (\alpha_{10} + \alpha_{14})c_1,$$
$$\xi_5 = \alpha_9 s_3 - \alpha_{10} s_1 + \alpha_{14} s_1, \quad \xi_6 = \alpha_6 s_2 + \alpha_{12} s_1. \tag{16.45}$$

From the first expression in Eq. (16.44) it follows that the pulse velocity is zero when $k = 0$ or π. In the former case we have the nonstaggered, stationary pulse solution while in the latter case, we have the staggered, stationary pulse solution. In particular, for $v = k = 0$, the pulse solution is given by

$$u_n = A \exp[-i(\omega t + \delta)] \operatorname{sech}[\beta(n + \delta_1)], \tag{16.46}$$

provided $2\gamma_2 C + \gamma_4 = 0$; $A^2(\gamma_1 C - \gamma_2 + \gamma_3/2) = \varepsilon S^2 C$; $\omega = 2\varepsilon(1 - C)$.

On the other hand, the exact moving kink solution to Eqs. (16.2) and (16.11) given by

$$u_n = A \exp[-i(\omega t - kn + \delta)] \tanh[\beta(n - vt + \delta_1)], \qquad (16.47)$$

exists provided

$$v\beta = 2\varepsilon s_1 T + \frac{4A^2 \xi_6 T^3}{(1+T^2)}, \quad \frac{(\omega - 2\varepsilon)}{A^2} = \frac{2\xi_1}{S^2} - \frac{2\xi_2}{T^2} + \frac{\xi_3}{T^2},$$

$$2\xi_6 + \xi_5(1+T^2) = 0, \quad \frac{\varepsilon s_1}{A^2} = -\frac{(\alpha_2 - \alpha_3)s_1}{T^2} - \frac{2\xi_6 T^2}{(1+T^2)},$$

$$2\xi_2 + \xi_4(1+T^2) = 0, \quad \frac{\varepsilon c_1}{A^2} = -\frac{\xi_1}{T^2} + \frac{\xi_2(1+2T^2-T^4)}{T^2(1+T^2)} - \frac{\xi_3(1+T^2)}{2T^2}.$$

$$(16.48)$$

For $k = 0$ we obtain the nonstaggered, stationary kink solution

$$u_n = A \exp[-i(\omega t + \delta)] \tanh[\beta(n + \delta_1)], \qquad (16.49)$$

provided $2\gamma_2 + \gamma_4(1+T^2) = 0$; $(\omega - 2\varepsilon)/A^2 = 2\gamma_1/S^2 - (2\gamma_2 - \gamma_3)/T^2$; $\varepsilon/A^2 = -\gamma_1/T^2 - \gamma_3(1+T^2)/(2T^2) + \gamma_2(1+2T^2-T^4)/[T^2(1+T^2)]$.

Unlike the JEF and the hyperbolic solutions, the moving as well as the stationary trigonometric solutions of Eqs. (16.2) and (16.11) exist even when all α_i are nonzero. In particular, the moving sine solution given by

$$u_n = A \exp[-i(\omega t - kn + \delta)] \sin[\beta(n - vt + \delta_1)], \qquad (16.50)$$

exists provided the following four relations are satisfied: $v\beta = -2\varepsilon \sin(\beta)s_1 - 2A^2 \sin^3(\beta)(\alpha_8 s_1 - \xi_5)$; $(\alpha_2 - \alpha_3)s_1 + 2\xi_6 \cos(\beta) - \alpha_8 s_1[4\sin^2(\beta) - 3] + \xi_5 = 0$; $\alpha_1 + 2(\xi_1 + \xi_4)\cos(\beta) + \xi_3 + 2\xi_2 \cos(2\beta) + 2\alpha_8 c_1 \cos(\beta)[1 - 4\sin^2(\beta)] = 0$; $\omega - 2\varepsilon + 2\varepsilon c_1 \cos(\beta) = A^2 \sin^2(\beta)[-2\xi_2 + \xi_3 - 6\alpha_8 c_1 \cos(\beta) + 2\xi_4 \cos(\beta)]$.

16.3.4 Stationary TI Solutions

Small-amplitude vibrational spectra calculated for stationary solutions to DNLS satisfying Eq. (16.3) always include a pair of zero-frequency eigenmodes reflecting the invariance with respect to the phase shift. Stationary TI solutions possess two additional zero-frequency modes in their linear spectra (the Goldstone translational modes) [3, 9, 11, 29]. Stationary solutions can be set in slow motion with the use of the TI eigenvectors whose amplitudes are proportional to propagation velocity (see, e.g., [9]). The accuracy of such slowly moving solutions increases with the decrease in the amplitude of the Goldstone translational mode, i.e., it increases with the decrease in propagation velocity.

Mobility of the bright and the dark solitons at small, as well as at finite velocities have been studied numerically, for the models supporting TI solutions, e.g., in [3, 9, 11]. TI coherent structures are not trapped by the lattice [3] and they can be accelerated by even a weak external field [33].

Properties of solitons in the DNLS with the nearest neighbor coupling in the nonlinear term [as in Eq. (16.11)] differ considerably from the classical DNLS with only α_1 nonzero where there is only on-site nonlinear coupling. For example, the TI dark solitons in case only α_2 and α_3 are nonzero do not survive the continuum limit, while in the classical DNLS they do [9]. On the other hand, in the classical DNLS, only highly localized on-site dark solitons are stable while the inter-site ones are unstable at any degree of discreteness [50, 51]. In the DNLS with only α_2 and α_3 nonzero, TI dark solitons can be robust, movable, and they can survive collisions with each other [9].

To illustrate the above-mentioned features of the stationary TI solutions, in Fig. 16.1 we show slowly moving, highly localized (a) bright and (b) colliding dark solitons (kink and antikink) in the nonintegrable lattices with $\varepsilon = 1/4$ and (a) $\alpha_2 = \alpha_3 = \alpha_{11} = 1/3$, with other $\alpha_i = 0$; (b) $\alpha_2 = \alpha_3 = 1/2$, with other $\alpha_i = 0$. Space–time evolution of $|u_n(t)|^2$ is shown and in both cases maximal $|u_n|^2$ is nearly equal to 1. To boost the solitons we used the zero-frequency Goldstone translational eigenmode with a small amplitude, which is proportional to the soliton velocity [9].

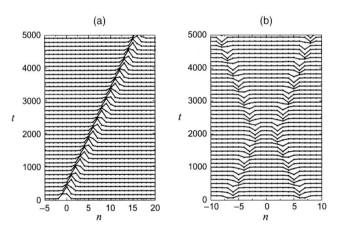

Fig. 16.1 Space-time evolution of $|u_n(t)|^2$ showing (**a**) moving bright and (**b**) moving and colliding dark solitons (kink and antikink) in nonintegrable lattices. To obtain the slowly moving solitons we used the stationary TI bright and dark soliton solutions supported by the corresponding ED models and boosted them applying the zero-frequency Goldstone translational eigenmode with a small amplitude, which is proportional to the soliton velocity. Parameters are given in the text. (After [9]; © 2007 IOP.)

16.3.5 Moving Bright Solitons

Properties of the moving bright solitons with the parameters satisfying (16.44) have been discussed in [11], in case $\alpha_{11} = \alpha_{12} = \alpha_{13} = \alpha_{14} = 0$. Here we reproduce some of the results of that work.

From (16.44) it follows that the moving bright soliton solution Eq. (16.43) exists, for example, in case only $\alpha_3, \alpha_5, \alpha_7$ are nonzero. While the first two relations in (16.44) are always valid [see Fig. 16.2 (a), (b) for the corresponding plots], the other relations and the constraint (16.12) take the form

$$\alpha_3 = \frac{1 - 2\alpha_5 s_1^2}{1 - 2c_1 C}, \quad \alpha_7 = 2(1 - \alpha_3 - \alpha_5), \quad A^2 = \frac{c_1 S^2 C}{1 + \alpha_3(c_1 C - 1) - 2\alpha_5 s_1^2}. \quad (16.51)$$

The number of constraints in this case is such that one has a free model parameter, say α_5, and pulse parameters k and β can change continuously within a certain domain [see Fig. 16.2 (c)]. It turns out that in this case, while the nonstaggered stationary pulse ($k = 0$) exists, the staggered stationary pulse ($k = \pi$) does not exist.

On the other hand, in case only $\alpha_2, \alpha_3, \alpha_5$ are nonzero we have the following constraints:

$$\alpha_3 = -\frac{\alpha_5 c_2}{2c_1 C}, \quad \alpha_2 = 1 - \alpha_3 - \alpha_5, \quad A^2 = \frac{c_1 S^2 C}{(\alpha_2 + \alpha_3) c_1 C + \alpha_5 c_2}. \quad (16.52)$$

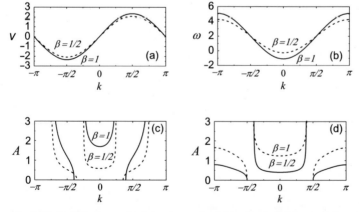

Fig. 16.2 (a) Velocity v, (b) frequency ω, and (c), (d) amplitude A of the pulse (bright soliton) as functions of the wavenumber parameter k at $\varepsilon = 1$ for the inverse width of the pulse $\beta = 1/2$ (*dashed lines*) and $\beta = 1$ (*solid lines*). The functions in (a), (b) are defined by the first two expressions in (16.44) and they do not depend on the model parameters α_i. To plot the amplitude A we set $\alpha_5 = 1$ and use (c) (16.51) and (d) (16.52). (After [11]; © 2007 IOP.)

The relation between pulse parameters and model parameters in this case is shown in Fig. 16.2 (d). In this case one has both nonstaggered and staggered stationary pulse solutions in case $k = 0$ and $k = \pi$, respectively.

In both cases, i.e., when Eqs. (16.51) and (16.52) are satisfied, it was found that the stationary bright solitons are generically stable.

The robustness of *moving* pulse solutions, in both these cases, was checked by observing the evolution of their velocity in a long-term numerical run for $\varepsilon = 1$. For pulses with amplitudes $A \sim 1$ and velocities $v \sim 0.1$ and for various model parameters supporting the pulse, $|\alpha_i| \sim 1$, we found that the pulse typically preserves its velocity with a high accuracy. Two examples of such simulations, one for the nonstaggered pulse and another one for the staggered pulse are given in Fig. 16.3 (a), (b) and (a'), (b'), respectively. In (a) and (a') we show the pulse configuration at $t = 0$ and in (b) and (b') the pulse velocity as a function of time for two different integration steps, $\tau = 5 \times 10^{-3}$ (solid lines) and $\tau = 2.5 \times 10^{-3}$ (dashed lines), while a numerical scheme with the accuracy $O(\tau^4)$ is employed.

In Fig. 16.3 (a,b) and (a',b') we give the numerical results for the pulse solutions given by Eqs. (16.51) and (16.52), respectively. The model characterized by Eq. (16.51) has one free parameter and we set $\alpha_5 = 1$. For the pulse parameters we set $\beta = 1$ and $k = 0.102102$. Then we find from the first two expressions in (16.44) and from (16.51) the pulse velocity $v = 0.239563$, frequency $\omega = -1.07009$, amplitude $A = 1.7087$, and the dependent model parameters $\alpha_3 = -0.473034$ and $\alpha_7 = 0.946068$. The model characterized by Eq. (16.52) has one free parameter and

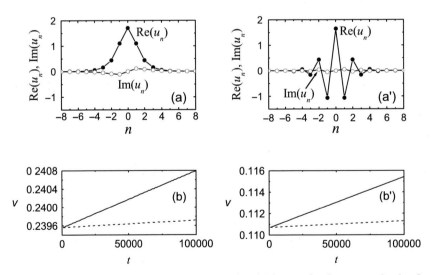

Fig. 16.3 (a) Nonstaggered moving pulse at $t = 0$ and (a') same for the staggered pulse. In (b) and (b') the long-term evolution of pulse velocity is shown for the corresponding pulses for the integration steps of $\tau = 5 \times 10^{-3}$ (*solid line*) and $\tau = 2.5 \times 10^{-3}$ (*dashed line*). Numerical scheme with an accuracy $O(\tau^4)$ is employed. In both models we find that the pulses preserve their velocity with the accuracy increasing with the increase in the accuracy of numerical integration. Parameters are given in the text. (After [11]; © 2007 IOP.)

we set $a_5 = 0.3$. For the pulse parameters we set $\beta = 1$ and $k = 3.09447$. Then we find from the first two expressions in (16.44) and from (16.52) the pulse velocity $v = 0.110719$, frequency $\omega = 5.08274$, amplitude $A = 1.65172$, and the dependent model parameters $\alpha_2 = 0.603116$ and $\alpha_3 = 0.0968843$.

In both cases, one can notice the linear increase in the pulse velocity with time, which is due to the numerical error, since the slope of the line decreases with the decrease in τ. The presence of perturbation in the form of rounding errors and integration scheme errors does not result in pulse instability within the numerical run. The velocity increase rate for the staggered pulse in (b') is larger than for the nonstaggered one in (b). This can be easily understood because the frequency of the staggered pulse is almost five times that of the nonstaggered one.

16.4 Conclusions and Future Challenges

In this contribution, we have given an overview of the recently reported exact stationary and moving solutions to nonintegrable discrete equations. Such solutions appear to be ubiquitous and they play an important role in our understanding of discrete nonlinear systems.

TI stationary solutions are potentially interesting for applications because they are not trapped by the lattice or, in other words, the PN barrier for them is exactly equal to zero. As a result, they can be accelerated by weak external fields. Such solutions possess the Goldstone translational mode, and thus, they can be boosted along this mode and can propagate at slow speed.

Exact moving solutions to discrete nonlinear equations are interesting in those cases where soliton mobility is an important issue. Such solutions indicate the "windows" in model and/or soliton parameters with enhanced mobility of solitons.

These studies open a number of new problems and research directions. Particularly, it would be interesting to look for the exact TI stationary or moving solutions in discrete systems other than DNLS and discrete Klein–Gordon equation. It would also be of interest to systematically examine the stability and other physical properties of the exact solutions to nonintegrable lattices. Finally, generalizing such approaches to higher dimensions and attempting to obtain analytical solutions in the latter context would constitute another very timely direction for future work.

Acknowledgments Our colleagues and friends D. Frantzeskakis, L. Hadžievski, P. G. Kevrekidis, A. Saxena, A. A. Sukhorukov, late S. Takeno, and N. Yoshikawa are greatly thanked for the collaboration on this topic.

References

1. Ablowitz, M.J., Ladik, J.F.: J. Math. Phys. **16**, 598 (1975)
2. Ablowitz, M.J., Ladik, J.F.: J. Math. Phys. **17**, 1011 (1976)
3. Dmitriev, S.V., Kevrekidis, P.G., Sukhorukov, A.A., Yoshikawa, N., Takeno, S.: Phys. Lett. A **356**, 324 (2006)

4. Melvin, T.R.O., Champneys, A.R., Kevrekidis, P.G., Cuevas, J.: Physica D **237**, 551 (2008)
5. Melvin, T.R.O., Champneys, A.R., Kevrekidis, P.G., Cuevas, J.: Phys. Rev. Lett. **97**, 124101 (2006)
6. Khare, A., Rasmussen, K.Ø., Samuelsen, M.R., Saxena, A.: J. Phys. A **38**, 807 (2005)
7. Khare, A., Rasmussen, K.Ø., Salerno, M., Samuelsen, M.R., Saxena, A.: Phys. Rev. E **74**, 016607 (2006)
8. Pelinovsky, D.E.: Nonlinearity **19**, 2695 (2006)
9. Dmitriev, S.V., Kevrekidis, P.G., Yoshikawa, N., Frantzeskakis, D.: J. Phys. A **40**, 1727 (2007)
10. Kevrekidis, P.G., Dmitriev, S.V., Sukhorukov, A.A.: Math. Comput. Simulat. **74**, 343 (2007)
11. Khare, A., Dmitriev, S.V., Saxena, A.: J. Phys. A **40**, 11301 (2007)
12. Maluckov, A., Hadžievski, L., Malomed, B.A.: Phys. Rev. E **77**, 036604 (2008)
13. Maluckov, A., Hadžievski, L., Stepić, M. Physica D **216**, 95 (2006)
14. Oxtoby, O.F., Barashenkov, I.V.: Phys. Rev. E **76**, 036603 (2007)
15. Quispel, G.R.W., Roberts, J.A.G., Thompson, C.J.: Physica D **34**, 183 (1989)
16. Hirota, R., Kimura, K., Yahagi, H.: J. Phys. A **34**, 10377 (2001)
17. Joshi, N., Grammaticos, B., Tamizhmani, T., Ramani, A. Lett. Math. Phys. **78**, 27 (2006)
18. Barashenkov, I.V., Oxtoby, O.F., Pelinovsky, D.E.: Phys. Rev. E **72**, 35602R (2005)
19. Dmitriev, S.V., Kevrekidis, P.G., Khare, A., Saxena, A.: J. Phys. A **40**, 6267 (2007)
20. Bender, C.M., Tovbis, A.: J. Math. Phys. **38**, 3700 (1997)
21. Dmitriev, S.V., Kevrekidis, P.G., Yoshikawa, N., Frantzeskakis, D.J.: Phys. Rev. E **74**, 046609 (2006)
22. Hennig, D., Rasmussen, K.O., Gabriel, H., Bulow, A.: Phys. Rev. E **54**, 5788 (1996)
23. Speight, J.M., Ward, R.S.: Nonlinearity **7**, 475 (1994)
24. Speight, J.M.: Nonlinearity **10**, 1615 (1997)
25. Speight, J.M.: Nonlinearity **12**, 1373 (1999)
26. Bogomol'nyi, E.B.: J. Nucl. Phys. **24**, 449 (1976)
27. Kevrekidis, P.G.: Physica D **183**, 68 (2003)
28. Cooper, F., Khare, A., Mihaila, B., Saxena, A.: Phys. Rev. E **72**, 36605 (2005)
29. Dmitriev, S.V., Kevrekidis, P.G., Yoshikawa, N.: J. Phys. A **38**, 7617 (2005)
30. Dmitriev, S.V., Kevrekidis, P.G., Yoshikawa, N.: J. Phys. A **39**, 7217 (2006)
31. Oxtoby, O.F., Pelinovsky, D.E., Barashenkov, I.V.: Nonlinearity **19**, 217 (2006)
32. Speight, J.M., Zolotaryuk, Y.: Nonlinearity **19**, 1365 (2006)
33. Roy, I., Dmitriev, S.V., Kevrekidis, P.G., Saxena, A.: Phys. Rev. E **76**, 026601 (2007)
34. Dmitriev, S.V., Khare, A., Kevrekidis, P.G., Saxena, A., Hadžievski, L.: Phys. Rev. E **77**, 056603 (2008)
35. Barashenkov, I.V., van Heerden, T.C.: Phys. Rev. E **77**, 036601 (2008)
36. Khare, A., Dmitriev, S.V., Saxena, A.: Exact Static Solutions of a Generalized Discrete ϕ^4 Model Including Short-Periodic Solutions (2007) arXiv:0710.1460.
37. Khare, A., Saxena, A.: J. Math. Phys. **47**, 092902 (2006)
38. Ross, K.A., Thompson, C.J.: Physica A **135**, 551 (1986)
39. Khare, A., Lakshminarayan A., Sukhatme, U.P.: Pramana (J. Phys.) **62**, 1201 (2004); math-ph/0306028.
40. Flach, S., Zolotaryuk, Y., Kladko, K.: Phys. Rev. E **59**, 6105 (1999)
41. Schmidt, V.H.: Phys. Rev. B **20**, 4397 (1979)
42. Jensen, M.H., Bak, P., Popielewicz, A.: J. Phys. A **16**, 4369 (1983)
43. Comte, J.C., Marquie, P., Remoissenet, M.: Phys. Rev. B **60**, 7484 (1999)
44. Kosevich, Yu.A.: Phys. Rev. Lett. **71**, 2058 (1993)
45. Chechin, G.M., Novikova, N.V., Abramenko, A.A.: Physica D **166**, 208 (2002)
46. Rink, B.: Physica D **175**, 31 (2003)
47. Shinohara, S.: J. Phys. Soc. Jpn. **71**, 1802 (2002)
48. Kosevich, Yu.A., Khomeriki, R., Ruffo, S.: Europhys. Lett. **66**, 21 (2004)

49. Abdullaev, F.Kh., Bludov, Yu.V., Dmitriev, S.V., Kevrekidis, P.G., Konotop, V.V.: Phys. Rev. E **77**, 016604 (2008)
50. Fitrakis, E.P., Kevrekidis, P.G., Susanto, H., Frantzeskakis, D.J.: Phys. Rev. E **75**, 066608 (2007)
51. Johansson, M., Kivshar, Yu.S.: Phys. Rev. Lett. **82**, 85 (1999)

Chapter 17
Solitary Wave Collisions

Sergey V. Dmitriev and Dimitri J. Frantzeskakis

17.1 Introduction and Setup

The well-known (see, e.g., [1] and references therein) elastic nature of the interaction among solitons in the completely integrable one-dimensional (1D) nonlinear Schrödinger (NLS) equation or in the completely integrable *Ablowitz–Ladik* (AL) lattice [2, 3] generally ceases to exist when perturbations come into play. This is due to the fact that, generally, perturbations are destroying the complete integrability and as a result many different effects in soliton interactions come into play. More specifically, in perturbed continuous (or discrete) NLS equations the outcome of the the collision process (i.e., the soliton trajectories and soliton characteristics) depends on the phase difference between two colliding solitons, emission of continuum radiation during soliton collisions, as well as the excitation of solitons' internal modes.

Here, we will discuss soliton collisions in the discrete NLS equation in the usual form,

$$i\dot{u}_n = -C\Delta_2 u_n - |u_n|^2 u_n. \qquad (17.1)$$

We will firstly discuss the case of *weak discreteness*, $C \gg 1$, which is a nearly integrable case: in the limit $C \to \infty$, Eq. (17.1) becomes the integrable 1D continuous NLS equation $i\partial_t u = -(1/2)\partial_x^2 u - |u|^2 u$, while the effect of a weak discreteness may be partially accounted for by the perturbation term proportional to $\partial_x^4 u$. The effect of such a weak discreteness on soliton collisions has been studied in Refs. [4, 5]. Then the case of *strong discreteness*, $C \sim 1$, will be discussed following the results reported in [6] for Eq. (17.1), and also the results reported in [7] for the following model:

$$i\dot{u}_n = -C\Delta_2 u_n - \delta|u_n|^2 u_n - \frac{1-\delta}{2}|u_n|^2(u_{n+1} + u_{n-1}) + \varepsilon|u_n|^4 u_n, \qquad (17.2)$$

S.V. Dmitriev (✉)
Institute for Metals Superplasticity Problems RAS, 450001 Ufa, Khalturina 39, Russia
e-mail: dmitriev.sergey.v@gmail.com

where δ, ε are two different perturbation parameters, providing tunable degree of nonintegrability. In particular, for $\delta = \varepsilon = 0$, Eq. (17.2) is reduced to the AL lattice [2, 3], which is integrable even in the case of strong discreteness. On the other hand, for $\varepsilon = 0$, Eq. (17.2) is reduced to the so-called *Salerno model* [8]. For $\varepsilon = 0$ and $\delta = 1$, Eq. (17.2) reduces to Eq. (17.1). Equation (17.2) has two integrals of motion, namely it conserves a modified norm and energy (Hamiltonian) [7]. For $\delta = \varepsilon = 0$ it supports the exact AL soliton solution [2, 3]

$$u_n(t) = \sqrt{2C} \sinh \mu \frac{\exp[ik(n-x) + i\alpha]}{\cosh[\mu(n-x)]}, \quad x = x_0 + \frac{2Ct}{\mu} \sinh \mu \sin k,$$

$$\alpha = \alpha_0 + 2Ct \left[\cos k \cosh \mu + \frac{k}{\mu} \sinh \mu \sin k - 1 \right], \tag{17.3}$$

where the parameters x_0 and α_0 are the initial coordinate and phase of the soliton, respectively, while the soliton's inverse width μ and the parameter k define the soliton's amplitude A and velocity V through the equations:

$$A = \sqrt{2C} \sinh \mu, \quad V = \frac{\sqrt{2C}}{\mu} \sinh \mu \sin k. \tag{17.4}$$

Finally, we review the results reported in the literature for the collisions of solitons in some physically relevant settings where discrete NLS equations are key models. These settings include optical waveguide arrays (in the optics context), and Bose–Einstein condensates (BECs) confined in optical lattices (in the atomic physics context).

17.2 Collisions in the Weakly Discrete NLS Equation

Examples of two-soliton collisions in the weakly discrete Eq. (17.1) (for $C = 15$) are presented in Fig. 17.1. Initial conditions were set employing the exact two-soliton solution to the integrable NLS equation [9, 10]. The out-of-phase collision in (a) and the in-phase collision in (b) are practically elastic, but they are very different in the sense that in (a) solitons repel each other and their cores do not merge at the collision point while in (b) the situation is opposite. The collision in (b) corresponds to the separatrix two-soliton solution to the integrable NLS equation [5]. Near-separatrix (nearly in-phase) collisions are strongly inelastic, as exemplified in (c), where the solitons emerge from the collision with different amplitudes and velocities. In (d), solitons' amplitudes after collision, $\tilde{A}_i$, are presented as the functions of the initial phase difference $\Delta\alpha_0$. Note the extreme sensitivity of the collision outcome to the initial phase difference $\Delta\alpha_0$ for the near-separatrix collisions ($\Delta\alpha_0 \approx 0$). Of particular importance is the fact that in the inelastic near-separatrix collisions in the regime of weak discreteness the energy given to the soliton's internal modes and to the radiation is negligible in comparison to the energy exchange between the

17 Solitary Wave Collisions

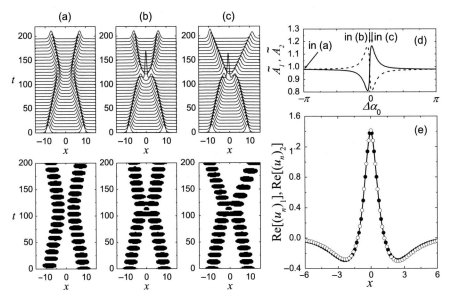

Fig. 17.1 Collisions in the weakly discrete NLS equation Eq. (17.1) at $C = 15$. *Upper panels* in (**a**)–(**c**) show $|u(x,t)|^2$, while *bottom panels* show the regions with $\text{Re}[u(x,t)] > 0.3$ in order to reveal the difference in the relative initial phase $\Delta\alpha_0$ of solitons: in (**a**), (**b**), and (**c**), $\Delta\alpha_0 = -\pi$, 0, and 0.1, respectively. (**d**) Soliton amplitudes after collision $\tilde{A}_i$ as functions of $\Delta\alpha_0$. The parameters of the solitons before the collision are $A_1 = A_2 = 0.98$, $V_1 = -V_2 = 0.05$. (**e**) Comparison of the real parts for two different exact two-soliton solutions to the NLS equation at the collision point. Imaginary parts are similarly close. The first solution (*dots*) has $A_1 = A_2 = 1$, $V_1 = -V_2 = 0.01$, while the second one (*circles*) has $A_1 = 1.1$, $A_2 = 0.9$, $V_1 = 0.0909$, $V_2 = -0.1111$. These two solutions have the same norm and momentum and $\Delta\alpha_0 = 0$. (After Ref. [5]; © 2002 APS.)

solitons [4]. This is the main feature of the so-called radiationless energy exchange (REE) effect in soliton collisions [11]. The REE in near-separatrix collisions can be understood by the fact that the profiles of two different two-soliton solutions to integrable NLS equation can be very close to each other at the collision point. An example is given in (e) by comparing the real parts of solutions with $A_1 = A_2 = 1$, $V_1 = -V_2 = 0.01$ (dots) and $A_1 = 1.1$, $A_2 = 0.9$, $V_1 = 0.0909$, $V_2 = -0.1111$ (circles). These two solutions have the same norm and momentum and $\Delta\alpha_0 = 0$. Their imaginary parts are similarly close. The presence of even weak perturbation can easily transform such close solutions one into another without violation of the conservation laws remaining in the weakly perturbed system.

For sufficiently small collision velocity, the REE effect can result in the fractal soliton scattering in the weakly discrete NLS equation [4, 5]. Fractal soliton scattering in the weakly perturbed NLS equation was explained qualitatively in the frame of a very simple model [4] and for the generalized NLS equation in the context of a more elaborate collective variable approach [12, 13], based on the method of Karpman and Solov'ev [14].

In Fig. 17.2a the solitons' velocities after collision, $\tilde{V}_i$, are shown as functions of the initial phase difference $\Delta\alpha_0$ for Eq. (17.1) at $C = 25$. Initial velocities and

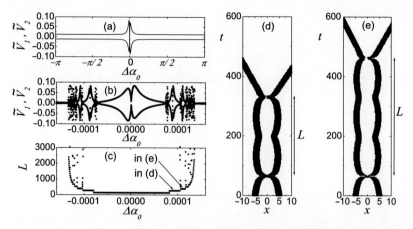

Fig. 17.2 Fractal soliton scattering in the weakly discrete NLS equation (17.1) at $C = 25$. (**a**) Soliton velocities after collision, $\tilde{V}_i$, as functions of initial phase difference $\Delta\alpha_0$. (**b**) Blowup of the region in (a) in the vicinity of $\Delta\alpha_0 = 0$. (**c**) Lifetime of the two-soliton bound states as function of $\Delta\alpha_0$. (**d**), (**e**) Examples of two-soliton bound states with different lifetime L. Regions of the x, t plane with $|u_n|^2 > 0.3$ are shown. The corresponding values of the initial phase difference $\Delta\alpha_0$ are indicated in (c). Initial soliton velocities are $V_1 = -V_2 = 0.012$ and initial amplitudes are $A_1 = A_2 = 1$

amplitudes are $V_1 = -V_2 = 0.012$ and $A_1 = A_2 = 1$, respectively. The collisions are inelastic in the vicinity of $\Delta\alpha_0 = 0$ where the solitons' velocities after collision differ considerably from their initial velocities. Blowup of the narrow region in the vicinity of $\Delta\alpha_0 = 0$ presented in (b) reveals a complex behavior of $\tilde{V}_i(\Delta\alpha_0)$. Smooth regions of these functions are separated by apparently chaotic regions. However, any chaotic region being expanded reveals the property of self-similarity at different scales (not shown in Fig. 17.2 but can be found in [4, 5]). The fractal soliton scattering can be explained through the following two facts: (i) in a weakly discrete system, the solitons attract each other with a weak force and (ii) the REE between colliding solitons is possible. As it is clearly seen from Fig. 17.2b, the chaotic regions appear where $\tilde{V}_i$ in smooth regions become zero. In these regions, the solitons after collision gain very small velocities so that they cannot overcome their mutual attraction and collide again. In the second collision, due to the momentum exchange, the solitons can acquire an amount of energy sufficient to escape each other, but there exists a finite probability to gain the energy below the escape limit. In the latter case, the solitons will collide for a third time, and so on. Physically, the multiple collisions of solitons can be regarded as the two-soliton bound state with certain lifetime L (two examples are given in Fig. 17.2d, e). The probability P of the bound state with the lifetime L was estimated to be $P \sim L^{-3}$ and this rate of decreasing of P with increase in L does not depend on the parameters of the colliding solitons [4].

Importantly, the REE effect in near-separatrix collisions has been predicted from the analysis of the two-soliton solution to the *unperturbed* integrable NLS equation [5] and thus, the precise form of the perturbation is not really important for the appearance of this effect. Moreover, it has been demonstrated that in the systems with more than one perturbation term, the collisions can be nearly elastic even in the vicinity of $\Delta\alpha_0 = 0$ when the effects of different perturbations cancel each other [5, 7].

17.3 Collisions in the Strongly Discrete NLS Equation

Let us now consider the strongly discrete NLS Eq. (17.1) with the coupling constant C now assumed to be a parameter of order $O(1)$. In such a case, soliton collisions may in principle be studied by means of a variational approximation (VA), as the ones used in various works (see, e.g., [15, 16]) to study discrete soliton solutions of the discrete nonlinear Schrödinger (DNLS) equation. However, in the case of soliton collisions under consideration, a direct application of VA may produce equations for the soliton parameters that could be very difficult to be expressed in an explicit form and, thus, to be treated analytically or numerically. A simple variant of VA was adopted in [6], where the variational ansatz was considered to be a combination of two *discrete spikes*, which was subsequently substituted into the *continuous* NLS equation, with the Lagrangian $\int_{-\infty}^{+\infty} \left[i(u^*\dot{u} - u\dot{u}^*) - |u_x|^2 + |u|^4 \right] dx$. Using this discreteness motivated ansatz in the continuum Lagrangian of the model, the resulting variational equations predicted that the collision of two solitons with large velocities leads to bounce, while the collision with small velocities gives rise to *merger* of the solitons.

The above result can be confirmed by means of systematic simulations. Here, following the analysis of [6], we use an initial condition for the DNLS equation (17.1) suggested by the AL model (see Eq. (17.3)), namely,

$$u_0 = B \operatorname{sech}\left[W^{-1}(n - x_1)\right] \exp\left[ia(n - x_1) + \left(\frac{i}{2}\right)\Delta\phi\right]$$
$$+ B \operatorname{sech}\left[W^{-1}(n - x_2)\right] \exp\left[-ia(n - x_2) - \left(\frac{i}{2}\right)\Delta\phi\right]. \quad (17.5)$$

It is clear that Eq. (17.5) is a superposition of two far separated pulses with common amplitude B and width W, initial phase difference $\Delta\phi$, and initial positions at $x_{1,2}$. The parameter a denotes the soliton wave number and, in fact, determines the initial soliton speed. Fixing the soliton width as, e.g., $W = 1$, and using a as a main control parameter, it is possible to study outcomes of the collision for several different values of the amplitude B, including $B = \sinh(1/W) \approx 1.1752$ (corresponding to the AL soliton), $B = 1$ (corresponding to the continuum NLS limit), and another smaller value, $B = 1/\sinh(1/W) \approx 0.851$. Although these values are not very different, the results obtained for them may differ dramatically, and they adequately represent the possible outcomes of the collision. Moreover, using the initial condition Eq. (17.5), it is also possible to study on-site (OS) and inter-site (IS) collisions (with the central point located, respectively, OS or at a midpoint between sites), varying the initial positions x_1 and x_2.

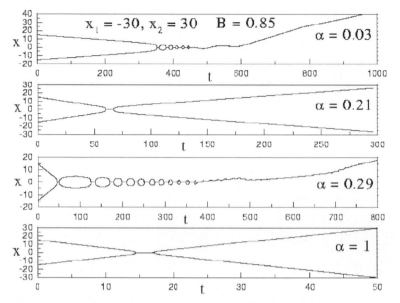

Fig. 17.3 The on-site collision, with $x_{1,2} = \pm 30$, for various values of the parameter a in the case of $B = 0.851$. The intervals of merger with spontaneous symmetry breaking are separated by regions of quasi-elastic collisions. In all the cases displayed herein $C = 0.5$

Figure 17.3 depicts several collision and merger scenarios that we now explain in some detail. Let us first consider the case of OS collisions, with $x_{1,2} = \pm 30$, and initial soliton amplitude $B = 1.175$ and 1 (below we will also consider that the solitons have the same phase, i.e., $\Delta\phi = 0$). In the former case ($B = 1.175$), the solitons cannot collide for $a < 0.550$ (as in this case these "taller" solitons encounter a higher Peierls–Nabarro [PN] barrier), they move freely and collide merging to a single pulse for $0.550 < a < 2.175$ (with multiple collisions, if a is close to the upper border of this interval), and they collide quasi-elastically for $a > 2.175$. In the latter case ($B = 1$), and for $0 < a < 0.7755$, the colliding solitons merge into a single pulse, while for $a > 0.7756$, the solitons undergo a quasi-elastic collision (as they separate after the collision). It is worth noting that these basic features of this phenomenology (for $B = 1$) are correctly predicted by the aforementioned VA devised in [6]. Nevertheless, some more peculiar characteristics can also be identified, since in the interval $0 < a < 0.7755$, there exist two subintervals, namely $0 < a < 0.711$, where the solitons fuse into one after a single collision, and $0.711 < a < 0.7755$, where the fusion takes place after multiple collisions. Finally, in the case of the smaller initial amplitude, $B = 0.851$, a new feature is found in intervals $a < 0.203$ and $0.281 < a < 0.3$. There, the solitons merge after multiple collisions, which is accompanied by strong *symmetry breaking* (SB): the resulting pulse moves to the left or to the right, at a well-defined value of the velocity, as is shown in Fig. 17.4. Between these intervals, i.e., at $0.203 < a < 0.281$, as well as in the case $a > 0.3$, the collisions are quasi-elastic.

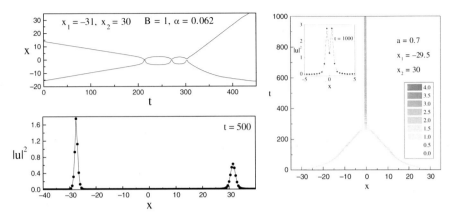

Fig. 17.4 *Left panel*: The inter-site collision, with $x_1 = -31$, $x_2 = 30$, for $a = 0.062$ and $B = 1$. Both trajectories of the colliding solitons (*top*) and their eventual profiles in terms of $|u|^2 \equiv |u_n|^2$ (*bottom*) are shown, with the latter picture illustrating the symmetry breaking. *Right panel*: The quarter-site collision ($x_1 = -29.5$, $x_2 = 30$) for $a = 0.7$ and $B = 1.175$. Shown are the soliton trajectories and the formation of an "M"-shaped merger (see inset). In all the cases displayed herein $C = 0.5$

It should be noted that the SB mechanism and the collision-induced momentum generation (recall that the lattice momentum is not conserved) were analyzed in [6]. As discussed in that work, the SB may be deterministic or spontaneous. The former one is accounted for by the location of the collision point relative to the lattice, and/or the phase shift between the solitons, while the momentum generated during the collision due to the phase shift was found to depend on the solitons' velocities. As far as spontaneous SB is concerned, the modulational instability of a quasi-flat plateau temporarily formed in the course of the collision was suggested as a possible explanation.

On the other hand, in the case of IS collisions (e.g., with $x_1 = -31$ and $x_2 = 30$), we expect a significant change in the phenomenology, as in this case the collision point is at a local maximum of the PN potential, while in the OS case it was at a local minimum. This important difference results in a strong reduction of the scale of the initial velocity (determining the different outcomes of the IS collisions), roughly by an order of magnitude, as compared to the OS case. Apart this reduction, most features of the phenomenology discussed above for the OS collisions can also be found in the case of IS collisions. Nevertheless, in the case of IS collisions with the intermediate value of the amplitude $B = 1$, and for $0.062 < a < 0.075$, lead to spontaneous SB, with mutual reflection (rather than merger) of the solitons after multiple collisions, see left panel of Fig. 17.4 (note that the collision results in a merger for $a < 0.061$ and $0.075 < a < 0.089$, while it is quasi-elastic for $a > 0.089$). Such a *multiple-bounce* window in the DNLS system, resembles a similar effect that was found for ϕ^4 kinks in [17, 18], but with an important difference that in the kink-bearing models, spontaneous SB is impossible. Finally, it should be mentioned that quarter-site collisions (corresponding, e.g., to $x_1 = -29.5$ and

$x_2 = 30$) lead also to qualitatively similar results: for the smaller amplitude value of $B = 0.851$ the collision results in the separation of solitons upon a single bounce with SB for all values of a, while for the larger amplitude value of $B = 1.175$ there exist windows of no collision (for $0 < a < 0.5$), formation of a static bound state, in the form of "M" (for $0.5 < a < 0.9$, see right panel of Fig. 17.4) and quasi-elastic collision with SB (for $a > 0.9$).

17.4 Strongly Discrete Nearly Integrable Case

Here, following [7], we discuss the weakly perturbed integrable AL system Eq. (17.2), in the regime of high discreteness setting $C = 0.78$ and small values for the perturbation parameters, δ and ε. In the simulations, initial conditions were set according to the exact AL soliton solution in Eq. (17.3).

In Fig. 17.5 the soliton amplitudes after collision, $\tilde{A}_i$, are shown as functions of the initial phase difference $\Delta\alpha_0$ for different coordinates of the collision point, x_c, with respect to the lattice: (a) $x_c = 0$ (OS collision), (b) $x_c = 0.25$, (c) $x_c = 0.5$ (IS collision), and (d) $x_c = 0.75$. It is readily seen that the collisions are inelastic only in the vicinity of $\Delta\alpha_0 = 0$, the situation typical for the weakly perturbed integrable systems. However, in contrast to the result presented in Fig. 17.1d for the weak discreteness, in the highly discrete case, as it was already described in Sect. 17.3, the collision outcome becomes extremely sensitive to the location of the collision point with respect to the lattice. For example, collisions of in-phase solitons ($\Delta\alpha_0 = 0$) are practically elastic for $x_c = 0$ and $x_c = 0.5$, while they are strongly inelastic

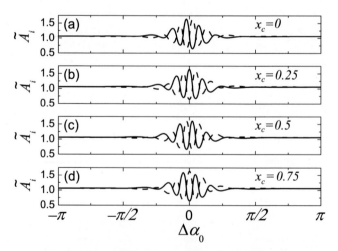

Fig. 17.5 Soliton amplitudes after collision, $\tilde{A}_i$, as the functions of the initial phase difference $\Delta\alpha_0$ for different coordinates of the collision point x_c with respect to the lattice: (a) $x_c = 0$ (on-site collision), (b) $x_c = 0.25$, (c) $x_c = 0.5$ (inter-site collision), and (d) $x_c = 0.75$. The soliton parameters before the collision are $\mu_1 = \mu_2 = 0.75$, $k_1 = -k_2 = 0.1$. The model parameters are $\delta = 0.04$, $\varepsilon = 0$, and $C = 0.78$

for $x_c = 0.25$ and $x_c = 0.75$. For any x_c, in the vicinity of $\Delta\alpha_0 = 0$, the collision outcome is extremely sensitive to small variations in $\Delta\alpha_0$, which is typical for the near-separatrix collisions.

We emphasize again that in the weakly perturbed integrable system, also in the case of strong discreteness, the dominant inelasticity effect is the REE between colliding solitons, while the radiation losses and the excitation of the soliton's internal modes are marginal even for the near-separatrix collisions with $\Delta\alpha_0 \approx 0$ [7].

The model Eq. (17.2) contains the perturbation parameters δ and ϵ (determining the strength of the cubic and quintic perturbation terms, respectively). In Fig. 17.6 we show the maximal degree of inelasticity of the collision as a function of δ at $\varepsilon = 0$ (a) and ε at $\delta = 0$ (b). The ordinate is the maximal (over x_c and $\Delta\alpha_0$) soliton amplitude after collision, $\tilde{A}_{\max}$. The discreteness parameter is $C = 0.78$. The soliton velocities and amplitudes before the collision are $V_1 = -V_2 = 0.137$ and $A_1 = A_2 = 1.05$, respectively.

The results presented in Fig. 17.6 reveal the asymmetry in the net inelasticity effect for positive and negative values of the perturbation parameters δ and ε. In (a) the asymmetry appears for $|\delta| > 0.02$ and in (b) for $|\varepsilon| > 0.0025$ and it is negligible for smaller values of perturbation parameters. This asymmetry can be explained through the influence of the soliton's internal modes that exist, as it is well-known, only if the perturbation parameter has the "right" sign [19]. To confirm this, we calculate the spectrum of small amplitude vibrations of the lattice containing a stationary soliton with frequency $\omega = 2C(\cosh\mu - 1)$ (for the chosen parameters $\omega = 0.4$). The spectrum includes the phonon band $\Omega = \pm[4C\sin^2(Q/\sqrt{8C}) + \omega]$, where Q and Ω are the phonon wave number and frequency, respectively, and it may include the frequencies of soliton's internal modes. The results are presented in Fig. 17.7. In (a) and (b) we show the bifurcation of the internal mode frequency, ω_{IM}, from the upper edge of the phonon spectrum, $\Omega_{\max}$, while in (c) and (d) from the lower edge of the spectrum, $\Omega_{\min}$. Particularly, we plot $\sqrt{\omega_{\text{IM}} - \Omega_{\max}}$ as a function of δ at $\varepsilon = 0$ (a), and ε at $\delta = 0$ (b), and also we plot $\sqrt{\Omega_{\min} - \omega_{\text{IM}}}$ as a function of δ at $\varepsilon = 0$ (c) and ε at $\delta = 0$ (d). Recall that $C = 0.78$.

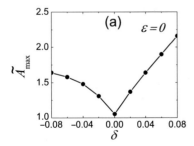
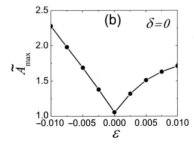

Fig. 17.6 Maximal (over x_c and $\Delta\alpha_0$) soliton amplitude after collision as a function of (**a**) δ at $\varepsilon = 0$ and (**b**) ε at $\delta = 0$. The discreteness parameter is $C = 0.78$. The soliton velocities before the collision are $V_1 = -V_2 = 0.137$ and the amplitudes before the collision are $A_1 = A_2 = 1.05$

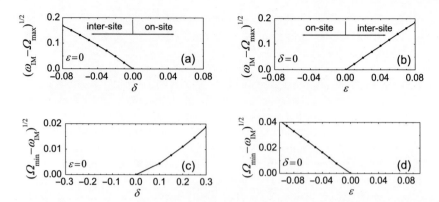

Fig. 17.7 Bifurcation of the soliton's internal mode frequency ω_{IM} from (**a, b**) the upper edge of the phonon spectrum, Ω_{max}, and (**c, d**) from the lower edge of the phonon spectrum, Ω_{min}. In (**a, c**) $\varepsilon = 0$, while in (**b, d**) $\delta = 0$. In all cases the stationary soliton has frequency $\omega = 0.4$ and the discreteness parameter is $C = 0.78$

The numerical results show that the solitons emerge from the collision bearing internal modes with frequencies corresponding to the lower edge of the spectrum and, thus, it can be concluded that these modes can influence the collision outcome but not the modes bifurcating from the upper edge.

Coming back to the asymmetry of the inelasticity of collisions with respect to the change of the sign of perturbation parameter (see Fig. 17.6), it can now be concluded that the net inelasticity effect is higher when the soliton's internal modes come into play. More precisely, at $\varepsilon = 0$ the collisions are more inelastic for $\delta > 0$ when the internal mode below the phonon band exists. Similarly, for $\delta = 0$ collisions are more inelastic for $\varepsilon < 0$, for the same reason.

We note in passing that a change of the sign of perturbation parameters switches the stable OS and IS configurations as indicated in Fig. 17.7a, b.

17.5 Role of Soliton's Internal Modes

As mentioned above, the REE is the dominant effect in soliton collisions in the weakly perturbed NLS or AL systems. However, if the perturbation is not small, the REE effect is mixed with radiation and possibly with excitation of the soliton's internal modes. In Sect. 17.4 we have already discussed the role of the soliton's internal modes and here we further elaborate on their role in the inelastic soliton collisions.

Particularly, we will now investigate if the energy exchange between the soliton's internal and translational modes is possible in the perturbed NLS equation. Such energy exchange plays an important role in the collisions among ϕ^4 kinks resulting in several nontrivial effects such as separation after multiple-bounce collisions [17, 18, 20–23]. In order to eliminate the influence of the location of the collision

17 Solitary Wave Collisions 321

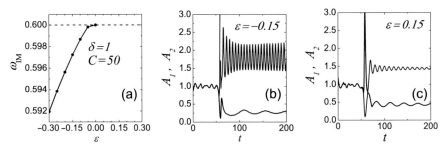

Fig. 17.8 (a) Bifurcation of the soliton's internal mode frequency ω_{IM} from the lower edge of the phonon spectrum as the function of perturbation parameter ε at $\delta = 1$, $C = 50$ for the stationary soliton with frequency $\omega = 0.6$. An internal mode exists only for $\varepsilon < 0$. Panels (**b**) and (**c**) show the amplitudes of solitons as a function of time for moderately perturbed systems with $\varepsilon = -0.15$ and 0.15, respectively. Solitons having initial amplitudes $A_1 = A_2 \approx 1$ and velocities $V_1 = -V_2 = 0.05$ collide at $t \approx 60$ and they emerge from the collision with different amplitudes. Moreover, in (b), high-amplitude internal modes with very long lifetime are excited, while in (c), they are not. Initial phase difference in both cases is $\Delta\alpha_0 = 0.12$

point with respect to the lattice we set in Eq. (17.2) $C = 50$ (extremely weak discreteness) and take $\delta = 1$, so that the cubic perturbation is absent and the only perturbation remains to be the quintic term with coefficient ε.

In Fig. 17.8a we show the bifurcation of the soliton's internal mode frequency ω_{IM} from the lower edge of the phonon spectrum $\Omega_{min} = \omega$ (where $\omega = 0.6$ is the soliton's frequency) as a function of ε. The internal mode exists only for $\varepsilon < 0$. The panels (b) and (c) of the same figure show the amplitudes of the solitons as a function of time for $\varepsilon = -0.15$ and 0.15, respectively. Solitons having initial amplitudes $A_1 = A_2 \approx 1$ and velocities $V_1 = -V_2 = 0.05$ collide at $t \approx 60$. The initial phase difference is $\Delta\alpha_0 = 0.12$.

Figure 17.8 clearly illustrates that in the case of $\varepsilon < 0$, when the internal mode exists, the collision is more inelastic than in the case of $\varepsilon > 0$. In addition to this, in the case of $\varepsilon < 0$ the solitons emerge from the collision bearing high-amplitude internal modes with very long lifetime, while in (c) such modes are not excited.

Now we focus on the case of $\varepsilon = -0.15$ and study the symmetric in-phase solitons ($\Delta\alpha_0 = 0$) with different velocity V for solitons with initial amplitudes $A_1 = A_2 = 1$. Relevant results are shown in Fig. 17.9. In (a) we plot the velocity of the solitons after collision, $\tilde{V}$, as a function of V. For the collision velocity $V > V^* \approx 0.42$, the solitons separate after the collision while for $V < V^*$ they merge. An example of collision with merger for the collision velocity $V = 0.4$ is given in (c) where only the particles with $|u_n|^2 > 0.2$ are shown. No separation "windows" in the region $V < V^*$, typical for the kink collisions in ϕ^4 model, are found. In (b), for the velocities $V < V^*$ we plot the maximal separation S_{max} of two solitons after the first collision as a function of V. If the energy exchange between the soliton's internal and translational modes took place, one would see the maxima of S_{max} at the resonant collision velocities, but nothing like that is observed. It should be pointed out that the high-amplitude internal modes *are* excited during the collision, similar

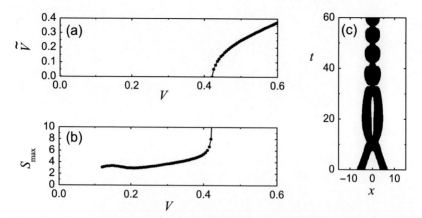

Fig. 17.9 Role of internal modes studied by colliding the symmetric in-phase solitons ($\Delta\alpha_0 = 0$) with velocity V in the moderately perturbed NLS equation Eq. (17.2) with $C = 50$ (extremely weak discreteness), $\varepsilon = -0.15$, $\delta = 1$. The initial soliton amplitudes are $A_1 = A_2 = 1$. (**a**) The velocity of solitons after collision, $\tilde{V}$, is shown as a function of V. (**b**) The maximal separation $S_{\max}$ of two solitons after the first collision is given as a function of V. (**c**) Example of collision for the collision velocity $V = 0.4$. Shown are the particles with $|u_n|^2 > 0.2$

to that shown in Fig. 17.8b but, in contrast to the ϕ^4 case, they do not significantly affect the translational motion of the solitons.

Another important note is the following: in the simulations presented in Fig. 17.9 we have excluded the REE effect by setting $\Delta\alpha_0 = 0$. For nonzero $\Delta\alpha_0$, in the vicinity of $\Delta\alpha_0 = 0$, the REE effect would completely change the collision outcome, and particularly, the soliton separation after multi-bounce collisions could be observed.

17.6 Solitary Wave Collisions in Physically Relevant Settings

In this section, we will discuss some physically relevant settings where solitary wave collisions in weakly discrete or discrete systems have been investigated. These include optical waveguide arrays in the nonlinear optics context, and BECs confined in periodic potentials, so-called optical lattices, in the atomic physics context.

First, in the context of nonlinear optics, in [24] (see also related work in [25]), the interaction of two solitary waves was studied experimentally in arrays of AlGaAs coupled optical waveguides. In this case, the relevant mathematical model is the discrete cubic NLS equation discussed above. The focus in this experimental work was on the case of the interaction of solitons with zero initial velocities (in terms of the nomenclature of the present work), and a result coinciding with our findings was that, in the limit of the zero collision velocity, two solitons with the zero phase difference always merge into one.

Soliton mobility [26] and collisions [27–29] were studied theoretically in the framework of a discrete NLS equation with a saturable nonlinearity. In particular, the nonlinear term in the DNLS equation was taken to be of the form

$-\beta u_n/(1 + I_n)$, (here $I_n \sim |u_n|^2$ is the normalized light peak intensity and $\beta =$ const.). This type of the discrete NLS equation is also called Vinetskii–Kukhtarev model (VK) [30], which is particularly relevant to waveguide arrays created in photorefractive crystals. The latter have attracted much attention (see, e.g., the reviews [31, 32]) as they offer real-time control of the waveguide array, as well as strong and tunable nonlinearity [33]. Note that for low-norm solutions the VK model is reduced to the cubic discrete NLS equation.

In [27] it was found that, similarly to the results of Sect. 17.2, low-norm solitons with small velocities merge and remain pinned, creating a breathing state, while for high enough velocities, the solitons are reflected. However, high-norm moving solutions are also allowed in the VK model [26, 29] (contrary to the cubic DNLS case, due to the potential vanishing of the PN barrier relevant to the VK case). When high-norm (power) solutions collide the effect of, so-called, breather creation can be observed [27, 29], consisting of a partial trapping of the energy of the incoming solitons, together with the reflection of the initial solitons. Note that this effect was previously found in the continuum counterpart of the model [34].

Symmetric collisions of two discrete breathers in the VK model were also investigated in [28]. The strong correlation of the collision properties and the parameters of colliding breathers (power, velocity, and phase difference), lattice parameters and position of the collision point was related to the internal structure of the colliding breathers and energy exchange with the phonon background. Several types of collision were observed in wide parameter space: elastic (quasi-elastic) OS (IS) collision, breather creation, fusion of colliding breathers, and creation of two asymmetric breathers (after IS collision).

Soliton collisions were also studied in the context of BECs trapped in a periodic potential created by the interference of optical beams, the so-called optical lattice [35]. In this case, the pertinent model describing the evolution of the BEC wave function is the continuous NLS (with a periodic external potential) equation or the discrete NLS equation, for a weak or strong (as compared to the system's chemical potential) optical lattice strength, respectively [36]. In the former case, collisions of the pertinent, so-called lattice solitons have been studied in various works (see, e.g., [37, 38]), while the transition from the continuous to discrete regime was studied in [39]. In this work, the outcome of the collision between two gap solitons was shown to serve as a measure of the discreteness imposed on the BEC by the optical lattice. Moreover, in [40], soliton collisions were studied in the framework of a strongly discrete NLS equation with a periodically time-modulated nonlinearity coefficient. This model describes a BEC confined in a strong optical lattice, whose interatomic interactions (effectively described by the nonlinear term in the NLS model) are controlled by time-periodic external fields, according to the so-called Feshbach resonance management technique [41]. Such a time-dependent variation of the nonlinearity was shown to assist the discrete soliton motion, and a study of soliton collisions revealed that there exist two different types of the interaction: elastic bounce, or bounce with mass transfer from one soliton to the other. It is relevant to note that, in contrast to the results of Sect. 17.3 (where the DNLS equation had constant coefficients), in the model analyzed in [40] a merger of colliding solitons into a standing one was not observed.

17.7 Conclusions

The above discussion of the effects observed in solitary wave collisions in 1D DNLS equation in the regimes of weak and strong discreteness for nearly integrable and nonintegrable cases can be summarized as follows.

For *nearly integrable systems*, i.e., for the integrable NLS equation perturbed by weak discreteness and for the weakly perturbed AL chain at any degree of discreteness, the inelasticity of collisions is solely due to the radiationless energy exchange between solitons with relatively small amount of radiation and almost no excitation of the soliton's internal modes. This is so because the radiationless energy exchange grows proportionally to the perturbation parameter [5, 11] while the radiation and excitation of the internal modes are the second-order effects [19].

The radiationless energy exchange happens in the vicinity of multisoliton separatrix solutions to the corresponding integrable equations [5]. Typical features of near-separatrix collisions in nearly integrable systems are as follows:

- Collisions are inelastic in a narrow window of parameters of colliding solitons while outside this window they are practically elastic. Examples are presented in Fig. 17.1d for the weakly discrete NLS equation and in Fig. 17.5 for the weakly perturbed AL lattice at high discreteness. Collisions are inelastic only for nearly in-phase (near separatrix) collisions with $\Delta\alpha_0 \sim 0$ [5].
- Near-separatrix collisions are naturally extremely sensitive to small variations in the collision phase $\Delta\alpha_0$ and, for highly discrete systems, to the location of the collision point with respect to the lattice (see Figs. 17.1d and 17.5).
- The fact that the inelasticity of soliton collisions for weakly perturbed systems increases linearly with the perturbation parameter is illustrated by Fig. 17.6. Panel (a) suggests that in the case $\varepsilon = 0$, the radiationless energy exchange effect is dominant within $|\delta| < 0.02$, while from panel (b), for $\delta = 0$, it is dominant for $|\varepsilon| < 0.0025$. For larger values of perturbation parameters, the soliton's internal modes start to affect the result of collision and the net inelasticity effect becomes asymmetric for positive and negative values of perturbation parameters.
- The radiationless energy exchange effect in near-separatrix collisions has been predicted from the analysis of the two-soliton solution to the *unperturbed* integrable NLS equation [5] and thus the actual type of perturbation is not important for the appearance of this effect. It has been also demonstrated that in the systems with more than one perturbation term, the collisions can be nearly elastic even in the vicinity of $\Delta\alpha_0 = 0$ when the effects of different perturbations cancel each other [5, 7].
- The inelasticity of collision increases with decrease in collision velocity. This feature is again related to the near-separatrix nature of the collision. Fast solitons spend a shorter time in the vicinity of separatrix during the collision and their properties are less affected than that of slow solitons.
- The radiationless energy exchange can be responsible for the fractal soliton scattering [4] if they collide with sufficiently small velocities, which is illustrated by the results presented in Fig. 17.2.

- The radiationless energy exchange effect is possible only if the number of parameters of the colliding solitons exceeds the number of conservation laws in the weakly perturbed system [42]. For example, in the weakly discrete Frenkel–Kontorova model, the conservation of energy (exact) and momentum (approximate) sets two constraints on the soliton parameters and, as a result, the two-kink collisions are practically elastic. A three-kink collision has one free parameter and radiationless energy exchange becomes possible [43]. Solitons in the NLS equation and AL chain have two parameters so that the two-soliton collisions are described by four parameters. If in the weakly perturbed NLS equation or AL lattice, the number of exact and approximate conservation laws is less than four (typically this is so) then the radiationless energy exchange is possible in two-soliton collisions.
- In the *moderately or strongly perturbed* systems the soliton collisions become even more complicated because in addition to the radiationless energy exchange the excitation of the soliton's internal modes and radiation become important and the net inelasticity effect is an admixture of these three effects.

    Particularly, for the strongly discrete, nonintegrable case, the merger of colliding solitons can be observed in certain range of collision velocities (see the results shown in Figs. 17.3 and 17.4). Another interesting effect of strong nonintegrability is the symmetry breaking effect described in Sect. 17.3. In the case of high discreteness, the collision outcome in a nonintegrable system is extremely sensitive to location of the collision point with respect to the lattice.
- When the soliton internal modes come into play, the radiationless energy exchange effect becomes more pronounced (compare panels (b) and (c) of Fig. 17.8). The soliton internal modes result in the asymmetry of the net inelasticity effect with respect to the change of sign of perturbation parameter, see Fig. 17.6. In Sect. 17.5 we have analyzed the influence of high-amplitude soliton's internal modes on the collision outcome for the NLS equation with moderate quintic perturbation in the absence of the radiationless energy exchange effect. We found that, in spite of the fact that the high-amplitude internal modes *are* excited during the collision, they do not significantly affect the translational motion of the solitons. This behavior contrasts that observed for the colliding ϕ^4 kinks
[17, 18, 20–23].

17.8 Future Challenges

Before closing, we would like to mention various interesting open problems that, in our opinion, deserve to be studied in more detail.

(i) In many physically relevant cases, 2D and 3D models describe realistic situations better than 1D models, but the solitary wave collisions in higher dimensions have been studied much less than in the 1D case.

(ii) Collisions between discrete vector solitons have not been studied in detail. In fact, although this issue was studied theoretically for waveguide arrays with the Kerr-type nonlinearity [44] (relevant experiments were reported in [45]), the case of vector soliton collisions in coupled discrete Vinetskii-Kukhtarev models has not been considered so far. This is an interesting direction, since relevant experimental [2] and theoretical [46] results have already appeared recently. Moreover, as per our previous remark, discrete vector soliton collisions in higher dimensional settings have not been studied in detail yet. Such studies would be particularly relevant in the context of multicomponent and spinor BECs [36].
(iii) The interplay between various mechanisms controlling the inelasticity of soliton collisions (e.g., radiationless energy exchange, internal modes, and radiation) is not fully understood yet even in 1D settings.
(iv) For nearly integrable models, the collision outcome depends on the number of exact and approximate conservation laws remaining in the system. However, so far, a detailed study on how a perturbation affects the conservation laws of an integrable equation is still missing.
(v) Some results presented here are not fully understood and certainly deserve a more careful consideration. For example, as concerns the findings of Sect. 17.5, it is worth noting the following. Contrary to what is observed in kink collisions in the ϕ^4 model, soliton collisions in continuum NLS equation with quintic perturbation do not reveal a noticeable interaction between the soliton's translational and internal modes. This observation should be better understood and explained.

Acknowledgments It is a pleasure and an honor to acknowledge the invaluable contribution of our colleagues and friends with whom we had a very pleasant and productive collaboration on this interesting and exciting topic: P. G. Kevrekidis, Yu. S. Kivshar, B. A. Malomed, A. E. Miroshnichenko, I. E. Papacharalampous, D. A. Semagin, T. Shigenari, A. A. Sukhorukov, and A. A. Vasiliev are greatly thanked.

References

1. Kivshar, Yu.S., Agrawal, G.P.: Optical Solitons: From Fibers to Photonic Crystals. Academic Press, San Diego (2003)
2. Ablowitz, M.J., Ladik, J.F.: J. Math. Phys. **16**, 598 (1975)
3. Ablowitz, M.J., Ladik, J.F.: J. Math. Phys. **17**, 1011 (1976)
4. Dmitriev, S.V., Shigenari, T.: Chaos **12**, 324 (2002)
5. Dmitriev, S.V., Semagin, D.A., Sukhorukov, A.A., Shigenari, T.: Phys. Rev. E **66**, 046609 (2002)
6. Papacharalampous, I.E., Kevrekidis, P.G., Malomed, B.A., Frantzeskakis, D.J.: Phys. Rev. E **68**, 046604 (2003)
7. Dmitriev, S.V., Kevrekidis, P.G., Malomed, B.A., Frantzeskakis, D.J.: Phys. Rev. E **68**, 056603 (2003)
8. Salerno, M.: Phys. Rev. A **46**, 6856 (1992)
9. Sukhorukov, A.A., Akhmediev, N.N.: Phys. Rev. Lett. **83**, 4736 (1999)

10. Gordon, J.P.: Opt. Lett. **8**, 596 (1983)
11. Frauenkron, H., Kivshar, v, Malomed, B.A.: Phys. Rev. E **54**, R2244 (1996)
12. Zhu, Y., Yang, J.: Phys. Rev. E **75**, 036605 (2007)
13. Zhu, Y., Haberman, R., Yang, J.: Phys. Rev. Lett. **100**, 143901 (2008)
14. Karpman, V.I., Solov'ev, V.V.: Physica D **3**, 142 (1981)
15. Malomed, B.A., Weinstein, M.I.: Phys. Lett. A **220**, 91 (1996)
16. Kaup, D.J.: Math. Comput. Simul. **69**, 322 (2005)
17. Campbell, D.K., Schonfeld, J.F., Wingate, C.A.: Physica D **9**, 1 (1983)
18. Campbell, D.K., Peyrard, M.: Physica D **18**, 47 (1986)
19. Kivshar, Yu.S., Pelinovsky, D.E., Cretegny, T., Peyrard, M.: Phys. Rev. Lett. **80**, 5032 (1998)
20. Anninos, P., Oliveira, S., Matzner, R.A.: Phys. Rev. D **44**, 1147 (1991)
21. Belova, T.I., Kudryavtsev, A.E.: Phys. Usp. **40**, 359 (1997)
22. Goodman, R.H., Haberman, R.: SIAM J. Appl. Dyn. Sys. **4**, 1195 (2005)
23. Goodman, R.H., Haberman, R.: Phys. Rev. Lett. **98**, 104103 (2007)
24. Meier, J., Stegeman, G.I., Silberberg, Y., Morandotti, R., Aitchison, J.S.: Phys. Rev. Lett. **93**, 093903 (2004)
25. Meier, J., Stegeman, G., Christodoulides, D., Morandotti, R., Sorel, M., Yang, H., Salamo, G., Aitchison, J., Silberberg, Y.: Opt. Express **13**, 1797 (2005)
26. Melvin, T.R.O., Champneys, A.R., Kevrekidis, P.G., Cuevas, J.: Phys. Rev. Lett. **97**, 124101 (2006)
27. Cuevas, J., Eilbeck, J.C.: Phys. Lett. A **358**, 15 (2006)
28. Maluckov, A., Hadžievski, Lj., Stepić, M.: Eur. Phys. J. B **53**, 333 (2006)
29. Melvin, T.R.O., Champneys, A.R., Kevrekidis, P.G., Cuevas, J.: Physica D **237**, 551 (2008)
30. Vinetskii, V.O., Kukhtarev, N.V.: Sov. Phys. Solid State **16**, 2414 (1975)
31. Fleischer, J.W., Bartal, G., Cohen, O., Schwartz, T., Manela, O., Freedman, B., Segev, M., Buljan, H., Efremidis, N.K.: Opt. Express **13**, 1780 (2005)
32. Chen, Z., Martin, H., Eugenieva, E.D., Xu, J., Yang, J.: Opt. Express **13**, 1816 (2005)
33. Efremidis, N.K., Sears, S., Christodoulides, D.N., Fleischer, J.W., Segev, M.: Phys. Rev. E **66**, 046602 (2002)
34. Królikowski, W., Luther-Davies, B., Denz, C.: IEEE J. Quantum Electron. **39**, 3 (2003)
35. Morsch, O., Oberthaler, M.K.: Rev. Mod. Phys. **78**, 179 (2006)
36. Kevrekidis, P.G., Frantzeskakis, D.J., Carretero-González, R. (eds.): Emergent Nonlinear Phenomena in Bose-Einstein Condensates: Theory and Experiment. Springer Series on Atomic, Optical, and Plasma Physics, vol. 45 2007
37. Sakaguchi, H., Malomed, B.A.: J. Phys. B: At. Mol. Opt. Phys. **37**, 1443 (2004)
38. Ahufinger, V., Sanpera, A.: Phys. Rev. Lett. **94**, 130403 (2005)
39. Dąbrowska, B.J., Ostrovskaya, E.A., Kivshar, Yu.S.: J. Opt. B: Quantum Semiclass. Opt. **6**, 423 (2004)
40. Cuevas, J., Malomed, B.A., Kevrekidis, P.G.: Phys. Rev. E **71**, 066614 (2005)
41. Kevrekidis, P.G., Theocharis, G., Frantzeskakis, D.J., Malomed, B.A.: Phys. Rev. Lett. **90**, 230401 (2003)
42. Kevrekidis, P.G., Dmitriev, S.V.: Soliton collisions. In: Scott, A. (ed.) Encyclopedia of Nonlinear Science, pp. 148–150. Routledge, New York (2005)
43. Miroshnichenko, A.E., Dmitriev, S.V., Vasiliev, A.A., Shigenari, T.: Nonlinearity **13**, 837 (2000)
44. Ablowitz, M.J., Musslimani, Z.H.: Phys. Rev. **65** 056618 (2002)
45. Meier, J., Hudock, J., Christodoulides, D.N., Stegeman, G., Silberberg, Y., Morandotti, R., Aitchison, J.S.: Phys. Rev. Lett. **91**, 143907 (2003)
46. Fitrakis, E.P., Kevrekidis, P.G., Malomed, B.A., Frantzeskakis, D.J.: Phys. Rev. E **74**, 026605 (2006)

Chapter 18
Related Models

Boris A. Malomed

18.1 Models Beyond the Standard One

18.1.1 Introduction

The model which plays the central role in this book is the discrete nonlinear Schrödinger (DNLS) equation, in one or two or even three dimensions (1D, 2D, and 3D, respectively). In fact, the DNLS equation is a paradigmatic model of nonlinear lattice dynamics, which provides for the framework allowing the study of basic effects in this area. In addition to that, the 1D and 2D DNLS equations find a direct physical realization as models of arrays of linearly coupled optical waveguides with the intrinsic Kerr nonlinearity, as first predicted in [1]. Later, 1D discrete solitons were created in the experiment performed in a 1D array built as a set of parallel semiconductor waveguides mounted on a common substrate [2]. More recently, 2D arrays were fabricated as a bundle of parallel waveguides written in a bulk sample of silica glass [3]. The creation of 2D discrete solitons in this setting was reported [4]. Moreover, 2D solitons were reported even in a disordered bundle built by means of this technology [5].

Another direct implementation of the DNLS equation is in the form of a model of a Bose–Einstein condensate (BEC) trapped in a very strong (deep) optical lattice (OL). The fact that the deep lattice splits the condensate into a set of weakly coupled droplets makes it possible to replace the respective Gross–Pitaevskii equation (GPE), with the cubic term accounting for collisions between atoms, by the discrete equation with the on-site cubic nonlinearity, see original works [6–10] and a review of the topic in [11]. This realization of the DNLS equation is relevant (unlike the situation in optics) in the 3D case as well. In a similar way, the 1D, 2D, and 3D versions of the DNLS equation may be used as models of crystals of the respective dimension, formed by microcavities which trap polaritons [12, 13].

B.A. Malomed (✉)
Department of Physical Electronics, School of Electrical Engineering, Faculty of Engineering, Tel Aviv University, Tel Aviv 69978, Israel
e-mail: malomed@eng.tau.ac.il

In addition to the standard DNLS equation, more general models of dynamical lattices are important to various applications, and are interesting in their own right, as specific examples of nonlinear dynamical systems. This chapter offers an overview of such generalized models. The emphasis in the presentation is made on basic theoretical results, obtained in them for 1D and 2D discrete solitons.

18.1.2 Anisotropic Inter-Site Couplings

The simplest nontrivial deviation from the standard DNLS model is provided by the 2D (or, generally speaking, 3D) discrete equation with anisotropic linear couplings between lattice sites [14]. The 2D version of the model is based on the following generalized DNLS equation:

$$i\dot{u}_{n,m} = -C\tilde{\Delta}u_{n,m} - |u_{n,m}|^2 u_{n,m}, \qquad (18.1)$$

where $u_{n,m}(t)$ is the complex 2D lattice field with integers n and m representing the lattice coordinates, the overdot stands for its time derivative (in the case of optical waveguides, the evolution variable is not time, but rather the distance along the waveguides), the on-site nonlinearity is assumed self-focusing, $C > 0$ is the lattice-coupling constant, and the anisotropic 2D lattice Laplacian is defined as follows:

$$\tilde{\Delta}u_{n,m} = \alpha\left(u_{n+1,m} + u_{n-1,m}\right) + u_{n,m+1} \\ + u_{n,m-1} - 2(1+\alpha)u_{n,m}, \qquad (18.2)$$

with positive coefficient α accounting for the degree of the anisotropy. Physical realizations of such a generalized model in terms of 2D arrays of linearly coupled optical waveguides or deep BEC-trapping OLs are straightforward.

Irrespective of the anisotropy, Eq. (18.1) conserves a fundamental dynamical invariant – the norm of the lattice field (alias power, which is an appropriate name for it in the model of arrays of optical waveguides):

$$N = \sum_{n,m} |u_{m,n}|^2. \qquad (18.3)$$

In fact, the norm defined in the same or similar way plays a crucial role in the analysis of all other models considered in this chapter.

18.1.3 Noncubic On-site Nonlinearities

A more conspicuous departure from the standard DNLS model is realized through the replacement of the simplest cubic (Kerr) on-site nonlinearity by more sophisticated forms of the nonlinear response. In particular, a lattice model with saturable nonlinear terms was introduced back in 1975 by Vinetskii and Kukhtarev [15]. Stable 1D bright solitons in this model were reported recently in [16–18] (2D discrete solitons in a model with the saturable nonlinearity were studied too [19]). Experimentally, lattice solitons supported by the saturable self-defocusing nonlinearity were created in an array of optical waveguides in a photovoltaic medium [20].

A simpler but, in some aspects, more interesting non-Kerr nonlinearity is represented by a combination of *competing* self-focusing cubic and self-defocusing quintic terms, which corresponds to the experimentally observed form of the nonlinear dielectric response in some optical materials [21–23] (earlier studies of bright solitons in the DNLS equation were dealing with on-site nonlinear terms of an arbitrary power, but without the competition between self-focusing and defocusing [24–26]). In the 1D continuum NLS equation with the cubic-quintic (CQ) nonlinearity, a family of exact stable soliton solutions is well known [27–30]. An intermediate step between that model and its discrete counterpart is a continuum equation combining the CQ nonlinearity and a periodic potential of the Kronig–Penney type, i.e., a periodic chain of rectangular potential wells. Numerous stable families of solitons were explored in both 1D [31–33] and 2D [34] versions of the latter equation (the 2D model, which features a "checkerboard" 2D potential, supports both fundamental and vortical solitons). The limit case of the Kronig–Penney potential composed of very deep and narrow potential wells amounts to the replacement of the continuum equation by its DNLS counterpart, whose 1D form is [35, 36]

$$i\dot{u}_n + C(u_{n+1} + u_{n-1} - 2u_n) + \left(2|u_n|^2 - |u_n|^4\right)u_n = 0, \qquad (18.4)$$

cf. Eq. (18.1). It is relevant to mention that, in addition to bright-soliton solutions to Eq. (18.4) that are considered below, stable dark solitons in the same model were studied too [37].

A more radical change of the on-site nonlinearity corresponds to lattice models of the second-harmonic generation, where the nonlinearity is quadratic ($\chi^{(2)}$), and two lattice fields are involved, representing the fundamental-frequency and second-harmonics (FF and SH) components. Originally, such a model was introduced to describe Fermi-resonance interface modes in multilayered systems based on organic crystals [38, 39]. The interest in $\chi^{(2)}$ lattices was boosted by the experimental realization of discrete $\chi^{(2)}$ solitons in nonlinear optics [40]. A variety of topics have been studied in this context, including the formation of 1D and 2D solitons [41–43], modulational instability (which was demonstrated experimentally) [44], $\chi^{(2)}$ photonic crystals [45], cavity solitons [46, 47], and multicolor localized modes [48].

A noteworthy result is the prediction of the (anisotropic) mobility of 2D solitons in the $\chi^{(2)}$ lattice [49], which is described by the following system of equations for the FF and SH lattice fields, $\psi_{m,n}(t)$ and $\phi_{m,n}(t)$:

$$i\dot{\psi}_{m,n} = -\left(C_1 \Delta \psi_{m,n} + \psi^*_{m,n}\phi_{m,n}\right), \qquad (18.5)$$

$$i\dot{\phi}_{m,n} = -(1/2)\left(C_2 \Delta \phi_{m,n} + \psi^2_{m,n} + k\phi_{m,n}\right), \qquad (18.6)$$

where the asterisk stands for the complex conjugation, Δ is the isotropic discrete Laplacian defined by Eq. (18.2) with $\alpha = 1$, real coefficients C_1 and C_2 are the FF and SH lattice-coupling constants, and k is a real mismatch parameter. The prediction of the soliton mobility in the framework of Eqs. (18.5) and (18.6) is noteworthy, as all 2D discrete solitons (unlike their 1D counterparts) are immobile in the DNLS lattice with the cubic ($\chi^{(3)}$) on-site nonlinearity. In the simulations, an initial quiescent 2D soliton was set in motion by the application of a *kick* to it, i.e., multiplying its FF and SH components by $\exp\left[i\left(S/C_{1,2}\right)(m\cos\theta + n\sin\theta)\right]$, where S and θ are the size and direction of the kick. The diagram of the resulting dynamical regimes is shown in Fig. 18.1, along with a similar diagram for solitons in the 1D version of the model (the "localization" region in Fig. 18.1 implies that the kicked soliton remains pinned). A noteworthy feature of the diagram for the 2D model is the dependence of the border between different types of the dynamical behavior on the direction of the kick, i.e., the *dynamical anisotropy* of the model.

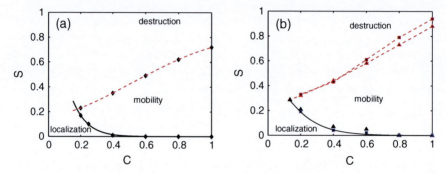

Fig. 18.1 Typical examples of diagrams in the plane of the coupling strength, $C = C_1 = C_2$, and kick size, S, showing outcomes of attempts to set quiescent solitons in motion in the 1D (**a**) and 2D (**b**) $\chi^{(2)}$ lattices (with mismatch $k = 0.25$), as per [49]. In the 2D case, *triangles and rectangles* present the results for two directions of the kick, $\theta = 20°$ and $\theta = 45°$, respectively

18.1.4 Nonlocal Coupling

Another extension of the DNLS equation deals with the character of the inter-site coupling: while in the standard model it is strictly linear and local (involving only nearest neighbors), various extensions can make it both nonlocal (where nonlocal herein will be assumed to mean that nonnearest neighbors are involved) and nonlinear. As concerns the nonlocality, the simplest possibility is to include the interaction with second-nearest neighbors (alias *higher order discrete diffraction*), which may be a natural additional ingredient in the above-mentioned physical realizations of the DNLS equation, in terms of arrays of optical waveguides and BEC trapped in deep OL potentials. The accordingly modified 1D version of the DNLS equation is [50]

$$i\dot{u}_n + C(u_{n+1} + u_{n-1} - 2u_n) + |u_n|^2 u_n$$
$$+ \mu(u_{n+2} + u_{n-2} - 2u_n) + i\kappa(u_{n+2} - u_{n-2}) = 0, \qquad (18.7)$$

where μ and κ are real and imaginary parts of the constant accounting for the linear coupling between next-nearest neighbors (NNN). While the ordinary coupling constant, C, is always defined as a real one, the NNN constant may be complex – for instance, because of a phase shift between fields in the respective array of optical waveguides. In [50] it was found that the addition of the NNN coupling may destabilize the usual on-site-centered solitons, if the imaginary part, κ, is large enough (in that case, the stationary soliton transforms itself into a persistent localized breather). Lattice models with more general long-range linear couplings were studied too. In particular, it was found that, if the strength of the nonlocal coupling in the 1D model decays with the distance between the sites, $|m - n|$, as $|m - n|^{-s}$, the family of on-site-centered discrete solitons features a region of *bistability* (which is absent in the standard DNLS equation), provided that power s is smaller than a critical value (in particular, $s = 3$, which corresponds to the dipole–dipole interactions, admits the bistability of on-site-centered solitons) [51, 52]. It may be relevant to mention that a very specific form of the 1D DNLS equation, with the linear coupling decaying as $\exp(-|m - n|)$ (the nonlocal linear interaction between lattice sites which depends on the distance between them as $\exp(-\alpha|m - n|)$, with $\alpha > 0$, is usually called the *Kac–Baker* potential coupling), and a logarithmic form of the on-site nonlinearity, admits exact solutions in the form of on-site *peakons*, i.e., solutions of the type of $u_n \sim \exp(-|n|)$ [53]. Effects of competition between local and long-range couplings in a 2D lattice model were studied in [54], where it was demonstrated that, if the strength of the nonlocal coupling and/or its range are large enough, the respective soliton-existence threshold [minimum value of norm (18.3) necessary for the creation of 2D discrete solitons] vanishes.

18.1.5 The Competition Between On-site and Inter-site Nonlinearities (The Salerno Model)

A famous example of the 1D discrete system with the nonlinear (cubic) coupling between nearest neighbors (and without the on-site nonlinearity) is the *Ablowitz–Ladik* (AL) model, which is integrable by means of the inverse scattering transform and, accordingly, admits a large number of exact solutions, such as moving discrete solitons [55]. Because the nonintegrable DNLS equation and the integrable AL equation (in the 1D case) converge to a common continuum limit, in the form of the ordinary nonlinear Schrödinger (NLS) equation, a combined model can be naturally introduced, which includes the cubic terms of both types (on-site and inter-site ones). Known as the *Salerno model* (SM) [56, 57], this combined system is based, in the 2D case, on the following discrete equation:

$$i\dot{u}_{n,m} = -C\left[(u_{n+1,m} + u_{n-1,m}) + \alpha\left(u_{n,m+1} + u_{n,m-1}\right)\right]$$
$$\times \left(1 + \mu \left|u_{n,m}\right|^2\right) - 2\nu \left|u_{n,m}\right|^2 u_{n,m}, \tag{18.8}$$

where real coefficients μ and ν account for the inter-site and on-site nonlinearities of the AL and DNLS types, respectively. As before, C stands here for the strength of the coupling between adjacent lattice sites, while α accounts for possible anisotropy of the 2D lattice, with $\alpha = 1$ and $\alpha = 0$ corresponding, respectively, to the isotropic 2D lattice and its 1D counterpart. Accordingly, the variation of α from 0 to 1 makes it possible to observe the *dimensionality crossover*, from 1D to 2D.

It has been demonstrated, chiefly by means of numerical methods, that the 1D version of Eq. (18.8) gives rise to static [58] and (sometimes) moving [59–62] solitons for any value of the DNLS coefficient, ν, and *positive* values of its AL counterpart, μ (collisions between moving 1D solitons were studied too). Static 2D solitons in the SM were also found, under the same condition, $\mu > 0$ [63–65]. Negative values of ν can be made positive by means of the standard staggering transformation, $u_{n,m} \rightarrow (-1)^n u_{n,m}$, and then one may fix $\nu \equiv +1$ by means of rescaling, $u_{n,m} \rightarrow u_{n,m}/\sqrt{\nu}$ (if $\nu \neq 0$). However, the sign of AL coefficient μ *cannot* be altered. In particular, the pure AL model ($\nu = 0$) with $\mu < 0$ supports no bright solitons.

Actually, the SM with $\mu < 0$ is a system with *competing nonlinearities*. It gives rise to soliton dynamics which is very different from that in the ordinary SM, with $\mu > 0$. For the 1D and 2D geometries, the SM with $\mu < 0$ was introduced, respectively, in [66, 67].

It is relevant to mention that, while the SM was originally proposed as a dynamical model in a rather abstract context, it has recently found direct physical realization, as a limit form of the GPE for the BEC of atoms with magnetic moments, trapped in a deep OL [68]. In that case, the evolution variable in Eq. (18.8) is time, the on-site nonlinearity accounts (as usual) for collisions between atoms, while the inter-site nonlinear terms take into account the dipole–dipole interactions, which are repulsive ($\mu < 0$) if the external magnetic field aligns the atomic moments perpendicular to the lattice. The moments may also be polarized parallel to the lattice. In the 1D setting, this will give rise to the attractive inter-site interactions, with $\mu > 0$, while in the 2D case such *in-plane* polarization corresponds to a complex anisotropic interaction, attractive along the direction of the moments and repulsive perpendicular to it. In the continuum 2D GPE with the same type of the anisotropic dipole–dipole interaction and repulsive local nonlinearity, a specific type of stable 2D solitons was predicted [69].

18.1.6 Semidiscrete Systems

A specific class of multicomponent systems includes a discrete component coupled to a continuum one. The simplest *semidiscrete* models of this type were elaborated, in the framework of the $\chi^{(2)}$ nonlinearity, in [70]. The general setting supporting

Fig. 18.2 A schematic of the optical setting which gives rise to the semidiscrete models with the $\chi^{(2)}$ nonlinearity. Reprinted from [70] with permission

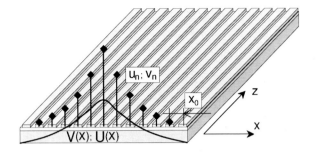

such a system is shown in Fig. 18.2. It includes a periodic array of waveguides, with spacing x_0, mounted on top of (or buried into) the slab waveguide. Both the array and the slab are assumed to be made of the same $\chi^{(2)}$ material. The description of this setting amounts to a system including a set of ordinary differential equations for the evolution of amplitudes of the optical fields in the discrete array, coupled to a partial differential equation for the propagation in the slab.

In fact, two distinct models may describe this semidiscrete system: *type A* includes a discrete FF component in the waveguide array, ϕ_n, coupled to a continuum SH component in the slab waveguide, $\Psi(\eta)$, where η is the transverse coordinate in the slab. Conversely, the *type-B* model combines a continuum FF component in the slab, $\Phi(\eta)$, and discrete SH components in the array, ψ_n. The respective systems of equations take the following form for type A (cf. Eqs. (18.5) and (18.6)):

$$i\frac{d\phi_n}{d\zeta} + \varrho(\phi_{n-1} + \phi_{n+1}) + \phi_n^* \Psi(\zeta, n) = 0, \tag{18.9}$$

$$i\frac{\partial \Psi}{\partial \zeta} + \frac{1}{2}\frac{\partial^2 \Psi}{\partial \eta^2} + \beta \Psi + \frac{1}{2}\sum_n \phi_n^2 \delta(\eta - n) = 0. \tag{18.10}$$

The model of type B is based on a different system:

$$i\frac{\partial \Phi}{\partial \zeta} + \frac{1}{2}\frac{\partial^2 \Phi}{\partial \eta^2} + \Phi^* \sum_n \psi_n \delta(\eta - n) = 0, \tag{18.11}$$

$$i\frac{d\psi_n}{d\zeta} + \beta\psi_n + \varrho(\psi_{n-1} + \psi_{n+1}) + \frac{1}{2}\Phi^2(\zeta, n) = 0. \tag{18.12}$$

Here, ζ stands for the propagation distance, β is the effective mismatch [similar to k in Eq. (18.6)], $\delta(\eta)$ is the ordinary Dirac delta function, and constant ϱ accounts for the coupling between adjacent waveguides in the array. It is sufficient to consider the case of $\varrho = 1$, which adequately represents the generic situation [70].

Basic types of mixed *discrete-continuum* solitons generated by models A and B, and their stability features will be presented in Section 5. It will be demonstrated that the stability of these solitons may be drastically different from that of their ordinary counterparts, both discrete and continuum ones.

18.2 The Anisotropic Two-Dimensional Lattice

18.2.1 Outline of Analytical and Numerical Methods

The presentation in this section follows [14]. Stationary solutions to Eq. (18.1) with anisotropic discrete Laplacian (18.2) are sought for in the usual form, $u_{n,m}(z) = U_{n,m} \exp(i\Lambda z)$, with real propagation constant Λ. This substitution leads to the stationary finite-difference equation, $\Lambda U_{n,m} = C\tilde{\Delta} U_{n,m} - |U_{n,m}|^2 U_{n,m}$. To develop a quasi-analytical variational approximation (VA) for 2D fundamental solitons, which correspond to real localized stationary solutions, one may use the fact that the stationary equation for real function $U_{n,m}$ can be derived from the Lagrangian,

$$L = \sum_{n,m} \left[U_{n,m+1} U_{n,m} + \alpha U_{n+1,m} U_{n,m} - \right.$$
$$\left. ((\Lambda/2C) - 1 - \alpha) U_{n,m}^2 + (1/4C) U_{n,m}^4 \right]. \tag{18.13}$$

For constructing solutions in the numerical form, a universal starting point is the anticontinuum (AC) limit corresponding to $C = 0$, i.e., an uncoupled lattice [71]. In the AC limit, all configurations are constructed as appropriate combinations of independent on-site states, which have either $U_{n,m} = \sqrt{\Lambda}$ with $\Lambda > 0$ (an obvious solution valid for $C = 0$) at a few *excited sites* or $U_{n,m} \equiv 0$ in the rest of the lattice. For $C > 0$, solution families are obtained by means of a numerical continuation in C. Then, the stability of the stationary solutions is analyzed numerically by linearizing Eq. (18.1) for small perturbations added to the solutions, i.e., taking

$$u_{n,m} = \left[u_{n,m}^{(0)} + \delta \cdot \left(a_{n,m} e^{\lambda t} + b_{n,m} e^{\lambda^* t} \right) \right] e^{i\Lambda t}, \tag{18.14}$$

where $a_{n,m}$ and $b_{n,m}$ represent infinitesimal eigenmodes of the perturbations, and λ is the respective eigenvalue (instability growth rate), which may be complex. Perturbations corresponding to purely real eigenvalues λ in Eq. (18.14) can be analyzed by means of the *Vakhitov–Kolokolov* (VK) criterion [72, 73]: the respective stability condition for a single-pulse soliton family characterized by the dependence $N(\Lambda)$, where N is the norm defined by Eq. (18.3), reduces to inequality $dN/d\Lambda > 0$.

The semi-analytical and numerical methods outlined above were applied, in different works, to all the models considered in this chapter. Basic results obtained for the anisotropic DNLS model are recapitulated in the rest of this section.

18.2.2 Fundamental Solitons and Vortices in the Anisotropic Model

Practically speaking, the only analytically tractable *ansatz* for the application of the VA to fundamental solitons in discrete systems is based on the following expression [26, 74]:

$$U_{n,m} = A \exp(-a|n| - b|m|), \tag{18.15}$$

with positive parameters a and b that determine the widths of the soliton in the two directions on the lattice, and arbitrary amplitude A. The ansatz can be made slightly more general, still keeping it in a tractable form, by allowing the field at the central point, $U_{0,0}$, to be an independent variational parameter, different from A [35]. The substitution of ansatz (18.15) in Eq. (18.13) makes it possible to calculate the corresponding effective Lagrangian and derive the respective variational equations, $\partial L/\partial N = \partial L/\partial a = \partial L/\partial b = 0$. In the case of a strongly anisotropic soliton, which is broad in one direction and narrow in the other, which corresponds to $a \ll 1$ and $b \gg 1$, the analysis yields the following relations between inverse widths a, b and norm N, and the soliton's propagation constant, Λ: $a = \sqrt{\Lambda/(3\alpha C)}$, $\sinh(b/2) = \sqrt{\Lambda/C}$, $N^2 = (4/3)C\alpha\Lambda$. The application of the VA criterion to this solution family demonstrates that it may be stable, as it obviously satisfies condition $dN/d\Lambda > 0$.

Figure 18.3 displays results obtained from numerical solution of the variational equations in the general case, together with direct numerical solutions for fundamental solitons. It also presents numerical results for the stability of the solitons, obtained through the computation of eigenvalues for perturbation modes. In particular, the figure demonstrates that the VA provides a good fit to the numerical solutions, and the stability border is *exactly* predicted by the VK criterion.

Further results of the numerical analysis are summarized in Fig. 18.4(a), which shows, as a function of anisotropy parameter α, the critical value of the coupling constant, ε_{cr}. It is defined so that the fundamental solitons with $\Lambda = 1$ are stable for $C \leq \varepsilon_{\text{cr}}$ and unstable for $C > \varepsilon_{\text{cr}}$. It is worthy to note that this dependence is well approximated by an empirical relation, $\epsilon_{\text{cr}} = 1/\sqrt{\alpha}$.

A family of vortex solitons (shaped as "vortex crosses", i.e., with zero field at the central site), with topological charge $S = 1$, was also investigated in the framework of the anisotropic model. The stability diagram for them is displayed in Fig. 18.4(b). As concerns another species of vortex solitons, in the form of "squares," i.e., without

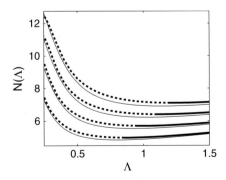

Fig. 18.3 The norm of fundamental solitons vs. propagation constant Λ for fixed coupling constant, $C = 1$, at several values of the anisotropy parameters: $\alpha = 1.5, 1.25, 1, 0.75$, from top to bottom. *Bold solid and dashed lines* represent stable and unstable numerical solutions, while the *thin lines* are obtained by means of the VA. Reprinted from [14] with permission

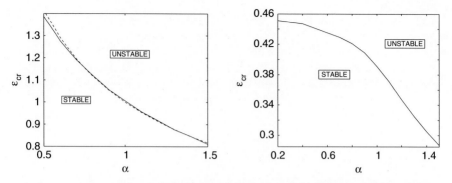

Fig. 18.4 (a) The critical value of the lattice-coupling constant, which separates stable and unstable fundamental solitons with $\Lambda \equiv 1$, vs. anisotropy parameter α, in the 2D DNLS model. The dashed line represents the empirical approximation, $\varepsilon_{\mathrm{cr}} = 1/\sqrt{\alpha}$. (b) The same stability border for vortex-cross solitons with topological charge $S = 1$. Reprinted from [14] with permission

the empty site at the center, an interesting result is that they are strongly destabilized even by a weak deviation from the isotropy, see further details in [14].

18.3 Solitons Supported by the Cubic-Quintic On-site Nonlinearity

In this section, the presentation follows [35, 36]. In the latter work, stationary solutions to Eq. (18.4) were looked for as $u_n(z) = U_n e^{ikn - i\mu z}$, where U_n is a real stationary lattice field which obeys equation

$$\mu U_n + C\left(U_{n+1} e^{ik} + U_{n-1} e^{-ik} - 2U_n\right) + 2U_n^3 - U_n^5 = 0, \quad (18.16)$$

where $-\mu$ is the propagation constant (the same as Λ in the previous section), and $k = 0$ or $k = \pi$ refer to *unstaggered* and *staggered* configurations, respectively. Two different types of both unstaggered or staggered soliton solutions can be found from Eq. (18.16): on-site- and inter-site-centered ones. Within the framework of the VA, the on-site soliton is approximated by the 1D version of ansatz (18.15), while the 1D ansatz for inter-site solitons is

$$U_{n,m} = A \exp\left(a/2 - a|n - 1/2|\right), \quad (18.17)$$

with $a > 0$, where it is implied that the lattice field attains its maximum values, A, at points $n = 0$ and 1, with the formal center of the soliton placed between them.

Global stability diagrams for fundamental unstaggered and staggered solitons in the parameter plane of (μ, C) are displayed in Figs. 18.5 and 18.6. A noteworthy peculiarity of the family of the unstaggered solitons, which occurs at $C < 0.15$, is the coexistence of two different species of the fundamental modes, with different

18 Related Models

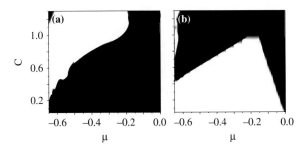

Fig. 18.5 Stability diagrams for fundamental unstaggered discrete solitons of the on-site-centered (**a**) and inter-site-centered (**b**) types, in the 1D model with the CQ on-site nonlinearity. *Black and white* areas depict stability and instability regions, respectively. The instability is accounted for by real eigenvalues. Reprinted from [36] with permission

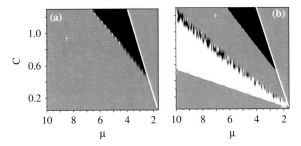

Fig. 18.6 The same as in Fig. 18.5, but for fundamental staggered solitons of the on-site (**a**) and inter-site (**b**) types. The regions of the stability, exponential instability (which is accounted for by real eigenvalues), and oscillatory instability (determined by complex eigenvalues) are depicted by *gray, black*, and *white* areas, respectively. Staggered solitons do not exist to the right of the *white lines* in both panels. Reprinted from [36] with permission

norms and different amplitudes, at a given value of μ. Both species of the on-site-centered unstaggered solitons are stable in the most part of the parameter plane, while the inter-site mode with a smaller amplitude is unstable. At $C > 0.15$, stability exchange between the on-site and inter-site solitons with equal norms takes place with the variation of μ.

In addition to the two species of fundamental unstaggered solitons whose stability is summarized in Fig. 18.5, many other species of unstaggered solitons, both symmetric and asymmetric ones, were found at small values of coupling constant C. Some of them are stable, as demonstrated by examples displayed in Fig. 18.7. With the increase of C, the number of branches of soliton solutions gradually decreases via a chain of bifurcations, so that the single exact stable soliton solution [27–30] survives in the continuum limit corresponding to $C \to \infty$. As shown in [35, 36], basic species of the unstaggered and staggered solitons (of both on-site- and inter-site-centered types) can be very accurately described by the VA based on *ansätze* (18.15) and (18.17).

Nonlinear dynamical effects in the 1D discrete CQ model, such as the long-time evolution of various kinds of unstable solitons, and mobility of stable ones, were also

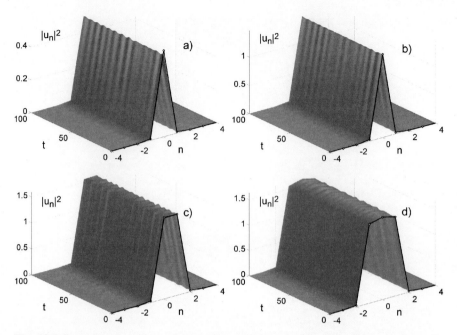

Fig. 18.7 Examples of stable evolution of perturbed on-site-centered (**a,b,d**) and inter-site-centered (**c**) solitons in the CQ model, found at $C = 0.1$ for $\mu = -0.6$. Reprinted from [35] with permission

studied in detail in [35, 36]. Most typically, unstable solitons rearrange themselves into localized breathers, although on-site-centered staggered solitons may be subject to strong instability, which completely destroys them. The application of a kick to stable unstaggered solitons reveals their robust mobility, while all staggered solitons investigated in [36] were found to be immobile.

18.4 Solitons in the Salerno Model with Competing Inter-site and On-site Nonlinearities

18.4.1 The 1D Model

The presentation in this section, for the 1D and 2D versions of the SM (Salerno model) based on Eq. (18.8), with $\nu = 1$ and $\mu < 0$, follows [66, 67], respectively. A necessary ingredient of the analysis of the 1D model is the continuum approximation. To introduce it, one can set $u(x, t) \equiv e^{2it}\Psi(x, t)$, and expand $\Psi_{n\pm 1} \approx \Psi \pm \Psi_x + (1/2)\Psi_{xx}$, where Ψ is treated as a function of the continuous coordinate x, which coincides with n when it takes integer values. The accordingly derived continuum counterpart of the 1D version of Eq. (18.8) (with $\nu \equiv 1$ and $\mu < 0$, as said above) is

18 Related Models

$$i\Psi_t = -2(1-|\mu|)|\Psi|^2 \Psi - \left(1 - |\mu||\Psi|^2\right)\Psi_{xx}. \tag{18.18}$$

Further, soliton solutions to Eq. (18.18) are sought as $\Psi = e^{-i\omega t}U(x)$. Straightforward analysis demonstrates that such solutions exist for $|\mu| < 1$, in a finite frequency band, $0 < \Omega \equiv \mu\omega/(1-|\mu|) \leq 1$. The soliton's amplitude, $A \equiv U(x=0)$, is smaller than $1/\sqrt{|\mu|}$ for $0 < \Omega < 1$. At the edge of the band, $\Omega = 1$ (i.e., at point $\omega = 1 - |\mu|^{-1}$), an exact solution can be found, in the form of a *peakon*,

$$U_{\text{peakon}} = \left(1/\sqrt{|\mu|}\right)\exp\left(-\sqrt{(1/|\mu|)-1}|x|\right), \tag{18.19}$$

whose amplitude attains the maximum possible value, $A_{\max} = 1/\sqrt{|\mu|}$, and the norm is $\pi^2/[6\sqrt{|\mu|}(1-|\mu|)]$. The existence of the peakon is a specific feature of the SM in the case of competing nonlinearities.

Numerical analysis of the stationary 1D version of the discrete equation (18.8) demonstrates that, similarly to its continuum counterpart, it has a family of regular soliton solutions, bordered by a discrete peakon. The boundary value, $\mu = \mu_p(\omega)$, at which the discrete peakon is found for given frequency, is roughly approximated by the prediction of the continuum limit, $\mu_p \approx 1/(1-\omega)$. However, unlike the continuum equation (18.18), the discrete model gives rise to another family of soliton solutions, which are found *beyond* the peakon limit (i.e., at $|\mu| > |\mu_p(\omega)|$), in the form of *cuspons*. There are no counterparts to cuspons among solutions of continuum equation (18.18). Moreover, discrete cuspons can be found even at $|\mu| > 1$, where the continuum limit does not admit any soliton solutions. A characteristic feature of the cuspon is that it is more strongly localized at the center than in the exponentially decaying tails. Examples of all the three species of the discrete solitons mentioned above are displayed in Fig. 18.8.

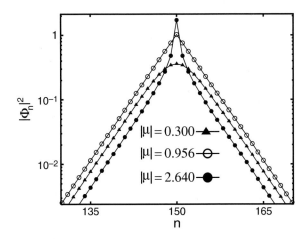

Fig. 18.8 Generic examples of three different types of discrete solitons, for $\omega = -2.091$, in the 1D Salerno model with competing nonlinearities, are shown on the logarithmic scale: a regular soliton at $\mu = -0.3$, a peakon at $\mu = -0.956$, and a cuspon at $\mu = -2.64$ (note that, in the latter case, $|\mu|$ exceeds 1). Reprinted from [66] with permission

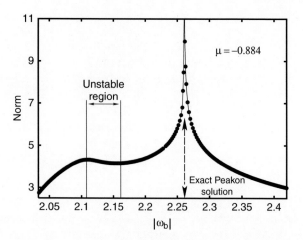

Fig. 18.9 The norm of the discrete solitons, in the 1D Salerno model with competing nonlinearities, vs. the soliton's frequency, for $\mu = -0.884$. Except for the explicitly marked unstable portion, the family is stable, including the entire cuspon subfamily, located to the right of the peakon. Reprinted from [66] with permission

The stability analysis for the discrete solitons in the SM with competing nonlinearities was performed through numerical computation of the respective eigenvalues, within the framework of the linearization of Eq. (18.8), and verified by direct simulations of the full equation. It was found that for the most part the soliton family is stable, except for a small portion, as shown in Fig. 18.9. In direct simulations, unstable solitons transform themselves into persistent localized breathers.

It is worthy to note that, as observed in Fig. 18.9, the stability and instability of regular discrete solitons precisely obeys the VK criterion, which, in the present case, takes the form of $dN/d\omega < 0$. On the other hand, the entire cuspon family is stable *contrary* to the VK condition (i.e., the criterion is valid for the regular solitons, but it does not apply to the cuspons).

Bound states of two discrete solitons were also investigated in this model, by means of the numerical continuation in μ, starting with the known bound states in the standard 1D DNLS equation, which corresponds to $\mu = 0$ [66]. The DNLS equation supports two different types of two-soliton bound states, in-phase and π-out-of-phase ones, which are represented, respectively, by symmetric and antisymmetric configurations, only the latter ones being stable [75, 76]. Bound states of both types can also be found in the SM with the competing nonlinearities, for different values of separation d between centers of the two solitons, see Fig. 18.10. A remarkable feature of this model is the *stability exchange* between the in-phase and out-of-phase bound states: with the increase of $-\mu$, precisely at point $\mu = \mu_p(\omega)$, where peakons exist for given soliton's frequency ω, the antisymmetric bound state loses its stability, while the symmetric one becomes stable. The character of the stability exchange implies that stable antisymmetric bound states are formed, at $|\mu| < |\mu_p(\omega)|$, by regular solitons, while stable symmetric complexes are built of two cuspons, at $|\mu| > |\mu_p(\omega)|$. Another notable feature of this effect is that, although the instability of unstable bound states is, naturally, weaker for larger d, the stability-exchange point *does not* depend on the separation between bound solitons.

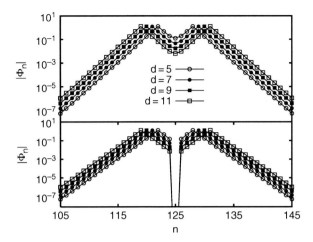

Fig. 18.10 The 1D Salerno model with the competing nonlinearities: Profiles of in-phase (*top*) and π-out-of-phase (*bottom*) bound states of two peakons, with different distances d between their centers, at $\omega = -3.086$ and $\mu = \mu_p(\omega) = -0.645$. The profiles are shown on the logarithmic scale, cf. Fig. 18.8. These two configurations are taken exactly at the point of the stability exchange between the in-phase and out-of-phase bound states, i.e., both are neutrally stable. Reprinted from [66] with permission

18.4.2 The 2D Model

In a rescaled form, the continuum limit of 2D equation (18.8) is (cf. Eq. (18.18))

$$i\Psi_t + \left(1 + \mu |\Psi|^2\right)\left(\Psi_{xx} + \Psi_{yy}\right) + 2[(1+\alpha)\mu + 1]|\Psi|^2 \Psi = 0. \quad (18.20)$$

Recall that, while α accounts for the anisotropy of the inter-site coupling in Eq. (18.8), the continuum limit is always tantamount to the isotropic equation. Similarly to its 1D counterpart, Eq. (18.20) gives rise to a family of regular axially symmetric solitons, in the form of $\Psi = e^{-i\omega t}U(r)$, where $r^2 \equiv x^2 + y^2$, with the amplitude bounded by $F(r=0) \le |\mu|^{-1/2}$, and the limiting case of $F(r=0) = |\mu|^{-1/2}$ corresponding to a *radial peakon*. However, the 2D peakon solution and the value of ω corresponding to it (for given $\mu < 0$) cannot be found in an exact form, unlike the 1D version of the model. It is also relevant to notice that, in both cases of positive and negative μ, the nonlinear dispersive term in Eq. (18.20), $\mu |\Psi|^2 (\Psi_{xx} + \Psi_{yy})$, prevents the onset of collapse in this equation.

As shown in Fig. (18.11), numerical solution of the stationary version of the 2D discrete equation (18.8) demonstrates the existence of the same two generic species of solitons which were found in the 1D counterpart of the model, i.e., regular discrete solitons and cuspons, which are separated by the peakon mode. The asymmetric variant of the model, with $\alpha \neq 1$, gives rise to an additional generic type of solutions, in the form of *semicuspons*, which feature the cuspon shape in one direction, and a regular shape in the other.

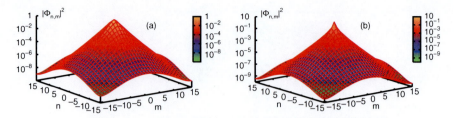

Fig. 18.11 Profiles of two discrete solitons found in the 2D isotropic ($\varepsilon = 1$) Salerno model with competing nonlinearities: (**a**) a regular soliton close to the peakon, at $\mu = -0.2$; (**b**) a cuspon, at $\mu = -0.88$. Reprinted from [67] with permission

The stability of the 2D discrete solitons was analyzed through the computation of the respective eigenvalues, which was verified by direct simulations of Eq. (18.8). Similar to the situation outlined above for the 1D version of the model, a small part of the family of regular 2D solitons was found to be unstable, while *all the 2D cuspons* are stable. The unstable solitons spontaneously transform themselves into persistent localized breathers. Another similarity to the 1D model can be observed through the comparison of the actual stability of the 2D solitons to the prediction of the VK criterion for them: the regular states exactly comply with the criterion, while the cuspons are stable contrary to it.

Also similar to what was found in the 1D model are fundamental properties of bound states of 2D fundamental solitons: out-of-phase and in-phase two-soliton complexes exchange their stability with the increase of $-\mu$ (recall that, in the usual 2D DNLS model, which corresponds to $\mu = 0$, only out-of-phase, alias antisymmetric, two-soliton bound states are stable [77]). As well as in the 1D version of the model, values of parameters at the stability-exchange point do not depend on the separation between the bound solitons (nor on the orientation of the bound state relative to the underlying 2D lattice). A new peculiarity, in comparison with the 1D model, is the fact that, exactly when they switch their stability, both the symmetric and antisymmetric bound states are built not of peakons, but of two cuspons.

A natural generalization of the 2D fundamental solitons are discrete vortices, which are well-known solutions in the ordinary 2D DNLS model [78]. A vortex is characterized by the phase circulation around its center, $\Delta\theta$, that must be a multiple of 2π, hence the vortex state may be labeled by the integer vorticity (topological charge) $S \equiv \Delta\theta/(2\pi)$. As mentioned above (in the context of the anisotropic version of the 2D DNLS model), two distinct types of lattice vortices can exist, on-site- and off-site-centered ones, alias *vortex crosses* and *vortex squares*. Examples of these two species of the vortex solitons found in the SM with competing nonlinearities ($\mu < 0$) are plotted in Fig. 18.12. In [67], vortex solitons in the SM were investigated for both $\mu < 0$ and $\mu > 0$.

The stability of both types of the vortices (crosses and squares) was identified through the computation of the full set of relevant eigenvalues for small perturbations, and verified in direct simulations. Stability regions for static vortices in parameter plane (μ, ω) exist, but they are narrow for the cross-shaped vortices, and

Fig. 18.12 Examples of discrete vortices with $S = 1$ in the isotropic ($\varepsilon = 1$) 2D Salerno model with the competing on-site and inter-site nonlinearities. Profiles of the real part of the "square"- and "cross"-shaped vortices are shown in the *top and bottom panels*, respectively. Both modes pertain to $\mu = -0.4$ and $C = 1$. Reprinted from [67] with permission

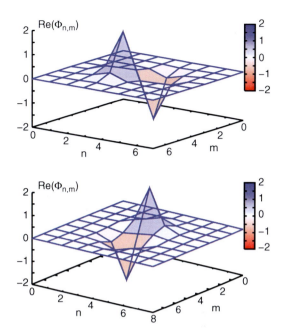

very narrow for the square-shaped ones. As in the usual 2D DNLS model [78], the destabilization of vortices occurs through the Hamiltonian Hopf bifurcation (i.e., through a quartet of complex eigenvalues). In the ordinary SM, with $\mu > 0$ (i.e., with noncompeting nonlinearities), unstable vortex solitons follow the same instability-development pattern as in the 2D DNLS equation [78], evolving into on-site-centered fundamental discrete solitons (ones with $S = 0$). However, an essentially novel finding is that unstable cross-shaped vortices in the model with $\mu < 0$ rearrange into very robust localized *vortical breathers*, which means that the topologically distinct vortex pattern *does not disappear* under the action of the instability but merely develops irregular modulations of its local amplitudes.

A similar instability-development pattern, which leads, in the model with $\mu < 0$, to the formation of robust vortical breathers, is demonstrated by unstable square-shaped vortices. A noteworthy peculiarity of the latter case is that the respective vortical breathers feature at least two distinct frequencies of intrinsic vibrations. One of them accounts for periodic transfer of the norm between four corners of the square vortex, following a path that can be schematically depicted by $(n_0, m_0) \to (n_0, m_0 + 1) \to (n_0 + 1, m_0 + 1) \to (n_0 + 1, m_0) \to (n_0, m_0)$. In addition to that, the amount of the norm circulating between the sites periodically varies in time, thus giving rise to the second frequency. Generally, in comparison with the vortical breathers generated by the unstable cross-shaped vortices, those emerging from unstable vortex squares feature much more regular internal dynamics.

Lastly, bound states of two cross-shaped vortices were also investigated in [67]. When $-\mu$ increases, a destabilizing bifurcation occurs in the equal-vorticity bound state, precisely at the same value of μ at which the instability of the corresponding out-of-phase bound state of two fundamental 2D solitons was found, as explained above.

18.5 One-Dimensional Solitons in the Semidiscrete System with the $\chi^{(2)}$ Nonlinearity

The semidiscrete models based on Eqs. (18.9) and (18.10), or (18.11) and (18.12) (recall these systems were defined as ones of types A and B, respectively), give rise to new specific families of composite solitons, as shown in work [70]. These solutions are sought for, respectively, in the following form:

$$\phi_n(\zeta) = e^{i\lambda\zeta}\bar{\phi}_n, \Psi(\eta,\zeta) = e^{2i\lambda\zeta}\bar{\Psi}(\eta); \tag{18.21}$$

$$\psi_n(\zeta) = e^{2i\lambda\zeta}\bar{\psi}_n, \Phi(\eta,\zeta) = e^{i\lambda\zeta}\bar{\Phi}(\eta), \tag{18.22}$$

where λ is the wavenumber. Similar to the ordinary discrete solitons, the composite ones can be odd or even: the odd modes are centered at a site of the discrete waveguide, whereas even solitons are inter-site-centered. Typical examples of odd and even composite solitons are displayed for both cases, A and B, in the left panels of Fig. 18.13.

The stability of the composite solitons can be analyzed by means of the VK criterion, in the form of $dP/d\lambda > 0$, where the total power, which includes contributions from both continuous and discrete components of the system, is

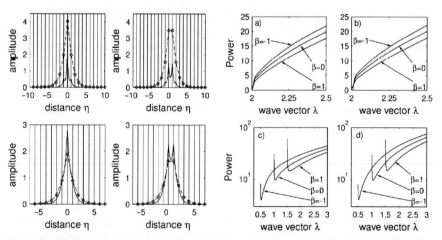

Fig. 18.13 *Left*: Profiles of composite solitons in the semidiscrete $\chi^{(2)}$ model. *Top and bottom panels* correspond to the solitons of types A and B, respectively, whereas *left and right subpanels* display odd and even solitons. *Vertical lines* designate the location of discrete waveguides. The mismatch is $\beta = 0$, and the soliton wavenumber is $\lambda = 3$ and $\lambda = 1.5$, in cases A and B, respectively. *Right*: The total power P of the families of composite solitons *vs*. wavenumber λ. Reprinted from [70] with permission

$$P(\lambda) = 2\int_{-\infty}^{+\infty} |\{\Psi, \Phi\}|^2 d\eta + \sum_n |\{\phi_n, \psi_n\}|^2, \tag{18.23}$$

in cases A and B, respectively. Curves $P(\lambda)$ for the four basic soliton families are displayed in the right panels of Fig. 18.13. Predictions of the VK criterion were completely corroborated by accurate computation of the respective stability eigenvalues.

As seen in the right panels of Fig. 18.13, the solitons exist (as might be expected) above the band of linear waves of the discrete subsystem, i.e., for $\lambda > 2$ in case A, and $\lambda > 1 + \beta/2$ in case B. Both odd and even solitons are stable in most of their existence domain. The latter result is significant, as even (inter-site-centered) solitons are never stable in the ordinary DNLS model. Further, a noteworthy property of the A-type solitons is that they exist up to $P = 0$, while the solitons of type B feature a cutoff in terms of P. In direct simulations, unstable composite solitons of the B-type decay into linear waves.

Twisted modes, built as antisymmetric bound states of two on-site-centered solitons, were also studied in [70]. It was found that the semidiscrete twisted states are stable in nearly the entire domain where they exist, on the contrary to continuous $\chi^{(2)}$ media, where twisted solitons exist too but are always unstable [79].

18.6 Conclusion and Perspectives

The discrete and semi-discrete models surveyed in this chapter feature a great variety of species of soliton states, and many different patterns of their dynamical behavior (especially, as concerns the stability). In most cases, their properties are notably different from those of the standard DNLS model.

A number of possibilities and problems which are suggested by these and allied models remain to be explored. First, none of them has been studied, as yet, in the 3D geometry. As concerns the model with the CQ on-site nonlinearity in 2D, some preliminary results were reported in [35]. As Fig. 18.14 illustrates, at sufficiently small values of the inter-site coupling constant (C), the model admits a large variety of coexisting stable and unstable discrete solitons. However, a systematic analysis of static (unstaggered and staggered) and mobile solitons in the 2D CQ model still has to be completed. The studies of this model may include the consideration of *semi-staggered* solitons, i.e., ones subjected to the staggering transformation only in one direction (similarly arranged *semi-gap solitons* were investigated in a BEC model based on the 2D GPE [80]).

It is known that a consistent derivation of the effective 1D dynamical equation for BEC trapped in a tight cigar-shaped confining potential leads to the *nonpolynomial* NLS equation [81–83]. If the cigar-shaped trap is combined with a deep OL potential acting in the axial direction, this gives rise to the DNLS equation with the nonpolynomial on-site nonlinearity,

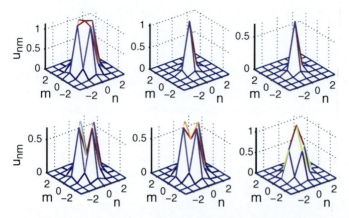

Fig. 18.14 A set of coexisting discrete solitons produced by the 2D version of Eqs. (18.4) and (18.16) for $\mu = -0.7$ and $C = 0.07$, as per [35]. The solitons in the *top and bottom rows* are stable and unstable, respectively. Reprinted from [35] with permission

$$i\frac{df_n}{dt} = -C(f_{n+1} + f_{n-1} - 2f_n) + \frac{1 - (3/2)g|f_n|^2}{\sqrt{1 - g|f_n|^2}} f_n, \qquad (18.24)$$

where constant $g > 0$ or $g < 0$ accounts for the attractive or repulsive interactions between atoms in the condensate. A unique feature of Eq. (18.24) is that, in the case of $g > 0$, this 1D discrete equation may give rise to *collapse*, when local density $|f_n|^2$ at any site attains the critical value, $1/g$. Soliton families supported by this new model were investigated very recently [84], and there remains to derive and explore a similar 2D discrete equation with a (different) nonpolynomial on-site nonlinearity, which would also serve as relevant BEC model.

As concerns 2D models with competing on-site and inter-site nonlinearity, a challenging problem is to construct fundamental solitons and vortices (if any) in the most interesting anisotropic version of such a model that would feature an attractive cubic inter-site coupling in one direction on the lattice, and a repulsive coupling in the other direction, while the on-site nonlinearity may be either repulsive or attractive. This configuration naturally corresponds to the dipolar BEC with atomic moments polarized in-plane [69].

Lastly, as concerns semidiscrete models for composite solitons, a natural extension of the $\chi^{(2)}$ model considered above is a two-component system with the $\chi^{(3)}$ nonlinearity that may describe an optical setting like the one shown above in Fig. 18.2, but made of a material with the ordinary $\chi^{(3)}$ (Kerr) nonlinearity. The respective system of equations for two optical fields, which corresponding to mutually orthogonal polarizations, can be derived in the following form:

$$i\frac{d\phi_n}{d\zeta} + \beta_d \phi_n + \phi_{n-1} + \phi_{n+1} + \phi_n |\phi_n|^2 + \kappa_d \phi_n |\Psi(\zeta, \eta = n)|^2 = 0, \qquad (18.25)$$

$$i\frac{\partial \Psi}{\partial \zeta} + \beta_c \Psi + \frac{1}{2}\frac{\partial^2 \Psi}{\partial \eta^2} + \Psi|\Psi|^2 + \kappa_c \Psi \sum_n |\phi_n|^2 \delta(\eta - n) = 0, \qquad (18.26)$$

where β_d and β_c are propagation constants, and coefficients κ_d and κ_c account for the nonlinear coupling between the fields. This model gives rise to its specific families of composite solitons [85]. It remains to extend this system, as well as its $\chi^{(2)}$ counterparts, based on Eqs. (18.9) and (18.10) or (18.11) and (18.12), into the 2D geometry.

References

1. Christodoulides, D.N., Joseph, R.I.: Opt. Lett. **13**, 794 (1988)
2. Eisenberg, H.S., Silberberg, Y., Morandotti, R., Boyd, A.R., Aitchison, J.S.: Phys. Rev. Lett. **81**, 3383 (1998)
3. Szameit, A., Blömer, D., Burghoff, J., Schreiber, T., Pertsch, T., Nolte, S., Tünnermann, A., Lederer, F.: Opt. Exp. **13**, 10552 (2005)
4. Szameit, A., Burghoff, J., Pertsch, T., Nolte, S., Tünnermann, A., Lederer, F.: Opt. Express **14**, 6055 (2006)
5. Pertsch, T., Peschel, U., Kobelke, J., Schuster, K., Bartelt, H., Nolte, S., Tünnermann, A., Lederer, F.: Phys. Rev. Lett. **93**, 053901 (2004)
6. Trombettoni, A., Smerzi, A.: Phys. Rev. Lett. **86**, 2353 (2001)
7. Alfimov, G.L., Kevrekidis, P.G., Konotop, V.V., Salerno, M.: Phys. Rev. E **66**, 046608 (2002)
8. Carretero-González, R., Promislow, K.: Phys. Rev. A **66**, 033610 (2002)
9. Cataliotti, F.S., Burger, S., Fort, C., Maddaloni, P., Minardi, F., Trombettoni, A., Smerzi, A., Inguscio, M.: Science **293**, 843 (2001)
10. Efremidis, N.K., Christodoulides, D.N.: Phys. Rev. A **67**, 063608 (2003)
11. Porter, M.A., Carretero-González, R., Kevrekidis, P.G., Malomed, B.A.: Chaos **15**, 015115 (2005)
12. Heebner, J.E., Boyd, R.W.: J. Mod. Opt. **49**, 2629 (2002)
13. Chak, P., Sipe, J.E., Pereira, S.: Opt. Lett. **28**, 1966 (2003)
14. Kevrekidis, P.G., Frantzeskakis, D.J., Carretero-González, R., Malomed, B.A., Bishop, A.R.: Phys. Rev. E **72**, 046613 (2005)
15. Vinetskii, V.O., Kukhtarev, N.V.: Sov. Phys. Solid State **16** (1975) 2414
16. Stepić, M., Kip, D., Hadžievski, L., Maluckov, A.: Phys. Rev E **69** (2004) 066618
17. Hadžievski, L., Maluckov, A., Stepić, M., Kip, D.: Phys. Rev. Lett. **93**, 033901 (2004)
18. Khare, A., Rasmussen, K.Ø., Samuelsen, M.R., Saxena, A.: J. Phys. A Math. Gen. **38**, 807 (2005)
19. Öster, M., Johansson, M., Eriksson, A.: Phys. Rev. E **73**, 046602 (2006)
20. Chen, F., Stepić, M., Rüter, C.E., Runde, D., Kip, D., Shandarov, V., Manela, O., Segev, M.: Opt. Exp. **13** 4314 (2005).
21. Smektala, F., Quemard, C., Couderc, V., Barthélémy, A.: J. Non-Cryst. Solids **274**, 232 (2000)
22. Boudebs, G., Cherukulappurath, S., Leblond, H., Troles, J., Smektala, F., Sanchez, F.: Opt. Commun. **219**, 427 (2003)
23. Zhan, C., Zhang, D., Zhu, D., Wang, D., Li, Y., Li, D., Lu, Z., Zhao, L., Nie, Y., J. Opt. Soc. Am. B **19**, 369 (2002)
24. Laedke, E.W., Spatschek, K.H., Turitsyn, S.K.: Phys. Rev. Lett. **73**, 1055 (1994)
25. Flach, S., Kladko, K., MacKay, R.S.: Phys. Rev. Lett. **78**, 1207 (1997)
26. Malomed, B.A., Weinstein, M.I.: Phys. Lett. A **220**, 91 (1996)
27. Pushkarov, Kh.I., Pushkarov, D.I., Tomov, I.V.: Opt. Quant. Electr. **11**, 471 (1979)
28. Pushkarov, Kh.I., Pushkarov, D.I.: Rep. Math. Phys. **17**, 37 (1980)

29. Cowan, S., Enns, R.H., Rangnekar, S.S., Sanghera, S.S.: Can. J. Phys. **64**, 311 (1986)
30. Herrmann, J.: Opt. Commun. **87**, 161 (1992)
31. Merhasin, I.M., Gisin, B.V., Driben, R., Malomed, B.A.: Phys. Rev. E **71**, 016613 (2005)
32. Wang, J., Ye, F., Dong, L., Cai, T., Li, Y.-P.: The cubic-quintic nonlinearity is combined with a 1D sinusoidal potential. Phys. Lett. A **339**, 74 (2005)
33. Abdullaev, F., Abdumalikov, A., Galimzyanov, R.: Phys. Lett. A **367**, 149 (2007)
34. Driben, R., Malomed, B.A., Gubeskys, A., Zyss, J.: Phys. Rev. E **76**, 066604 (2007)
35. Carretero-González, R., Talley, J.D., Chong, C., Malomed, B.A.: Physica D **216**, 77 (2006)
36. Maluckov, A., Hadžievski, L., Malomed, B.A.: Phys. Rev. E **77**, 036604 (2008)
37. Maluckov, A., Hadžievski, L., Malomed, B.A.: Phys. Rev. E **76**, 046605 (2007)
38. Agranovich, V.M., Dubovskii, O.A., Orlov, A.V.: Solid State Commun. **72**, 491 (1989)
39. Dubovskii, O.A., Orlov, A.V.: Phys. Solid State **41**, 642 (1999)
40. Iwanow, R., Stegeman, G., Schiek, R., Min, Y., Sohler, W.: Phys. Rev. Lett. **93**, 113902 (2004)
41. Darmanyan, S., Kobyakov, A., Lederer, F.: Phys. Rev. E **57**, 2344 (1998)
42. Konotop, V.V., Malomed, B.A.: Phys. Rev. B **61**, 8618 (2000)
43. Malomed, B.A., Kevrekidis, P.G., Frantzeskakis, D.J., Nistazakis, H.E., Yannacopoulos, A.N.: Phys. Rev. E **65**, 056606 (2002)
44. Iwanow, R., et al.: Opt. Express **13**, 7794 (2005)
45. Sukhorukov, A.A., et al.: Phys. Rev. E **63**, 016615 (2001)
46. Egorov, O., Peschel, U., Lederer, F.: Phys. Rev. E **71**, 056612 (2005)
47. Egorov, O., Peschel, U., Lederer, F.: Phys. Rev. E **72**, 066603 (2005)
48. Molina, M.I., Vicencio, R.A., Kivshar, Y.S.: Phys. Rev. E **72**, 036622 (2005)
49. Susanto, H., Kevrekidis, P.G., Carretero-González, R., Malomed, B.A., Frantzeskakis, D.J.: Phys. Rev. Lett. **99**, 214103 (2007)
50. Kevrekidis, P.G., Malomed, B.A., Saxena, A., Bishop, A.R., Frantzeskakis, D.J.: Physica D **183**, 87 (2003)
51. Gaididei, Yu.B., Mingaleev, S.F., Christiansen, P.L., Rasmussen, K.Ø.: Phys. Rev. E **55**, 6141 (1997)
52. Rasmussen, K.Ø., Christiansen, P.L., Johansson, M., Gaididei, Yu.B., Mingaleev, S.F.: Physica D **113**, 134 (1998)
53. Comech, A., Cuevas, J., Kevrekidis, P.G.: Physica D **207**, 137 (2005)
54. Kevrekidis, P.G., Gaididei, Yu.B., Bishop, A.R., Saxena, A.: Phys. Rev. E **64**, 0666606 (2001)
55. Ablowitz, M.J., Ladik, J.F.: J. Math. Phys. **17**, 1011 (1976)
56. Salerno, M.: Phys. Rev. A **46**, 6856 (1992)
57. Salerno, M.: Encyclopedia of Nonlinear Science. Scott, A. (ed.), p. 819. Routledge, New York (2005)
58. Cai, D., Bishop, A.R., Grønbech-Jensen, N.: Phys. Rev. E **53**, 4131 (1996)
59. Cai, D., Bishop, A.R., Grønbech-Jensen, N.: Phys. Rev. E **56**, 7246 (1997)
60. Dmitriev, S.V., Kevrekidis, P.G., Malomed, B.A., Frantzeskakis, D.J.: Phys. Rev. E **68**, 056603 (2003)
61. Gómez-Gardeñes, J., Falo, F., Floría, L.M.: Phys. Lett. A **332**, 213 (2004)
62. Gómez-Gardeñes, J., Floría, L.M., Peyrard, M., Bishop, A.R.: Chaos **14**, 1130 (2004)
63. Christiansen, P.L., Gaididei, Yu.B., Rasmussen, K.Ø., Mezentsev, V.K., Juul Rasmussen, J.: Phys. Rev B **54**, 900 (1996)
64. Christiansen, P.L., Gaididei, Yu.B., Mezentsev, V.K., Musher, S.L., Rasmussen, K.Ø., Juul Rasmussen, J., Ryzhenkova, I.V., Turitsyn, S.K.: Phys. Scripta **67**, 160 (1996)
65. Gómez-Gardeñes, J., Floría, L.M., Bishop, A.R.: Physica D **216**, 31 (2006)
66. Gomez-Gardeñes, J., Malomed, B.A., Floría, L.M., Bishop, A.R.: Phys. Rev. E **73**, 036608 (2006)
67. Gomez-Gardeñes, J., Malomed, B.A., Floría, L.M., Bishop, A.R.: Phys. Rev. E **74**, 036607 (2006)
68. Li, Z.D., He, P.B., Li, L., Liang, J.Q., Liu, W.M.: Phys. Rev. A **71**, 053611 (2005)
69. Tikhonenkov, I., Malomed, B.A., Vardi, A.: Phys. Rev. Lett. **100**, 090406 (2008)

70. Panoiu, N.C., Osgood, R.M., Malomed, B.A.: Opt. Lett. **31**, 1097 (2006)
71. MacKay, R.S., Aubry, S.: Nonlinearity **7**, 1623 (1994)
72. Vakhitov, M.G., Kolokolov, A.A.: Sov. J. Radiophys. Quantum Electr. **16**, 783 (1973)
73. Bergé, L.: Phys. Rep. **303**, 260 (1998)
74. Weinstein, M.I.: Nonlinearity **12**, 673 (1999)
75. Kapitula, T., Kevrekidis, P.G., Malomed, B.A.: Phys. Rev. E **63**, 036604 (2001)
76. Pelinovsky, D.E., Kevrekidis, P.G., Frantzeskakis, D.J.: Physica D **212**, 1 (2005)
77. Kevrekidis, P.G., Malomed, B.A., Bishop, A.R.: J. Phys. A Math. Gen. **34**, 9615 (2001)
78. Malomed, B.A., Kevrekidis, P.G.: Phys. Rev. E **64**, 026601 (2001)
79. Buryak, A.V., Di Trapani, P., Skryabin, D.V., Trillo, S.: Phys. Rep. **370**, 63 (2002)
80. Baizakov, B.B., Malomed, B.A., Salerno, M.: Eur. Phys. J. **38**, 367 (2006)
81. Salasnich, L.: Laser Phys. **12**, 198 (2002)
82. Salasnich, L., Parola, A., Reatto, L.: Phys. Rev. A **65**, 043614 (2002)
83. Salasnich, L., Parola, A., Reatto, L.: Phys. Rev. A **66**, 043603 (2002)
84. Maluckov, A., Hadžievski, L., Malomed, B.A., Salasnich, L.: Phys. Rev. A **78**, 013616 (2008)
85. Panoiu, N.-C., Malomed, B.A., Osgood, R.M.: Phys. Rev. A **78**, 013801 (2008)

Chapter 19
DNLS with Impurities

Jesús Cuevas and Faustino Palmero

19.1 Introduction

The past few years have witnessed an explosion of interest in discrete models and intrinsic localized modes (discrete breathers or solitons) that has been summarized in a number of recent reviews [1–3]. This growth has been motivated by numerous applications of nonlinear dynamical lattice models in areas as broad and diverse as the nonlinear optics of waveguide arrays [4], the dynamics of Bose–Einstein condensates in periodic potentials [5, 6], micro-mechanical models of cantilever arrays [7], or even simple models of the complex dynamics of the DNA double strand [8]. Arguably, the most prototypical model among the ones that emerge in these settings is the discrete nonlinear Schrödinger (DNLS) equation, the main topic of this book.

While DNLS combines two important features of many physical lattice systems, namely nonlinearity and periodicity, yet another element which is often physically relevant and rather ubiquitous is disorder. Localized impurities are well known in a variety of settings to introduce not only interesting wave-scattering phenomena [9], but also to create the possibility for the excitation of impurity modes, which are spatially localized oscillatory states at the impurity sites [10, 11]. Physical applications of such phenomena arise, e.g., in superconductors [12, 13], in the dynamics of the electron–phonon interactions [14, 15], in the propagation of light in dielectric superlattices with embedded defect layers [16] or in defect modes arising in photonic crystals [17, 18].

In the context of the DNLS, there has been a number of interesting studies in connection to the interplay of the localized modes with impurities. Some of the initial works were either at a quasicontinuum limit (where a variational approximation could also be implemented to examine this interplay) [19] or at a more discrete level but with an impurity in the coupling [20] (see also in the latter setting

---

J. Cuevas (✉)
Grupo de Física No Lineal, Departamento de Física Aplicada I, Escuela Universitaria Politécnica.
Universidad de Sevilla. C/ Virgen de África, 7, 41011 Sevilla, Spain
e-mail: jcuevas@us.es

the more recent studies of a waveguide bend [21, 22] and the boundary defect case of [23]). More recently the experimental investigations of [24, 25] motivated the examinations of linear and nonlinear defects in a DNLS context [26–28].

The aim of this chapter is to summarize the properties of the DNLS equation in the presence of a linear impurity, as shown previously in [29]. Our first aim is to present the full bifurcation diagram of the stationary localized modes in the presence of the impurity and how it is drastically modified in comparison to the case of the homogeneous lattice. The relevant bifurcations are quantified whenever possible even analytically, in good agreement with our full numerical computations. A second aim is to show the outcome of the interaction of an incoming solitary wave with the linear impurity.

The DNLS equation with the defect can be written as

$$i\dot{u}_n + \gamma |u_n|^2 u_n + \epsilon(u_{n+1} + u_{n-1}) + \alpha_n u_n = 0, \quad (n = 1\ldots N) \quad (19.1)$$

where α_n allows for the existence of local, linear inhomogeneities. Hereafter, we consider a single point defect, thus $\alpha_n = \alpha \delta_{n,0}$, that can be positive (attractive impurity) or negative (repulsive impurity). In general, the presence of an on-site defect would affect the nearest-neighbor coupling, and Eq. (19.1) should be modified to take this effect into account, as in [30]. This inhomogeneity in the coupling, however, can be avoided using different techniques, for example, in nonlinear waveguide arrays, changing slightly the separation between the defect waveguide and its nearest neighbors, as in the case of [31]. We will assume here that the coupling parameter ϵ is independent of the site and positive.

Note that the defocusing case can be reduced, under the staggering transformation $u_n \to (-1)^n u_n$, to the previous one with opposite sign of the impurity α. Also, under the transformation $u_n \to u_n e^{2i\epsilon t}$, Eq. (19.1) can be written in the standard form

$$i\dot{u}_n + |u_n|^2 u_n + C \Delta u_n + \alpha_n u_n = 0, \quad (19.2)$$

In what follows, we use the form given by Eq. (19.1).

19.2 Stationary Solutions

In this part, we look for stationary solutions with frequency Λ, $u_n(t) = e^{i\Lambda t} v_n$, and the stationary analog of Eq. (19.1) then reads

$$-\Lambda u_n + \epsilon(u_{n+1} + u_{n-1}) + u_n^3 + \alpha_n u_n = 0. \quad (19.3)$$

19.2.1 Linear Modes

Some of the properties of solitons are related to the characteristics of linear localized modes. These modes arise when an inhomogeneity appears and can be obtained from the linearized form (around the trivial solution $u_n = 0, \forall\, n$) of Eq. (19.3). In this case, and considering an inhomogeneity located at the first site of the chain and with periodic boundary conditions, the problem reduces to solving the eigenvalue problem

$$\begin{bmatrix} \alpha & \epsilon & 0 & . & . & \epsilon \\ \epsilon & 0 & \epsilon & 0 & . & 0 \\ 0 & \epsilon & 0 & \epsilon & . & . \\ . & . & . & . & . & . \\ . & . & . & \epsilon & 0 & \epsilon \\ \epsilon & . & . & . & \epsilon & 0 \end{bmatrix} \begin{bmatrix} v_0 \\ v_1 \\ . \\ . \\ v_{N-2} \\ v_{N-1} \end{bmatrix} = \Lambda \begin{bmatrix} v_0 \\ v_1 \\ . \\ . \\ v_{N-2} \\ v_{N-1} \end{bmatrix}, \qquad (19.4)$$

that is, a particular case of the eigenvalue problem studied in [32]. There it was shown that, if $\alpha \neq 0$, the solution corresponds to $N-1$ extended modes and an impurity-localized mode. Also, if N becomes large, the frequencies of extended modes are densely distributed in the interval $\Lambda \in [-2\epsilon, 2\epsilon]$ and the localized mode can be approximated by

$$v_n = s^n v_0 \left[\left(\frac{\alpha}{2\epsilon} + \beta\right)^{-n} + s^N \left(\frac{\alpha}{2\epsilon} + \beta\right)^{n-N} \right], \quad \Lambda = 2s\epsilon\beta, \quad \beta \equiv \sqrt{\frac{\alpha^2}{4\epsilon^2} + 1} \qquad (19.5)$$

with $s = \text{sign}(\alpha)$ and v_0 an arbitrary constant. Note that for $\alpha > 0$ ($\alpha < 0$) the localized mode has an in-phase (staggered) pattern. In Fig. 19.1 we depict the linear

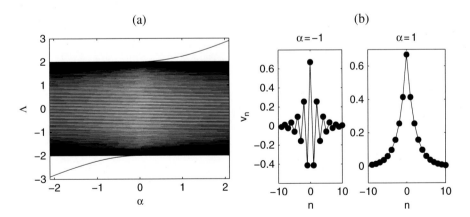

Fig. 19.1 (a) Linear modes spectrum as function of impurity parameter α. Periodic boundary conditions are considered. (b) Examples of the profiles of the impurity modes. The impurity is located at $n = 0$. (*left*) Profile for $\alpha = -1$; (*right*) profile for $\alpha = 1$. In all cases $N = 200$ and $\epsilon = 1$. Reprinted from [29] with permission

mode spectrum as a function of the inhomogeneity parameter α and examples of the profiles of the ensuing localized modes.

19.2.2 Bifurcations

In order to explore the existence and stability of the nonlinear stationary states described by Eq. (19.3), we have used the well-known technique based on the concept of continuation from the anticontinuum (AC) limit using a Newton–Raphson algorithm. Also, a standard linear stability analysis of these stationary states has been performed.

In the homogeneous lattice case of $\alpha = 0$, fundamental stationary modes are well known to exist and be centered either on a lattice site or between two adjacent lattice sites. The site-centered solitary waves are always stable, while the inter-site-centered ones are always unstable [33].

In order to study the effects of the inhomogeneity on the existence and properties of localized modes, we have performed a continuation from the homogeneous lattice case of $\alpha = 0$. We found that if α increases ($\alpha > 0$, attractive impurity case), the amplitude of the stable on-site mode decreases, while if α decreases ($\alpha < 0$, repulsive impurity case), in general, the stable on-site soliton localized at the impurity merges with the unstable inter-site-centered one localized between the impurity and its neighboring site (beyond some critical value of $|\alpha|$) and the resulting state becomes unstable. Notice that, at heart, the latter effect is a pitchfork bifurcation as the on-site mode collides with both the inter-site mode centered to its right and the one centered to its left.

In Fig. 19.2 we show a typical bifurcation scenario where, for fixed values of Λ and ϵ, we depict the mode power P corresponding to different on-site and inter-site localized modes as a function of impurity parameter α. If we denote as n_0 the site of the impurity, when $\alpha > 0$ increases, we found that the unstable inter-site soliton localized at $n = 0.5$ disappears in a saddle-node bifurcation with the stable on-site soliton localized at $n = 1$. Also, if we continue this stable mode, when α decreases, and for a given value $\alpha = \alpha_c < 0$, it also disappears together with the unstable mode localized at $n = 1.5$ through a saddle-node bifurcation. If we increase again the impurity parameter, this unstable mode localized at $n = 1.5$ bifurcates with the stable site mode localized at $n = 2$ for a critical value of parameter $\alpha = \alpha'_c > 0$ through a saddle-node bifurcation again, and it could be possible to continue this bifurcation pattern until a site k, where the value of site k increases with the value of ϵ and Λ parameters. This scenario is similar to the one found in previous studies with different kinds of impurities [21, 27] and appears to be quite general. It should be noted that when the coupling parameter increases, more bifurcations take place, in a narrower interval of power P and impurity parameter α values.

Some of the particularly interesting experimentally tractable suggestions that this bifurcation picture brings forward are the following:

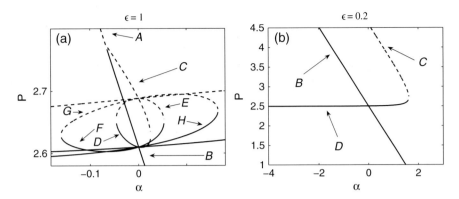

Fig. 19.2 Bifurcation diagram of stable (*solid line*) and unstable (*dashed line*) nonlinear modes. Shown is the power P as a function of the impurity parameter α. In all cases $N = 100$ and $\Lambda = 2.5$. The branch designation is as follows: (**A**) Unstable soliton centered at the impurity ($n = 0$), (**B**) stable on-site soliton centered at $n = 0$, (**C**) Unstable inter-site soliton centered at $n = 0.5$, (**D**) stable on-site soliton at $n = 1$, (**E**) unstable inter-site soliton at $n = 1.5$, (**F**) stable on-site soliton at $n = 2$, (**G**) unstable inter-site soliton at $n = 2.5$, and (**H**) stable on-site soliton at $n = 3$. The stable on-site mode located at the impurity, in the homogeneous case ($\alpha = 0$), disappears for a coupling value of $\epsilon \simeq 1.25$ due to resonances with the phonon band. Reprinted from [29] with permission

- A localized mode centered at the impurity may be impossible for sufficiently large attractive impurities (because the amplitude of the mode may decrease to zero), while it may be impossible to observe also in the defocusing case due to the instability induced by the pitchfork bifurcation with its neighboring inter-site configurations.
- A localized on-site mode centered at the neighborhood of the impurity should not be possible to localize for sufficiently large impurity strength both in the attractive *and* in the repulsive impurity case.

19.2.3 Invariant Manifold Approximation

This subsection shows a more detailed study of the bifurcation between the on-site nonlinear mode centered at the impurity and its inter-site and one-site neighbor. From the discussion of the previous section we can determine for a given value of the coupling parameter ϵ, the corresponding critical value of impurity parameter $\alpha = \alpha_c$. Note that this bifurcation takes place only if α is negative (repulsive impurity). In case of α positive (attractive impurity), the inter-site solution disappears in a saddle-node bifurcation with the on-site wave centered at the site next to the impurity at $\alpha = \alpha'_c$. In these cases, via an analysis of invariant manifolds of the DNLS map, and following the method developed in Sect. 4.1.4 of [34], some approximate analytical expressions corresponding to this bifurcation point can be obtained. This method is sketched below.

The difference equation (19.3), for $\alpha = 0$, can be recast as a 2D real map by defining $y_n = v_n$ and $x_n = v_{n-1}$ [35]:

$$\begin{cases} x_{n+1} = y_n \\ y_{n+1} = (\Lambda y_n - y_n^3)/\epsilon - x_n. \end{cases} \quad (19.6)$$

For $\Lambda > 2$, the origin $x_n = y_n = 0$ is hyperbolic and a saddle point. Consequently, there exists a 1D stable ($W^s(0)$) and a 1D unstable ($W^u(0)$) manifolds emanating from the origin in two directions given by $y = \lambda_\pm x$, with

$$\lambda_\pm = \frac{\Lambda \pm \sqrt{\Lambda^2 - 4\epsilon^2}}{2\epsilon}. \quad (19.7)$$

These manifolds intersect in general transversally, yielding the existence of an infinity of homoclinic orbits. Each of their intersections corresponds to a localized solution. Fundamental solitons (i.e., on-site and inter-site solitons) correspond to the primary intersection points, i.e., those emanating from the first homoclinic windings. Each intersection point defines an initial condition (x_0, y_0), that is (v_{-1}, v_0), and the rest of the points composing the soliton are determined by application of the map (19.6) and its inverse. Figure 19.3 shows an example of the first windings of the manifolds. Intersections corresponding to fundamental solitons are labeled as follows: (1) is the on-site soliton centered at $n = 0$, (2) is the inter-site soliton centered at $n = 0.5$, and (3) is the on-site soliton centered at $n = 1$.

The effect of the inhomogeneity is introduced as a linear transformation of the unstable manifold $A(\alpha)W^u(0)$ with $A(\alpha)$ given by

$$A(\alpha) = \begin{bmatrix} 1 & 0 \\ -\alpha/C & 1 \end{bmatrix}. \quad (19.8)$$

When $\alpha > 0$, the unstable manifold moves downwards, changing the intersections between the transformed unstable manifold and the stable manifold to points $1'$, $2'$, and $3'$ (see Fig. 19.3). For $\alpha = \alpha_c$, both manifolds become tangent. Thus, for $\alpha > \alpha_c'$ intersections $3'$ and $2'$ are lost, that is, for $\alpha = \alpha_c'$ the breathers centered at $n = 1$ and $n = 0.5$ experience a tangent bifurcation. On the contrary, if $\alpha < 0$, intersections $1'$ and $2'$ are lost when $|\alpha| > |\alpha_c|$, leading to a bifurcation between the breathers centered at $n = 0.5$ and $n = 0$.

A method for estimating $\alpha_c(\Lambda)$ and $\alpha_c'(\Lambda)$ is based on a simple approximation of $W^u(0)$. Let us consider a cubic approximation W^u_{app} of the local unstable manifold of Fig. 19.3, parametrized by $y = \lambda x - c^2 x^3$, with $\lambda \equiv \lambda_+$. The coefficient c depends on Λ and C and need not be specified in what follows (a value of c suitable when λ is large is computed in [36]). We have

$$y = \lambda_0 x - c^2 x^3 \quad (19.9)$$

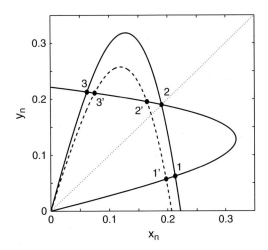

Fig. 19.3 First winding of the homoclinic tangle of the map (19.6). *The dashed line* corresponds to the linear transformed unstable manifold when $\alpha = 0$. Labels 1, 2, 3 (1′, 2′, 3′) correspond to fundamental solitons for $\alpha = 0$ ($\alpha \neq 0$). Reprinted from [29] with permission

on the curve $A(\alpha)W^u_{\text{app}}$, where $\lambda_0 = \lambda - \alpha/C$. By symmetry we can approximate the local stable manifold using the curve W^s_{app} parametrized by

$$x = \lambda y - c^2 y^3. \tag{19.10}$$

The curves $A(\alpha)W^u_{\text{app}}$ and W^s_{app} become tangent at (x, y) when in addition

$$\left(\lambda - 3c^2 x^2\right)\left(\lambda_0 - 3c^2 y^2\right) = 1. \tag{19.11}$$

In order to compute α_c and α'_c as a function of Λ, or, equivalently, the corresponding value of λ_0 as a function of λ, one has to solve the nonlinear system (19.9), (19.10) and (19.11) with respect to x, y, λ_0, which yields a solution depending on λ. Instead of using λ it is practical to parametrize the solutions by $t = y/x$. This yields

$$x = \frac{1}{c\sqrt{2}}\left(t + \frac{1}{t^3}\right)^{1/2}, \quad y = \frac{t}{c\sqrt{2}}\left(t + \frac{1}{t^3}\right)^{1/2}; \quad \lambda_0 = \frac{3}{2}t + \frac{1}{2t^3}, \quad \lambda = \frac{3}{2t} + \frac{1}{2}t^3.$$

Since $\lambda + \lambda^{-1} = \Lambda/\epsilon$ it follows that

$$t^4 - 2\lambda t + 3 = 0, \tag{19.12}$$

$$\alpha = \frac{\epsilon}{2}\left(t - \frac{1}{t}\right)^3. \tag{19.13}$$

As this system of equations has two real positive solutions, given a value of Λ, one can approximate α_c and α'_c by the values of α given by Eqs. (19.12)–(19.13). Despite the fact that it gives precise numerical results in a certain parameter range,

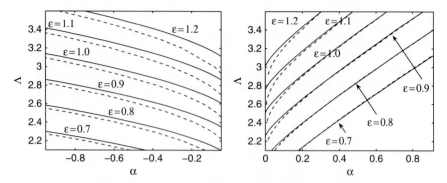

Fig. 19.4 Bifurcation loci corresponding to the bifurcation between the on-site-localized mode at the impurity ($n = 0$) and its neighbor inter-site breather ($n = 0.5$) (*left panel*), and to the bifurcation between the on-site-localized mode next to the impurity ($n = 1$) and its neighbor inter-site breather ($n = 0.5$) (*right panel*), for different values of parameter ϵ. *Dashed lines* correspond to numerical results and *continuous lines* to approximate analytical calculations. Reprinted from [29] with permission

the approximation (19.12)–(19.13) is not always valid. Indeed, the parameter regime $\Lambda < 5\epsilon/2$ is not described within this approximation (see [36]).

Figure 19.4 shows the comparison between the exact numerical and the approximate analytical results. For a fixed value of the coupling parameter ϵ, the critical value of the frequency increases with $|\alpha|$.

19.3 Interaction of a Moving Soliton with a Single Impurity

Early studies of the DNLS had shown that discrete solitary waves in the DNLS can propagate along the lattice with a relatively small loss of energy [37], and more recent work suggests that such (almost freely) propagating solutions might exist, at least for some range of control parameters [38–40]; nevertheless, genuinely traveling single-hump solitary wave solutions are not present in the DNLS, but only in variants of that model such as the ones with saturable nonlinearity [41–43].

In this section we deal with the interaction of propagating (with only weak radiative losses) localized modes with the impurity. Thus, we consider a nonlinear localized mode, far enough from the impurity, of frequency Λ, and perturb it by adding a thrust q to a stationary breather v_n [44], so that

$$u_n(t = 0) = v_n e^{iqn}. \tag{19.14}$$

This is similar in spirit to the examination of [26], although we presently examine both attractive and repulsive impurities. In what follows, $\Lambda = 2.5$ and $\epsilon = 1$; a similar scenario emerges for other values of Λ.

In general, if q is large enough, the soliton moves with a small loss of radiation. We have calculated, as a function of parameters q and α, the power and energy that

19 DNLS with Impurities

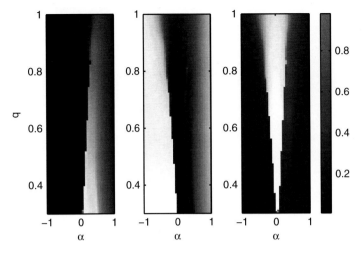

Fig. 19.5 Power trapping (*left*), reflection (*center*), and transmission (*right*) coefficients as function of impurity parameter α and initial thrust q. In all cases $\epsilon = 1$. Reprinted from [29] with permission

remain trapped by the impurity, or get reflected and transmitted along the chain, and determined the corresponding coefficients of trapping, reflection, and transmission, defined as the corresponding fraction of power (energy). Figure 19.5 summarizes the relevant numerical results.

We can essentially distinguish four fundamental regimes:

(a) *Trapping.* If parameters q and α are small enough, and the impurity is attractive, nearly all the energy remains trapped at the impurity, and only a small fraction of energy is lost by means of phonon radiation. An example of this phenomenon is shown in Fig. 19.6 (left). In this case, the central power (power around the impurity) before the collision is nearly zero. When the localized mode reaches the impurity, the former loses power as phonon radiation and remains trapped. The analysis of the Fourier spectrum of this trapped breather, carried out after the initial decay and at an early stage of the evolution, shows a frequency close to the initial soliton frequency, as shown in Fig. 19.6 (right). We have observed that, in general, this frequency is slightly smaller than that of the incident soliton, and, in consequence, it has even smaller energy (in absolute value) and power than the corresponding nonlinear mode with the frequency of incident soliton. In this particular case, corresponding to $q = 0.3$ and $\alpha = 0.2$, the initial incident wave (after perturbation) has power $P = 2.61$ and energy $E = -5.40$ and the stationary mode, trapped at the impurity, with the same frequency, has $P = 2.17$ and $E = -4.73$. Thus, the incident breather can activate this nonlinear mode, and nearly all energy and norm remain trapped.

(b) *Trapping and reflection.* If the impurity is attractive, but strong enough, some fraction of energy remains trapped by the impurity, but a considerable amount of it is reflected. The reflected excitation remains localized. This case is similar to

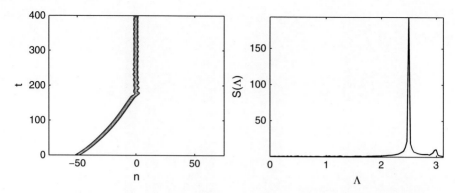

Fig. 19.6 Trapping: Contour plot corresponding to the power of the soliton P as function of site n and time t (*left panel*) and Fourier power spectrum of the trapped soliton calculated soon after the collision (*right panel*). The parameters are $\alpha = 0.2$, $q = 0.3$, $\Lambda = 2.5$, $\epsilon = 1$ and the impurity is located at $n = 0$. Reprinted from [29] with permission

the previous one, but now the incident traveling structure has enough energy and norm to excite a stationary mode centered at the impurity, remaining localized and giving rise to a reflected pulse. A typical case is shown in Fig. 19.7 that corresponds to $q = 0.6$ and $\alpha = 1.0$. The incident wave has power and energy $P = 2.61$ and $E = -4.79$, and the stationary nonlinear mode centered at the impurity, with the same frequency, $P = 0.76$ and $E = -1.79$. When the incident breather reaches the impurity, it excites the nonlinear mode, and, after losing some power, part of it remains localized, and another part is reflected. Also, in our numerical simulations, we have detected, as in the previous case, that the frequency of the remaining trapped mode is slightly lower than that of the incident breather, so it has even smaller power than the corresponding nonlinear mode with the frequency of incident soliton.

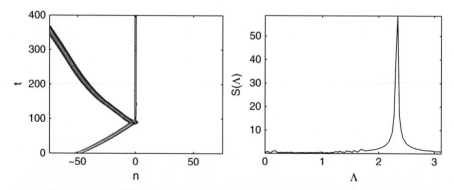

Fig. 19.7 Trapping and reflection: Contour plot corresponding to the power of soliton P as a function of site n and time t (*left panel*) and Fourier power spectrum of the trapped soliton calculated soon after the collision (*right panel*). The parameters are $\alpha = 1.0$, $q = 0.6$, $\Lambda = 2.5$, $\epsilon = 1$, and the impurity is located at $n = 0$. Reprinted from [29] with permission

In general, we have found that a necessary condition to trap energy and power by the impurity is the existence of a nonlinear localized mode centered at the impurity, with similar frequency, and energy (in absolute value) and power smaller than that of the corresponding incident soliton.

(c) *Reflection with no trapping.* Here, we have to distinguish two cases. If the impurity is repulsive, and q small enough, neither trapping nor transmission occurs. Instead, all energy is reflected, and the traveling nonlinear excitation remains localized. In this case, as shown in Fig. 19.8 (left), the incident wave has no energy and power to excite the localized mode. In a typical case, i.e., $\Lambda = 2.5$, $q = 0.6$, and $\alpha = -0.5$, the incident soliton has energy and power $E = -4.79$ and $P = 2.61$, and the nonlinear localized mode on the impurity with the same frequency $E = -8.038$ and $P = 3.77$. No trapping phenomenon occurs, and the pulse is reflected.

On the other hand, if the impurity is attractive and strong enough, i.e., $q = 0.7$, $\Lambda = 2.5$, and $\alpha = 2.0$, the frequency of the soliton is smaller than the one corresponding to the linear impurity mode ($\Lambda_L \simeq 2.82$), and all the energy is reflected. This is in accordance with the necessity of a nonlinear localized mode at the impurity site in order for the trapping to occur.

(d) *Transmission with no trapping.* If $|\alpha|$ is small enough, and q high enough, transmission with no trapping occurs, as shown in Fig. 19.8 (right). There exists a critical value of $q = q_c > 0$ such that, if $q > q_c$, the incident soliton crosses through the impurity. The value of q_c grows with $|\alpha|$. In the case where $q < q_c$, if $\alpha < 0$, reflection with no trapping occurs, while if $\alpha > 0$, trapping with no reflection phenomenon takes place.

Our results related to trapping, reflection, and transmission phenomena are in agreement with some results recently obtained, using a different approach, in a

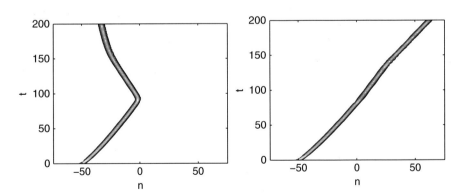

Fig. 19.8 Reflection with no trapping (*left panel*) corresponding to parameters $\alpha = -0.5$, $q = 0.6$ and $\Lambda = 2.5$, and transmission with no trapping (*right panel*) corresponding to parameters $\alpha = 0.1$, $q = 0.7$, and $\Lambda = 2.5$. In both cases we represent a contour plot corresponding to the power of the soliton P as function of site n and time t, $\epsilon = 1$ and the impurity is located at $n = 0$. Reprinted from [29] with permission

similar system [26]. In this work, where approximate discrete moving solitons with fixed amplitude are generated using a continuous approximation, the authors study the trapping process by a linear and a nonlinear attractive impurity. In the former framework, trapping can be explained by means of resonances with the linear localized mode. In our case, where nonlinear effects become stronger, the phenomena are related to resonances with a nonlinear localized mode.

Finally, a very interesting phenomenon occurs when the parameter α is repulsive and small enough (in absolute value). In this case, the solitary wave can be reflected or transmitted depending on its velocity. Also, when it is reflected, our numerical tests show that its velocity is similar to its incident velocity. Thus, if we consider the soliton as a "quasiparticle," the effect of the impurity is similar to the effect of a potential barrier. To determine this potential barrier for a given value of parameter α, we have used a method similar to the one described by [45]. We have considered different values of the thrust parameter q corresponding to the reflection regime, and determine, for each value, the turning point, $X(q)$. Thus the energy of the barrier for this value of q is defined as the difference between the energy of the moving soliton and the stationary state of the same frequency far from the impurity. It can be written as $V(q) = C \sin(q/2)|P(q/2)|$, with $P(q) = i \sum_n (\psi_n^* \psi_{n+1} - \psi_n^* \psi_{n-1})$ being the lattice momentum, as defined in [46]. Results are shown in Fig. 19.9, which exhibits, as expected, an irregular shape, whose origin lies in the nonuniform behavior of the translational velocity due to the discreteness of the system.

On the other hand, if the parameter α is small enough, and positive (attractive), the solitary wave faces a potential "well" and can be trapped if its translational energy is small or, if the translational energy is high enough, it may be transmitted, losing energy that remains trapped by the impurity, and decreasing its velocity.

To sum up, we have examined in detail for both impurity cases (attractive and repulsive) the interaction of the impurity with a moving localized mode initiated away from it. The principal regimes that we have identified as a function of the

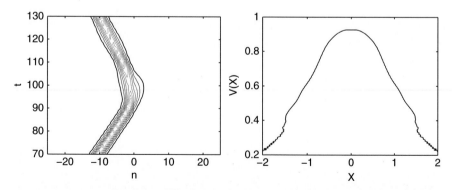

Fig. 19.9 Contour plot of the phenomenon of reflection of a soliton corresponding to thrust parameter $q = 0.6$ (*left panel*). Potential barrier calculated as described in the text (*right panel*). In both cases $\alpha = -0.2$, $\epsilon = 1$, $\Lambda = 2.5$, and the impurity is located at $n = 0$. Reprinted from [29] with permission

impurity strength (and sign) and initial speed are trapping, partial trapping and partial reflection, pure reflection and pure transmission. In general, if the impurity is repulsive, and the speed small enough, the wave is always reflected. If the impurity strength (in absolute value) is small enough and the speed is high enough, then transmission can take place. On the other hand, if the impurity is attractive, trapping can occur, and if the speed is high enough two different effects are observed: for small values of α, transmission takes place and for high values of α, the trapping is accompanied by a partial reflection (for intermediate values of α, the trapping is pure). If the impurity is attractive and sufficiently strong, the frequency of the soliton is smaller than the one corresponding to the linear localized impurity mode and the wave is reflected.

The above-described scenario is slightly different to that shown in [26], where small-amplitude solitons are considered. Contrary to the high-amplitude solitons considered in [29] where the trapping was shown to be originated by the excitation of a nonlinear localized mode centered at the impurity site, in [26], the trapping is due to excitations of linear localized modes, as the frequency of small-amplitude solitons are close to the phonon band. Thus, the scenario observed for attractive impurities is the following: for small impurity strength, the soliton is transmitted; above a critical value of α, the soliton is trapped partially, with the reflected fraction increasing with the impurity strength.

19.4 Comparison with Other Related Models

The findings herein, while presented for a linear impurity, are representative of other models including DNLS lattices with nonlinear impurities and Klein–Gordon (KG) lattices. We briefly expose hereafter the main similarities and differences between them.

19.4.1 Nonlinear Impurities

Stationary solitons with a quintic nonlinear impurity were considered in [27]:

$$i\dot{u}_n + |u_n|^2 u_n + \epsilon(u_{n+1} + u_{n-1} - 2u_n) + \alpha \delta_{n,0} |u_n|^4 u_n = 0. \tag{19.15}$$

In that work, only attractive impurities were considered. The bifurcation diagram of solitons close to the impurity site is similar to the one found for linear impurities (see Fig. 19.2). By means of a variational approximation the value of the impurity strength at which the branches corresponding to solitons centered at $n = 0$ and $n = 0.5$ merge. The dependence of this critical value with the frequency is

$$\alpha_c(\Lambda) = -\frac{16\pi^4 \epsilon^3}{\Lambda^4} \exp\left(-\pi^2 \sqrt{\frac{\epsilon}{\Lambda}}\right). \tag{19.16}$$

This dependence was approximated in the linear impurity case by means of an invariant manifold approximation (see Sect. 19.2.3).

The interaction of small-amplitude moving solitons with nonlinear impurities has been briefly considered in [26], where, contrary to (19.15), a nonlinear cubic impurity was considered:

$$i\dot{u}_n + (1 + \alpha\delta_{n,0})|u_n|^2 u_n + \epsilon(u_{n+1} + u_{n-1}) = 0. \tag{19.17}$$

The observed scenario for attractive impurities was the following: for small α, the soliton is transmitted and, above a threshold value, the soliton is trapped without reflection. If the impurity is increased, the soliton is totally reflected, and, surprisingly, for relatively high values of α, the soliton is trapped again.

19.4.2 Comparison with Klein–Gordon Breathers

The interaction of moving localized in KG chains (called discrete breathers) with a point inhomogeneity in the substrate potential was considered in [47, 48]. The model equation is given by

$$\ddot{u}_n + V'_n(u_n) + C(2u_n - u_{n+1} - u_{n-1}) = 0 \tag{19.18}$$

with $V(u_n) = (1 + \alpha\delta_{n,0})(\exp(-u_n) - 1)^2/2$ being the Morse substrate potential (for different potentials or kinds of impurities, the reader is referred to [49]). In this setting, for $\alpha > 0$ ($\alpha < 0$) the impurity is repulsive (attractive) contrary to the DNLS case shown throughout this chapter.

The observed regimes for KG breathers are qualitatively equivalent to those of DNLS solitons. An important analogy in both settings is the necessary condition for the trapping; i.e., the energy of the moving localized mode must be higher (in absolute value) than that of the stationary localized mode centered at the impurity (with the same frequency of the moving soliton/breather). This result for KG lattices was established in [47, 48, 50].

It is worth mentioning the study performed in [34] where many existence conditions are established for stationary KG breathers in inhomogeneous lattices based on a center manifold approach. The latter work also predicted and illustrated gap breathers, that is breathers whose frequency lies in the gap left in the phonon band by the linear localized mode when departing from it. These structures, however, do not exist in DNLS lattices.

19.5 Summary and Future Challenges

In this chapter, the existence and stability of discrete solitons close to a local inhomogeneity in a 1D DNLS lattice have been studied. A systematic study of the interaction of a moving discrete soliton with that local inhomogeneity has been performed. Finally, a brief comparison of these results with other related settings, such as DNLS lattices with nonlinear impurities or KG lattices, was given.

Further development of this direction of research could include the consideration of saturable nonlinearities which can describe nonlinear waveguides made of photorefractive materials. This kind of nonlinearity may enhance the mobility of solitary waves in isotropic 2D lattices [51], whereas moving solitons only can take place in anisotropic 2D lattices for cubic lattices [52]. A similar study to that shown in this section could be done in both settings. Another interesting direction could be the inclusion of two or more local inhomogeneities and the examination of the interplay between them.

References

1. Aubry, S.: Physica D **103**, 201, (1997)
2. Flach, S., Willis, C.R.: Phys. Rep. **295**, 181 (1998)
3. Hennig, D., Tsironis, G.: Phys. Rep. **307**, 333 (1999)
4. Christodoulides, D.N., Lederer, F., Silberberg, Y.: Nature **424**, 817 (2003)
5. Morsch, O., Oberthaler, M.: Rev. Mod. Phys. **78**, 179 (2006)
6. Kevrekidis, P.G., Frantzeskakis, D.J.: Mod. Phys. Lett. B **18**, 173 (2004)
7. Sato, M., Hubbard, B.E., Sievers, A.J.: Rev. Mod. Phys. **78**, 137 (2006)
8. Peyrard, M.: Nonlinearity **17**, R1 (2004)
9. Maradudin, A.A.: Theoretical and Experimental Aspects of the Effects of Point Defects and Disorder on the Vibrations of Crystal. Academic Press, New York (1966)
10. Lifschitz, I.M.: Nuovo Cimento Suppl. **3**, 716 (1956)
11. Lifschitz, I.M., Kosevich, A.M.: Rep. Progr. Phys. **29**, 217 (1966)
12. Andreev, A.F.: JETP Lett. **46**, 584 (1987)
13. Balatsky, A.V.: Nature (London) **403** 717 (2000)
14. Molina, M.I., Tsironis, G.P.: Phys. Rev. B **47**, 15330 (1993)
15. Tsironis, G.P., Molina, M.I., Hennig, D.: Phys. Rev. E **50**, 2365 (1994)
16. Lidorikis, E., Busch, K., Li, Q., Chan, C.T., Soukoulis, C.M.: Phys. Rev. B **56**, 15090 (1997)
17. Jin, S.Y., Chow, E., Hietala, V., Villeneuve, P.R., Joannopoulos, J.D.: Science **282**, 274 (1998)
18. Khazhinsky, M.G., McGurn, A.R.: Phys. Lett. A **237**, 175 (1998)
19. Forinash, K., Peyrard, M., Malomed, B.: Phys. Rev. E **49**, 3400 (1994)
20. Krolikowski, W., Kivshar, Yu.S.: J. Opt. Soc. Am. B **13**, 876 (1996)
21. Kivshar, Yu.S., Kevrekidis, P.G., Takeno, S., Phys. Lett. A **307**, 287 (2003)
22. Agrotis, M., Kevrekidis, P.G., Malomed, B.A.: Math. Comp. Simul. **69**, 223 (2005)
23. Longhi, S.: Phys. Rev. E **74**, 026602 (2006)
24. Peschel, U., Morandotti, R., Aitchison, J.S., Eisenberg, H.S., Silberberg, Y.: Appl. Phys. Lett., **75**, 1348 (1999)
25. Morandotti, R., Eisenberg, H.S., Mandelik, D., Silberberg, Y., Modotto, D., Sorel, M., Stanley, C.R., Aitchison, J.S.: Opt. Lett. **28**, 834 (2003)
26. Morales-Molina, L., Vicencio, R.A.: Opt. Lett. **31**, 966 (2006)
27. Kevrekidis, P.G., Kivshar, Yu.S., Kovalev, A.S.: Phys. Rev. E **67**, 046604 (2003)
28. Smirnov, E., Rüter, C.E., Stepić, M., Shandarov, V., Kip, D.: Opt. Express **14**, 11248 (2006)
29. Palmero, F., Carretero-González, R., Cuevas, J., Kevrekidis, P.G., Królikowski, W.: Phys. Rev. E **77**, 036614 (2008)
30. Trompeter, H., Peschel, U., Pertsch, T., Lederer, F., Streppel, U., Michaelis, D., Bräuer, A.: Opt. Express **11**, 3404 (2003)

31. Morandotti, R., Peschel, U., Aitchison, J.S., Eisenberg, H.S., Silberberg, Y.: Phys. Rev. Lett. **83**, 2726 (1999)
32. Palmero, F.J., Dorignac, F., Eilbeck, J.C., Römer, R.A.: Phys. Rev. B **72**, 075343 (2005)
33. Kevrekidis, P.G., Rasmussen, K.Ø., Bishop, A.R.: Int. J. Mod. Phys. B **15**, 2833 (2001)
34. James, G., Sánchez-Rey, B., Cuevas, J.: Rev. Math. Phys. **21**, 1 (2009)
35. Qin, W.X., Xiao, X.: Nonlinearity **20**, 2305 (2007)
36. Cuevas, J., James, G., Malomed, B.A., Kevrekidis, P.G., Sánchez-Rey, B.: J. Nonlinear Math. Phys. **15**, 134 (2008)
37. Eilbeck, J.C.: Computer Analysis for Life Science Progress and Challenges. In Kawabata, C., Bishop, A.R. (Eds.) Biological and Synthetic Polymer Research, pp. 12–21. Ohmsha, Tokyo (1986)
38. Feddersen, H.: Nonlinear Coherent Structures in Physics and Biology. In M. Remoissenet, M. Peyrard (Eds.) Lecture Notes in Physics, vol. 393, pp. 159–167. Springer, Heidelberg (1991)
39. Eilbeck, J.C., Johansson, M.: Localization and Energy Transfer in Nonlinear Systems. Vázquez, L., MacKay, R.S., Zorzano, M.P. (Eds.), pp. 44–67. World Scientific, Singapore (2003)
40. Gómez-Gardeñes, J., Floría, L.M., Peyrard, M., Bishop, A.R.: Chaos **14**, 1130 (2004)
41. Melvin, T.R.O., Champneys, A.R., Kevrekidis, P.G., Cuevas, J.: Phys. Rev. Lett. **97**, 124101 (2006)
42. Oxtoby, O.F., Barashenkov, I.V.: Phys. Rev. E **76**, 036603 (2007)
43. Melvin, T.R.O., Champneys, A.R., Kevrekidis, P.G., Cuevas, J.: Physica D **237**, 551 (2008)
44. Cuevas, J., Eilbeck, J.C.: Phys. Lett. A **358**, 15 (2006)
45. Cuevas, J., Kevrekidis, P.G.: Phys. Rev. E **69**, 056609 (2004)
46. Papacharalampous, I.E., Kevrekidis, P.G., Malomed, B.A., Frantzeskakis, D.J.: Phys. Rev. E **68**, 046604 (2003)
47. Cuevas, J., Palmero, F., Archilla, J.F.R., Romero, F.: J. Phys. A **35**, 10519 (2002)
48. Cuevas, J., Palmero, F., Archilla, J.F.R., Romero, F.: Theor. Math. Phys. **137**, 1406 (2003)
49. Cuevas, J.: PhD Dissertation. University of Sevilla.
50. Alvarez, A., Romero, F.R., Archilla, J.F.R., Cuevas, J., Larsen, P.V.: Eur. Phys. J. B **51**, 119 (2006)
51. Vicencio, R., Johansson, M.: Phys. Rev. E **73**, 046602 (2006)
52. Gómez-Gardeñes, J., Floría, L.M., Bishop, A.R.: Physica D **216**, 31 (2006)

Chapter 20
Statistical Mechanics of DNLS

Panayotis G. Kevrekidis

20.1 Introduction

In this section, we focus our attention on the long-time asymptotics of the system, aiming to understand the dynamics from the viewpoint of statistical mechanics. The study of the thermalization of the lattice for $T \geq 0$ was performed analytically as well as numerically in [1]. In that work, a regime in phase space was identified wherein regular statistical mechanics considerations apply, and hence, thermalization was observed numerically and explored analytically using regular, grand-canonical, Gibbsian equilibrium measures. However, the nonlinear dynamics of the problem renders permissible the realization of regimes of phase space which would formally correspond to "negative temperatures" in the sense of statistical mechanics. The novel feature of these states was found to be that the energy spontaneously localizes in certain lattice sites forming breather-like excitations. Returning to statistical mechanics, such realizations are not possible (since the Hamiltonian is unbounded, as is seen by a simple scaling argument similar to the continuum case studied in [2]) unless the grand-canonical Gibbsian measure is refined to correct for the unboundedness. This correction was argued in [1] to produce a discontinuity in the partition function signaling a phase transition which was identified numerically by the appearance of the intrinsic localized modes (ILMs).

In our presentation below, we first elaborate on the semianalytical calculations of [1]. We then present direct numerical simulation results, supporting the theoretically analyzed scenario. We conclude our discussion with a number of more recent results, including a generalization of the considerations to higher dimensions and/or nonlinearity exponents, as well as other classes of related nonlinear lattices.

P.G. Kevrekidis (✉)
University of Massachusetts, Amherst, MA, 01003, USA
e-mail: kevrekid@math.umass.edu

20.2 Theoretical Results

To present the analysis of [1], it is convenient to use a slightly modified (yet equivalent, up to rescaling and gauge transformations) form of the DNLS

$$i\dot{u}_n + (u_{n+1} + u_{n-1}) + v|u_n|^2 u_n = 0. \tag{20.1}$$

In order to study the statistical mechanics of the system, we calculate the classical grand-canonical partition function $\mathcal{Z}$. Using the canonical transformation $u_n = \sqrt{A_n}\exp(i\phi_n)$, the Hamiltonian expressed as

$$H = \sum_n \left[(u_n^* u_{n+1} + u_n u_{n+1}^*) + \frac{v}{2}|u_n|^4 \right] \tag{20.2}$$

becomes

$$H = \sum_n 2\sqrt{A_n A_{n+1}}\cos(\phi_n - \phi_{n+1}) + \frac{v}{2}\sum_n A_n^2. \tag{20.3}$$

The partition function in this setting can be expressed as

$$\mathcal{Z} = \int_0^\infty \int_0^{2\pi} \prod_n d\phi_n dA_n \exp[-\beta(H + \mu P)], \tag{20.4}$$

where the multiplier μ is analogous to the chemical potential introduced to ensure conservation of the squared l^2 norm $P = \sum_n |u_n|^2$. Straightforward integration over the phase variable ϕ_n yields

$$\mathcal{Z} = (2\pi)^N \int_0^\infty \prod_n dA_n I_0(2\beta\sqrt{A_n A_{n+1}}) \times$$
$$\exp\left[-\beta \sum_n \left(\frac{v}{4}(A_n^2 + A_{n+1}^2) + \frac{\mu}{2}(A_n + A_{n+1})\right)\right]. \tag{20.5}$$

This integral can be evaluated exactly in the thermodynamic limit of a large system ($N \to \infty$) using the eigenfunctions and eigenvalues of the transfer integral operator [3, 4],

$$\int_0^\infty dA_n \kappa(A_n, A_{n+1}) y(A_n) = \lambda\, y(A_{n+1}), \tag{20.6}$$

where the kernel κ is

$$\kappa(x, z) = I_0\left(2\beta\sqrt{xz}\right)\exp\left[-\beta\left(\frac{v}{4}(x^2 + z^2) + \frac{\mu}{2}(x + z)\right)\right]. \tag{20.7}$$

Similar calculations were performed for the statistical mechanics of the ϕ^4 field [3, 4], and for models of DNA denaturation [5]. The partition function can, thus, be obtained as $\mathcal{Z} \simeq (2\pi\lambda_0)^N$, as $N \to \infty$ where λ_0 is the largest eigenvalue of the operator. From this expression the usual thermodynamic quantities such as the free energy, F, or specific heat can be calculated. More importantly, for our purposes, the averaged energy density, $h = \langle H \rangle / N$, and the average excitation norm, $a = \langle P \rangle / N$, can be found as

$$a = -\frac{1}{\beta\lambda_0}\frac{\partial\lambda_0}{\partial\mu}, \quad h = -\frac{1}{\lambda_0}\frac{\partial\lambda_0}{\partial\beta} - \mu a. \tag{20.8}$$

The average excitation norm a can also be calculated as

$$a = (1/\mathcal{Z})\int_0^\infty \prod_n dA_n\, A_n \exp[-\beta(H + \mu P)], \tag{20.9}$$

where the integral again can be calculated using the transfer integral technique [3, 4] and yields $a = \int_0^\infty y_0^2(A) A\, dA$, where y_0 is the normalized eigenfunction corresponding to the largest eigenvalue λ_0 of the kernel κ (Eq. (20.5)). This shows that $p(A) = y_0^2(A)$ is the probability distribution function (PDF) for the amplitudes A. Subsequently, λ_0, y_0 were obtained numerically in [1]. However, two limits ($\beta \to \infty$ and $\beta \to 0$) can also be explored analytically. In particular, the minimum of the Hamiltonian is realized by a plane wave, $u_n = \sqrt{a}\exp im\pi$, whose energy density is $h = -2a + va^2/2$. This relation defines the zero temperature (or the $\beta = \infty$) line. For the high temperature limit, $\beta \ll 1$, the modified Bessel function in the transfer operator can be approximated to leading order, by unity which, in turn, reduces the remaining eigenvalue problem to the approximate solution,

$$y_0(A) = \frac{1}{\sqrt{\lambda_0}}\exp\left[-\frac{\beta}{4}\left(vA^2 + 2\mu A\right)\right]. \tag{20.10}$$

Using this approximation and enforcing the constraint $\beta\mu = \gamma$ (where γ remains finite as we take the limits $\beta \to 0$ and $\mu \to \infty$), one can obtain $h = v/\gamma^2$ and $a = 1/\gamma$. Thus, we get $h = va^2$ at $\beta = 0$.

Figure 20.1 depicts (with thick lines) the two parabolas in (a, h)-space corresponding to the $T = 0$ and $T = \infty$ limits. Within this region all considerations of statistical mechanics in the grand-canonical ensemble are applicable and there is a one-to-one correspondence between (a, h) and (β, μ). Thus, within this range of parameter space the system thermalizes in accordance with the Gibbsian formalism. However, the region of the parameter space that is experimentally (numerically) accessible is actually wider since it is possible to initialize the lattice at any energy density h and norm density a above the $T = 0$ line in an infinite system.

A statistical treatment of the remaining domain of parameter space was accomplished in [1] by introducing formally negative temperatures. However, the

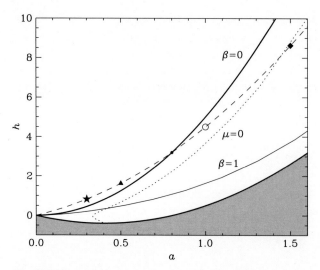

Fig. 20.1 From [1]: parameter space (a, h), where the *shaded area* is inaccessible. The *thick lines* represent, respectively, the $\beta = \infty$ and $\beta = 0$ cases and thus bound the Gibbsian regime. The *dashed line* represents the $h = 2a + \frac{v}{2}a^2$ curve along which the reported numerical simulations are performed (pointed by the symbols). The *dotted line* shows the locus of points where the chemical potential vanishes

partition function (20.4) is clearly not suited for that purpose since the constraint expressed in the grand-canonical form fails to bound the Hamiltonian of Eq. (20.2) from above. In all the alternative approaches of the study of negative temperatures a finite system of size N was thus considered. As suggested in [2] the grand-canonical ensemble can be realized using the modified partition function $\mathcal{Z}'(\beta, \mu') = \int \exp(-\beta(H + \mu' P^2)) \prod_n du_n du_n^*$, but this introduces long-range coupling and μ' will have to be of order $1/N$. Now β can be negative since $H + \mu' P^2$ can be seen to be bounded from above when $\mu' < -v/2N$. The important consequence of this explicit modification of the measure is a jump discontinuity in the partition function, associated with a phase transition. More explicitly, if one starts in a positive T, thermalizable (in the Gibbsian sense) state in phase space with $h > 0$, and continuously varies the norm, then one will, inevitably, encounter the $\beta = 0$ parabola. Hence, in order to proceed in a continuous way, a discontinuity has to be assigned to the chemical potential. This discontinuity will destroy the analyticity of the partition function as the transition line is crossed, and will indicate a phase transformation according to standard statistical mechanics.

In order to characterize the dynamics of both phases (above and below the $\beta = 0$ line) and to verify that the system relaxes to a thermalized state, numerical experiments were performed in [1]. The parameters (a, h) were restricted to the dashed line of Fig. 20.1, choosing an appropriate perturbed phonon as initial condition. The modulational instability of the latter [6] naturally gives rise to localized states. For these initial conditions, the important question is whether relaxation to equilibrium

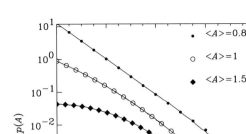

Fig. 20.2 From [1]: probability distribution of $A = |\psi|^2$ for three cases under (and on) the transition line. *The solid lines* show the results of simulations and the symbols are given by the transfer operator. Curves are vertically shifted to facilitate visualization

is really achieved and whether different qualitative behavior is indeed observed on the two sides of the $\beta = 0$ line.

Figure 20.2 shows three typical examples of what can be observed when the energy-norm density point lies below the $\beta = 0$ line (the symbols refer to Fig. 20.1). Since the initial condition is modulationally unstable, the energy density forms small localized excitations but their lifetime is not very long and, rapidly, a stationary distribution of the amplitudes A_n is reached (Fig. 20.2). Hence an equilibrium state is reached as predicted by means of the transfer-operator method.

The scenario is found to be very different when the energy and norm densities are above the $\beta = 0$ line. A rapid creation of ILMs due to the modulational instability is again observed and is accompanied by thermalization of the rest of the lattice. Once created however, these localized excitations remained mostly pinned and because the internal frequency increased with amplitude their coupling with the small-amplitude radiation was very small. This introduces a new time scale in the thermalization process necessitating symplectic integration for as long as $10^6 - 10^7$ time units in order to reach a stationary PDF. This was also qualitatively justified by the effective long-range interactions, introduced in the modified partition function, which produce stronger memory effects as one observes regimes in phase space which are further away from the transition line.

Typical distribution functions of the amplitudes are shown in Fig. 20.3. The presence of high-amplitude excitations is directly seen here. The positive curvature of the PDF at small amplitudes clearly indicates that the system evolves in a regime of negative temperature and the appearance of the phase transformation is signaled in the dynamics by the appearance of the strongly localized, persistent in the long-time asymptotics, ILMs.

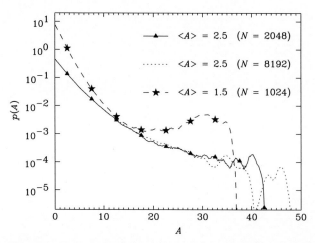

Fig. 20.3 From [1]: distribution of $A = |\psi|^2$ for parameters (h, a) above the transition line (*triangles* and *stars* as in Fig. 20.1.)

20.3 Recent Results

More recently, the statistical understanding of the formation of localized states and of the asymptotic dynamics of the DNLS equation has been addressed in the works of [7, 8].

The analysis of the former paper presented a complementary viewpoint to that of [1] which attempted for the first time to address the thermodynamics of the states within the localization regime. Assuming small-amplitude initial conditions, [7] argued that the phase space of the system can be divided roughly into two weakly interacting domains, one of which corresponds to the low-amplitude fluctuations (linear or phonon modes), while the other consists of the large-amplitude, localized mode nonlinear excitations. A remarkable feature of that work is that based on a simple partition of the energy $H = H_< + H_>$ and of the norm $P = P_< + P_>$, into these two broadly (and also somewhat loosely) defined fractions, one smaller than a critical threshold (denoted by $<$) and one larger than a critical threshold (denoted by $>$); it allows to compute thermodynamic quantities such as the entropy in this localization regime. In particular, one of the key results of [7] is that for a partition of K sites with large-amplitude excitations and $N - K$ sites with small-amplitude ones, it derives an expression for the total entropy (upon computing $S_<$, $S_>$, and a permutation entropy due to the different potential location of the K and $N - K$ sites). This expression reads

$$S = N\ln\Omega + \frac{P_>^2}{2E_>}\ln\Gamma, \tag{20.11}$$

where $\Omega = (4P_<^2 - E_<^2)/(4A_<(N - K))$ and $\Gamma = 2P_<N/P_>E_<$, while $K = P_>^2/(2E_>)$. While some somewhat artificial assumptions are needed to arrive at the result of Eq. (20.11) [such as the existence of a cutoff amplitude radius R in phase space], nevertheless, the result provides a transparent physical understanding of the localization process. The contributions to the entropy stem from the fluctuations [first term in Eq. (20.11)] and from the high-amplitude peaks (second term in the equation). However, typically the contribution of the latter in the entropy is negligible, while they can absorb high amounts of energy. The underlying premise is that the system seeks to maximize its entropy by allocating the ideal amount of energy $H_<$ to the fluctuations. Starting from an initial energy $H_<$, this energy is decreased in favor of localized peaks (which contribute very little to the entropy). The entropy would then be maximized if eventually a single peak was formed, absorbing a very large fraction of the energy while consuming very few particles. Nevertheless, practically, this regime is not reached "experimentally" (i.e., in the simulations). This is because of the inherent discreteness of the system which leads to a pinning effect of large-amplitude excitations which cannot move (and, hence, cannot eventually merge into a single one) within the lattice. Secondly, the growth of the individual peaks, as argued in [7], stops when the entropy gain due to energy transfer to the peaks is balanced by the entropy loss due to transfer of power. While placing the considerations of [7] in a more rigorous setting is a task that remains open for future considerations, this conceptual framework offers considerable potential for understanding the (in this case argued to be infinite, rather than negative, temperature) thermal equilibrium state of coexisting large-amplitude localized excitations and small-amplitude background fluctuations.

On the other hand, the work of [8] extended the considerations of the earlier work of [1] to the generalized DNLS model of the form

$$i\dot{u}_n + (u_{n+1} + u_{n-1}) + |u_n|^{2\sigma} u_n = 0. \tag{20.12}$$

Our analytical considerations presented above are directly applicable in this case as well, yielding a partition function

$$\mathcal{Z} = (2\pi)^N \int_0^\infty \prod_n dA_n I_0(2\beta\sqrt{A_n A_{n+1}}) \times \tag{20.13}$$

$$\exp\left[-\beta \sum_n A_n \left(\mu + A_n^\sigma/(\sigma+1)\right)\right]. \tag{20.14}$$

This can be again directly evaluated in the high-temperature limit, where the modified Bessel function is approximated as $I_0 \approx 1$, as

$$\mathcal{Z} = (2\pi)^N \frac{1}{(\beta\mu)^N}\left(1 - \frac{\beta\Gamma(\sigma+1)}{(\beta\mu)^{\sigma+1}}\right), \tag{20.15}$$

with Γ denoting the Γ function. As a result, in this case,

$$a = \frac{1}{\beta\mu} - \frac{\Gamma(\sigma+1)}{\mu(\beta\mu)^{\sigma+1}} \qquad (20.16)$$

$$h = \frac{\Gamma(\sigma+1)}{(\beta\mu)^{\sigma+1}} \qquad (20.17)$$

and the relation between the energy density h and the norm density a is generalized from the $\sigma = 1$ limit as

$$h = \Gamma(\sigma+1)a^{\sigma+1}, \qquad (20.18)$$

encompassing the parabolic dependence of that limit as a special case. It was once again confirmed in the setting of [8] that crossing this limit of $\beta = 0$ results in the formation of the large-amplitude, persistent few-site excitations. Another interesting observation made in [8] is that in this limit of $\beta = 0$, the coupling terms are inactive, hence, if additional dimension(s) are added to the problem, these do *not* affect the nature of the critical curve of Eq. (20.18). In that sense, the role of the dimensionality is different than the role of σ [with the latter being evident in Eq. (20.18]. This is to be contradistincted with the situation regarding the excitation thresholds or the thresholds for collapse, as discussed in Sect. III.2, whereby the dimensionality d and the exponent σ play an equivalent role, since it is when their product exceeds a critical value (in particular for $d\sigma \geq 2$) that such phenomena arise.

Finally, in the work of [8], the connection of these DNLS considerations with the generally more complicated Klein–Gordon (KG) models was discussed. Much of the above-mentioned phenomenology, as argued in [7], is critically particular to NLS-type models, due to the presence of the second conserved quantity, namely of the l^2 norm; this feature is absent in the KG lattices, where typically only the Hamiltonian is conserved. [8] formalizes the connection of DNLS with the KG lattices, by using the approximation of the latter via the former through a Fourier expansion whose coefficients satisfy the DNLS up to controllable corrections. Within this approximation, they connect the conserved quantity of the KG model to the ones of the DNLS model approximately reconstructing the relevant transition (to formation of localized states) criterion discussed above. However, in the KG setting this only provides a guideline for the breather formation process, as the conservation of the norm is no longer a true but merely an approximate conservation law. This is observed in the dynamical simulations of [8], where although as the amplitude remains small throughout the lattice the process is well described by the DNLS formulation, when the breathers of the KG problem grow, they violate the validity of the DNLS approximation and of the norm conservation; thus, a description of the asymptotic state and of the thermodynamics of such lattices requires further elucidation that necessitates a different approach. This is another interesting and important problem for future studies.

References

1. Rasmussen, K.Ø., Cretegny, T., Kevrekidis, P.G., Grønbech-Jensen, N.: Phys. Rev. Lett., **84**, 3740 (2000)
2. Lebowitz, J.L., Rose, H.A., Speer, E.R., J. Stat. Phys., **50**, 657 (1988).
3. Krumhansl, J.A., Schrieffer, J.R.: Phys. Rev. B, **11**, 3535 (1975)
4. Scalapino, D.J., Sears, M., Ferrell, R.A.: Phys. Rev. B, **6**, 3409 (1972)
5. Dauxois, T., Peyrard, M., Bishop, A.R.: Physica, **66D**, 35 (1993)
6. Kivshar, Yu.S., Peyrard, M.: Phys. Rev. A, **46**, 3198 (1992)
7. Rumpf, B.: Phys. Rev E **69**, 016618 (2004)
8. Johansson, M., Rasmussen, K.Ø.: Phys. Rev. E **70**, 066610 (2004)

Chapter 21
Traveling Solitary Waves in DNLS Equations

Alan R. Champneys, Vassilis M. Rothos and Thomas R.O. Melvin

21.1 Introduction

The existence of localized traveling waves (sometimes called "moving discrete breathers" or "discrete solitons") in discrete nonlinear Schrödinger (DNLS) lattices has shown itself to be a delicate question of fundamental scientific interest (see e.g [1]). This interest is largely due to the experimental realization of solitons in discrete media, such as waveguide arrays [2] optically induced photorefractive crystals [3] and Bose–Einstein condensates coupled to an optical lattice trap [4]; see Chapter 8 for more details. The prototypical equation that emerges to explain the experimental observations is the DNLS model of the form

$$i\dot{u}_n(t) = \frac{u_{n+1}(t) - 2u_n(t) + u_{n-1}(t)}{h^2} + F(u_{n+1}(t), u_n(t), u_{n-1}(t)), \quad (21.1)$$

where the integer $n \in \mathbb{Z}$ labels a 1D array of lattice sites, with spacing h. Alternatively h^2 can be thought of as representing the inverse coupling strength between adjacent sites. The nonlinear term F can take a number of different forms:

DNLS equation

$$F_{\text{DNLS}} = |u_n|^2 u_n,$$

Ablowitz–Ladik (AL) model [5]

$$F_{\text{AL}} = |u_n|^2 (u_{n+1} + u_{n-1}),$$

Salerno model [6]

$$F_{\text{S}} = 2(1-\alpha)F_{\text{DNLS}} + \alpha F_{\text{AL}},$$

A.R. Champneys (✉)
Department of Engineering Mathematics, University of Bristol, UK
e-mail: a.r.champneys@bristol.ac.uk

cubic-quintic DNLS

$$F_{3-5} = (|u_n|^2 + \alpha |u_n|^4) u_n,$$

saturable DNLS

$$F_{\text{sat}} = \frac{u_n}{1 + |u_n|^2},$$

generalized cubic DNLS equation

$$\begin{aligned}F_{g3} &= \alpha_1 |u_n|^2 u_n + \alpha_2 |u_n|^2 (u_{n+1} + u_{n-1}) + \alpha_3 u_n^2 (\bar{u}_{n+1} + \bar{u}_{n-1}) \\ &\times \alpha_4 (|u_{n+1}|^2 + |u_{n-1}|^2) u_n + \alpha_5 (\bar{u}_{n+1} u_{n-1} + u_{n+1} \bar{u}_{n-1}) u_n \\ &+ \alpha_6 (u_{n+1}^2 + u_{n-1}^2) \bar{u}_n + \alpha_7 u_{n+1} u_{n-1} \bar{u}_n \\ &+ \alpha_8 (|u_{n+1}|^2 u_{n+1} + |u_{n-1}|^2 u_{n-1}) + \alpha_9 (u_{n+1}^2 \bar{u}_{n-1} + u_{n-1}^2 \bar{u}_{n+1}) \\ &+ \alpha_{10} (|u_{n+1}|^2 u_{n-1} + |u_{n-1}|^2 u_{n+1}),\end{aligned} \quad (21.2)$$

where $(\bar{\cdot})$ is used to represent complex conjugation. Note that when $\alpha_1 = 2(1 - \alpha_2)$, $\alpha_2 \in \mathbb{R}$, and $\alpha_j = 0$ for $3 \leq j \leq 10$, the nonlinear function F_{g3} reduces to the Salerno nonlinearity F_S.

Stationary localized solutions to (21.1) of the form $u(n, t) = e^{-i\omega t} U(n)$ abound in such models under quite general hypotheses on the function F, and indeed one can pass to the continuum limit $h \to 0$, $x = nh$ and find the corresponding solutions to continuum NLS equations of the form

$$i\dot{u} = u_{xx} + f(|u|^2) u, \quad (21.3)$$

for appropriate nonlinear functions f. The continuum model (21.3) possesses Galilean invariance and so localized solutions can be found that move with a range of different wavespeeds v, including arbitrarily low speeds $v \to 0$. However, the corresponding discrete problem (21.1) is such that each stationary localized mode has its maximum intensity centered on either a lattice site or halfway between two sites. These site-centered and off-site-centered modes typically have different energies, leading to a so-called Peierls–Nabarro (PN) energy barrier that, in general, prevents localized excitations moving with small wavespeeds without shedding radiation. But could localized waves exist at finite wavespeeds?

One negative result in this direction by Gómez–Gardeñes et al. [7] for the Salerno nonlinearity F_S shows that starting from the AL limit $\alpha = 1$ where the equation is completely integrable, traveling localized waves acquire nonvanishing tails as soon as parameters deviate from the integrable limit. Hence exponentially localized fundamental (single humped) localized traveling waves cannot be constructed. While these results settled a long-standing controversy, see e.g. [8–11], they are also somewhat unsatisfactory since they do not give conditions under which moving discrete breathers might exist for generic, nonintegrable lattices. In fact, following [12], we shall show below that there *are* truly localized traveling waves in the Salerno model that do not shed radiation, which can be found in continuation from the AL soliton only by including enough parameters.

Mathematically, the shedding of radiation is due to a large number of resonances in the spectrum of the linear operator around the localized mode when one moves to a traveling frame with small wavespeed. To minimize the number of these radiation modes it is necessary to look for solutions that travel with a *finite* wavespeed [13], that is, in an appropriate region of parameter space where the minimum wavespeed is bounded away from zero. However, posing DNLS equations in a traveling frame gives rise to differential advance-delay equations which are notoriously hard to analyze. Progress in this area has been made by developing a Mel'nikov method around existing solution families [14] or by using a pseudo-spectral method to transform the advance-delay equation into a large system of algebraic equations [9, 15]. Alternatively, looking for small-amplitude (but nonzero wavespeed) solutions bifurcating from the rest state involves a beyond-all-order computation of the so-called Stokes constants [16] which measure the splitting of the stable and unstable manifolds [17]. All these methods are reviewed more carefully in Sect. 21.2 below.

The rest of this chapter is outlined as follows: Section 21.2 contains a mathematical formulation of the central problem under investigation. After introduction of a careful parametrization of possible localized traveling waves, Sect. 21.2.1 goes on to show how to pose the resulting advance-delay equation as a spatial dynamical system. Study of the linear part of this system leads to an argument that finite wavespeeds are required in order to find truly localized traveling waves in a persistent way. Sections 21.2.2 and 21.2.3 then show how to apply center-manifold and normal-form techniques to this dynamical system near a special zero-dispersion point. This leads in Sects. 21.2.4 and 21.2.5 to a brief discussion of more technical results based on Mel'nikov methods near the zero-point and Stokes constant elsewhere in parameter space. The section ends with a brief discussion of how pseudospectral methods can be used to transform the advance-delay equation into a large system of algebraic equations suitable for a numerical path-following investigation. Section 21.3 reviews results from the literature that were obtained using the array of mathematical methods introduced in Sect. 21.2. The results are presented for three particular nonlinearities; saturable F_{sat}, Salerno F_S, and the generalized cubic F_{g3}. Finally, Sect. 21.4 draws conclusions and points to open questions.

21.2 Mathematical Formulation

We address here the existence of traveling wave solutions to (21.1) in which we set $h = 1$ without loss of generality, subsuming this parameter into the form of F. That is, we seek solutions in the form

$$u_n(t) = \phi(z)e^{-i\beta n - i\omega t}, \qquad z = n - vt, \qquad (21.4)$$

where ω, v, and β are real-valued parameters, while $z \in \mathbb{R}$, $\phi \in \mathbb{C}$, and $\lim_{|z| \to \infty} \phi(z) = 0$. Only two parameters ($\omega$ and v) are truly independent, since the term $e^{-i\beta n}$ can be included in the definition of $\phi(z)$. A number of methods can be used to normalize the parameter β uniquely. For example, one can require [1] that the exponential decay of $\phi(z)$ be real-valued: $\lim_{|z| \to \infty} e^{\kappa |z|} \phi(z) = \phi_\infty$, where $\kappa \in \mathbb{R}$ and $\phi_\infty \in \mathbb{C}$. The tail analysis of the traveling wave solutions (21.4) shows that parameters β and κ are uniquely defined in terms of ω and v. In particular, as we shall see, the traveling wave solutions (21.4) are exponentially decaying, if they exist, only if ω and v lie inside the region $\mathcal{K}$ represented by the white region in Fig. 21.2 below.

21.2.1 Spatial Dynamics Formulation

Let $\Gamma = l^2(\mathbb{Z}; \mathbb{C})$ denote the Hilbert space of all square-summable bi-infinite complex-valued sequences; for any $u, v \in \Gamma$ and $\lambda \in \mathbb{C}$ we denote by λu, $\bar{u}$, $|u|$, uv and $u + v$ the sequences $(\lambda u_n)_{n \in \mathbb{Z}}$, $(\bar{u}_n)_{n \in \mathbb{Z}}$, $(|u_n|)_{n \subset \mathbb{Z}}$, $(u_n v_n)_{n \subset \mathbb{Z}}$, and $(u_n + v_n)_{n \in \mathbb{Z}}$, respectively. As usual, the norm of $u \in \Gamma$ is represented by $\|u\|$.

The ansatz (21.4) for traveling wave solutions reduces the discrete NLS equation (21.1) to a differential advance-delay equation of the form:

$$-iv\phi'(z) = \phi_+ e^{-i\beta} + \phi_- e^{i\beta} - (2+\omega)\phi + \epsilon^2 F\left(\phi, \phi_+ e^{-i\beta}, \phi_- e^{i\beta}\right), \qquad (21.5)$$

where $\phi_\pm(z) = \phi(z \pm 1)$ and we have rescaled the equation so that a parameter ϵ^2 appears in front of the nonlinearity.

First, consider the linear properties of the Eq. (21.5). The dispersion curve

$$\omega = \omega(k) = -vk + 2(\cos(\beta - k) - 1) \qquad (21.6)$$

is shown on Fig. 21.1 for (a) $v = 0$ and (b) $v = 0.5$, when $\beta = 0$. The wave spectrum resides on the segment $\omega \in [-4, 0]$ in the case $v = 0$ and on the line $\omega \in \mathbb{R}$ in the case $v \neq 0$. As a result, traveling wave solutions with $v \neq 0$ must have resonances with the wave spectrum.

Bifurcation of small-amplitude traveling wave solutions may occur from quadratic points of the dispersion relation $\omega = \omega(k)$, when the double root $k = 0$ for the Fourier mode splits into two imaginary values $k = \pm i\kappa$ for the exponentially decaying tails. In parameter space (ω, v), this could imply that the bifurcation occurs at points for which $\kappa = 0$ into the parameter region, $\mathcal{K}$, say, represented by $\kappa > 0$. From (21.6) we find that there is a one-to-one mapping between (ω, v) and (κ, β) in such a parameter region:

$$\omega = 2\cos\beta \cosh\kappa - 2, \qquad v = \frac{2\sin\beta \sinh\kappa}{\kappa}. \qquad (21.7)$$

21 Traveling Solitary Waves in DNLS Equations

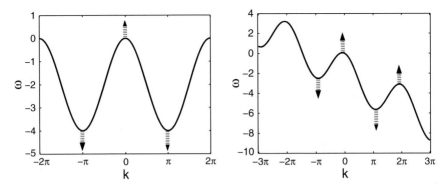

Fig. 21.1 Dispersion curves $\omega = \omega(k)$ for (**a**) $v = 0$ and (**b**) $v = 0.5$, when $\beta = 0$. The *vertical arrows* show directions of possible bifurcations of traveling wave solutions

If other roots $k = k_n$ are present for the same values of ω and v, this bifurcation becomes complicated by the fact that the resonant Fourier modes with $k = k_n$ coexist with the tail solution as $|z| \to \infty$. Such modes would represent small-amplitude radiation modes or *phonons* which would typically lead to the collapse of any localized solutions. The region $\mathcal{K}$ is given by the exterior of the gray wedge-shaped region in Fig. 21.2. The region where additional Fourier modes exist within $\mathcal{K}$ is given by the dark shaded region. Note that the set of points at which small-amplitude bifurcation might occur is thus the boundary of the gray wedge (corresponding to $\kappa = 0$) which can be parametrized by β lying between the two depicted values β_0 and β_1.

For general points $\kappa = 0$, a normal form can be derived that governs the small amplitude behavior. However, one finds that this normal form is integrable to all powers of ϵ, such that persistence or nonpersistence of traveling wave solutions can

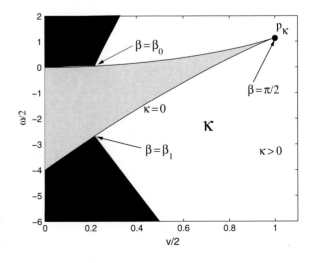

Fig. 21.2 The bifurcation curve $\kappa = 0$ is represented in the (ω, v)-plane for $\epsilon = 1$ as the boundary of the *gray wedge-shaped region*. Localized solutions of the differential advance-delay equation (21.5) may bifurcate into the region $\mathcal{K}$ which is shaded white. For $\beta < \beta_0$ and $\beta > \beta_1$, more than one radiation mode exists

only be studied by analyzing exponentially small in ϵ, beyond-all-orders correction terms. However, for the special point p_K: $(\omega, v) = (2, 2)$, which corresponds to $\beta = \pi/2$, the persistence problem can be studied at third order in ϵ, by deriving a (generically) nonintegrable polynomial normal form that can be analyzed using regular asymptotics, as we shall show in Sect. 21.2.2 below. Before doing so, it is helpful to introduce a bit more notation.

Instead of treating the system (21.5) directly, following Pelinovsky and Rothos [18] we shall adopt a "spatial dynamics" point of view by rewriting (21.5) as an (infinite–dimensional) evolution problem in the spatial coordinate z. For this purpose we introduce the new coordinate $p \in [-1, 1]$ and the vector $\mathbf{u} = \mathbf{u}(z, p) = (u_1, u_2, u_3, u_4)^T$ defined by $u_1 = \phi(z)$, $u_2 = \phi(z + p)$, $u_3 = \bar{\phi}(z)$, $u_4 = \bar{\phi}(z + p)$, which we assume to lie within the Banach space

$$\mathcal{D} = \left\{ \mathbf{u} \in \mathbb{C}^4, \ \mathbf{u} \in C^1(\mathbb{R}, [-1, 1]) : \ u_2(z, 0) = u_1(z), \ u_4(z, 0) = u_3(z) \right\}.$$

We use the notation $\delta^{\pm 1}$ to represent the difference operators $\delta^{\pm 1} \mathbf{u}(z, p) = \mathbf{u}(z, \pm 1)$. Within this formulation, the differential advance-delay equation (21.5) can be written in vector form as

$$-iv\mathcal{J}\frac{d\mathbf{u}}{dz} = \mathcal{L}\mathbf{u} + \epsilon^2 \mathcal{M}(\mathbf{u}), \tag{21.8}$$

where $\mathcal{J} = \text{diag}(1, 1, -1, -1)$,

$$\mathcal{L} = \begin{pmatrix} -(2+\omega) & e^{-i\beta}\delta^{+1} + e^{i\beta}\delta^{-1} & 0 & 0 \\ 0 & -iv\dfrac{\partial}{\partial p} & 0 & 0 \\ 0 & 0 & -(2+\omega) & e^{i\beta}\delta^{+1} + e^{-i\beta}\delta^{-1} \\ 0 & 0 & 0 & iv\dfrac{\partial}{\partial p} \end{pmatrix}, \tag{21.9}$$

is the linear operator that maps $\mathcal{D}$ into $\mathcal{H} = \left\{ \mathbf{F} \in \mathbb{C}^4, \mathbf{F} \in C^0(\mathbb{R}, [-1, 1]) \right\}$ continuously, where $\delta^{\pm 1} \mathbf{u}(z, p) = \mathbf{u}(z, \pm 1)$, and the nonlinear operator

$$\mathcal{M} = (F(u_1, e^{-i\beta}\delta^{+1}u_2, e^{i\beta}\delta^{-1}u_2), 0, \bar{F}(u_3, e^{i\beta}\delta^{+1}u_4, e^{-i\beta}\delta^{-1}u_4), 0)^T.$$

In this spatial dynamics context, we can look at the dispersion relation in a different way. Roots of the dispersion relation

$$D(p; \kappa, \beta) \equiv \cos\beta(\cosh p - \cosh\kappa) + i\sin\beta\left(\sinh p - \frac{\sinh\kappa}{\kappa}p\right) = 0,$$

correspond to eigenvalues of the linearized dynamical system. Using the theory of *embedded solitons* [19] a condition to find isolated *branches* of truly localized

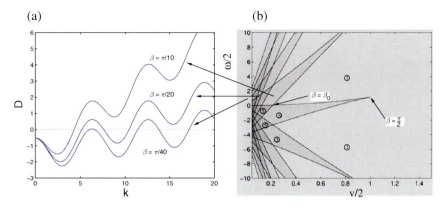

Fig. 21.3 (a) The graph of $D(ik; \kappa, \beta)$ versus k for $\kappa = 1$ and for different values of β. (b) The (ω, v) parameter plane divided into domains corresponding to different numbers of pairs of eigenvalues of the spatial-dynamics problem. This number of roots is indicated by the depth of shading and also the number indicated in the *small circles*. *Arrows* indicate the equivalence of these numbers with the number of real-valued roots of corresponding dispersion curves in panel a. As $v \to 0$ the number of real roots increases to infinity

solutions inside $\mathcal{K}$ is that there should only be one pair of pure imaginary eigenvalues; see Fig. 21.3. This leads to exactly the white region in Fig. 21.2, because we can interpret the multiple roots of the dispersion curves as where the number of pairs of imaginary eigenvalues of the spatial dynamical system increases.

21.2.2 Center Manifold Reduction

The technique of analyzing small-amplitude bifurcation of localized solutions near the special point $p_\mathcal{K}$ is to use a reduction of the infinite-dimensional dynamical system (21.8) to a low-dimensional system. The technique is similar to how one analyzes bifurcations in finite dimensions, where one considers the problem reduced to the *center manifold* that is tangent to the central eigenspace (that is, the generalized eigenspace containing all eigenvalues with zero real part). At $p_\mathcal{K}$, the center eigenspace of the system (21.8) includes only the zero eigenvalue $\lambda = 0$, but this has algebraic multiplicity *six* and geometric multiplicity *two*. The two eigenvectors of the kernel of $\mathcal{L}$ are $\mathbf{u}_0 = (1, 1, 0, 0)^T$, $\mathbf{w}_0 = (0, 0, 1, 1)^T$ and the four eigenvectors of the generalized ker$\mathcal{L}$ are $\mathbf{u}_1 = (0, p, 0, 0)^T$, $\mathbf{w}_1 = (0, 0, 0, p)^T$, $\mathbf{u}_2 = \frac{1}{2}(0, p^2, 0, 0)^T$, $\mathbf{w}_2 = \frac{1}{2}(0, 0, 0, p^2)^T$. Near $p_\mathcal{K}$, it is natural to rescale parameters so that

$$\omega = 2 + \epsilon^2 \Omega, \qquad v = 2 + \epsilon^2 V. \qquad (21.10)$$

With this rescaling, the nonlinear problem (21.8) takes the explicit form

$$-2i\mathcal{J}\frac{d\mathbf{u}}{dz} = \mathcal{L}_0\mathbf{u} + \epsilon^2\mathcal{N}(\mathbf{u}), \tag{21.11}$$

where the operator $\mathcal{L}_0 = \mathcal{L}(p_K, \beta = \pi/2)$ and the perturbation vector $\mathcal{N}(\mathbf{u})$ is given explicitly as $\mathcal{N} = \mathcal{M}(p_K, \beta = \pi/2) + \mathcal{M}_1$ where

$$\mathcal{M}_1 = (-\Omega u_1 + iVu_1', iV(\partial_z u_2 - \partial_p u_2), -\Omega u_3 - iVu_3', -iV(\partial_z u_4 - \partial_p u_4))^T.$$

To form a reduced problem on the center manifold, we apply the decomposition

$$\mathbf{u}(z) = \mathbf{u}_c(z) + \epsilon^2 \mathbf{u}_h(z), \tag{21.12}$$

where

$$\mathbf{u}_c(z) = A(z)\mathbf{u}_0 + B(z)\mathbf{u}_1 + C(z)\mathbf{u}_2 + \bar{A}(z)\mathbf{w}_0 + \bar{B}(z)\mathbf{w}_1 + \bar{C}(z)\mathbf{w}_2$$

is the projection onto the center eigenspace and $\mathbf{u}_h(z)$ is the projection onto the space spanned by the rest of the spectrum of the operator $\mathcal{L}_0$:

$$-2i\mathcal{J}\frac{d\mathbf{u}_h}{dz} - \mathcal{L}_0\mathbf{u}_h = \mathbf{F}_h \equiv \frac{1}{\epsilon^2}\left[2i\mathcal{J}\frac{d\mathbf{u}_c}{dz} + \mathcal{L}_0\mathbf{u}_c\right] + \mathcal{N}(\mathbf{u}_c + \epsilon^2\mathbf{u}_h). \tag{21.13}$$

Here, the function $\mathbf{F}_h$ can be assumed to act on a certain function space of vectors $\mathbf{u}$ that decay exponentially quickly as $|z| \to \infty$. Within this function space it can be shown that there exists a continuous map $\mathcal{A}_\epsilon$ such that $\mathbf{u}_h = \mathcal{A}_\epsilon \mathbf{u}_c$, see [18] for details. That is, there is a one-to-one correspondence between localized solutions $\mathbf{u}_c$ in the center manifold and solutions to the full system. The reduced system on the center manifold can then be written in the form [18]

$$\frac{d}{dz}\begin{pmatrix}\mathbf{x}\\\bar{\mathbf{x}}\end{pmatrix} = \mathcal{L}_c\begin{pmatrix}\mathbf{x}\\\bar{\mathbf{x}}\end{pmatrix} + \epsilon^2\begin{pmatrix}\mathbf{R}(\mathbf{x})\\\bar{\mathbf{R}}(\mathbf{x})\end{pmatrix} + O\left(\epsilon^2\|g\| + \epsilon^2\|\mathbf{u}_h\|\right), \tag{21.14}$$

where $\mathbf{x} = (A, B, C)^T \in \mathbb{C}^3$, the vector function $\mathbf{R} : \mathbb{C}^3 \mapsto \mathbb{C}^3$ is given by

$$\mathbf{R}(\mathbf{x}) = 3i\left(\frac{1}{20}, 0, -1\right)^T(g(A, B, C) - \Omega A + iVB) \tag{21.15}$$

and

$$g_{\text{DNLS}} = |A|^2 A, \quad g_{\text{AL}} = -2i|A|^2 B, \quad g_{\text{sat}} = A/(1 + |A|^2) \tag{21.16}$$

and the linear operator $\mathcal{L}_c$ corresponds to the central part of the operator $\mathcal{L}_0$ spanned by the 6D generalized eigenspace of the zero eigenvalue.

21.2.3 Normal Form Equations Near the Zero-Dispersion Point

Equation (21.14) for the center manifold reduction can be further reduced by using near-identity transformations of the form

$$\mathbf{x} = \boldsymbol{\xi} + \epsilon^2 \Phi(\boldsymbol{\xi}), \tag{21.17}$$

where $\Phi: \mathbb{C}^3 \mapsto \mathbb{C}^3$ contains strictly nonlinear terms that are designed to remove the *nonresonant* terms in the center manifold equations. The resulting normal form equations take the form

$$\frac{d}{dz}\begin{pmatrix}\boldsymbol{\xi}\\ \bar{\boldsymbol{\xi}}\end{pmatrix} = \mathcal{L}_c\begin{pmatrix}\boldsymbol{\xi}\\ \bar{\boldsymbol{\xi}}\end{pmatrix} + \epsilon^2\begin{pmatrix}\mathbf{P}(\boldsymbol{\xi})\\ \bar{\mathbf{P}}(\boldsymbol{\xi})\end{pmatrix}, \tag{21.18}$$

where the nonlinear vector function $\mathbf{P} : \mathbb{C}^3 \mapsto \mathbb{C}^3$ contains only the essential resonant terms. That is,

$$\mathcal{D}\mathbf{P}(\boldsymbol{\xi})\mathcal{L}_c^*\boldsymbol{\xi} = \mathcal{L}_c^*\mathbf{P}(\boldsymbol{\xi}), \tag{21.19}$$

where $\mathcal{L}_c^*$ is the adjoint operator and $\mathcal{D}$ is the Jacobian derivative. Introducing $\boldsymbol{\xi} = (a, b, c)^T$, the normal-form so obtained can be written in the form [18]

$$\frac{da}{dz} = b + ia\mathcal{P}_1\left(|a|^2, I, J, \epsilon^2\right), \tag{21.20}$$

$$\frac{db}{dz} = ib\mathcal{P}_2\left(|a|^2, I, J, \epsilon^2\right) + \mathcal{Q}_1(a, b, c), \tag{21.21}$$

$$\frac{dc}{dz} = ic\mathcal{P}_3(|a|^2, I, J, \epsilon^2) + \mathcal{Q}_2(a, b, c), \tag{21.22}$$

where $I = a\bar{b} - \bar{a}b$, $J = a\bar{c} - \bar{a}c$ and $\mathcal{P}_1, \mathcal{P}_2, \mathcal{P}_3, \mathcal{Q}_1, \mathcal{Q}_2$ are polynomial functions of their arguments.

Now, a key observation made in [18] is that the analysis of solutions to (21.20), (21.22) and (21.22) is greatly simplified by introducing a further transformation of variables and parameters in order to write the system as a single third-order equation for a complex scalar Φ. Specifically, one obtains an equation of the form

$$\frac{i}{3\epsilon^2}\Phi''' - iV\Phi' + \Omega\Phi = h(\Phi, \Phi', \Phi'', \Phi''') + O(4), \tag{21.23}$$

where h contains pure cubic terms and $O(4)$ represents terms that are fourth-order or higher in Φ or its derivatives. Now, the third-order differential equation (21.23) is equivalent to the traveling-wave equation for the so-called third-order nonlinear Schrödinger equation that has been studied by a number of authors, e.g., [20, 21]. In particular, there is a localized embedded soliton solution of this equation provided certain conditions on the nonlinear term h are satisfied, most importantly that there should be a zero coefficient of the term $|\Phi|^2\Phi$.

To determine whether localized solutions bifurcate for a particular lattice equation (21.1), one therefore has to carry out the center manifold reduction and near-identity (21.17) to compute the specific coefficients of the nonlinear terms in h. Specifically, in [18] it was found for the DNLS and AL equations that

$$h_{\text{DNLS}} = |\Phi|^2\Phi + \frac{1}{140}\left(6|\Phi|^2\Phi'' - 2\Phi^2\bar{\Phi}'' + (\Phi')^2\bar{\Phi} - 3|\Phi'|^2\Phi\right), \quad (21.24)$$

$$h_{\text{AL}} = -2i|\Phi|^2\Phi' + \frac{i}{100}\left(4\Phi\Phi''\bar{\Phi}' - 2\Phi\Phi'\bar{\Phi}'' - 2\Phi'\Phi''\bar{\Phi} + |\Phi|^2\Phi''' - \Phi^2\bar{\Phi}'''\right). \quad (21.25)$$

This therefore proves the existence of a continuous (two-parameter) family of single-humped traveling wave solutions in the third-order derivative NLS equation, when it is derived from the integrable AL nonlinearity F_{AL}. Whereas there are no single-humped solutions in the third-order derivative NLS equation when it is derived from the usual cubic DNLS nonlinearity F_{DNLS} with pure on-site interactions.

More generally, for the generalized cubic nonlinearity F_{g3} Pelinovsky [22] derived the condition

$$\alpha_1 + 2\alpha_4 - 2\alpha_5 - 2\alpha_6 + \alpha_7 = 0. \quad (21.26)$$

on the parameters α_i in order for the $|\Phi|^2\Phi$ term to vanish in the normal form nonlinearity h. Under this constraint then, it can be proved rigorously that the generalized DNLS equation supports a branch of single-humped traveling waves that bifurcates into $\mathcal{K}$ from the special point $p_\mathcal{K}$.

21.2.4 Mel'nikov Calculations for Generalized Cubic DNLS

Furthermore, if (21.26) holds, $\alpha_2 + 2\alpha_8 - 2\alpha_9 \neq 0$ and

$$\text{either } \alpha_3 - \alpha_8 - \alpha_9 + \alpha_{10} = 0 \text{ or } \alpha_2 + 3\alpha_3 - \alpha_8 - 5\alpha_9 + 3\alpha_{10} = 0, \quad (21.27)$$

the relevant third-order ordinary differential equation (21.23) can be shown to reduce to the integrable Hirota or Sasa–Satsuma equations respectively (see [22]) which admit two-parameter families of traveling solutions in (κ, β). Note that the Hirota family includes the AL case where $\alpha_2 = 1$ and $\alpha_i = 0$ for all $i \neq 1$. In all these cases, one has therefore localized solutions throughout an open parameter region within $\mathcal{K}$. But these integrable cases are somewhat special, because, since the center manifold for parameter values within $\mathcal{K}$ is 2D, then one should generically expect localized solutions, if they occur at all, to exist along curves in the (κ, β)-plane.

21 Traveling Solitary Waves in DNLS Equations

A starting point to find such curves is the observation in [22] that if

$$(\alpha_2, \alpha_3) \in \mathbb{R}^2, \quad \text{with } \alpha_2 > \alpha_3, \quad \text{and } \alpha_j = 0, \quad \text{for } j = 1 \text{ and } j = 4\ldots 10, \tag{21.28}$$

then the explicit solution

$$\Phi(Z) = \frac{\sinh \kappa}{\sqrt{\alpha_2 - \alpha_3}} \operatorname{sech}(\kappa Z) \tag{21.29}$$

solves the differential advance-delay Eq. (21.5) exactly for $\kappa > 0$ and $\beta = \pi/2$. A natural question to ask then is the persistence of this explicit solution as one adds nonlinear terms that break the constraint (21.28). Pelinovsky et al. [14] studied this question using Mel'nikov theory. That is, one computes the splitting distance between the stable and unstable manifolds of the origin as one adds the additional nonlinear terms as perturbations. Intuitively, provided the constraint (21.28) is satisfied, then one should expect that a curve in the (κ, β)-plane persists. The key to Mel'nikov theory is to study the linearized variational equations evaluated around the explicit solution. The splitting distance is then projected onto the space of bounded solutions of these variational equations.

Without going into the details here, a study of the form of the variational equations shows that there are two cases. If $\alpha_1 = 0$ and $\alpha_4 = \alpha_6$ and $\alpha_7 = 2\alpha_5$ [which we note automatically satisfies the constraint (21.26)], then a localized solution continues to exist exactly along the straight line $\beta = \pi/2$. If these additional constraints are not met, the curve of parameter values along which a localized solution exists shifts to a local neighborhood of this line.

The linear variational equations around the exact solution are also precisely what is required to understand the temporal stability of the localized solutions when launched as initial conditions to the initial-value problem. Approximate arguments in [14] indicate that localized modes should be stable, just as the corresponding solitons are to the third-order NLS equation [19].

21.2.5 Beyond-All-Orders Asymptotic Computation

It is also possible for solutions to bifurcate from $\kappa = 0$ for $\beta \neq \pi/2$. In this case it can be shown that the normal form one can derive describes the bifurcation of a saddle-center equilibrium from a pure center in a reversible system. This problem has been studied by many authors, see for example, the book [23] and it is well known that the integrable normal form does not provide the leading-order expression to the Mel'nikov integral. In general, solutions that persist into $\kappa > 0$ for some β-value $\neq \pi/2$ have nonvanishing tails that are exponentially small in κ. Thus, the question of the persistence of the localized solution to the normal form becomes a question that can only be answered by considering the beyond-all-orders terms in the normal form expansion.

The technique of beyond-all-orders asymptotic expansions for problems of this type was developed by Tovbis *et al.* [16] and first applied to DNLS equations by Oxtoby and Barashenkov [17]. One seeks a solution as a regular series solution to (21.5), in which the radiation tail appears only beyond all orders of the asymptotic expansion in powers of κ. By analyzing the outer expansion near a singularity in the complex z plane and by rescaling dependent and independent variables with a blow-up technique [16], the amplitude of the radiation tail can be measured at the leading-order approximation. This in turn leads to a recurrence relation to determine the splitting between the stable and unstable manifolds of the saddle-center point. This splitting is known as a Stokes constant, and a zero of this constant implies the true persistence of a single-humped localized solution.

21.2.6 Numerical Continuation Using Pseudospectral Methods

Localized solutions to the advance-delay equation (21.5) can easily be sought numerically using a pseudospectral method originally proposed by Eilbeck *et al.* [9, 15]; see also [24, 25] for similar results for discrete sine-Gordon lattices. The pseudospectral method is tantamount to making a truncated Fourier series expansion of solutions $\phi(z)$ to (21.5) on a long finite interval $[-L/2, L/2]$. A particular choice of expansion terms can be made that exploits the underlying symmetry of the localized solutions we seek, namely by choosing even real functions and odd imaginary functions

$$\phi(z) = \sum_{j=1}^{N} a_j \cos\left(\frac{\pi j z}{L}\right) + i b_j \sin\left(\frac{\pi j z}{L}\right), \qquad (21.30)$$

where $a_j, b_j \in \mathbb{R}$ are the coefficients of the Fourier series. Substituting the expansion (21.30) in (21.5) at the series of *collocation* points

$$z_i = \frac{Li}{2(N+1)}, \qquad i = 1, \ldots, 2N$$

gives a system of $2N$ nonlinear algebraic equations for the unknown coefficients a_j, b_j, which can be solved using globally convergent root-finding methods (for example, the Powell hybrid method [26]). Once a solution is found, this can be continued in a single parameter using a numerical path-following scheme built around Newton's method, for example, the code AUTO [27].

The solutions found using the pseudospectral approximation will generally be weakly nonlocal solitary waves that exhibit nonzero oscillatory tails. To find waves with zero tails we need to add an extra condition – a signed measure of the amplitude of the tail – and seek zeros of this function. Given the symmetry assumed by the expansion (21.30), a good choice of such a tail function is

$$\Delta = \text{Im}\left(\phi\left(\frac{L}{2}\right)\right), \tag{21.31}$$

which measures the amplitude of the imaginary part of the tail of a solution of period L. To see why this is a good choice, it is helpful to regard the numerical solutions to (21.5) as being made up of two parts: an exponentially localized core and a nonvanishing oscillatory background. At a sufficient distance from the center of the soliton the core part will be zero due to its exponential localization. Because of the way $\phi(z)$ has been approximated in (21.30), we know that the real part of $\phi(z)$ is odd around $z = L/2$ and the imaginary part is even. Moreover, within parameter region $\mathcal{K}$, where there is only one phonon branch, a small amplitude tail will be purely sinusoidal. Therefore (21.31) is a pure measure of the amplitude of the tail.

Figure 21.4 shows results from Melvin et al. [28] on computation of various branches of solutions to (21.5) for the saturable nonlinearity F_{sat} with varying discreteness parameter

$$\varepsilon = 1/\sqrt{\epsilon},$$

while keeping other parameters fixed. Note that zeros of Δ are indeed found to be values at which the tail vanishes to numerical accuracy. Moreover, since Δ is a *signed* measure of the tail amplitude, we can note a topological distinction between branches of the nonzero tail solutions that do contain a zero (they are u- or n-shaped in the figure) compared with those that do contain a zero (they are s-shaped). Hence we can assert with confidence that we have found true zeros. Moreover, since

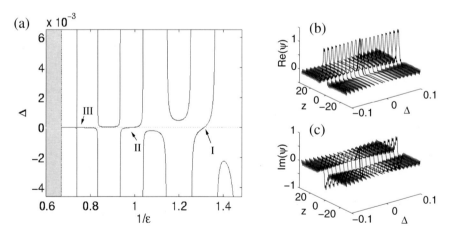

Fig. 21.4 Reprinted from [28] with permission. (a) Continuation of weakly localized solutions (with nonzero oscillatory tails) to (21.5) for the saturable nonlinearity F_{sat} against $\varepsilon = 1/\sqrt{\epsilon}$ for $v = 0.7$, $\omega = -0.5$ and $L = 60$ showing three zeros of Δ at $\varepsilon \approx 0.76, 1.02, 1.36$. The *shaded region* represents the spectral band where any embedded solitons would be of co-dimension 2. (**b, c**) Continuation of branch with second zero of Δ at $\varepsilon \approx 1.02$. for $c = 0.7$, $\Lambda = 0.5$, $L = 60$. (b) $Re(\phi)$, (c) $Im(\phi)$

solution branches that contain truly localized solutions cross the Δ axis transversely, the condition $\Delta = 0$ can be added to the list of algebraic equations and another parameter freed. In so doing, we can use path-following to trace branches of localized solutions in two parameters, for example, in the (ω, v)-plane.

21.3 Applications

21.3.1 Saturable DNLS

Melvin et al. [13, 28] considered localized modes in the DNLS model (21.1) with the saturable nonlinearity F_{sat} by continuation of zeros of Δ defined in (21.31). Figure 21.5 shows several solution branches. Note that the numerics becomes unreliable in Fig. 21.5 at the upper end of the branch labeled I, because it terminates by reaching the small-amplitude limit. Here, beyond-all-orders asymptotics applies and the function Δ in Fig. 21.4 becomes remarkably flat, so that zeros cannot be detected accurately. However this is precisely the realm where bifurcation can be predicted by zeros of the Stokes constant. Such an isolated zero in this model was found by Oxtoby and Barashenkov [17].

Note the open circles for zero wavespeed in Fig. 21.5. These correspond to where a generalized PN barrier [13] between the on-site and off-site stationary solutions precisely vanishes. It has been conjectured that these would be "transparent points" where localized traveling waves form. However, the computations show that this is not the case. Each branch in some sense "points" to one of these transparent points, but it terminates when it hits the "multi-phonon band" where there are extra branches of linear waves. Numerically, in this region we find that all approximately

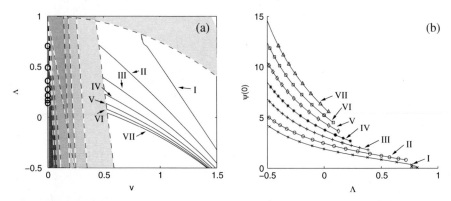

Fig. 21.5 Reprinted from [28] with permission. (**a**) Continuation of the three solitary wave branches found in Fig. 21.4 with varying $\Lambda = 1 - \omega/2$ and v for fixed $\varepsilon = 1$, and four other branches. *Open circles* for $v = 0$ correspond to points where the generalized PN barrier vanishes (see text for details.) (**b**) Amplitude of solutions $\phi(z = 0)$ along these seven continuation branches

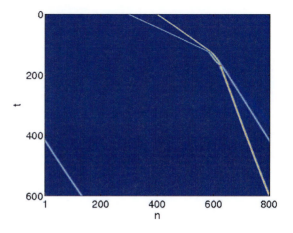

Fig. 21.6 Reprinted from [28] with permission. Interaction of two soliton solutions with $\varepsilon = 1$ and $\Lambda = 1 - \omega/2 = 0.5$. The branch I soliton is initially centered on site $n = 300$ with $v = 1.00926$ and the branch II soliton is centered on site $n = 400$ with $v = 0.67725$

localized solutions to the advance-delay equation (21.5) generate nonzero oscillatory or quasi-periodic tails. In some sense, then these multiphonon bands form an alternative to the PN barrier that prevents any low wavespeed localized waves.

Spectral computations in [28] suggest that the localized waves presented in Fig. 21.4 are temporally stable. These results are backed up by simulation results. If one provides a strong perturbation then radiation is shed for a finite time, as the wave relaxes to a new solution on the same branch, with a different internal frequency ω and speed v. Small internal oscillations of the core of the soliton may remain for long times, but radiation does not appear to be continuously shed and the resulting localized wave persists for a long time. The results of Fig. 21.6 show that these coherent structures can even survive collision with one another.

21.3.2 Generalized Cubic DNLS

The Mel'nikov analysis for the generalized cubic nonlinearity F_{g3} as mentioned in Sect. 21.2.4 was backed up by using the pseudospectral continuation method by Pelinovsky *et al.* [14]. The results are reproduced here in Figs. 21.7 and 21.8. The coefficients α_i were chosen to satisfy

$$\alpha_1 = 2\alpha_4 = 2\alpha_5 = 2\alpha_6, \quad \alpha_7 = \alpha_9 = 0, \quad \alpha_{10} = \alpha_8, \tag{21.32}$$

subject to the normalization

$$\alpha_2 + \alpha_3 + 4\alpha_6 + 2\alpha_8 = 1.$$

The conditions (21.32) were shown in [22, 29] to lead to a precise vanishing of the PN barrier. That is, stationary solutions exist that are translational invariant, or alternatively a continuous family of waves exists that interpolates between the site-centered and the off-site-centered localized modes. Nevertheless, as we have already argued, localized waves with infinitesimal speed cannot occur due to the

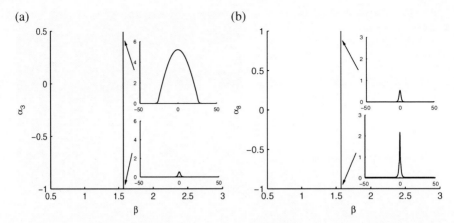

Fig. 21.7 Persistence of the localized solution for $\kappa = 1$ and $\beta = \pi/2$ versus α_3 for $(\alpha_6, \alpha_8) = (0, 1)$ (*left*) and α_8 for $(\alpha_3, \alpha_6) = (-1, 0)$ (*right*). The insets show the single-humped profiles of the localized solution. Reprinted from [14] with permission

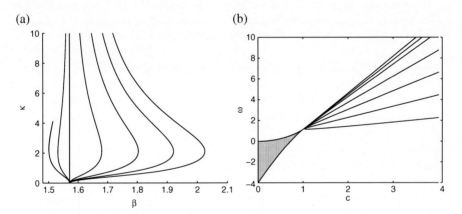

Fig. 21.8 Persistence of solutions for $(\alpha_3, \alpha_8) = (-1, 1)$ as κ is varied on the (β, κ)-plane (*left*) and on the (v, ω)-plane (*right*). Different curves correspond to different values of $\alpha_6 = 0.5, 0.25, 0, -0.5, -1, -1.5, -2$ from *left to right* on the *left panel* and from *top to bottom* on the *right panel*. *The shaded area* in the *right panel* indicates the boundary of the existence domain at $\kappa = 0$ and $\beta \in [0, \pi]$. Reprinted from [14] with permission

multiphonon bands. Instead, given the constraint (21.26) we get a regular bifurcation of small-amplitude localized waves from the point p_K at wavespeed $v = 2$.

The results of the numerical continuation for varying β and κ are shown in Figs. 21.7 and 21.8, for different groups of the parameters α_i. In the former figure, we see that the solitary traveling waves persist precisely along the curve $\beta = \pi/2$, whereas in the latter they persist along a curve for which β varies. This is precisely as predicted by the theory outlined in Sect. 21.2.4.

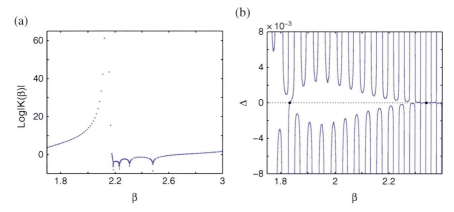

Fig. 21.9 After [12]. Existence of traveling wave solutions in the Salerno model for $\alpha = 0.65$ computed via calculation of (**a**) the Stokes constant $K(\alpha; \beta)$ ($\kappa = 0$) and (**b**) the radiation tail amplitude Δ ($\kappa = 0.5$). A number of points where localized waves exist are found, at $\beta \approx 2.182, 2.237, 2.308$, and 2.48 using the Stokes constant method and at $\beta \approx 1.834, 2.34$ from measuring the radiation tail amplitude (indicated by *dots*)

21.3.3 The Salerno Model

In [12], Melvin *et al.* considered a combination of Stokes constant computation and numerical continuation to explore the existence of localized solitary waves in the Salerno model (21.1) with $F = F_S$ as the *three* parameters, α, β, and κ vary. Figure 21.9 shows the correspondence between zeros of the Stokes constant $K(\alpha; \beta)$, where there are four distinct zeros for $\kappa = 0$, and zeros of the numerical tail coefficient Δ, where there are two zeros for $\kappa = 0.5$, in the case $\alpha = 0.65$. The discrepancy in the number of zeros between the Stokes constant calculations and the pseudospectral computations is due to the fact that some of zeros of $K(\alpha; \beta)$ move to the domain $\beta > \beta_1 \approx 13\pi/14$ for nonzero values of κ and hence do not generate bifurcations of nontrivial zeros of Δ.

We would expect the zeros of Δ to approach the zeros of $K(\alpha; \beta)$ as κ is reduced toward zero. However, as we have already noted, continuation into the beyond-all-orders limit $\kappa \to 0$ is problematic. It is a much easier task to compute curves $\Delta = 0$ for a fixed nonzero value of κ in the (α, β)-plane, the results of which are depicted as dashed curves in the left panel of Fig. 21.10. All curves for finite κ have a fold point at the maximum value of α for some value of $\beta < \pi/2$, which approaches the point $(\alpha, \beta) = (1, \pi/2)$ as $\kappa \to 0$. For β less than this fold point the solutions split into multiple-humped solutions as shown in panel (b). Also plotted as solid lines in the figure are lines corresponding to four zeros of the Stokes constant $K(\alpha; \beta)$ which can each be seen to approach the point $(\alpha, \beta) = (1, \pi/2)$. In results not shown here, the dashed curves corresponding to zeros of Δ are found to approach these solid lines for $\beta > \pi/2$ as κ becomes smaller.

These numerical results can be summarized in the topological sketch in Fig. 21.11. When $\kappa = 0$, only the single-humped solutions for $\beta > \pi/2$ and $\alpha > 0.5$ exist,

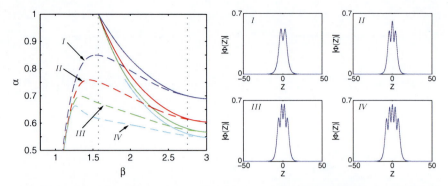

Fig. 21.10 Continuation of zeros of the Stokes constants (*solid lines*) along with the corresponding branches of solutions with zero radiation tail $\Delta = 0$ for $\kappa = 0.5$ (*dashed lines*). For $\beta < \pi/2$ the solutions are multihumped. The number of humps corresponding to each branch (I–IV) is shown in the *right panels*, with $\alpha = 0.5$ for all plots and $\beta =$ I 1.3267717242, II 1.2057587752, III 1.2167118387, and IV 1.1752123732

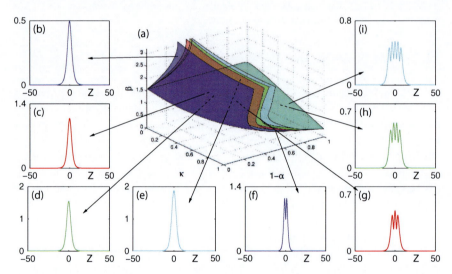

Fig. 21.11 (a) Existence sketch of the first four "sheets" of solutions in $(\kappa, 1-\alpha, \beta)$-parameter space. (**b**)–(**i**) Profiles of solutions from different sheets showing single-humped solutions (**b**)–(**e**) for the upper surfaces and multihump solutions (**f**)–(**i**) on the lower surface. Shadings on plots (b)–(i) match the solution sheet they originate from as shown in Fig. 21.10

corresponding to the upper parts of each solution sheet, but as soon as κ is nonzero, extra branches of multihumped solutions for $\beta < \pi/2$ form, corresponding to the lower parts of the sheets. For $\kappa > 0$, the fold point joining these two branches moves in the negative α direction in the (α, β)-plane, such that the only place the sheet of solutions approaches the point of the integrable AL model ($\alpha = 1$) is at the co-dimension-three point $\kappa = 0$, $\beta = \pi/2$. Behind the sheet of solutions shown, in the negative α direction, a number of other solution sheets exist, and one might conjecture that there are indeed infinitely many such sheets. Further numerical integration results presented in [12] suggest that the single-humped solutions are stable as solutions to the initial-value problem, whereas perturbed multi-humped solutions continuously shed radiation and decay.

21.4 Conclusion

In this chapter we have illustrated that localized waves in 1D DNLS lattices cannot in general move without shedding radiation. However, for *finite* wavespeeds one might expect to find one-parameter families of these waves, such that isolated values of the internal frequency ω are selected for each wavespeed v. It seems though that certain extra ingredients are required. First, the wavespeed should be sufficiently high that we are in the "embedded soliton" parameter region where there is only one branch of the linear dispersion relation. Second, it seems necessary to have some kind of competing nonlinearity. It would seem that, despite previous suggestive numerical results [9], there are no truly localized traveling waves for the pure DNLS equation nonlinearity F_{DNLS}. But the saturable nonlinearity F_{sat} when expanded as a Taylor series has a competition between cubic, quintic, septic, etc., terms, and this leads to the large (possibly infinite) number of branches of localized waves, the first seven of which are reported in Fig. 21.5. The vanishing of a generalized PN barrier seems to act as an indication of the existence of such branches, although they cannot actually exist for small wavespeeds due to the different "barrier" caused by the presence of infinitely many phonon bands in the limit $v \to 0$. Preliminary computations suggest that there is just one such vanishing point and consequent localized branch for the case of a mere cubic-quintic nonlinearity F_{3-5}.

We have also shown how it is sometimes possible to use local bifurcation theory to predict small-amplitude persistence of the localized waves, for nonzero velocity, but only close to the special point p_K. Furthermore we require special relations to hold between the parameters in the nonlinear terms, such as for the AL lattice with $F = F_{AL}$ and for the more general cubic nonlinearity F_{g3} under the constraint (21.26). Otherwise, to find small-amplitude traveling waves requires beyond-all-orders asymptotic analysis.

A particular difficulty that is not present for the continuous NLS equations when trying to find localized traveling waves is that moving to a traveling frame results in an advance-delay equation, which has an infinite-dimensional spectrum and cannot be solved as an initial-value problem. Nevertheless, pseudospectral methods are well suited to the numerical detection and continuation of the parameter values at which the localized waves exist.

There are many open questions that remain. First, largely speaking, we have not dealt with the issue of stability of the localized traveling waves computed. Nevertheless, in several results reviewed here there is strong numerical evidence to suggest the orbital stability of at least single-humped versions of these waves. A rigorous theory of stability remains to be developed. Second, it would be interesting to see whether these localized traveling waves are experimentally realizable (perhaps in one of the experimental realizations of DNLS equations reviewed in Chapter 8). Also, whether truly localized traveling breathers can occur in FPU-type lattices (see, e.g., [30]) is a much more challenging question than that addressed here. In such lattices, the envelope of the excitation sits on a temporally periodic state that is not a result of the trivial rotational symmetry as it is for DNLS-type lattices. Finally, a huge challenge is to seek analog of localized traveling states in more general lattices perhaps in two or three spatial dimensions and with different coupling topologies.

Acknowledgments We are grateful to Dimitry Pelinovsky (McMaster University) and Panos Kevrekidis (University of Massachusetts, Amherst) for useful discussions and for allowing us to reproduce their material in this chapter. Thanks also go to Panos for his considered editorial comments on an earlier draft.

References

1. Flach, S., Kladko, K.: Moving discrete breathers? Physica D **127**, 61–72 (1999)
2. Christodoulides, D.N., Lederer, F., Silderberg, Y., Discretizing light behaviour in linear and nonlinear waveguide lattices. Nature **424**, 817–823 (2003)
3. Fleischer, J.W., Carmon, T., Segev, M., Efremidis, N.K., Christodoulides, D.N.: Observation of discrete solitons in optically induced real time waveguide arrays. Phys. Rev. Lett. **90**, art.no. 023902 (2003)
4. Kevrekidis, P.G., Rasmussen, K.O., Bishop, A.R.: The discrete nonlinear Schrödinger equation: a survey of recent results. Int. J. Mod. Phys. B **15**, 2833–2900 (2001)
5. Ablowitz, M.J., Ladik, J.F., Nonlinear differential-difference equations and Fourier analysis. J. Math. Phys. **17**, 1011–1018 (1976)
6. Salerno, M.: Quantum deformations of the discrete nonlinear Schrdinger equation. Phys. Rev. A **46**, 6856–6859 (1992)
7. Gómez-Gardeñes, J., Floria, L.M., Peyrard, M., Bishop, A.R.: Nonintegrable Scroödinger discrete breathers. Chaos **14**, 1130–1147 (2004)
8. Remoissenet, M., Peyrard, M. (eds.): Nonlinear Coherent Structures in Physics and Biology. Springer-Verlag Berlin (1991)
9. Duncan, D.B., Eilbeck, J.C., Feddersen, H., Wattis, J.A.D.: Solitons on lattices. Physica D **68**, 1–11 (1993)
10. Ablowitz, M.J., Musslimani, Z.H., Discrete spatial solitons in a diffraction-managed nonlinear waveguide array: a unified approach. Physica D **184**, 276–303 (2003)
11. Ablowitz, M.J., Musslimani, Z.H., Biondini, G.: Methods for discrete solitons in nonlinear lattices. Phys. Rev. E **65**, 026602 (2002)
12. Melvin, T.R.O., Champneys, A.R., Pelinovsky, D.E.: Discrete Travelling Solitons in the Salerno Model, Preprint. University of Bristol, Bristol (2008)
13. Melvin, T.R.O., Champneys, A.R., Kevrekidis, P.G., Cuevas, J.: Radiationless traveling waves in saturable nonlinear Schrödinger lattices. Phys. Rev. Lett. **97**, 124101 (2006)
14. Pelinovsky, D.E., Melvin,T.R.O., Champneys, A.R.: One-parameter localized traveling waves in nonlinear Schrödinger lattices. Physica D **236**, 22–43 (2007)
15. Eilbeck, J.C., Flesch, R.: Calculation of families of solitary waves on discrete lattices. Phys. Lett. A **149**, 200–202 (1990)

16. Tovbis, A., Tsuchiya, M., Jaffé, C., Exponential asymptotic expansions and approximations of the unstable and stable manifolds of singularly perturbed systems with the Hénon map as an example. Chaos **8**, 665–681 (1998)
17. Oxtoby, O.F., Barashenkov, I.V.: Moving solitons in the discrete nonlinear Schödinger equation. Phys. Rev. E **76** 036603 (2007)
18. Pelinovsky, D.E., Rothos, V.M.: Bifurcations of travelling wave solutions in the discrete NLS equations. Physica D **202**, 16-3-6, (2005)
19. Yang, J.: Stable embedded solitons. Phys. Rev. Lett. **91**, 143903 (2003)
20. Yang, J., Akylas, T.R.: Continuous families of embedded solitons in the third-order nonlinear Schrödinger equation. Stud. Appl. Math. **111**, 359–375 (2003)
21. Pelinovsky, D.E., Yang, J.: Stability analysis of embedded solitons in the generalized third-order NLS equation. Chaos **15**, 037115, (2005)
22. Pelinovsky, D.E.: Translationally invariant nonlinear Schrödinger lattices. Nonlinearity **19**, 2695–2716 (2006)
23. Lombardi, E.: Oscillatory Integrals and Phenomena Beyond All Orders; With Applications to Homoclinic Orbits in Reversible Systems. Springer-Verlag, Berlin (2000)
24. Savin, A.V., Zolotaryuk, Y., Eilbeck, J.C.: Moving kinks and nanopterons in the nonlinear Klein-Gordon Lattice. Phys. D **138**, 267 (2000)
25. Aigner, A.A., Champneys, A.R., Rothos, V.M.: A new barrier to the existence of moving kinks in Frenkel – Kontorova lattices. Physica D **186**(3–4), 148–170 (2003)
26. Powell, M.J.D.: A hybrid method for nonlinear algebraic equations. In: Numerical Methods for Nonlinear Algebraic Equations. Gordon and Breach (1970)
27. Doedel, E.J., Champneys, A.R., Fairgrieve, T.R., Kuznetsov, Y.A., Sandstede, B., Wang, X.J.: Auto97 Continuation and Bifurcation Software for Ordinary Differential Equations. (1997) ftp://ftp.es.concordia.ca/directory/doedel/auto
28. Melvin, T.R.O., Champneys, A.R., Kevrekidis, P.G., Cuevas, J.: Travelling solitary waves in the discrete Schödinger equation with saturable nonlinearity; existence stability and dynamics. Physica D **237**, 551–567 (2008)
29. Dmitriev, S.V., Kevrekidis, P.G., Sukhorukov, A.A., Yoshikawa, N., Takeno, S.: Discrete nonlinear Schrödinger equations free of the Peierls-Nabarro potential. Phys. Lett. A **356**, 324–332 (2006)
30. Iooss, G.: Travelling waves in the Fermi – Pasta – Ulam lattice. Nonlinearity **13**, 849–866 (2000)

Chapter 22
Decay and Strichartz Estimates for DNLS

Atanas Stefanov

22.1 Introduction

In the study of the continuous Schrödinger equation

$$i\partial_t u(t,x) = \Delta u + F(t,x), \qquad (t,x) \in \mathbf{R}^1 \times \mathbf{R}^d, \tag{22.1}$$

one deals with a time-evolution problem, which does not improve smoothness with time. This simple but very fundamental observation (in sharp contrast with the better behaved parabolic evolution) has necessitated radically different approaches to the standard questions of local and global well-posedness, persistence of smoothness, stability of localized structures, etc. Indeed, tackling these questions took some time, and in fact the first rigorous mathematical results did not appear until the late 1970s. Let us explain some of the ingredients that were crucial for understanding these issues. First, for $F = 0$ in (22.1), we have the representation

$$u(t,x) = e^{it\Delta}u_0 := \frac{1}{(4\pi it)^{d/2}} \int_{\mathbf{R}^d} e^{i|x-y|^2/(4t)} u_0(y)dy.$$

From that, we immediately get $\|u(t,\cdot)\|_{L_x^\infty} \leq (4\pi t)^{-d/2} \|u_0\|_{L_x^1}$. That is, the L^∞ norm of the solution decays at the rate of $t^{-d/2}$. Since we have conservation of charge $\|u(t,\cdot)\|_{L^2} = \|u_0\|_{L^2}$, it follows that the Schrödinger evolution "spreads" the conserved energy around as time goes by.[1] In a surprisingly easy way,[2] one can show that the $L^1 \to L^\infty$ decay and the L^2 conservation imply the so-called Strichartz estimates. Before we state them precisely, we introduce the mixed Lebesgue spaces with norm

---

A. Stefanov (✉)
The University of Kansas, Lawrence, KS, USA
e-mail: stefanov@math.ku.edu

---

[1] This is sometimes expressed as "the uncertainty of the system increases as time goes by."
[2] But this was not realized until the seminal work of Ginibre and Velo [1]. The work of Strichartz was about estimating the Fourier restriction operator to cones and paraboloid, which after dualization argument implies the Strichartz estimates for $q = r$ only, see Theorem 1.

$$\|f\|_{L^q_t L^r_x} := \left(\int_0^\infty \left(\int_{\mathbf{R}^d} |f(t,x)|^r dx \right)^{q/r} dt \right)^{1/q}.$$

The following theorem has been proved by many people in different situations, but the main names that stand out are those of Strichartz [2], Ginibre-Velo [1], and Keel-Tao [3].

Theorem 1. *Let both (q,r) and $(\tilde{q},\tilde{r})$ be Strichartz pairs, that is, $q,r \geq 2$, $2/q + d/r = d/2$, and $(q,r,d) \neq (2,\infty,2)$. Then*

$$\|e^{it\Delta} f\|_{L^q_t L^r_x} \leq C \|f\|_{L^2} \tag{22.2}$$

$$\left\| \int e^{-it\Delta} F(t,\cdot) dt \right\|_{L^2} \leq C \|F\|_{L^{\tilde{q}'}_t L^{\tilde{r}'}_x}. \tag{22.3}$$

$$\left\| \int_0^t e^{i(t-s)\Delta} F(s,\cdot) ds \right\|_{L^2} \leq C \|F\|_{L^{\tilde{q}'}_t L^{\tilde{r}'}_x},$$

where we use the notation $q' = q/(q-1)$. Equivalently, for the solution of (22.1),

$$\|u\|_{L^q L^r} \leq C \left(\|u(0)\|_{L^2} + \|F\|_{L^{\tilde{q}'}_t L^{\tilde{r}'}_x} \right).$$

Moreover the range of the exponents in (22.2) and (22.3) is sharp.

Theorem 1 was instrumental in understanding the local and global behavior of continuous nonlinear Schrödinger equations and systems. In essence, the approach is to show existence of a (local) solution by a contraction argument in $L^q L^r$-type-spaces, as in Theorem 1. This argument works well to produce global solutions, if their initial data are small (in an appropriate sense). Global solutions for large data can then be obtained (whenever they exist) either by means of conservation laws or by some more sophisticated methods ("I-method," Bourgain and Fourier splitting methods, etc.), which also rely to a significant degree on the Strichartz estimates. For a thorough treatment of these issues, see the excellent book of Tao [4].

We would like to point out to some other, more recent developments, which use the same circle of ideas. In the study of asymptotic stability of solitons for continuous NLS, one linearizes around the special solution and considers matrix Schrödinger equations in the form

$$i\partial_t v + Hv = F(t,v), \tag{22.4}$$

where $v = (v, \bar{v})$, $F(t,v)$ is at least quadratic in v,

$$H = \begin{pmatrix} -\Delta + V & W \\ -W & \Delta - V \end{pmatrix},$$

and V, W are exponentially decaying and smooth potentials. On the spectral side, one has the typical absolutely continuous (hereafter denoted as a.c.) spectrum, which spans a finite co-dimension space. The remaining dimensions represent a nontrivial eigenspace (and possibly resonances) at zero mode and at least another point in the pure point spectrum. The study of small solutions[3] clearly necessitates an analog of Theorem 1 for the free generators away from eigenvalues,[4] namely $e^{itH} P_{a.c.}$. It has been shown in [5] that under suitable decay and smoothness assumptions on V, W, one has Strichartz estimates as in Theorem 1. On the other hand, we should point out that the lack of a clear spectral picture for H as well as the high number of degrees of freedom,[5] especially in the cases $d = 2, 3$, leaves many fascinating natural questions and conjectures unresolved.

22.2 Decay and Strichartz Estimates for the Discrete Schrödinger and Klein–Gordon Equation

The author of this section and the author of the main part of the book have initiated a program in [6] to extend these results to the case of the DNLS. In this section, we will present these results as well as some applications. Introduce the discrete Laplacian

$$\Delta_d u(k) = \sum_{j=1}^{d} [u(k+e_j) + u(k-e_j) - 2u(k)].$$

Theorem 2. *(Theorem 3, [6]) For the free discrete Schrödinger equation $i\partial_t u(t) + \Delta_d u = 0$, one has the decay estimate*

$$\|u(t)\|_{l^\infty} \leq C <t>^{-d/3} \|u(0)\|_{l^1}. \qquad (22.5)$$

As a consequence, for the inhomogeneous equation $iu'_n(t) + \Delta_d u_n + F_n(t) = 0$, there are the Strichartz estimates

$$\|u(t)\|_{L^q l^r} \leq C \left(\|u(0)\|_{l^2} + \|F(t)\|_{L^{\tilde{q}'} l^{\tilde{r}'}} \right), \qquad (22.6)$$

where $q, r \geq 2$, $1/q + d/(3r) \leq d/6$ and $(q, r, d) \neq (2, \infty, 3)$. Both the decay estimate (22.5) and the Strichartz estimates (22.6) are sharp.

Remark. Note that the decay rate $t^{-d/3}$ (and consequently the Strichartz estimates) is strictly smaller than the usual $t^{-d/2}$ decay that one has for the continuous analog.

---

[3] In this context, v is a variable which stands for the deviation of the solution from a suitable modification of the soliton and hence orbital/asymptotic stability (22.4) requires one to control the growth of v or $\|v\| \to 0$.
[4] and resonances in some cases.
[5] i.e., high-dimensional eigenspaces.

Let us explain why we get this rate of decay $t^{-d/3}$. For the solution of the free equation, there is the formula

$$u_n(t) = \sum_{m \in \mathbf{Z}^d} u_m(0) \int_{[0,1]^d} e^{-4it \sum_{j=1}^d \sin^2(\pi k_j)} e^{2\pi i (m-n) \cdot k} dk. \quad (22.7)$$

Thus

$$\|e^{it\Delta_d}\|_{l^1 \to l^\infty} = \sup_{m,n \in \mathbf{Z}^d} \left| \int_{[0,1]^d} e^{-4it \sum_{j=1}^d \sin^2(\pi k_j)} e^{2\pi i (m-n) \cdot k} dk \right|$$

and hence, noting that the integration over $[0, 1]^d$ splits in d integrals over $[0, 1]$, and after elementary change of variables, it remains to show

$$\sup_{s \in \mathbf{R}^1} \left| \int_0^1 e^{-it(\sin^2(x) - sx)} dx \right| \leq C \min(1, |t|^{-1/3}). \quad (22.8)$$

The estimate now follows from the Vander Corput lemma, since the phase function for $s = 1$ is $\varphi(x) = \sin^2(x) - x$ and one can check that $\varphi'(\pi/4) = \varphi''(\pi/4) = 0$, hence the estimate (22.8). In [6], there is a rigorous proof that this is sharp as well as numerical evidence, which shows that this rate is indeed the best possible.

For the Klein–Gordon model $\partial_t^2 u(t) - \Delta_d u + u + F(t) = 0$, we have a similar result, which has the unfortunate restriction to one-space dimension.[6]

Theorem 3. *(Theorem 5, [6]) For the solutions of the 1D homogeneous discrete Klein–Gordon equation one has the decay estimates*

$$\|u(t)\|_{l^\infty} \leq C <t>^{-1/3} (\|u(0)\|_{l^1} + \|u_t(0)\|_{l^1}). \quad (22.9)$$

For the solutions of the inhomogeneous equation, one has the Strichartz estimates with $(q, r) \geq 2$, $1/q + 1/(3r) \leq 1/6$. That is

$$\|u(t)\|_{L^q l^r} \leq C(\|u(0)\|_{l^2} + \|F(t)\|_{L^{q'} l^{r'}}).$$

As in the Schrödinger case, the 1D case is sharp both in the decay and in the Strichartz estimates statements.

Remark. One could still obtain a decay rate in the form $\|u(t)\|_{l^\infty} \leq C t^{-1/3}$ for the solutions of the Klein–Gordon equation in $\mathbf{Z}^d$, but this is clearly far from the conjectured optimal decay of $t^{-d/3}$.

In the next few sections, we will present some immediate corollaries of these theorems.

---

[6] This is due to the fact that the Vander Corput lemma, which is used to estimate the oscillatory integrals that arise is not quite sharp in dimensions $d \geq 2$. On the other hand, for DNLS, due to the nature of the dispersion relation, one can eventually reduce to d 1D oscillatory integrals.

22.2.1 A Spectral Result for Discrete Schrödinger Operators

In the continuous case, it is well known that the Schrödinger operator $H = -\Delta + V$ may (and in general does) support eigenvalues if V is not nonnegative everywhere. For small potential V, this question is more subtle and its answer depends on the dimension. Namely, for $d = 1, 2$, even very small potentials may support eigenvalues[7] [7], while in $d \geq 3$, this is not the case, as it may be seen from the CLR inequalities

$$|\sigma_{p.p.}(-\Delta + V)| \leq C_d \int_{\mathbf{R}^d} |V^-(x)|^{d/2} dx.$$

Here $V^- = \min(V(x), 0)$ is the negative part of the potential.

For the discrete Schrödinger operators, there are partial results in this direction in the case of $d = 1, 2$, namely that weak coupling *always* generates eigenvalues. This is in [8] for the case $d = 1$ and in [9] for $d = 2$. In [6], we have proved that similar to the continuous case, in high dimensions small potentials do not generate eigenvalues.

Theorem 4. *(Theorem 4, [6]) Let $d \geq 4$ and $Hu(n) = -\Delta_d u(n) + V(n)u(n)$. Then there exists $\varepsilon > 0$, so that whenever $\|V\|_{l^{d/3}(\mathbf{Z}^d)} \leq \varepsilon$, the eigenvalue problem*

$$-\Delta_d u_n + V_n u_n = \lambda u_n \tag{22.10}$$

has no solution $u \in l^2(\mathbf{Z}^d)$ for any λ.

The proof of Theorem 4 is simple, so we include it to illustrate the power of Theorem 2. Note that this proof just fails (by the failure of the Strichartz estimates at the endpoint $q = 2, r = \infty, d = 3$) to give the case of $d = 3$, which seems to be open at the moment of this writing.

Proof. It is clear that u is an eigenstate for (22.10) corresponding to an eigenvalue λ if and only if $\{e^{it\lambda} u_n\}$ is a solution to $i u'_n(t) + (-\Delta_d + V_n) u_n = 0$ with initial data $\{u_n\}$. Assume that a solution to (22.10) exists. Apply the Strichartz estimates in a time interval $(0, T)$ with $q = \tilde{q} = 2, r = \tilde{r} = 2d/(d-3)$. We get

$$\|u\|_{L^2(0,T)l^{2d/(d-3)}} \leq C\|u\|_{l^2} + C\|Vu\|_{L^2(0,T)l^{2d/(d+3)}} \leq$$
$$\leq C\|u\|_{l^2} + C\|V\|_{l^{d/3}} \|u\|_{L^2(0,T)l^{2d/(d-3)}} \leq$$
$$\leq C\|u\|_{l^2} + C\varepsilon \|u\|_{L^2(0,T)l^{2d/(d-3)}}.$$

If we assume $\varepsilon : C\varepsilon < 1/2$, we deduce from the last inequality that

$$\|u(t)\|_{L^2(0,T)l^{2d/(d-3)}} \leq 2C\|u\|_{l^2}$$

---

[7] In fact, eigenvalues exist whenever $\int V dx \leq 0$.

for every $T > 0$. This is a contradiction, since $u_n(t) = e^{it\lambda} u_n$ and therefore $\|u(t)\|_{L^2(0,\infty)l^{2d/(d-3)}} = \infty$.

22.2.2 Application to Excitation Thresholds

Consider now the DNLS

$$i\partial_t u + \Delta_d u \pm |u|^{2\sigma} u = 0 \qquad (22.11)$$

M. Weinstein [10] has proved that for $\sigma \geq 2/d$, one has an energy excitation threshold for (22.11), i.e., that there exists $\varepsilon = \varepsilon(d)$, so that every standing wave solution $\{e^{i\Lambda t}\phi_n\}$ must satisfy $\|\phi\|_{l^2} \geq \varepsilon$. In the same paper, he has also conjectured that for sufficiently small solutions, one has $\lim_{t\to\infty} \|u(t)\|_{l^p} = 0$ for all $p \leq \infty$. M. Weinstein has also shown in the same paper the complementary statement that if $\sigma < 2/d$, then there are arbitrarily small standing wave solutions, in particular solutions with $\|u(t)\|_{l^p} = const \neq 0$. In [6], we verify this conjecture in dimensions $d = 1, 2$. We have

Theorem 5. *(Theorem 7, [6]) Let $\sigma > 2/d$ and $d = 1, 2$. There exists an $\varepsilon = \varepsilon(d)$, so that whenever $\|u(0)\|_{l^{(8+2d)/(d+7)}} \leq \varepsilon$, one has for all $p : 2 \leq p \leq (8+2d)/(d+1)$,*

$$\|u(t)\|_{l^p} \leq C t^{-d(p-2)/(3p)} \|u(0)\|_{p'}. \qquad (22.12)$$

which is the generic rate of decay for the free solutions. Note $\lim_{t\to\infty} \|u(t)\|_{l^p} = 0$ for any $p > 2$, for small data. Also, no standing wave solutions are possible under the smallness assumptions outlined above.

We also establish the Weinstein conjecture for the 1D discrete Klein–Gordon equation

$$\partial_{tt} u - \Delta_1 u + u \pm |u|^{2\sigma} u = 0. \qquad (22.13)$$

Theorem 6. *(Theorem 9, [6]) Let $\sigma > 2$. There exists an ε, so that whenever $\|u(0)\|_{l^{5/4}} \leq \varepsilon$, $\|\partial_t u(0)\|_{l^{5/4}} \leq \varepsilon$, one has for all $p : 2 \leq p \leq 5$,*

$$\|u(t)\|_{l^p} \leq C t^{-(p-2)/(3p)} \|u(0)\|_{p'}. \qquad (22.14)$$

In particular, there are no small standing wave solutions to the discrete Klein–Gordon equation (22.13).

22.3 Decay and Strichartz Estimates for the Discrete Schrödinger Equation Perturbed by a Potential

In [11], the authors have considered, among other things, the question for the time decay of $e^{itH} P_{a.c.}(H)$, where $H = -\Delta_1 + V$, where V is a potential supported at finitely many sites. A decay estimate in the form

$$\|e^{itH} P_{a.c.}(H)\|_{l_\sigma^2 \to l_{-\sigma}^2} \leq Ct^{-3/2} \tag{22.15}$$

was established under the assumption $\sigma > 7/2$. This was achieved by means of an explicit formula for the free resolvent[8] and then using it to establish the limiting absorption principle and subsequently a Puiseux expansion at the spectral edges. In a very recent follow up to this paper [12], a similar result was established in two spatial dimensions. In it, the authors had to rely on estimates for the free resolvent[9] rather than on an exact formula, which makes the analysis much harder.

It is clear however that in the applications (especially when one linearizes around solitons), one must consider (decaying) potentials with infinite supports. In [13], D. Pelinovsky and A. Stefanov have addressed this case, but only for the 1D discrete Schrödinger equation. Let us present a quick summary of the results and the methods in this chapter.

We first show the *limiting absorption principle for potentials with infinite supports*. More precisely,

Theorem 7. *Fix $\sigma > 1/2$ and assume that $V \in l_{2\sigma}^1$. The resolvent $R(\lambda) = (-\Delta + V - \lambda)^{-1}$, defined for $\lambda \in \mathbf{C} \setminus [0, 4]$ as a bounded operator on l^2, satisfies*

$$\sup_{\varepsilon \downarrow 0} \|(-\Delta + V - \omega \pm \varepsilon)^{-1}\|_{l_\sigma^2 \to l_{-\sigma}^2} < \infty. \tag{22.16}$$

for any fixed $\omega \in (0, 4)$. As a consequence, there exist $R^\pm(\omega) = \lim_{\varepsilon \downarrow 0} R(\omega \pm i\varepsilon)$ in the norm of $B(l_\sigma^2, l_{-\sigma}^2)$.

The other technical tool is obtaining a Puiseux expansion at the spectral edges, which however requires that $H = -\Delta_1 + V$ does not support a resonance at zero. This looks somewhat different in each situation, so we give a precise technical definition.

Definition 1. $V \in l_1^1$ *is called a generic potential if no solution ψ_0 of equation $(-\Delta + V)\psi_0 = 0$ exists in $l_{-\sigma}^2$ for $1/2 < \sigma \leq 3/2$.*

As one expects, such a condition is generically satisfied with respect to V. Also, this condition guarantees a nice Taylor expansion (up to order one) for the resolvent $R_V(w)$ as $w \sim 0$.

---

[8] which unfortunately only holds in one spatial dimension.
[9] which features a hyperbolic point inside the a.c. spectrum, which one has to control separately.

Theorem 8. *(Theorem 3, [13]) Fix $\sigma > 5/2$ and assume that $V \in l_s^1$ for all $s < 2\sigma - 1$ and that $H = -\Delta_1 + V$ does not support resonance at zero. Then, there exists a constant C depending on V, so that*

$$\left\| e^{itH} P_{a.c.}(H) \right\|_{l_\sigma^2 \to l_{-\sigma}^2} \leq Ct^{-3/2}. \tag{22.17}$$

If one is interested in the boundedness of $e^{itH} P_{a.c.}(H)$ on $l^1 \to l^\infty$ (which is a key element of any asymptotic analysis for the DNLS and which has not been addressed in [11] and [12]), one needs to study the detailed behavior of the Jost solutions for the potential V.

Theorem 9. *(Theorem 4, [13]) Fix $\sigma > 5/2$ and assume that $V \in l_s^\infty$ for $s < 2\sigma - 1$ and that V is generic in the sense of Definition 1. Then, there exists a constant C depending on V, so that*

$$\left\| e^{itH} P_{a.c.}(H) \right\|_{l^1 \to l^\infty} \leq Ct^{-1/3} \tag{22.18}$$

Remark. In both Theorems 5 and 9, it is possible to obtain a slightly worst decay rate without the nonzero resonance assumption, as it has been done in the continuous case.

The idea of the proof of Theorem 9 is as follows: We split the spectral projection $P_{a.c.}(H)$ in a portion close to the spectral edges ($\omega = 0, 4$) and away from it. For a smooth cutoff χ_0, supported close to $\theta = 0, 4$, the piece $e^{itH} P_{a.c.} \chi_0(H)$ is given by the formula

$$[e^{itH} P_{a.c.} \chi_0(H) u]_n = \frac{i}{\pi} \sum_m u_m \int_{-\pi}^{\pi} e^{it(2 - 2\cos\theta)} \chi_0(\theta) \frac{\psi_m^+(\theta) \psi_n^-(\theta) \sin\theta}{W(\theta)} d\theta,$$

where $\psi_m^\pm$ are the Jost solutions and $W(\theta)$ is the Wronskian.[10] This is estimated by the following lemma, which gives the needed bound for the kernel of this operator.

Lemma 1. *Assume $V \in l_2^1$ and $W(0) \neq 0$. Then, there exists $C > 0$ such that*

$$\sup_{n<m} \left| \int_{-\pi}^{\pi} e^{it(2-2\cos\theta)} \chi_0(\theta) \frac{\psi_m^+(\theta) \psi_n^-(\theta) \sin\theta}{W(\theta)} d\theta \right| \leq Ct^{-1/2} \tag{22.19}$$

Note that the actual decay

$$\left\| e^{itH} P_{a.c.} \chi_0(H) \right\|_{l^1 \to l^\infty} \leq Ct^{-1/2}.$$

matches the continuous case.

---

[10] The nonresonance condition basically amounts to $W(0) \neq 0$.

The proof of Lemma 1 proceeds via a regularity estimate[11] for $f_n^\pm : \psi_n^\pm = e^{\mp n} f_n^\pm$. Namely, we establish that if $V \in l_2^1$, then for some small θ_0 (and where we have chosen $supp \chi_0 \subset (0, \theta_0)$)

$$\sup_{\theta: \in [-\theta_0, \theta_0]} \left(\|\partial_\theta f^+(\theta)\|_{l^\infty([0,\infty))} + \|\partial_\theta f^+(\theta)\|_{l^\infty([0,\infty))} \right) < \infty.$$

This is enough to perform an integration by parts argument (Vander Corput lemma), which implies (22.19).

For the piece $e^{itH} P_{a.c.}(1 - \chi_0)(H)$, we have

Lemma 2. *Fix $\sigma > 5/2$ and assume that $V \in l_s^1$ for all $s < 2\sigma - 1$. Then, there exists $C > 0$ such that*

$$\left\| \int_{-\pi}^{\pi} e^{it(2-2\cos\theta)} \chi(\theta) \operatorname{Im} R(2 - 2\cos\theta) \sin\theta \, d\theta \right\|_{l^1 \to l^\infty} \leq C t^{-1/3} \quad (22.20)$$

for any $t > 0$.

The proof of Lemma 2 goes through the Born series representation

$$e^{itH} P_{a.c.}(1 - \chi_0)(H) = \int_{-\pi}^{\pi} e^{it(2-2\cos\theta)} (1 - \chi_0(\theta)) \operatorname{Im} R(2 - 2\cos\theta) \sin\theta \, d\theta =$$

$$= \int_{-\pi}^{\pi} e^{it(2-2\cos\theta)} (1 - \chi_0(\theta)) \operatorname{Im} R_0(2 - 2\cos\theta) \sin\theta \, d\theta +$$

$$+ \int_{-\pi}^{\pi} e^{it(2-2\cos\theta)} (1 - \chi_0(\theta)) \operatorname{Im} R_0(2 - 2\cos\theta) V R_0(2 - 2\cos\theta) \sin\theta \, d\theta +$$

$$+ \int_{-\pi}^{\pi} e^{it(2-2\cos\theta)} (1 - \chi_0(\theta)) \operatorname{Im} R_0(2 - 2\cos\theta) G_V(\theta) R_0(2 - 2\cos\theta) \sin\theta \, d\theta,$$

where $G_V(\theta) := V R_V(2 - 2\cos\theta) V$. The first and the second expressions above are explicit[12] via the explicit representation of $R_0(\omega)$. Hence, these are estimated easily via the Vander Corput lemma. For the third integral, notice that the perturbed resolvent appears only sandwiched between two copies of V and so, we apply the limited absorption bounds for good estimates of $G_m(\theta)$. In fact, to apply the Vander Corput machinery, we need and show the following estimate

$$\sup_{\theta \in [-\pi, \pi]} \sum_m |G_m(\theta)| + \left| \frac{d}{d\theta} G_m(\theta) \right| \leq C \|V\|_{l_\sigma^2}^2 \|f\|_{l^1}.$$

---

[11] Note that in the case $V = 0$, $f_n \equiv 1$.
[12] In fact the first integral represents the free discrete Schrödinger equation, which was addressed in [6].

22.3.1 Spectral Theoretic Results for 1D Schrödinger Operators

By the limiting absorption principle (Theorem 7), one deduces immediately that $\sigma(-\Delta_1+V)$ does not have a singular component and by the Weyl's theorem consists of absolutely continuous part $\sigma_{a.c.}(-\Delta_1+V) = \sigma_{a.c.}(-\Delta_1) = [0,4]$ and (potentially infinite) number of eigenvalues, which are outside $[0,4]$. By the results of Killip and Simon,[13] [8], if $V \neq 0$, there is always a portion of the spectrum which lies outside $[0,4]$. That is $\sigma_{p.p.}(-\Delta_1+V) \neq \emptyset$. However

Theorem 10. *(Lemma 1, [13]) Fix $\sigma > 5/2$ and assume that $V \in l_s^1$ all $s < 2\sigma - 1$ and that V is generic in the sense of Definition 1. Then, the discrete spectrum of H is finite-dimensional and located in the two segments $(\omega_{\min}, 0) \cup (4, \omega_{\max})$, where*

$$\omega_{\min} = \min_{n \in \mathbf{Z}}\{0, V_n\}, \qquad \omega_{\max} = \max_{n \in \mathbf{Z}}\{4, 4+V_n\}$$

The proof uses essentially the analyticity of the perturbed resolvent, which is a consequence of the theory built to address the decay estimates.

22.4 Challenges and Open Problems

There are several natural and outstanding problems, related to the material presented in the previous sections. We will try to organize them in the order of appearance, although it is possible that some of these problems are related in some fashion.

22.4.1 Does Weak Coupling Allow Eigenvalues in Three Dimensions?

In relation to Theorem 4, we have the question of existence of eigenvalues in the regime of weak coupling in $d = 3$. As we have already mentioned, the continuous case has been resolved completely. Following this analogy, we can form the following conjecture

Conjecture 1. Show that weak coupling in 3D does not generate eigenvalues. That is, for sufficiently small potentials V, the associated Schrödinger operator $H = -\Delta + V : l^2(\mathbf{Z}^3) \to l^2(\mathbf{Z}^3)$ does not have point spectrum.

---

[13] This is stated for Jacobi matrices but translates into this statement for discrete Schrödinger operators of the form $-\Delta_1 + V$.

22.4.2 CLR-Type Bounds for Discrete Schrödinger Operators and Related Issues

A related, more subtle question is whether one can bound the number of eigenvalues by some quantity depending on the potential (for example, CLR-type bounds). This is typically impossible whenever weak coupling generates eigenvalues, but even in 1D, it seems to be an open question whether for arbitrarily small potential one can generate arbitrarily large number of eigenvalues. We will tentatively conjecture that this is possible.

Conjecture 2. Show that for each $\varepsilon > 0$, there exists a finitely supported potential $V^\varepsilon : \sup_n |V_n^\varepsilon| < \varepsilon$, so that $|\sigma_{p.p.}(-\Delta_1 + V^\varepsilon)| > \varepsilon^{-1}$.

This is even more relevant for high dimensions $d \geq 4$ (and in the case $d = 3$, if the weak-coupling conjecture above is true), where we know that the natural obstacle to such bounds is not present. In such cases, it may be possible to obtain some variant of the CLR bounds on the number of eigenvalues of $H = -\Delta_d + V$, for $d \geq 3$. Thus, we state a tentative

Conjecture 3. Show CLR-type bounds for the number of eigenvalues of $H = -\Delta_d + V$ in $d \geq 3$.

22.4.3 Show Analogs of Theorems 8, 9, 10 in Higher Dimensions

This is self-explanatory, but the goal is to extend the spectral and decay results of the 1D dynamics [13] to multidimensions, as the relevant applications (which are naturally posed in high dimensions) require.

We should also mention that in reality one will need to show more general statements of the type of Theorems 8, 9, 10 for the matrix-valued non-self adjoint Hamiltonians (whose continuous analogs were discussed in Sect. 22.1)

$$H = \begin{pmatrix} -\Delta_d + V & W \\ -W & \Delta_d - V \end{pmatrix}.$$

This will have to be done parallel to a spectral theoretic study of these operators, because it seems that even the most basic questions regarding the spectrum of such operators H (even in the case of special H arising from linearization around solitons) remain rather unclear. The discussion in this Sect. 22.4.3 is really a program that is currently under active investigation by the authors Kevrekidis and Pelinovsky to address the question for asymptotic stability of special solutions.

22.4.4 Asymptotic Stability and Nucleation

In analogy with the continuous case, one seems to need the ingredients of the previous section (plus some hard analysis to analyze the modulation equations) to prove the asymptotic stability of solitons. We state the relevant

Conjecture 4. Show asymptotic stability of all orbitally stable solitons for the focusing 1D DNLS.

A related problem of slightly different nature is the solitary wave formation process, where one seeks to understand the asymptotic behavior of the DNLS evolution with initial data of the form $A\delta_{n,0}$ (or more ambitiously with multiple initially excited sites). As is shown numerically in [14], such a solution relaxes to a discrete soliton ϕ^Λ, where $\Lambda = \Lambda(A)$. Notice that this only happens if $|A|$ is above some threshold value; below that, the solution tends asymptotically to zero. As of now, the relation between A and Λ remains elusive, but besides that, the issues that still need to be addressed are a rigorous proof that this actually happens as well as the actual rate of convergence at which u_A approaches ϕ^Λ.

Conjecture 5. Show that there is a member of the discrete soliton family ϕ^Λ such that for A larger than the threshold value in some relevant norm $\|u_A(t) - \phi^\Lambda\| \leq t^{-1/3}$.

References

1. Ginibre, J., Velo, G.: Smoothing properties and retarded estimates for some dispersive evolution equations. Comm. Math. Phys. **123**, 535–573 (1989)
2. Strichartz, R.: Restrictions of fourier transforms to quadratic surfaces and decay of solutions of wave equations. Duke Math. J. **44**, 705–714 (1977)
3. Keel, M., Tao, T.: Endpoint Strichartz Estimates. Amer. J. Math. **120**, 955 (1998)
4. Tao, T.: Nonlinear dispersive equations. Local and global analysis. CBMS Regional Conference Series in Mathematics, vol. 106. American Mathematical Society, Providence, RI (2006)
5. Rodnianski, I., Schlag, W., Soffer, A.: Dispersive analysis of charge transfer models. Comm. Pure Appl. Math. **58**, 149–216 (2005)
6. Stefanov, A., Kevrekidis, P.: Asymptotic behavior of small solutions for the discrete nonlinear Schrödinger and Klein-Gordon equations. Nonlinearity **18**, 1841–1857 (2005)
7. Simon, B.: The bound states of weakly coupled Schrödinger operators in one and two dimensions. Ann. Phys. (N.Y.) **97**, 279-288 (1976)
8. Killip, R., Simon, B.: Sum rules for Jacobi matrices and their applications to spectral theory. Ann. Math. **158**(2), 253–321 (2003)
9. Damanik, D., Hundertmark, D., Killip, R., Simon, B.: Variational estimates for discrete Schrödinger operators with potentials of indefinite sign. Comm. Math. Phys. **238**, 545–562 (2003)
10. Weinstein, M.I.: Excitation thresholds for nonlinear localized modes on lattices. Nonlinearity **12**, 673–691 (1999)
11. Komech, A., Kopylova, E., Kunze, M.: Dispersive estimates for 1D discrete Schrödinger and Klein-Gordon equations. Appl. Anal. **85**, 1487–1508 (2006)
12. Komech, A., Kopylova, E., Vainberg, B.R.: On dispersive properties of discrete 2D Schrödinger and Klein-Gordon equations. J. Funct. Anal. **254**, 2227–2254 (2008)
13. Pelinovsky, D., Stefanov, A.: On the spectral theory and dispersive estimates for a discrete Schrödinger equation in one dimension. J. Math. Phys. **49**, 113501 (2008)
14. Kevrekidis, P.G., Espinola-Rocha, J., Drossinos, Y.: Dynamical barrier for the nucleation of solitary waves in discrete lattices. Phys. Lett. A **372**, 2247–2253 (2008)

Index

A
Ablowitz–Ladik, 15, 236, 237, 250, 269, 293, 379

B
BEC, *see* Bose–Einstein condensate
Bloch functions, 4
Bloch oscillations, 4, 176, 179, 184, 273
Bose–Einstein condensate, 4, 11, 12, 99, 143, 150, 153, 176, 182, 255, 277, 278, 289, 312, 322, 326, 329, 332, 353, 379
Bright soliton, 52, 117, 118, 143, 179, 181, 184, 185, 221, 226, 228, 231, 298, 303, 306–308, 331, 334

C
Collisions
    of eigenvalues, 35, 39, 41, 86, 90, 111, 112, 122, 126, 130, 133, 136, 138, 140, 164, 165, 167
    soliton–soliton, 52, 278, 288, 305, 311, 312, 315, 322, 334, 393
Continuous nonlinear Schrödinger equation, *see* NLS
    cross-phase modulation, 160, 167, 178
Cross-phase modulation, 160, 167, 178

D
Dark soliton, 117, 118, 120, 123, 125, 126, 131, 179, 214, 221, 226, 227, 229, 231, 296, 305, 331
Defocusing nonlinearity, 5, 11, 28, 116, 117, 124, 131, 140, 146, 151, 175, 181, 183, 206, 214, 218, 222, 224, 225, 228, 235, 245, 261, 264, 265, 272, 298, 331, 354, 357
Diffraction management, 3, 176, 179, 279
Discrete nonlinear Schrödinger equation, *see* DNLS

Discrete self-trapping (equation), 7, 249
Dispersion management, 281, 282
DNLS, 3, 6, 11, 55, 99
    derivation, 3
    saturable, *see* Saturable nonlinearity
Domain wall, 123, 186
Dynamics, 7, 153, 176, 181, 186, 205, 207, 211, 214, 236, 237, 242, 243, 277, 280, 281, 285, 287, 334, 345, 353, 382

E
Equation, 3, 117, 128, 176, 180, 182, 185, 230, 231, 281, 282, 284, 333, 347, 380, 402
Evans function, 20–22, 24
Exceptional discretizations, 293
Experiments, 3, 4, 99, 117, 118, 126, 133, 152, 158, 160, 163, 175, 179, 180, 182, 184, 186, 235, 245, 254, 256, 259, 270, 284, 322, 326, 329, 331, 354, 379

F
Feshbach resonance, 278, 323
Few-lattice-site, 249, 255
Focusing nonlinearity, 3, 5, 11, 27, 28, 56, 116, 117, 124, 140, 144, 146, 152, 175, 183, 206, 216, 222, 224, 225, 227, 228, 235, 236, 243, 261, 264, 265, 271, 298, 330, 331, 412
Fredholm alternative, 69, 123

G
Galilean invariance, 13, 380
Gap soliton, 3, 4, 118, 129, 176, 179, 181, 184, 185, 187, 259, 265, 272, 323
Gap vortex, 129, 131, 181

413

G

Gauge invariance, 12, 14, 16, 22, 27, 66, 68, 70, 72, 77, 82, 96, 125, 163, 194, 202
GPE, *see* Gross–Pitaevskii Equation
Gross–Pitaevskii Equation, 289, 329

H

Hamiltonian–Hopf bifurcation, 39, 41, 88, 91, 122, 126, 131, 208, 215, 345
Harmonic potential, *see* Parabolic potential
Hexagonal lattice, 95, 97, 206, 213, 274
Honeycomb lattice, *see* Hexagonal lattice

I

Impurity, 92, 94, 353, 355, 357, 360, 365
Invariance
    Galilean, *see* Galilean invariance
    gauge, *see* Gauge invariance
    phase, *see* Gauge invariance
    translational, *see* Translational invariance

L

Linear stability, 101, 108, 118–120, 122, 124, 125, 129, 132, 156, 158, 161, 191, 192, 194
Lyapunov–Schmidt reduction, 42, 44, 61, 73, 100, 103, 132, 167, 256

M

Map approach, 221
Mobility, 274, 278, 305, 308, 322, 331, 340, 347, 367
Modulational instability, 4, 143, 145, 149, 150, 159, 176, 179, 184, 235, 251, 317, 331, 372
Multibreathers, *see* Multipulses
Multicomponent
    DNLS, 147, 153, 158, 192, 334
    vortices, 170
Multipulses, 4, 19, 28, 34, 48, 56, 60, 66, 99, 118, 127, 133, 182, 221, 226, 229, 231, 243, 245, 395, 396

N

NLS, 3, 4, 11, 12, 13, 15, 16, 48, 61, 66, 90, 117, 128, 143, 144, 145, 167, 176, 180, 182, 185, 235, 236, 243, 281, 282, 283, 284, 293, 294, 296, 298, 311–315, 320, 322–325, 331, 333, 347, 376, 380, 382, 388, 389, 397, 402
Nonlinear Schrödinger equation, *see* NLS
Nonlocal coupling, 298, 332
Numerical methods, 193, 194, 199, 336, 390

O

One-dimensional DNLS, 11
Optical lattice, 4, 5, 99, 117, 150, 153, 175, 176, 183, 185, 186, 255, 277, 289, 312, 322, 323, 329, 330, 332, 334, 347, 379
Oscillatory instability, 85, 87, 89, 125, 167, 208, 215, 217, 339

P

Parabolic potential, 150, 184
Peierls–Nabarro (barrier/potential), 3, 18, 21, 49, 51, 159, 176, 179, 267, 277, 283, 285, 293, 380, 392, 393, 397
Periodic potential, 4, 8, 151, 176, 180, 185, 295, 322, 323, 331, 353
Phase invariance, *see* Gauge invariance
Photorefractive crystals, 3, 92, 117, 133, 177, 176, 179, 181, 186, 219, 235, 271, 290, 323, 367, 379
PN, *see* Peierls–Nabarro (barrier/potential)
Potential
    harmonic, *see* Parabolic potential
    localized, *see* Impurity
    optical lattice, *see* Optical lattice
    periodic
        see also Optical lattice
    quadratic, *see* Parabolic potential

S

Saturable nonlinearity, 117, 126, 133, 180, 265, 290, 295, 297, 322, 331, 360, 367, 380, 391, 392
Scattering length, 183, 278
    time-varying
        see also Feshbach resonance
Self-phase modulation, 167, 178
Soliton
    bright, *see* Bright soliton
    dark, *see* Dark soliton
    gap, *see* Gap soliton
    staggered, *see* Staggered soliton
Soliton interactions, 13, 48, 49, 51, 129, 227, 270, 293, 311, 322, 323, 393
Soliton scattering, 313, 324, 353
SPM, *see* Self-phase modulation
Staggered soliton, 124, 222, 228, 231, 245, 265, 270, 303, 307, 308, 338, 339, 355
Staggering transformation, 28, 117, 123, 131, 140, 224, 228, 245, 278, 334, 354
Statistical mechanics, 369, 374
Surface soliton, 259, 260, 262, 267, 268, 270

Index 415

T
Three-dimensional DNLS, 99
Topological charge, 91, 95, 100, 128, 129, 131, 139, 167, 182, 222, 284, 337, 344
Translational invariance, 12, 16, 18, 19, 21, 22, 99, 293, 393
Travelling solitons, *see* Mobility
Two-dimensional DNLS, 55

V
Variational approach (VA), 30, 34, 48, 57, 156, 159, 279, 281, 282, 283, 315, 316, 336, 338, 353, 365
Vortex, 4, 56, 60, 61, 63, 65, 67, 70, 71, 73, 77, 82, 87, 90, 92, 95, 97, 100, 107, 108, 112, 118, 128, 129, 131, 133, 138, 139, 140, 159, 167, 169, 176, 180, 182, 208, 213, 214, 215, 222, 236, 259, 268, 273, 289, 336
gap, *see* Gap vortex
multicomponent, *see* Multicomponent; vortices

W
Wannier function (WF), 5, 255
Waveguide arrays, 3, 7, 117, 147, 152, 153, 175, 176, 181, 235, 255, 256, 259, 261, 265, 267, 268, 270, 277, 290, 312, 322, 326, 329, 330, 332, 335, 346, 353, 354, 367, 379

X
XPM, *see* Cross-phase modulation